RF Passive Network Design and Synthesis for Mobile Communications

Volume 1

For a listing of recent titles in the
Artech House Microwave Library,
turn to the back of this book.

RF Passive Network Design and Synthesis for Mobile Communications

Volume 1

Peter V. Wright

ARTECH HOUSE
BOSTON | LONDON
artechhouse.com

Library of Congress Cataloging-in-Publication Data
A catalog record for this book is available from the U.S. Library of Congress.

British Library Cataloguing in Publication Data
A catalog record for this book is available from the British Library.

ISBN 13: 978-1-68569-089-2

Cover design by Joi Garron

685 Canton St.
Norwood, MA

10 9 8 7 6 5 4 3 2 1

To my family, who kept me going

Contents

Preface

Without RF engineering, life as we know it today would be impossible. Most people are familiar with radar that allows them to fly safely across continents in less time than it took their ancestors to travel to the next town. They take cellphones for granted that allow them to talk to and see loved ones on the other side of the globe in real time. They rely heavily on microwave ovens to quickly cook or heat up their food after a long day's work. When they go to the dentist, or if they break a bone, or have some other internal medical issue, they assume that their care provider will have some non-invasive way to peer inside their body to diagnose what is going on. They are not surprised when doors at their local supermarket open automatically for them as they approach, nor that stores have inventory control tags on merchandise to prevent theft. Many pet owners also pay to have microchips embedded in their loved ones to help ensure their safe return should they go astray. Anyone who has flown is only too familiar with security screening devices they must walk through before boarding the flight to help protect their safety. All of these applications rely on RF technology. As we move into the future, with the rapid growth of the Internet of Things (IoT) and self-driving cars, our dependence on such technology will become ever greater.

The field of RF engineering encompasses an enormously wide domain of design and analysis. Included within it are both passive and active devices, which are frequently required to operate in unison. The devices may have operational frequencies ranging from a few hertz to those in the terahertz range, increasingly the focus of recent efforts seeking ever higher data rates, or even higher into the optical range. However, despite the seemingly wide divergence of RF technologies and applications, they all operate under the same physical laws and thus many of the same design and analytic analysis approaches are widely shared across the field.

Beginning as a microwave engineer, during my career I worked as a surface-acoustic-wave (SAW) designer, a non-destructive-testing (NDT) engineer in the oil industry, an RF front-end module designer, a power-amplifier designer, and a filter designer in the cellphone industry. Over the years, I had to develop many analytical tools to aid my design challenges. Invariably, to be sure that I fully understood the problem, I began from first principles to develop the underlying equations for a circuit that would best achieve the requirements for the device. This book is an attempt to share those fundamental design analyses, and the synthesis approaches derived from them, with the practicing RF engineer.

In the early years of RF engineering, an engineer had little alternative to developing the basic circuit equations by themselves to predict the performance of a design, to best ensure it would meet the design goals. Computers were limited in their capabilities and RF design tools were rudimentary.

Nowadays, in contrast, there are several excellent, very powerful, though expensive, design tools available to aid the practicing RF engineer. Two such examples are Advanced Design System® from Keysight and Microwave Office® from AWR Corporation. These allow even a relatively inexperienced RF engineer to design a circuit with a high degree of confidence that if realized, the circuit will perform much as predicted. The design tools also include very capable optimizers for improving the performance of a selected circuit configuration. However, invariably, such optimizations are very time consuming.

Without a doubt, modern RF software tools are absolutely required for the design of today's complex and compact RF devices. In particular, they are essential for incorporating the effects of parasitic and 3D interaction effects into the final design. However, a complete dependence on advanced software tools, especially in the early conceptual stages of RF circuit design, can result in designs that are non-optimal in performance, larger, more expensive, have lower production yield, and higher cost than one conceived from a more first-principles approach.

Under the constraint of time-pressure, RF engineers approaching a new design all too frequently seek to "leverage" an existing circuit design to meet the new requirements. With the advanced software design tools, this may consist of simply adding a few components to an existing design and letting the software reoptimize the circuit. This approach frees the engineer from having to spend time on any basic circuit analysis or having to consider any new fundamentally different circuit topologies that might be advantageous. Unfortunately, this design approach does not guarantee that the resulting solution is optimal for the requirements. To achieve a more optimal solution would require alternate circuit topologies be built and their performance evaluated in the software tool. Time constraints typically preclude this course of action.

A principal objective of this book is to detail how analytic design approaches can be used for rapid design optimization and design of RF circuits. Formulae, algorithms, and analysis and data display techniques are presented for use by the practicing RF design engineer. Not only do these techniques allow for a rapid comparison to be made of the performance of alternate RF circuit topologies, but they also impart to the designer a deeper understanding of the critical design trade-offs that can be made for each of the solutions. The optimum circuit topology selected can then either be implemented directly, or if further refinements are needed due to, for example, parasitic interactions, used as a starting point for a more complex RF design tool for final optimization.

The use of analytic solutions for RF circuit design has fallen out of favor in recent times as design engineers have increasingly shied away from writing equations and performing detailed circuit analysis. Instead, designers have become increasingly dependent on software packages to do the circuit design and optimization for them. This book is an attempt to change that by providing techniques that are relatively simple to apply and can help the designer become more creative by exploring circuit innovations that might otherwise have been time-prohibitive to consider.

Motivation for This Book

Practicing RF engineers are constantly under pressure to quickly wrap up their current design and move on quickly to the next. The next project will typically have design specifications very close to those of a previous design. The fastest design approach is then to reuse the old design configuration and simply reoptimize it in an off-the shelf computer-aided design (CAD) package. While this approach may frequently yield a conforming design, alternative configurations with advantageous characteristics in terms of performance, size, and cost may be never be considered.

To take full advantage of the many analyses presented in this book, it is important to understand the prime goal of the work. Its purpose is not to provide analytic design solutions as an alternative to full 3D CAD radio frequency (RF) design suites. Quite the opposite! The information is intended to complement the latter. With the solutions, it is envisaged that the engineer can use them to quickly evaluate critical performance parameters of alternate architectures and thereby select the best configuration for the application. To this end, the frequency ranges and element dissipative factors used for the majority of design examples are kept largely constant throughout the book. Extensive performance plots are also provided to allow the reader to quickly visually compare trade-offs between design scenarios.

It was my intention when beginning the book to simply collate the accumulation of analytic design solutions I had worked on during my career. However, it quickly expanded to be much more than that. While the design equations I had used in practice were in a large part valid, providing them in book form forced me to return to first principles and provide rigorous justification for them. To my surprise, in many instances this threw up the possibilities of new circuit architectures and applications that had not been obvious previously. In addition, rigorous circuit analyses threw new light on some well-known circuit architectures, for example Chireix and Doherty, that I had previously viewed as largely disadvantageous. The rigorous analyses pointed the way to new configurations that could overcome many of the previous disadvantages. Such welcome revelations occurred frequently throughout the book. As a result, many novel circuit architectures are provided that offer significant performance advantages to the practicing engineer.

Overview of This Book

This work contains design approaches and formulae intended to aid the practicing RF engineer in designing high-performance RF components for mobile applications. In particular, the focus is on RF couplers, combiners, splitters, and cellular power amplifiers. The approach followed is based on using lumped-element models, incorporating all important parasitics, to obtain analytically derived closed-form expressions for the behavior of the network. These solutions, which can be instantly updated with changing inputs, typically give greater insight into the behavior of a network than is obtained by simply trying to optimize the performance of a network in an EM software simulation program. Rather than an alternative to the latter, these analytic tools are best exploited in the initial design stages by employing them to quickly explore the capabilities of alternate architectures before embarking on a lengthy EM design optimization.

CHAPTER 1

[ABCD] Parameters: Key Relationships

Before commencing the analysis of any radio frequency (RF) circuit, the engineer must decide which set of mathematical parameters will be used to represent the electrical characteristics of the network. Typically, there are multiple options available, each of which is equally capable of representing the network. The choice made is usually based on the personal preferences of the designer, the software tools available, the dominant circuit architecture, and the number of network ports and interconnections. In general, the designer should choose the set of electrical parameters that most simplifies the design task.

In this and following chapters, we shall summarize the salient characteristics of several of the most commonly employed network parameters used for characterization. We begin with a set that is restricted to two-port networks, but that offers great simplification for the series-cascading of multiple networks. A complex network may often be reduceable to an equivalent two-port, after all internal interconnections and terminations are applied. In those cases, these parameters can also be used to concisely express the key network transfer characteristics.

[*ABCD*], or matrix [*M*], parameters are restricted to representing two-port networks. However, frequently, even complex RF networks can comprise an interconnection of such two-port sections. As [*ABCD*] parameters are extremely user-friendly and convenient manipulating two-port sections, their use is encouraged wherever possible. [*ABCD*] parameters are particularly useful in calculations of impedance matching, power transfer calculations, and the design of filter networks.

The circuit variables defined for an [*ABCD*] description of a two-port network are shown in Figure 1.1. Note that the input current I_1 flows into the network, while the output current I_2 flows out of the network. This is contrary to almost all other network matrix characterizations in which the power-flow vectors are assigned the same sense (i.e., either in or out) on all ports. However, for [*ABCD*] parameters this choice greatly facilitates series cascading multiple two-port network elements.

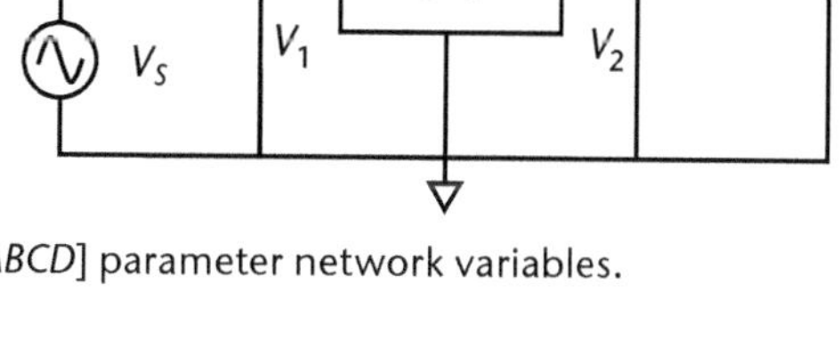

Figure 1.1 Two-port [*ABCD*] parameter network variables.

The network is represented by the following matrix equation:

$$\begin{bmatrix} V_1 \\ I_1 \end{bmatrix} = \begin{bmatrix} A & B \\ C & D \end{bmatrix} \begin{bmatrix} V_2 \\ I_2 \end{bmatrix} \tag{1.1}$$

that is,

$$V_1 = AV_2 + BI_2 \tag{1.2}$$

$$I_1 = CV_2 + DI_2 \tag{1.3}$$

with

$$A = \left.\frac{V_1}{V_2}\right|_{I_2=0} \qquad B = \left.\frac{V_1}{I_2}\right|_{V_2=0}$$
$$C = \left.\frac{I_1}{V_2}\right|_{I_2=0} \qquad D = \left.\frac{I_1}{I_2}\right|_{V_2=0} \tag{1.4}$$

and inversion formulas are

$$V_2 = \frac{DV_1 - BI_1}{AD - BC} \tag{1.5}$$

$$I_2 = \frac{-CV_1 + AI_1}{AD - BC} \tag{1.6}$$

In the following section, formulas are provided for deriving key circuit electrical characteristics from the network [*ABCD*] parameters. Subsequently, a list of [*ABCD*] parameters is provided for network topologies that are commonly used in RF design. Following that, parameter expressions are provided for both a series or a parallel cascade of two-port networks, and also those to account for a nonzero voltage on the reference ground plane. The latter is particularly useful for including a circuit a series impedance element, whether intended or parasitic, in the ground path of an active device.

1.1 Some Useful [*ABCD*] Relationships

If the network is reversed, the matrix equation becomes

$$\begin{bmatrix} V_1 \\ I_1 \end{bmatrix} = \begin{bmatrix} \dfrac{D}{AD-BC} & \dfrac{B}{AD-BC} \\ \dfrac{C}{AD-BC} & \dfrac{A}{AD-BC} \end{bmatrix} \begin{bmatrix} V_2 \\ I_2 \end{bmatrix} \tag{1.7}$$

Since, for a reciprocal network,

$$AD - BC = 1 \tag{1.8}$$

for a reversed reciprocal network

$$\begin{bmatrix} V_1 \\ I_1 \end{bmatrix} = \begin{bmatrix} D & B \\ C & A \end{bmatrix} \begin{bmatrix} V_2 \\ I_2 \end{bmatrix} \tag{1.9}$$

Hence, for a symmetric network,

$$A \Leftrightarrow D \tag{1.10}$$

If the external source and load impedances are Z_S and Z_L, respectively,

Input impedance
$$Z_{\text{in}} = \frac{V_1}{I_1} = \frac{AZ_L + B}{CZ_L + D} \tag{1.11}$$

where V_1 is the voltage at the input terminal 1, and I_1 is current flowing into terminal 1, with load Z_L on output terminal 2.

Output impedance
$$Z_{\text{out}} = \frac{V_2}{I_2} = \frac{DZ_S + B}{CZ_S + D} \tag{1.12}$$

where V_2 is the voltage at the output terminal 2, and I_2 is current flowing into terminal 2, with load Z_S on output terminal 1.

If the network is driven by a source voltage V_S, having a series impedance Z_S, and the output of the network is terminated by an impedance Z_L, then the output voltage V_2 on port 2, across the load, is given by

$$\frac{V_2}{V_S} = \frac{Z_L}{AZ_L + B + Z_S(CZ_L + D)} \tag{1.13}$$

Alternatively, V_2 may be expressed in terms of the input voltage at port 1, V_1, as

$$\frac{V_2}{V_1} = \frac{Z_L}{AZ_L + B} \tag{1.14}$$

where

$$\frac{V_1}{V_S} = \frac{AZ_L + B}{AZ_L + B + Z_S(CZ_L + D)} \tag{1.15}$$

Input reflection coefficient on port 1 is given by

$$\Gamma_1 = \frac{Z_{\text{in}} - Z_S}{Z_{\text{in}} + Z_S} = \frac{AZ_L + B - Z_S(CZ_L + D)}{AZ_L + B + Z_S(CZ_L + D)} \tag{1.16}$$

The transmission coefficient T of a network, is the ratio of the amplitude of the complex transmitted wave V_2, to that of the incident wave V_i. For the two-port network

$$V_1 = (1 + \Gamma_1)V_i \tag{1.17}$$

thus, from (1.16),

$$\frac{V_1}{V_i} = \frac{2(AZ_L + B)}{AZ_L + B + Z_S(CZ_L + D)}$$

Hence, from (1.14)

$$T = \frac{V_2}{V_i} = \frac{2Z_L}{AZ_L + B + Z_S(CZ_L + D)} \tag{1.18}$$

The current entering the input port 1 is

$$\frac{I_1}{V_S} = \frac{CZ_L + D}{AZ_L + B + Z_S(CZ_L + D)} \tag{1.19}$$

The DC component is

$$\left(\frac{I_1}{V_S}\right)_{\mathrm{DC}} = \Re\left(\frac{CZ_L + D}{AZ_L + B + Z_S(CZ_L + D)}\right) \tag{1.20}$$

1.2 Common Two-Port Network [*M*] Parameters

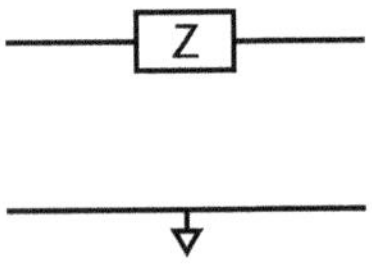

$$[M] = \begin{bmatrix} 1 & Z \\ 0 & 1 \end{bmatrix} \tag{1.21}$$

Figure 1.2 Series impedance.

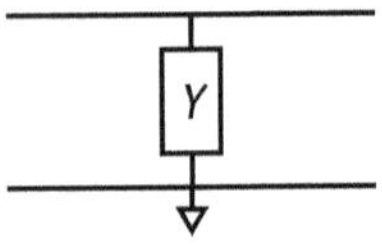

$$[M] = \begin{bmatrix} 1 & 0 \\ Y & 1 \end{bmatrix} \tag{1.22}$$

Figure 1.3 Shunt admittance.

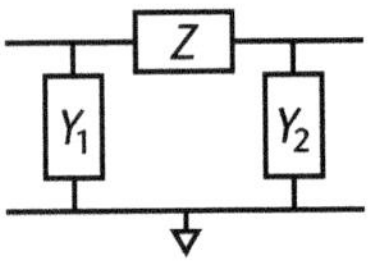

$$[M] = \begin{bmatrix} 1 + Y_2Z & Z \\ Y_1 + Y_2(1 + Y_1Z) & 1 + Y_1Z \end{bmatrix} \tag{1.23}$$

Figure 1.4 π-network.

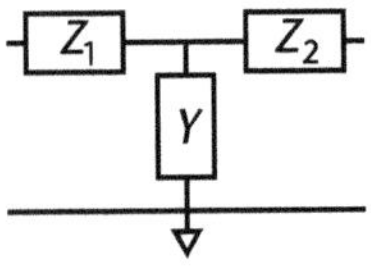

$$[M] = \begin{bmatrix} 1 + YZ_1 & Z_1 + Z_2(1 + YZ_1) \\ Y & 1 + YZ_2 \end{bmatrix} \tag{1.24}$$

Figure 1.5 *T* network.

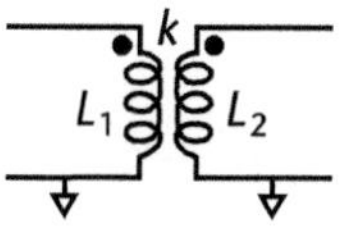

$$[M] = \begin{bmatrix} \dfrac{\sqrt{L_1/L_2}}{k} & \dfrac{j\omega\left(1-k^2\right)\sqrt{L_1L_2}}{k} \\ \dfrac{1}{j\omega k\sqrt{L_1L_2}} & \dfrac{\sqrt{L_2/L_1}}{k} \end{bmatrix} \tag{1.25}$$

Figure 1.6 Coupled inductors.

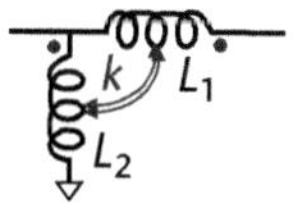

$$[M] = \begin{bmatrix} \dfrac{L_2}{L_2 + k\sqrt{L_1L_2}} & \dfrac{j\omega\left(1-k^2\right)L_1L_2}{L_2 + k\sqrt{L_1/L_2}} \\ \dfrac{1}{j\omega\left(L_2 + k\sqrt{L_1L_2}\right)} & 1 + \dfrac{L_1 + k\sqrt{L_1/L_2}}{L_2 + k\sqrt{L_1/L_2}} \end{bmatrix} \tag{1.26}$$

Figure 1.7 Coupled-inductors auto-transformer.

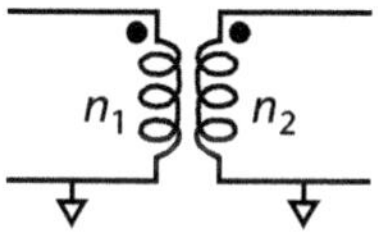

$$[M] = \begin{bmatrix} \dfrac{n_1}{n_2} & 0 \\ 0 & \dfrac{n_2}{n_1} \end{bmatrix} \tag{1.27}$$

Figure 1.8 Transformer.

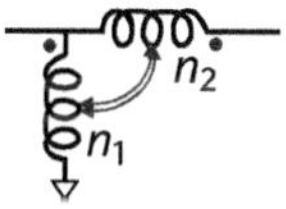

$$[M] = \begin{bmatrix} \dfrac{1}{1 + n_2/n_1} & 0 \\ 0 & 1 + n_2/n_1 \end{bmatrix} \tag{1.28}$$

Figure 1.9 Ideal auto-transformer.

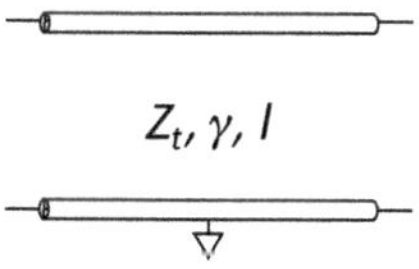

$$[M] = \begin{bmatrix} \cosh(\gamma l) & Z_t \sinh(\gamma l) \\ \sinh(\gamma l)/Z_t & \cosh(\gamma l) \end{bmatrix} \tag{1.29}$$

For the lossless case

$$[M] = \begin{bmatrix} \cos\phi & jZ_t \sin\phi \\ j\sin\phi/Z_t & \cos\phi \end{bmatrix} \tag{1.30}$$

Figure 1.10 Transmission line.

In (1.29) and (1.30), Z_t = transmission-line impedance. The complex propagation constant $\gamma = \alpha + j(2\pi/\lambda)$, where α = dissipative coefficient, λ = wavelength, and l = length of transmission line in wavelengths. The phase length of the transmission line is represented by $\phi = 2\pi l/\lambda$.

As a further aid to the designer, a compendium of useful [*ABCD*] formulae for additional common two-port RF topologies is given in Appendix A. In Appendix B, formulae are provided for converting between the various sets of two-port network parameters.

1.3 Interconnection of Two [*ABCD*] Matrices

Figure 1.11 shows a series cascade of two [*ABCD*] networks.

Cascaded parameters are

$$[M'] = \begin{bmatrix} A_1A_2 + B_1C_2 & A_1B_2 + B_1D_2 \\ C_1A_2 + D_1C_2 & C_1B_2 + D_1D_2 \end{bmatrix} \tag{1.31}$$

Figure 1.12 shows a parallel connection of two [*ABCD*] networks.

Parallel parameters are

$$[M'] = \begin{bmatrix} A' & B' \\ C' & D' \end{bmatrix} \tag{1.32}$$

where

$$\begin{aligned} A' &= \frac{(A_1B_2 + A_2B_1)}{(B_1 + B_2)} \\ B' &= \frac{B_1B_2}{(B_1 + B_2)} \\ C' &= \frac{((B_1 + B_2)(C_1 + C_2) - (A_1 + A_2)(D_1 + D_2))}{(B_1 + B_2)} \\ D' &= \frac{(B_1D_2 + B_2D_1)}{(B_1 + B_2)} \end{aligned} \tag{1.33}$$

1.4 Modified Matrix Coefficients with Ground Impedance

In practical RF circuits, it frequently happens that the ground reference in the device or module is not at the same potential as the external system ground. This is typically attributable to parasitic resistance or reactance in the connection between the device and system ground planes. If the device [*M*] matrix coefficients are known with reference to the device ground, the modified coefficients referenced to the external ground can be determined from the formulas below.

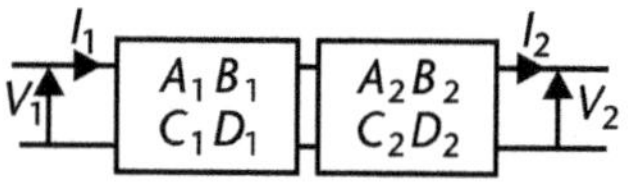

Figure 1.11 Cascaded [*ABCD*] networks.

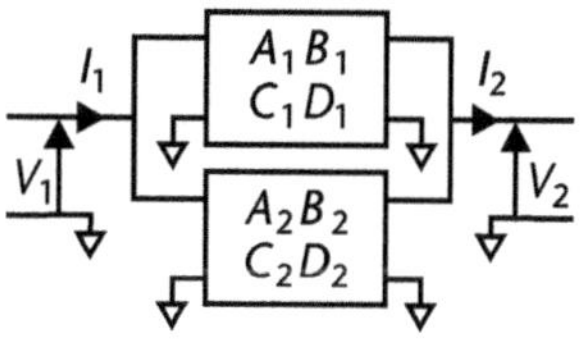

Figure 1.12 Parallel [*ABCD*] networks.

Let the parasitic impedance between the device and external ground planes be represented by Z_g, as shown in Figure 1.13.

Modified parameters $[M']$ are:

$$[M'] = \begin{bmatrix} A' & B' \\ C' & D' \end{bmatrix} \tag{1.34}$$

where

$$\begin{aligned} A' &= (A + CZ_g)/(CZ_g + 1) \\ B' &= B - Z_g(A-1)(D-1)/(CZ_g + 1) \\ C' &= (C)/(CZ_g + 1) \\ D' &= (D + CZ_g)/(CZ_g + 1) \end{aligned} \tag{1.35}$$

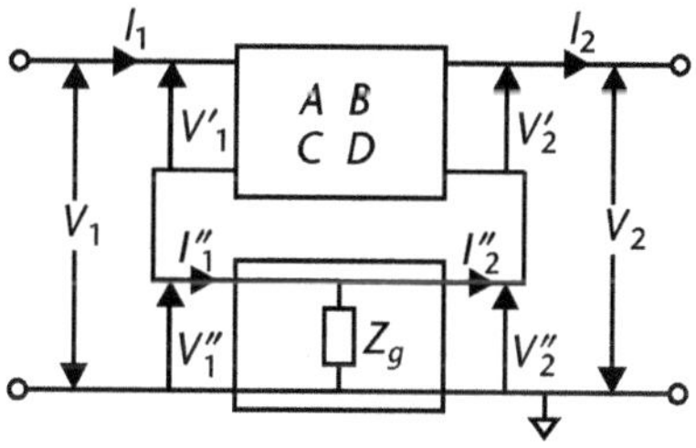

Figure 1.13 Network with ground impedance.

CHAPTER 2

S-Parameters: Key Relationships

Unlike [$ABCD$] parameters, scattering (S) parameters can be used to characterize networks with an unlimited number of ports. They are by far and away the most widely used electrical parameters for characterizing RF networks. The reason for this is that they are based on normalized power flow variables. For a high-frequency RF network, it is typically very difficult to directly measure the currents and voltages at the network ports, while power flow measurements are relatively straightforward. Thus, rather than direct current and voltage parameters, the S-parameters are based on power-based normalized voltages at each of the ports. The normalized input voltages are customarily denoted by a_n, while the normalized output voltages are denoted by b_n. These voltages are normalized to characteristic impedances on each of the ports, which may be different from one another. However, it is common practice to choose 50Ω as the default normalization impedance for all ports of an RF network.

Given the dominance of S-parameters in RF circuit analysis, a wider range of formulae for differing network configurations than in those for other network parameters is presented in this chapter.

A network with n ports and the S-parameter voltage variables and scattering matrix is shown schematically in Figure 2.1.

The network is represented by the scattering equations:

$$\begin{bmatrix} b_1 \\ \cdot \\ \cdot \\ b_n \end{bmatrix} = \begin{bmatrix} S_{11} & \cdot & \cdot & S_{1n} \\ \cdot & \cdot & \cdot & \cdot \\ \cdot & \cdot & \cdot & \cdot \\ S_{n1} & \cdot & \cdot & S_{nn} \end{bmatrix} \begin{bmatrix} a_1 \\ \cdot \\ \cdot \\ a_n \end{bmatrix} \tag{2.1}$$

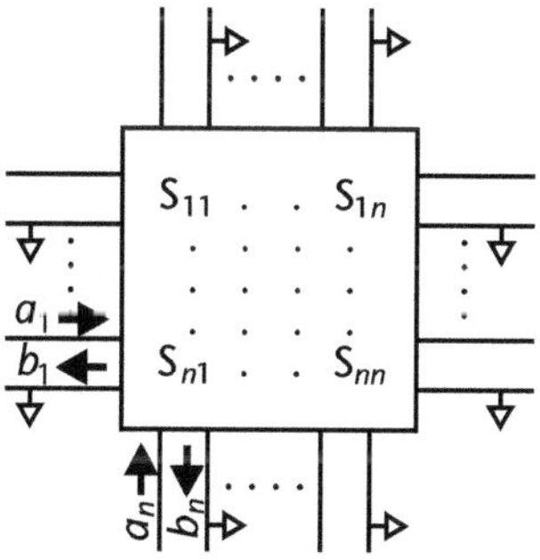

Figure 2.1 Definition of multiport scattering parameters.

that is,

$$
\begin{aligned}
b_1 &= S_{11}a_1 \cdots\cdots + S_{1n}a_n \\
&\;\cdot \quad\;\; \cdot \qquad\quad \cdot \\
&\;\cdot \quad\;\; \cdot \qquad\quad \cdot \\
b_n &= S_{n1}a_1 \cdots\cdots + S_{nn}a_n
\end{aligned}
\tag{2.2}
$$

The variables a_n and b_n, at each port, are related to the voltages and currents at the port by the following relations

$$V_n = a_n + b_n \tag{2.3}$$

$$I_n = (a_n - b_n)/\underline{Z}_n \tag{2.4}$$

where $\underline{Z}_n$ = characteristic reference impedance at port n.

Thus,

$$a_n = (V_n + I_n\underline{Z}_n)/2 \tag{2.5}$$

$$b_n = (V_n - I_n\underline{Z}_n)/2 \tag{2.6}$$

S-parameters are extensively used for representing linear RF networks because of the ease with which they can be measured on the bench and because they can be used to characterize a network with an unlimited number of ports. However, they are generally very unwieldly to manipulate directly mathematically. S-parameter formulae for even relatively simple network interconnections are typically very complex. This is not an issue when using RF commercial analysis software tools to manipulate the networks. However, it typically makes them unattractive to use as the basis for the development of analytic network synthesis solutions.

2.1 Some Useful S-Parameter Relationships

Some useful S-parameter relationships are the following. If the network is reciprocal

$$S_{mn} = S_{nm} \tag{2.7}$$

In addition, if the network is lossless

$$
\begin{bmatrix} S_{11} & \cdot & \cdot & S_{1n} \\ \cdot & \cdot & \cdot & \cdot \\ \cdot & \cdot & \cdot & \cdot \\ S_{n1} & \cdot & \cdot & S_{nn} \end{bmatrix} \cdot \begin{bmatrix} S^*_{11} & \cdot & \cdot & S^*_{1n} \\ \cdot & \cdot & \cdot & \cdot \\ \cdot & \cdot & \cdot & \cdot \\ S^*_{n1} & \cdot & \cdot & S^*_{mn} \end{bmatrix}^T = [1] \tag{2.8}
$$

where $[\ ]^T$ represents the transpose of the matrix, and [1] represents the unitary matrix, that is,

$$\begin{aligned} |S_{11}|^2 + |S_{12}|^2 \cdots\cdots + |S_{1n}|^2 &= 1 \\ \cdot \qquad \cdot \qquad \cdot \quad &= 1 \\ \cdot \qquad \cdot \qquad \cdot \quad &= 1 \\ |S_{1n}|^2 + |S_{2n}|^2 \cdots\cdots + |S_{nn}|^2 &= 1 \end{aligned} \tag{2.9}$$

and

$$\begin{aligned} S_{11}S_{12}^* + S_{12}S_{22}^* \cdots\cdots\cdots\cdots + S_{1n}S_{n2}^* &= 0 \\ \cdot \qquad \cdot \qquad \cdot \quad &= 0 \\ \cdot \qquad \cdot \qquad \cdot \quad &= 0 \\ S_{n1}S_{1(n-1)}^* + S_{n2}S_{2(n-1)}^* \cdots + S_{nn}S_{n(n-1)}^* &= 0 \end{aligned} \tag{2.10}$$

Input reflection coefficient at port n

$$\Gamma_n = b_n / a_n \tag{2.11}$$

Input impedance at port n

$$(Z_{\text{in}})_n = \frac{a_n + b_n}{a_n - b_n} \underline{Z}_n \tag{2.12}$$

Output voltage at port n

$$V_n = (1 + \Gamma_n) b_n \tag{2.13}$$

where Γ_n = load reflection coefficient at port n.

If V_S = voltage driving source on port n and Z_S = series source impedance, as in Figure 2.2,

$$V_S = \frac{\underline{Z}_n + Z_S}{\underline{Z}_n} a_n + \frac{\underline{Z}_n - Z_S}{\underline{Z}_n} b_n \tag{2.14}$$

or with

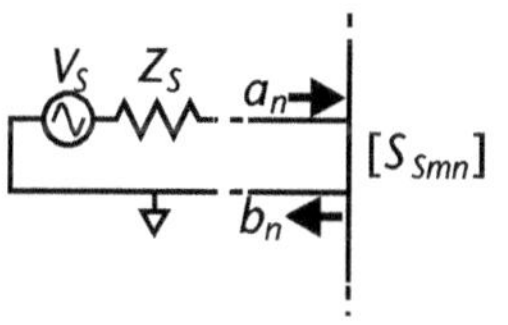

Figure 2.2 Voltage driving source on port n of a multiport network.

$$\Gamma_S = \frac{Z_S - \underline{Z}_n}{Z_S + \underline{Z}_n} \tag{2.15}$$

$$V_S = \frac{2}{1 - \Gamma_S}(a_n - \Gamma_S b_n) \tag{2.16}$$

2.2 Changing Port Normalization Impedances

The S-parameters for an RF network may be based on differing normalization impedances on each of the ports. They may be chosen, for example, to match the loading impedances on the ports. Occasionally, it may be desirable to change the normalization impedance on one or more of the network ports. This is not straightforward as the S-parameters are all interdependent. Thus, changing the normalization impedance on only one port can result in a change in all the S-parameters.

The formulae for changing the normalization impedance on port P of a network from $\underline{Z}_P$ to $\tilde{\underline{Z}}_P$ are given below.

The network S-parameters are modified from $S_{mn} = \tilde{S}_{mn}$, where

For $m = P$, $n \neq P$,

$$\tilde{S}_{Pn} = \frac{2S_{Pn}}{1 + \underline{Z}_P/\tilde{\underline{Z}}_P - S_{PP}\left(1 - \underline{Z}_P/\tilde{\underline{Z}}_P\right)} \tag{2.17}$$

For $m = n = P$,

$$\tilde{S}_{PP} = \frac{1 - \underline{Z}_P/\tilde{\underline{Z}}_P - S_{PP}\left(1 + \underline{Z}_P/\tilde{\underline{Z}}_P\right)}{1 + \underline{Z}_P/\tilde{\underline{Z}}_P - S_{PP}\left(1 - \underline{Z}_P/\tilde{\underline{Z}}_P\right)} \tag{2.18}$$

For $m \neq P$, $n \neq P$,

$$\tilde{S}_{mn} = S_{mn} + \frac{S_{mP}S_{Pn}\left(1 - \underline{Z}_P/\tilde{\underline{Z}}_P\right)}{1 + \underline{Z}_P/\tilde{\underline{Z}}_P - S_{PP}\left(1 - \underline{Z}_P/\tilde{\underline{Z}}_P\right)} \tag{2.19}$$

For $m \neq P$, $n = P$,

$$\tilde{S}_{mP} \frac{2S_{mP}\underline{Z}_P/\tilde{\underline{Z}}_P}{1 + \underline{Z}_P/\tilde{\underline{Z}}_P - S_{PP}\left(1 - \underline{Z}_P/\tilde{\underline{Z}}_P\right)} \tag{2.20}$$

These formulae yield the modified scattering parameters of a multiport network when the normalization impedance is changed on one of the ports, $n = P$. If it is desired to change the normalization impedances on more than one port, the above formulae should be applied sequentially to the ports in question.

2.3 Some Useful Two-Port S-Parameter Relationships

Complex RF networks are typically characterized by S-parameter representations comprising multiple ports. However, it is frequently the case that, for the final analysis, the network is reduced to a two-port representation by terminating the remaining ports with impedance networks. The characteristics of the network for transforming an input signal on one of the ports to an output signal at the second port can then be evaluated. Among the key characteristics of interest are the network input and output impedances, gain, and phase or group delay through the network.

The S-parameter matrix for a two-port is reduced to

$$\begin{bmatrix} b_1 \\ b_2 \end{bmatrix} = \begin{bmatrix} S_{11} & S_{12} \\ S_{21} & S_{22} \end{bmatrix} \begin{bmatrix} a_1 \\ a_2 \end{bmatrix} \tag{2.21}$$

corresponding to scattering equations

$$b_1 = S_{11}a_1 + S_{12}a_2 \tag{2.22}$$

$$b_2 = S_{21}a_1 + S_{22}a_2 \tag{2.23}$$

where the normalization impedances on ports 1 and 2 are $\underline{Z}_1$ and $\underline{Z}_2$, respectively.

If the network is reversed,

$$\begin{bmatrix} b_1 \\ b_2 \end{bmatrix} = \begin{bmatrix} S_{22} & S_{21} \\ S_{12} & S_{11} \end{bmatrix} \begin{bmatrix} a_1 \\ a_2 \end{bmatrix} \tag{2.24}$$

For a two-port network, with port 1 as input, and source voltage V_S and output on port 2,

$$V_{\text{in}} = a_1 + b_1 \tag{2.25}$$

$$I_{\text{in}} = (a_1 - b_1)/\underline{Z}_1 \tag{2.26}$$

Defining

$$\Gamma_S = \frac{Z_S - \underline{Z}_1}{Z_S + \underline{Z}_1} \quad \text{and} \quad \Gamma_L = \frac{Z_L - \underline{Z}_2}{Z_L + \underline{Z}_2} \tag{2.27}$$

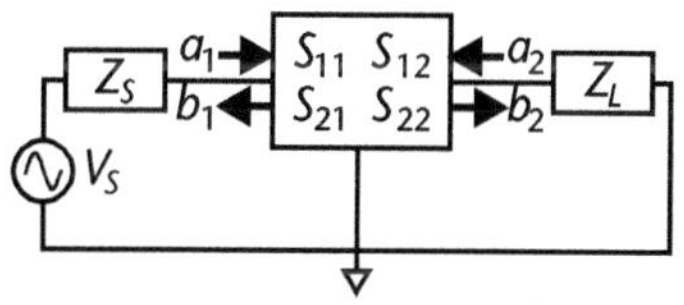

Figure 2.3 Two-port S-parameter network variables.

$$a_1 = \frac{1-\Gamma_S}{2}\left(\frac{1-\Gamma_L S_{22}}{(1-\Gamma_S S_{11})(1-\Gamma_L S_{22})-\Gamma_S\Gamma_L S_{12}S_{22}}\right)V_S \tag{2.28}$$

$$b_1 = \left(S_{11} + \frac{\Gamma_L S_{12}S_{21}}{1-\Gamma_L S_{22}}\right)a_1 \tag{2.29}$$

$$a_2 = \frac{\Gamma_L S_{21}}{1-\Gamma_L S_{22}}a_1 \tag{2.30}$$

$$b_2 = \frac{S_{21}}{1-\Gamma_L S_{22}}a_1 \tag{2.31}$$

$$V_L = \frac{(1+\Gamma_L)S_{21}}{1-\Gamma_L S_{22}}a_1 \tag{2.32}$$

$$Z_{\text{in}} = \left(\frac{2}{1-S_{11}-\Gamma_L S_{12}S_{21}/(1-\Gamma_L S_{22})}\right)\underline{Z}_1 \tag{2.33}$$

$$\frac{V_{\text{in}}}{V_S} = \frac{((1+S_{11})(1-\Gamma_L S_{22})+\Gamma_L S_{12}S_{21})\underline{Z}_1}{(Z_1+Z_S+S_{11}(Z_1-Z_S))(1-\Gamma_L S_{22})+\Gamma_L S_{12}S_{21}(\underline{Z}_1-Z_S)} \tag{2.34}$$

$$Z_{\text{out}} = \left(\frac{2}{1-S_{22}-\Gamma_S S_{12}S_{21}/(1-\Gamma_S S_{11})}\right)\underline{Z}_2 \tag{2.35}$$

$$\frac{V_L}{V_S} = \frac{S_{21}(1-\Gamma_S)(1+\Gamma_L)}{2(1-\Gamma_S S_{11})(1-\Gamma_L S_{22})-\Gamma_S\Gamma_L S_{12}S_{21}} \tag{2.36}$$

2.4 Common Two-Port Network S-Parameters

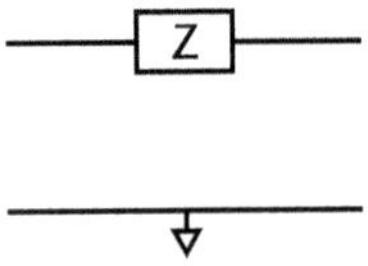

$$[S] = \begin{bmatrix} \dfrac{Z+\underline{Z}_2-\underline{Z}_1}{Z+\underline{Z}_1+\underline{Z}_2} & \dfrac{2\underline{Z}_1}{Z+\underline{Z}_1+\underline{Z}_2} \\ \dfrac{2\underline{Z}_2}{Z+\underline{Z}_1+\underline{Z}_2} & \dfrac{Z+\underline{Z}_1-\underline{Z}_2}{Z+\underline{Z}_1+\underline{Z}_2} \end{bmatrix} \tag{2.37}$$

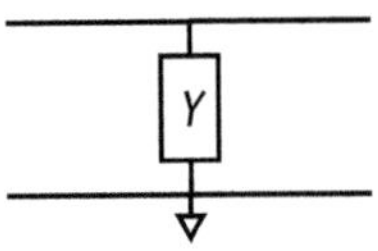

$$[S] = \begin{bmatrix} \dfrac{\underline{Z}_2-(1+Y\underline{Z}_2)\underline{Z}_1}{\underline{Z}_1+\underline{Z}_2+Y\underline{Z}_1\underline{Z}_2} & \dfrac{2\underline{Z}_1}{\underline{Z}_1+\underline{Z}_2+Y\underline{Z}_1\underline{Z}_2} \\ \dfrac{2\underline{Z}_2}{\underline{Z}_1+\underline{Z}_2+Y\underline{Z}_1\underline{Z}_2} & \dfrac{\underline{Z}_1-(1+Y\underline{Z}_1)\underline{Z}_2}{\underline{Z}_1+\underline{Z}_2+Y\underline{Z}_1\underline{Z}_2} \end{bmatrix} \tag{2.38}$$

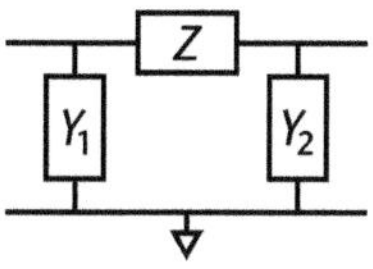

$$[S] = \begin{bmatrix} \dfrac{\overline{S}_{11}}{D} & \dfrac{\overline{S}_{12}}{D} \\ \dfrac{\overline{S}_{21}}{D} & \dfrac{\overline{S}_{22}}{D} \end{bmatrix} \quad (2.39)$$

where

$$\begin{aligned} \overline{S}_{11} &= -\left(Y_1 + Y_2 - (1 + Y_2 Z)/\underline{Z}_1 + (1 + Y_1 Z)\underline{Z}_2 + Z\left(Y_1 Y_2 - 1/\underline{Z}_1\underline{Z}_2\right)\right) \\ \overline{S}_{12} &= 2/\underline{Z}_2 \\ \overline{S}_{21} &= 2/\underline{Z}_1 \\ \overline{S}_{22} &= -\left(Y_1 + Y_2 + (1 + Y_2 Z)/\underline{Z}_1 - (1 + Y_1 Z)\underline{Z}_2 + Z\left(Y_1 Y_2 - 1/\underline{Z}_1\underline{Z}_2\right)\right) \\ D &= Y_1 + Y_2 + (1 + Y_2 Z)/\underline{Z}_1 + (1 + Y_1 Z)\underline{Z}_2 + Z\left(Y_1 Y_2 + 1/\underline{Z}_1\underline{Z}_2\right) \end{aligned}$$

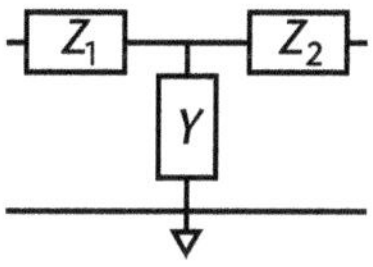

$$[S] = \begin{bmatrix} \dfrac{\overline{S}_{11}}{D} & \dfrac{\overline{S}_{12}}{D} \\ \dfrac{\overline{S}_{21}}{D} & \dfrac{\overline{S}_{22}}{D} \end{bmatrix} \quad (2.40)$$

where

$$\begin{aligned} \overline{S}_{11} &= (1 + Z_2/\underline{Z}_2)/Y\underline{Z}_1 - (1 - Z_1/\underline{Z}_1)/Y\underline{Z}_2 - (1 - Z_1/\underline{Z}_1)(1 + Z_2/\underline{Z}_2) \\ \overline{S}_{12} &= 2/Y\underline{Z}_2 \\ \overline{S}_{21} &- 2/Y\underline{Z}_1 \\ \overline{S}_{22} &= (1 + Z_1/\underline{Z}_1)/Y\underline{Z}_2 - (1 - Z_2/\underline{Z}_2)/Y\underline{Z}_1 - (1 + Z_1/\underline{Z}_1)(1 - Z_2/\underline{Z}_2) \\ D &= (1 + Z_2/\underline{Z}_2)/Y\underline{Z}_1 + (1 + Z_1/\underline{Z}_1)/Y\underline{Z}_2 + (1 + Z_1/\underline{Z}_1)(1 + Z_2/\underline{Z}_2) \end{aligned}$$

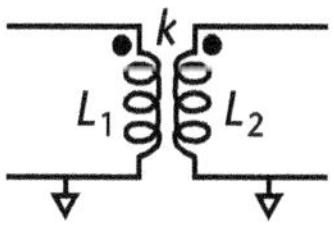

$$[S] = \begin{bmatrix} \dfrac{\overline{S}_{11}}{D} & \dfrac{\overline{S}_{12}}{D} \\ \dfrac{\overline{S}_{21}}{D} & \dfrac{\overline{S}_{22}}{D} \end{bmatrix} \quad (2.41)$$

where

$$
\begin{aligned}
\overline{S}_{11} &= j\omega\left(L_1/\underline{Z}_1 - L_2/\underline{Z}_2\right) - 1 - \omega^2\left(1-k^2\right)L_1L_2/\underline{Z}_1\underline{Z}_2 \\
\overline{S}_{12} &= 2j\omega k\sqrt{L_1L_2}/\underline{Z}_2 \\
\overline{S}_{21} &= 2j\omega k\sqrt{L_1L_2}/\underline{Z}_1 \\
\overline{S}_{22} &= j\omega\left(L_2/\underline{Z}_2 - L_1/\underline{Z}_1\right) - 1 - \omega^2\left(1-k^2\right)L_1L_2/\underline{Z}_1\underline{Z}_2 \\
D &= j\omega\left(L_1/\underline{Z}_1 + L_2/\underline{Z}_2\right) + 1 - \omega^2\left(1-k^2\right)L_1L_2/\underline{Z}_1\underline{Z}_2
\end{aligned}
$$

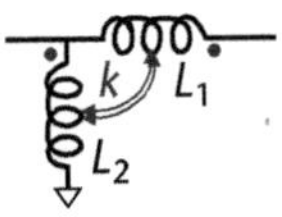

$$
[S] = \begin{bmatrix} \dfrac{\overline{S}_{11}}{D} & \dfrac{\overline{S}_{12}}{D} \\ \dfrac{\overline{S}_{21}}{D} & \dfrac{\overline{S}_{22}}{D} \end{bmatrix} \tag{2.42}
$$

where

$$
\begin{aligned}
\overline{S}_{11} &= -\left(1 + \omega^2\left(1-k^2\right)L_1L_2/\underline{Z}_1\underline{Z}_2 + j\omega\left(\left(L_1 + L_2 + 2k\sqrt{L_1L_2}\right)/\underline{Z}_2 - L_2/\underline{Z}_1\right)\right) \\
\overline{S}_{12} &= 2j\omega\left(L_2 + k\sqrt{L_1L_2}\right)/\underline{Z}_2 \\
\overline{S}_{21} &= 2j\omega\left(L_2 + k\sqrt{L_1L_2}\right)/\underline{Z}_1 \\
\overline{S}_{22} &= -\left(1 + \omega^2\left(1-k^2\right)L_1L_2/\underline{Z}_1\underline{Z}_2 - j\omega\left(\left(L_1 + L_2 + 2k\sqrt{L_1L_2}\right)/\underline{Z}_2 - L_2/\underline{Z}_1\right)\right) \\
D &= 1 - \omega^2\left(1-k^2\right)L_1L_2/\underline{Z}_1\underline{Z}_2 + j\omega\left(\left(L_1 + L_2 + 2k\sqrt{L_1L_2}\right) / \underline{Z}_2 + L_2/\underline{Z}_1\right)
\end{aligned}
$$

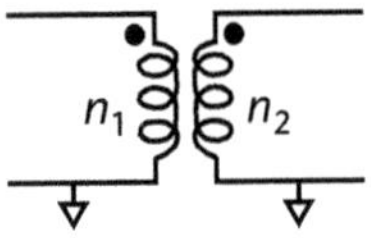

$$
[S] = \begin{bmatrix} \dfrac{n_1^2\underline{Z}_2 - n_2^2\underline{Z}_1}{n_2^2\underline{Z}_1 + n_1^2\underline{Z}_2} & \dfrac{2n_1n_2\underline{Z}_1}{n_2^2\underline{Z}_1 + n_1^2\underline{Z}_2} \\ \dfrac{2n_1n_2\underline{Z}_2}{n_2^2\underline{Z}_1 + n_1^2\underline{Z}_2} & \dfrac{n_2^2\underline{Z}_1 - n_1^2\underline{Z}_2}{n_2^2\underline{Z}_1 + n_1^2\underline{Z}_2} \end{bmatrix} \tag{2.43}
$$

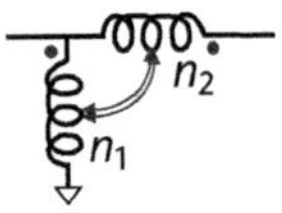

$$
[S] = \begin{bmatrix} \dfrac{\underline{Z}_2/\underline{Z}_1 + \left(1 + n_2/n_1\right)^2}{\underline{Z}_2/\underline{Z}_1 - \left(1 + n_2/n_1\right)^2} & -\dfrac{2\left(1 + n_2/n_1\right)}{\underline{Z}_2/\underline{Z}_1 - \left(1 + n_2/n_1\right)^2} \\ \dfrac{2\left(1 + n_2/n_1\right)\underline{Z}_2/\underline{Z}_1}{\underline{Z}_2/\underline{Z}_1 - \left(1 + n_2/n_1\right)^2} & -\dfrac{\underline{Z}_2/\underline{Z}_1 + \left(1 + n_2/n_1\right)^2}{\underline{Z}_2/\underline{Z}_1 - \left(1 + n_2/n_1\right)^2} \end{bmatrix} \tag{2.44}
$$

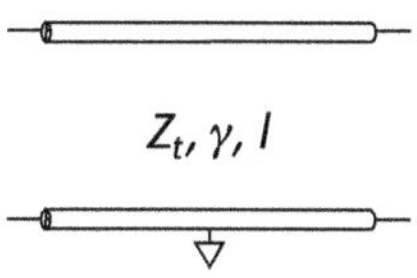

$$[S] = \begin{bmatrix} \frac{\bar{S}_{11}}{D} & \frac{\bar{S}_{12}}{D} \\ \frac{\bar{S}_{21}}{D} & \frac{\bar{S}_{22}}{D} \end{bmatrix} \tag{2.45}$$

where

$$\bar{S}_{11} = \underline{Z}_t(\underline{Z}_2 - \underline{Z}_1)\cosh(\gamma l) + (\underline{Z}_t^2 - \underline{Z}_1\underline{Z}_2)\sinh(\gamma l)$$
$$\bar{S}_{12} = 2\underline{Z}_1\underline{Z}_t$$
$$\bar{S}_{21} = 2\underline{Z}_2\underline{Z}_t$$
$$\bar{S}_{22} = \underline{Z}_t(\underline{Z}_1 - \underline{Z}_2)\cosh(\gamma l) + (\underline{Z}_t^2 - \underline{Z}_1\underline{Z}_2)\sinh(\gamma l)$$
$$D = \underline{Z}_t(\underline{Z}_1 + \underline{Z}_2)\cosh(\gamma l) + (\underline{Z}_t^2 + \underline{Z}_1\underline{Z}_2)\sinh(\gamma l)$$

where $\underline{Z}_t$ = transmission-line impedance. The complex propagation constant $\gamma = \alpha + j(2\pi/\lambda)$, where α = dissipative coefficient, λ = wavelength, and l = length of transmission line in wavelengths.

2.5 Coupled-Inductor Multiport S-Parameters

Coupled inductors are an important component and building block essential to the functioning of many RF communications circuits. They are used principally for phase splitting/combining of two RF signals or for impedance matching. A simplified schematic representation for such a component is shown in Figure 2.4.

Such a pair of coupled inductors is often referred to as a transformer. However, it should be noted that the coupled-inductor pair is characterized by two inductance values L_1 and L_2 and not by a turns ratio as would typically be used for a true transformer. This difference can be seen in the two-port examples in Figures 1.6 and 1.8. Viewing a coupled-inductor pair as a transformer can lead to many unrealistic expectations as is discussed in detail in Section 12.1.

Because coupled-inductor pairs are such an important component of RF circuits, it is worthwhile to characterize their S-parameter representations in the common topologies in which they are used.

As shown in Figure 2.4, in its most general form, a coupled-inductor pair constitutes a four-port network. Due to reciprocity and top-to-bottom symmetry of

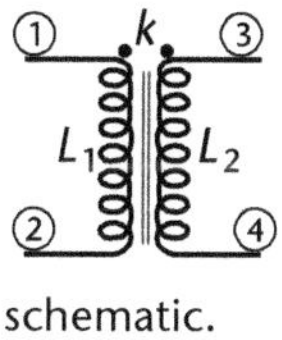

Figure 2.4 Coupled-inductor four-port schematic.

the network, the number of independent S-parameters required to represent the network reduces from 16 to 6. These are S_{11}, S_{12}, S_{13}, S_{14}, S_{33}, and S_{34}. The four-port S-parameter matrix for a coupled-inductor pair takes the form

$$\begin{bmatrix} b_1 \\ b_2 \\ b_3 \\ b_4 \end{bmatrix} = \begin{bmatrix} S_{11} & S_{12} & S_{13} & S_{14} \\ S_{12} & S_{11} & S_{14} & S_{13} \\ S_{13} & S_{14} & S_{33} & S_{34} \\ S_{14} & S_{13} & S_{34} & S_{33} \end{bmatrix} \cdot \begin{bmatrix} a_1 \\ a_2 \\ a_3 \\ a_4 \end{bmatrix} \tag{2.46}$$

or, writing the equations out explicitly,

$$\begin{aligned} b_1 &= S_{11}a_1 + S_{12}a_2 + S_{13}a_3 + S_{14}a_4 \\ b_2 &= S_{12}a_1 + S_{11}a_2 + S_{14}a_3 + S_{13}a_4 \\ b_3 &= S_{13}a_1 + S_{14}a_2 + S_{33}a_3 + S_{34}a_4 \\ b_4 &= S_{14}a_1 + S_{13}a_2 + S_{34}a_3 + S_{33}a_4 \end{aligned} \tag{2.47}$$

With the following definitions:

$$\begin{aligned} Z_0 &= \text{common four-port normalization impedance} \\ k &= \text{mutual coupling coefficient} \\ Z_{L1} &= R_{L1} + j\omega L_1 \\ Z_{L2} &= R_{L2} + j\omega L_2 \end{aligned} \tag{2.48}$$

The individual S-parameters can be expressed in terms of the circuit elements as

$$\begin{aligned} S_{11} &= \frac{(2Z_0 + Z_{L2})Z_{L1} + (\omega k)^2 L_1 L_2}{(2Z_0 + Z_{L1})(2Z_0 + Z_{L2}) + (\omega k)^2 L_1 L_2} \\ S_{12} &= \frac{2Z_0(2Z_0 + Z_{L2})}{(2Z_0 + Z_{L1})(2Z_0 + Z_{L2}) + (\omega k)^2 L_1 L_2} \\ S_{13} &= \frac{2j\omega k\sqrt{L_1 L_2}Z_0}{(2Z_0 + Z_{L1})(2Z_0 + Z_{L2}) + (\omega k)^2 L_1 L_2} \\ S_{14} &= \frac{-2j\omega k\sqrt{L_1 L_2}Z_0}{(2Z_0 + Z_{L1})(2Z_0 + Z_{L2}) + (\omega k)^2 L_1 L_2} \\ S_{33} &= \frac{(2Z_0 + Z_{L2})Z_{L2} + (\omega k)^2 L_1 L_2}{(2Z_0 + Z_{L1})(2Z_0 + Z_{L2}) + (\omega k)^2 L_1 L_2} \\ S_{34} &= \frac{2Z_0(2Z_0 + Z_{L1})}{(2Z_0 + Z_{L1})(2Z_0 + Z_{L2}) + (\omega k)^2 L_1 L_2} \end{aligned} \tag{2.49}$$

Another, more restricted, application of the coupled-inductor configuration is its use as a three-port. In this case, one of the ports of the generalized network is grounded, as shown in Figure 2.5.

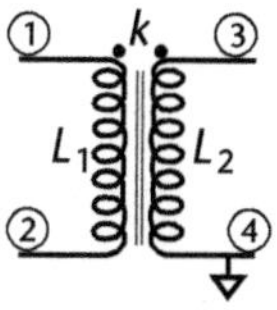

Figure 2.5 Coupled-inductor three-port schematic.

This architecture is commonly employed for differential combining or splitting and its use is considered in detail in Chapter 14.

Due to reciprocity and top-to-bottom symmetry of the network, in this configuration, the network requires only five independent S-parameters to represent it. In terms of the four-port S-parameters, given in (2.49), the modified S-parameter matrix $[S']$ for the three-port configuration is

$$\begin{bmatrix} b_1 \\ b_2 \\ b_3 \end{bmatrix} = \begin{bmatrix} S'_{11} & S'_{12} & S'_{13} \\ S'_{12} & S'_{11} & S'_{23} \\ S'_{13} & S'_{23} & S'_{33} \end{bmatrix} \cdot \begin{bmatrix} a_1 \\ a_2 \\ a_3 \end{bmatrix} \tag{2.50}$$

or, writing the equations out explicitly,

$$\begin{aligned} b_1 &= S'_{11}a_1 + S'_{12}a_2 + S'_{13}a_3 \\ b_2 &= S'_{12}a_1 + S'_{11}a_2 + S'_{23}a_3 \\ b_3 &= S'_{13}a_1 + S'_{23}a_2 + S'_{33}a_3 \end{aligned} \tag{2.51}$$

From circuit analysis, the three-port S-parameters can be derived in terms of thc four-port S-paramctcrs as:

$$\begin{aligned} S'_{11} &= S_{11} - \frac{S_{13}^2}{1+S_{33}} \\ S'_{12} &= S_{12} + \frac{S_{13}^2}{1+S_{33}} \\ S'_{13} &= \frac{2S_{13}}{1+S_{33}} \\ S'_{23} &= \frac{-2S_{13}}{1+S_{33}} \\ S'_{33} &= \frac{3S_{33}-1}{1+S_{33}} \end{aligned} \tag{2.52}$$

A still more restricted use of the four-port coupled-inductor network is it use as a two-port. In this case, the four-port network is grounded on two ports, as shown in Figure 2.6.

Applications of this inline coupler configuration for PA output matching are examined in Chapter 12.

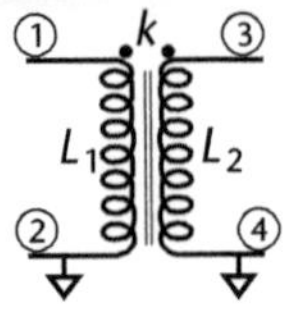

Figure 2.6 Coupled-inductor two-port schematic.

For this configuration, only three independent S-parameters are required to represent the network. The modified two-port S-parameter network matrix $[S']$ is

$$\begin{bmatrix} b_1 \\ b_3 \end{bmatrix} = \begin{bmatrix} S'_{11} & S'_{13} \\ S'_{13} & S'_{33} \end{bmatrix} \cdot \begin{bmatrix} a_1 \\ a_3 \end{bmatrix} \tag{2.53}$$

or, writing the equations out explicitly,

$$\begin{aligned} b_1 &= S'_{11}a_1 + S'_{13}a_3 \\ b_1 &= S'_{13}a_1 + S'_{33}a_3 \end{aligned} \tag{2.54}$$

From circuit analysis, the two-port S-parameters can be derived in terms of the four-port S-parameters as:

$$\begin{aligned} S'_{11} &= \frac{(1-3S_{11})(1+S_{33})+3S_{13}^2}{S_{13}^2-(1+S_{11})(1+S_{33})} \\ S'_{13} &= \frac{-4S_{13}}{S_{13}^2-(1+S_{11})(1+S_{33})} \\ S'_{33} &= \frac{(1+S_{11})(1-3S_{33})+3S_{13}^2}{S_{13}^2-(1+S_{11})(1+S_{33})} \end{aligned} \tag{2.55}$$

When a coupled-inductor pair is incorporated in a compact RF layout, for optimum circuit performance, it is generally desirable that the mutual magnetic coupling factor be made as large as possible. Doing this requires that the two inductors are placed in the closest possible proximity to each other. The two inductors are typically implemented by conductor traces on the substrate or printed circuit board (pcb). These traces may be on the same layer, as in Figure 12.2, or superposed on adjacent layers. In either case, with the inductors in close proximity, in addition to the magnetic coupling, there will be a distributed capacitive coupling. This undesired capacitive coupling between the two coils can significantly impact the performance of the coupler, especially at higher frequencies.

Adding capacitive cross-coupling into the model for the generalized four-port coupler of Figure 2.4 is extremely challenging. However, closed-form analytic solutions can be obtained for a more restricted case, which are helpful in understanding the impact of cross-capacitance on a coupled-inductor performance. For the restricted case $L_1 = L_2$, the terminating external impedances on all ports are Z_0. A coupled-inductor network with equal-value inductors is an important building block in many circuits. These include shunted-inductor lattice couplers (SILCs) (Chapter

1 of Volume 2 of this series), branch-line couplers (Chapter 2 of Volume 2 of this series), Wilkinson couplers (Chapter 8 of Volume 2 of this series), and others. A simplified schematic for such a coupler is shown in Figure 2.7.

The S-parameters for the four-port network are again represented by (2.46) and (2.47). The following variables are defined for the analysis:

Definitions:

L = inductance of top and bottom inductors
Q_L = inductor quality factors
$R_L = \omega_0 L/Q_L$ = series resistance of each inductor
$Z_L = R_L + j\omega L$
k = inductors' mutual coupling coefficient
C = total cross-capacitance between inductors
Q_C = capacitor quality factors
$G_C = \omega_0 C/Q_C$ = capacitive parallel conductance
$Y_C = G_C + j\omega C$

Determination of the S-parameters is most easily carried out using an incremental model for the coupler. The incremental schematic to be used for analysis is shown in Figure 2.8.

The generalized four-port coupled-inductor network with $L_1 = L_2 = L$ is represented by the distributed schematic shown in Figure 2.8. The external terminating impedances on all four-ports are Z_0.

Here, δL is the incremental inductance per unit length and δC is the incremental capacitance. In terms of the overall inductor values, these incremental values are given by $\delta L - L/l$, and $\delta C - C/l$, where l − length of the inductor. The analysis of the coupled-inductor network is most readily carried out using an even-odd mode analysis. For this, the boundary conditions applied are a source voltage V_S on port 1, with zero excitation on port 3. This excitation is then split into an odd-mode drive with $+V/2$ on port 1 and $-V/2$ on port 3 and an even-mode drive with $+V/2$ applied to both ports. The two drive scenarios are shown schematically in Figure 2.9.

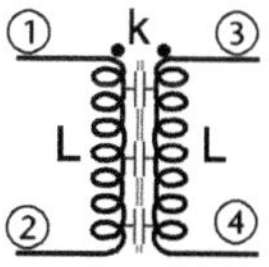

Figure 2.7 Equal coupled inductors with cross-capacitance.

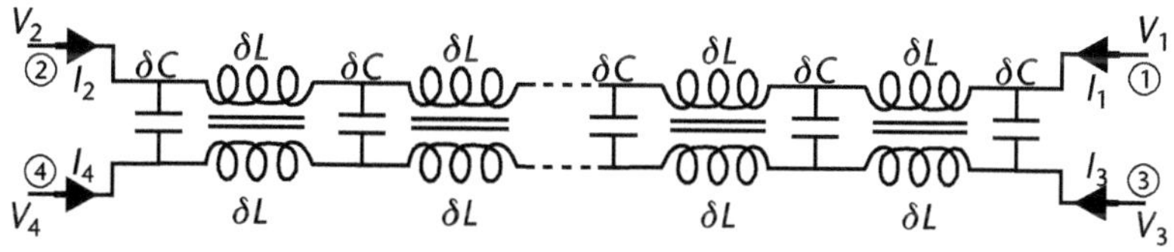

Figure 2.8 Distributed four-port coupled-inductor model with cross-capacitance.

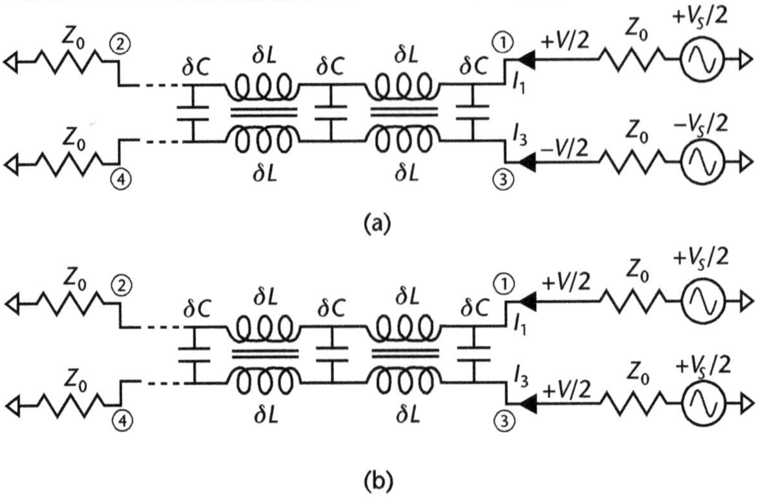

Figure 2.9 Even-odd mode drive on ports 1 and 3 of equal coupled inductors: (a) odd mode, and (b) even mode.

The net transfer coefficients defining the network are then obtained by linear superposition.

For the odd-mode drive, the equivalent network for both the upper and lower inductors is as shown in Figure 2.10. For the even-mode drive, the equivalent network for both the upper and lower inductors is as shown in Figure 2.11.

For the odd-mode drive, the network is equivalent to a transmission line. The $[ABCD]$ matrix coefficients for such a transmission line, with lossy elements are determined from (1.29), as:

$$[M] = \begin{bmatrix} \cosh\phi & Z_l \sinh\phi \\ \dfrac{\sinh\phi}{Z_l} & \cosh\phi \end{bmatrix} \tag{2.56}$$

where line impedance

$$Z_l = \sqrt{\frac{(Z_L - j\omega kL)}{2Y_C}} \tag{2.57}$$

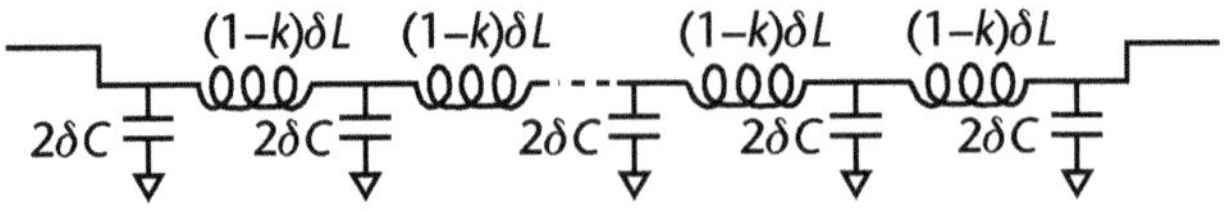

Figure 2.10 Odd-mode network schematic for equal coupled inductors with capacitance.

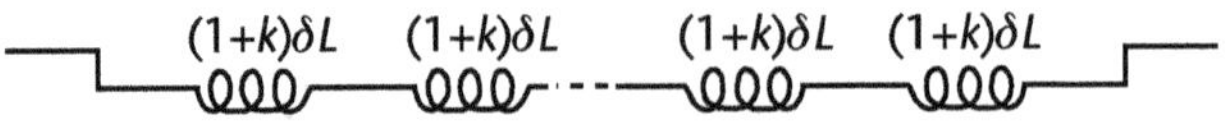

Figure 2.11 Even-mode network schematic for equal coupled inductors with capacitance.

propagation coefficient $$\gamma = \frac{\sqrt{(Z_L - j\omega kL)2Y_C}}{l} \tag{2.58}$$

and

$$\phi = \gamma l = \sqrt{(Z_L - j\omega kL)2Y_C} \tag{2.59}$$

From analysis, we determine on port 1

$$\frac{V_1}{V_S/2} = \frac{Z_l\left(Z_0 \cosh\phi + Z_l \sinh\phi\right)}{2Z_0Z_l \cosh\phi + \left(Z_0^2 + Z_l^2\right)\sinh\phi} \tag{2.60}$$

and on port 2

$$\frac{V_2}{V_S/2} = \frac{Z_0Z_l}{2Z_0Z_l \cosh\phi + \left(Z_0^2 + Z_l^2\right)\sinh\phi} \tag{2.61}$$

For the even-mode drive, circuit analysis yields:

$$\frac{V_1}{V_S/2} = \frac{Z_0 + R_L + j\omega(1+k)L}{2Z_0 + R_L + j\omega(1+k)L} \tag{2.62}$$

$$\frac{V_2}{V_S/2} = \frac{Z_0}{2Z_0 + R_L + j\omega(1+k)L} \tag{2.63}$$

Hence, for excitation on port 1, the net voltages are

$$\frac{V_1}{V_S/2} = \frac{Z_0 + R_L + j\omega(1+k)L}{2Z_0 + R_L + j\omega(1+k)L} + \frac{Z_l\left(Z_0 \cosh\phi + Z_l \sinh\phi\right)}{2Z_0Z_l \cosh\phi + \left(Z_0^2 + Z_l^2\right)\sinh\phi} \tag{2.64}$$

$$\frac{V_2}{V_S/2} = \frac{Z_0}{2Z_0 + R_L + j\omega(1+k)L} + \frac{Z_0Z_l}{2Z_0Z_l \cosh\phi + \left(Z_0^2 + Z_l^2\right)\sinh\phi} \tag{2.65}$$

$$\frac{V_3}{V_S/2} = \frac{Z_0 + R_L + j\omega(1+k)L}{2Z_0 + R_L + j\omega(1+k)L} - \frac{Z_l\left(Z_0 \cosh\phi + Z_l \sinh\phi\right)}{2Z_0Z_l \cosh\phi + \left(Z_0^2 + Z_l^2\right)\sinh\phi} \tag{2.66}$$

$$\frac{V_4}{V_S/2} = \frac{Z_0}{2Z_0 + R_L + j\omega(1+k)L} - \frac{Z_0Z_l}{2Z_0Z_l \cosh\phi + \left(Z_0^2 + Z_l^2\right)\sinh\phi} \tag{2.67}$$

Assuming the normalization impedance Z_0 to be pure real, as is invariably the case, for the S-parameters, we have

$$V_1 = \left(1 + S_{11}\right)V_S/2 \tag{2.68}$$

thus

$$S_{11} = \frac{V_1}{V_S/2} - 1 \tag{2.69}$$

$$S_{12} = \frac{V_2}{V_S/2} \tag{2.70}$$

$$S_{13} = \frac{V_3}{V_S/2} \tag{2.71}$$

$$S_{14} = \frac{V_4}{V_S/2} \tag{2.72}$$

Substituting from (2.64) to (2.67), the S-parameter expressions for the equal-valued, coupled-inductor network, accounting for cross-coupling capacitance are determined as

$$\begin{aligned} S_{11} &= S_{22} = S_{33} = S_{44} \\ &= \frac{-Z_0}{2Z_0 + R_L + j\omega(1+k)L} + \frac{Z_l(Z_0\cosh\phi + Z_l\sinh\phi)}{2Z_0Z_l\cosh\phi + (Z_0^2 + Z_l^2)\sinh\phi} \end{aligned} \tag{2.73}$$

$$\begin{aligned} S_{12} &= S_{21} = S_{34} = S_{43} \\ &= \frac{Z_0}{2Z_0 + R_L + j\omega(1+k)L} + \frac{Z_0Z_l}{2Z_0Z_l\cosh\phi + (Z_0^2 + Z_l^2)\sinh\phi} \end{aligned} \tag{2.74}$$

$$\begin{aligned} S_{13} &= S_{31} = S_{24} = S_{42} \\ &= \frac{Z_0 + R_L + j\omega(1+k)L}{2Z_0 + R_L + j\omega(1+k)L} - \frac{Z_l(Z_0\cosh\phi + Z_l\sinh\phi)}{2Z_0Z_l\cosh\phi + (Z_0^2 + Z_l^2)\sinh\phi} \end{aligned} \tag{2.75}$$

$$\begin{aligned} S_{14} &= S_{41} = S_{23} = S_{32} \\ &= \frac{Z_0}{2Z_0 + R_L + j\omega(1+k)L} - \frac{Z_0Z_l}{2Z_0Z_l\cosh\phi + (Z_0^2 + Z_l^2)\sinh\phi} \end{aligned} \tag{2.76}$$

2.6 Interconnection of Two Two-Port S-Parameter Networks

Figure 2.12 shows a series cascade of two two-port S-parameter networks.

Denoting S-parameters resulting from the cascade of the two two-port networks by $\tilde{S}_{mn}$, we have

$$\tilde{S}_{11} = S_{111} + \frac{S_{112}S_{121}S_{211}}{1 - S_{122}S_{211}} \tag{2.77}$$

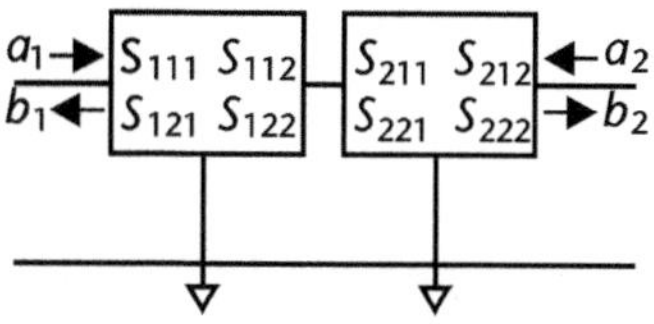

Figure 2.12 Cascaded two-port S-parameter networks.

$$\tilde{S}_{12} = \frac{S_{112}S_{212}}{1 - S_{122}S_{211}} \tag{2.78}$$

$$\tilde{S}_{21} = \frac{S_{121}S_{221}}{1 - S_{122}S_{211}} \tag{2.79}$$

$$\tilde{S}_{22} = S_{222} + \frac{S_{122}S_{212}S_{221}}{1 - S_{122}S_{211}} \tag{2.80}$$

Figure 2.13 shows a parallel connection of two two-port S-parameter networks with the following equalities:

$$D_1 = (1 + S_{111})(1 + S_{122}) - S_{112}S_{121} \tag{2.81}$$

$$D_2 = (1 + S_{211})(1 + S_{222}) - S_{212}S_{221} \tag{2.82}$$

$$\begin{aligned} E = {} & D_1D_2 - 2(S_{111} + S_{122})D_2 - 2(S_{211} + S_{222})D_1 \\ & + 4\left((1 + S_{111})(1 + S_{222}) + (1 + S_{211})(1 + S_{222}) - S_{112}S_{221} - S_{212}S_{221}\right) \end{aligned} \tag{2.83}$$

Denoting S-parameters resulting from the cascade of the two two-port networks by $\tilde{S}_{mn}$, we have

$$\tilde{S}_{11} = -1 + \frac{2\left(2(1 + S_{111})D_2 + 2(1 + S_{211})D_1 - D_1D_2\right)}{E} \tag{2.84}$$

$$\tilde{S}_{12} = \frac{4(S_{111}D_2 + S_{212}D_1)}{E} \tag{2.85}$$

$$\tilde{S}_{21} = \frac{4(S_{121}D_2 + S_{221}D_1)}{E} \tag{2.86}$$

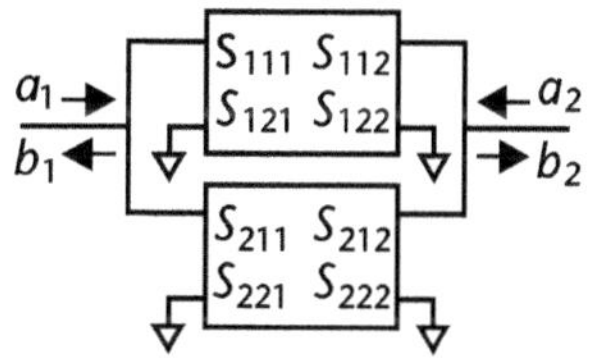

Figure 2.13 Two parallel two-port S-parameter networks.

$$\tilde{S}_{22} = -1 + \frac{2\left(2(1+S_{122})D_2 + 2(1+S_{222})D_1 - D_1D_2\right)}{E} \quad (2.87)$$

Figure 2.14 shows a stacked connection of two two-port S-parameter networks with the following equalities:

$$D_1 = (1 - S_{111})(1 - S_{122}) - S_{112}S_{121} \quad (2.88)$$

$$D_2 = (1 - S_{211})(1 - S_{222}) - S_{212}S_{221} \quad (2.89)$$

$$\begin{aligned} E = {} & D_1D_2 + 2(S_{111} + S_{122})D_2 + 2(S_{211} + S_{222})D_1 \\ & + 4\left((1 - S_{122})(1 - S_{211}) + (1 - S_{111})(1 - S_{222}) - S_{112}S_{221} - S_{121}S_{212}\right) \end{aligned} \quad (2.90)$$

Denoting S-parameters resulting from the cascade of the two two-port networks by $\tilde{S}_{mn}$, we have

$$\tilde{S}_{11} = 1 - \frac{2\left(2(1-S_{111})D_2 + 2(1-S_{211})D_1 - D_1D_2\right)}{E} \quad (2.91)$$

$$\tilde{S}_{12} = \frac{4(S_{112}D_2 + S_{212}D_1)}{E} \quad (2.92)$$

$$\tilde{S}_{21} = \frac{4(S_{121}D_2 + S_{221}D_1)}{E} \quad (2.93)$$

$$\tilde{S}_{22} = 1 - \frac{2\left(2(1-S_{122})D_2 + 2(1-S_{222})D_1 - D_1D_2\right)}{E} \quad (2.94)$$

Figure 2.15 shows a parallel-series connection of two two-port S-parameter networks with the following equalities:

$$D_1 = (1 + S_{111})(1 - S_{122}) + S_{112}S_{121} \quad (2.95)$$

$$D_2 = (1 + S_{211})(1 - S_{222}) + S_{212}S_{221} \quad (2.96)$$

$$\begin{aligned} E = {} & D_1D_2 - 2(S_{111} - S_{122})D_2 - 2(S_{211} - S_{222})D_1 \\ & + 4\left((1 + S_{111})(1 - S_{222}) + (1 + S_{211})(1 - S_{122}) + S_{112}S_{221} + S_{121}S_{212}\right) \end{aligned} \quad (2.97)$$

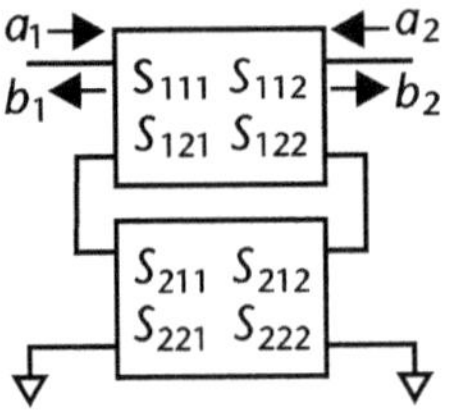

Figure 2.14 Two stacked two-port S-parameter networks.

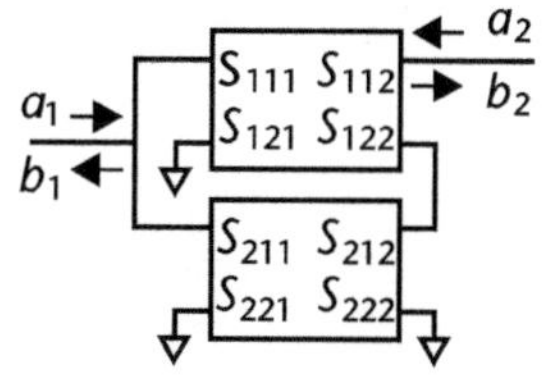

Figure 2.15 Parallel-series connection of two, two-port S-parameter networks.

Denoting S-parameters resulting from the cascade of the two two-port networks by $\tilde{S}_{mn}$, we have

$$\tilde{S}_{11} = -1 + \frac{2\left(2(1+S_{111})D_2 + 2(1+S_{211})D_1 - D_1D_2\right)}{E} \quad (2.98)$$

$$\tilde{S}_{12} = \frac{4(S_{112}D_2 + S_{212}D_1)}{E} \quad (2.99)$$

$$\tilde{S}_{21} = \frac{4(S_{121}D_2 + S_{221}D_1)}{E} \quad (2.100)$$

$$\tilde{S}_{22} = 1 - \frac{2\left(2(1-S_{122})D_2 + 2(1-S_{222})D_1 - D_1D_2\right)}{E} \quad (2.101)$$

Figure 2.16 shows a series-parallel connection of two two-port S-parameter networks with the following equalities:

$$D_1 = (1-S_{111})(1+S_{122}) + S_{112}S_{121} \quad (2.102)$$

$$D_2 = (1-S_{211})(1+S_{222}) + S_{212}S_{221} \quad (2.103)$$

$$\begin{aligned} E = {} & D_1D_2 + 2(S_{111}-S_{122})D_2 + 2(S_{211}-S_{222})D_1 \\ & + 4\left((1-S_{111})(1+S_{222}) + (1+S_{122})(1-S_{211}) + S_{112}S_{221} + S_{121}S_{212}\right) \end{aligned} \quad (2.104)$$

Denoting S-parameters resulting from the cascade of the two two-port networks by $\tilde{S}_{mn}$, we have

$$\tilde{S}_{11} = 1 - \frac{2\left(2(1-S_{111})D_2 + 2(1-S_{211})D_1 - D_1D_2\right)}{E} \quad (2.105)$$

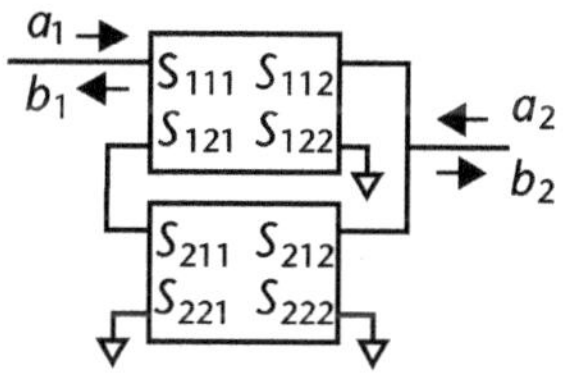

Figure 2.16 Series-parallel connection of two, two-port S-parameter networks.

$$\tilde{S}_{12} = \frac{4(S_{112}D_2 + S_{212}D_1)}{E} \quad (2.106)$$

$$\tilde{S}_{21} = \frac{4(S_{121}D_2 + S_{221}D_1)}{E} \quad (2.107)$$

$$\tilde{S}_{22} = -1 + \frac{2(2(1 + S_{122})D_2 + 2(1 + S_{222})D_1 - D_1D_2)}{E} \quad (2.108)$$

Figure 2.17 shows a two-port with a direct input-output connection across the network.

Denoting the resulting S-parameters of the two-port network by $\tilde{S}_{mn}$, we have

$$\tilde{S}_{11} = \tilde{S}_{22} = \frac{S_{11}S_{22} - (1 - S_{12})(1 - S_{21})}{2 + S_{11} + S_{22} - S_{12} - S_{21}} \quad (2.109)$$

$$\tilde{S}_{12} = \tilde{S}_{21} = \frac{(1 + S_{11})(1 + S_{22}) - S_{12}S_{21}}{2 + S_{11} + S_{22} - S_{12} - S_{21}} \quad (2.110)$$

Figure 2.18 shows a two-port comprising a direct line with terminated two-port shunt to ground.

The reflection coefficient on output port of the two-port is

$$\Gamma_L = \frac{Z_L - \underline{Z}_2}{Z_L + \underline{Z}_2} \quad (2.111)$$

where $\underline{Z}_2$ is the characteristic impedance on port 2 of the network.

Denoting the resulting S-parameters of the two-port network by $\tilde{S}_{mn}$, we have

$$\tilde{S}_{11} = \tilde{S}_{22} = \frac{S_{11} - 1 + \Gamma_L(S_{22} - S_{11}S_{22} + S_{12}S_{21})}{S_{11} + 3 - \Gamma_L(3S_{22} + S_{11}S_{22} - S_{12}S_{21})} \quad (2.112)$$

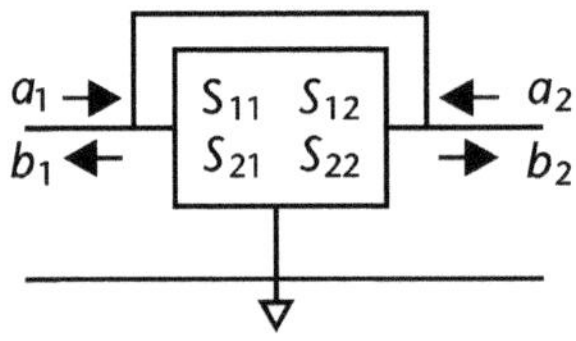

Figure 2.17 Two-port with input-output short S-parameters.

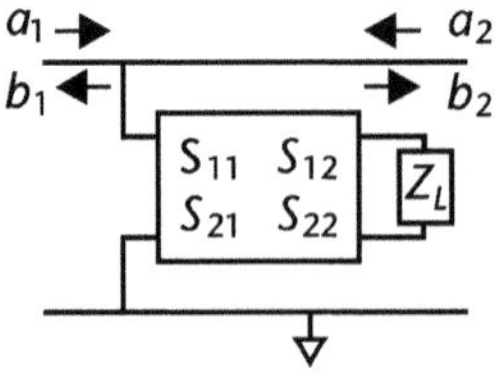

Figure 2.18 Terminated two-port in parallel with through-line S-parameters.

$$\tilde{S}_{12} = \tilde{S}_{21} = 1 + \tilde{S}_{11} \tag{2.113}$$

Figure 2.19 shows a terminated two-port connected in series with direct line.

$$\tilde{S}_{11} = \tilde{S}_{22} = \frac{S_{11} + 1 - \Gamma_L\left(S_{22} + S_{11}S_{22} - S_{12}S_{21}\right)}{3 - S_{11} - \Gamma_L\left(3S_{22} - S_{11}S_{22} + S_{12}S_{21}\right)} \tag{2.114}$$

$$\tilde{S}_{12} = \tilde{S}_{21} = 1 - \tilde{S}_{11} \tag{2.115}$$

Figures 2.20(a, b) show the schematic exchange of port 1 with the ground of a two-port network. Such a transformation could be used, for example, to convert the S-parameters of a transistor, measured in common-emitter configuration, to common-base, as shown in Figures 2.20(c, d).

Defining

$$D = (1 + S_{11})(1 - S_{22}) + (2 - S_{12})(2 - S_{21}) \tag{2.116}$$

and, denoting the resulting S-parameters of the two-port network by $\tilde{S}_{mn}$, we have

$$\tilde{S}_{11} = -1 + \frac{4(1 + S_{11})}{D} \tag{2.117}$$

$$\tilde{S}_{12} = 2\left(1 - \frac{2(2 - S_{21})}{D}\right) \tag{2.118}$$

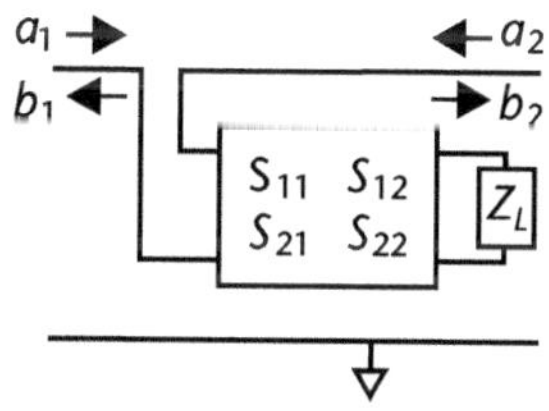

Figure 2.19 Terminated two-port in series with through-line S-parameters.

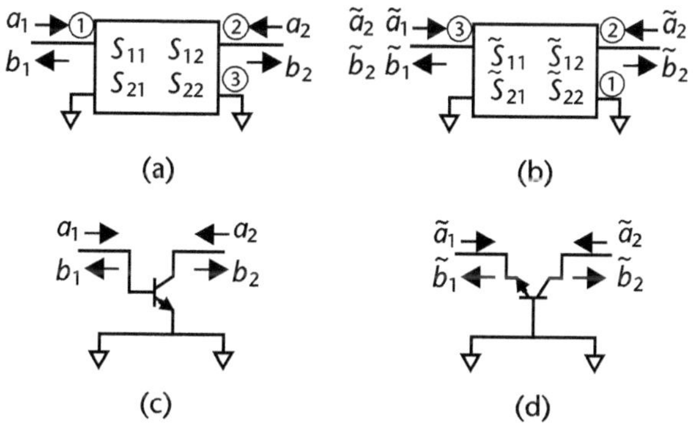

Figure 2.20 Exchanging port 1 and ground port S-parameters: (a), (b), (c), and (d).

$$\tilde{S}_{21} = 2\left(1 - \frac{2(2 - S_{12})}{D}\right) \quad (2.119)$$

$$\tilde{S}_{22} = 1 - \frac{4(1 - S_{22})}{D} \quad (2.120)$$

The transformations for exchanging port 2 with ground follow trivially from above.

2.7 S-Parameter Reduction of a Terminated Three-Port to a Two-Port Network

When a three-port S-parameter network is terminated on one-port by a complex impedance, the resulting network reduces to an equivalent two-port network. This section presents the modified S-parameters of the resultant two-port network in terms of the original network parameters. Figure 2.21 shows the S-parameter representation of a three-port network with a complex termination impedance, Z_L, on port 3.

The reflection coefficient of the load on port 3 is defined by

$$\Gamma_L = \frac{Z_L - \underline{Z}_3}{Z_L + \underline{Z}_3} \quad (2.121)$$

where $\underline{Z}_3$ is the characteristic impedance on port 3 of the network.

Denoting the S-parameters, the resulting two-port network by $\tilde{S}_{mn}$, we find

$$\tilde{S}_{11} = S_{11} + \frac{\Gamma_L S_{13} S_{31}}{1 - \Gamma_L S_{33}} \quad (2.122)$$

$$\tilde{S}_{12} = S_{12} + \frac{\Gamma_L S_{13} S_{32}}{1 - \Gamma_L S_{33}} \quad (2.123)$$

$$\tilde{S}_{21} = S_{21} + \frac{\Gamma_L S_{23} S_{31}}{1 - \Gamma_L S_{33}} \quad (2.124)$$

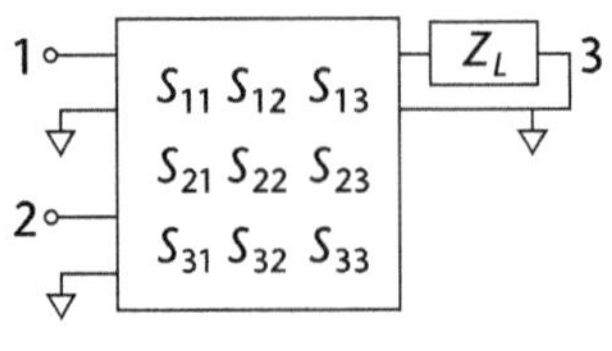

Figure 2.21 Terminated three-port S-parameter network.

$$\tilde{S}_{22} = S_{22} + \frac{\Gamma_L S_{23} S_{32}}{1 - \Gamma_L S_{33}} \tag{2.125}$$

2.8 S-Parameter Reduction of a Terminated Four-Port to a Three-Port Network

When a four-port S-parameter network is terminated on one-port by a complex impedance, the resulting network reduces to an equivalent three-port. This section presents the modified S-parameters of the resultant three-port network in terms of the original network parameters. Figure 2.22 shows the S-parameter representation of a four-port network with the complex termination impedance, Z_L, on port 4.

The reflection coefficient of the load on port 4 is defined by

$$\Gamma_4 = \frac{Z_L - \underline{Z}_4}{Z_L + \underline{Z}_4} \tag{2.126}$$

where $\underline{Z}_4$ is the characteristic impedance on port 4 of the network.

Denoting the S-parameters, the resulting three-port network by $\tilde{S}_{mn}$, we find

$$\tilde{S}_{11} = S_{11} + \Gamma_4 S_{14} S_{41} / (1 - \Gamma_4 S_{44}) \tag{2.127}$$

$$\tilde{S}_{12} = S_{12} + \Gamma_4 S_{14} S_{42} / (1 - \Gamma_4 S_{44}) \tag{2.128}$$

$$\tilde{S}_{13} = S_{13} + \Gamma_4 S_{14} S_{43} / (1 - \Gamma_4 S_{44}) \tag{2.129}$$

$$\tilde{S}_{21} = S_{21} + \Gamma_4 S_{24} S_{41} / (1 - \Gamma_4 S_{44}) \tag{2.130}$$

$$\tilde{S}_{22} = S_{22} + \Gamma_4 S_{24} S_{42} / (1 - \Gamma_4 S_{44}) \tag{2.131}$$

$$\tilde{S}_{23} = S_{23} + \Gamma_4 S_{24} S_{43} / (1 - \Gamma_4 S_{44}) \tag{2.132}$$

$$\tilde{S}_{31} = S_{31} + \Gamma_4 S_{34} S_{41} / (1 - \Gamma_4 S_{44}) \tag{2.133}$$

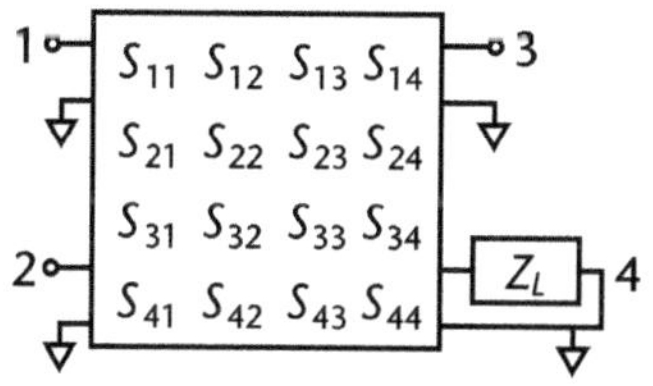

Figure 2.22 Terminated four-port S-parameter network.

$$\tilde{S}_{32} = S_{32} + \Gamma_4 S_{34} S_{42} / (1 - \Gamma_4 S_{44}) \quad (2.134)$$

$$\tilde{S}_{33} = S_{33} + \Gamma_4 S_{34} S_{43} / (1 - \Gamma_4 S_{44}) \quad (2.135)$$

2.9 Useful Three-Port Formulae

Three-port networks are very common and find many applications in RF circuits. Examples include couplers, power splitters and combiners, and diplexers. In this chapter, some useful formulae are presented to aid in their analysis and design. Figure 2.23 shows key circuit parameters appropriate to analysis of a three-port network characterized by scattering parameters.

Three-port S-parameter equations are:

$$\begin{bmatrix} b_1 \\ b_2 \\ b_3 \end{bmatrix} = \begin{bmatrix} S_{11} & S_{12} & S_{13} \\ S_{21} & S_{22} & S_{23} \\ S_{31} & S_{32} & S_{33} \end{bmatrix} \cdot \begin{bmatrix} a_1 \\ a_2 \\ a_3 \end{bmatrix} \quad (2.136)$$

that is,

$$b_1 = S_{11} a_1 + S_{12} a_2 + S_{13} a_3 \quad (2.137)$$

$$b_2 = S_{21} a_1 + S_{22} a_2 + S_{23} a_3 \quad (2.138)$$

$$b_3 = S_{31} a_1 + S_{32} a_2 + S_{33} a_3 \quad (2.139)$$

From circuit analysis, we have

$$V_1 = a_1 + b_1 \quad (2.140)$$

$$V_2 = a_2 + b_2 \quad (2.141)$$

$$V_3 = a_3 + b_3 \quad (2.142)$$

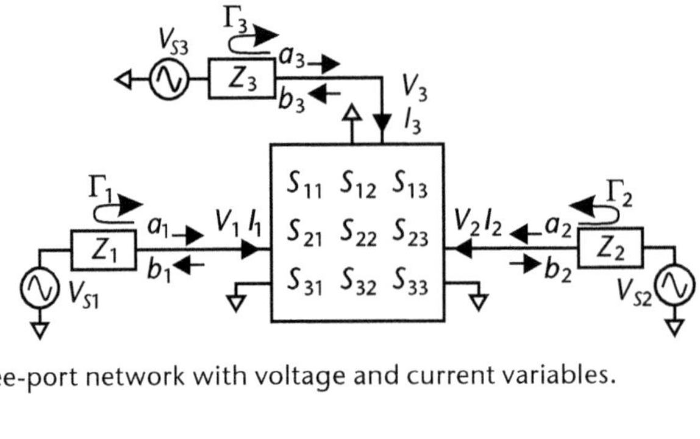

Figure 2.23 Three-port network with voltage and current variables.

and

$$I_1 = (a_1 - b_1)/\underline{Z}_1 \tag{2.143}$$

$$I_2 = (a_2 - b_2)/\underline{Z}_2 \tag{2.144}$$

$$I_3 = (a_3 - b_3)/\underline{Z}_3 \tag{2.145}$$

where $\underline{Z}_1$, $\underline{Z}_2$, and $\underline{Z}_3$ are the normalization impedances on ports 1, 2, and 3, respectively.

If the external impedances on the three ports are Z_1, Z_2, and Z_3, the loading reflection coefficients are given by

$$\begin{aligned} \Gamma_1 &= \frac{Z_1 - \underline{Z}_1}{Z_1 + \underline{Z}_1} \\ \Gamma_2 &= \frac{Z_2 - \underline{Z}_2}{Z_2 + \underline{Z}_2} \\ \Gamma_3 &= \frac{Z_3 - \underline{Z}_3}{Z_3 + \underline{Z}_3} \end{aligned} \tag{2.146}$$

While three-port networks are extremely useful, some care must be taken when working with them, on the bench or analytically. The principal difficulty is that, from simple analysis, it can be shown that the three ports of a lossless network cannot be matched simultaneously, which implies that the following equation cannot be realized in any physical lossless reciprocal network

$$S_{11} = S_{22} = S_{33} = 0 \tag{2.147}$$

Although a lossless three-port network cannot be simultaneously matched on all ports, this does not mean that power transfer through the network must incur some mismatch losses. In fact, power transfer through the network can occur with great efficiency, save only for dissipative losses, so what is the consequence of not being able to implement a three-port network that satisfies (2.147)?

Consider a network with two ports arbitrarily terminated. At a given frequency, it will generally be possible to conjugately match the input at the third port to an arbitrary source impedance. Power will then flow from the source to the two loads with no mismatch loss, despite the fact that at the two loaded ports there would not exist a conjugate match between the loads and the network. Due to the latter, if all external impedances were left unchanged and one of the two previously loaded ports now driven instead, there would be power loss due to mismatch at that port.

In the case that a three-port is used as a coupler, two ports, rather than one, are simultaneously driven. Again, this function can be accomplished in the three-port without any mismatch loss on the two ports or output if it is designed appropriately. It is simply necessary to imagine the three-port scenario in the previous paragraph being operated in reverse. By reciprocity, if the input power on one port of the network is split efficiently to the two alternate ports, then simultaneously applying corresponding output voltages to the said output ports will result in the network efficiently combining the two inputs at the previous input port. Note that, in this case, the two input ports strongly interact. If either of the ports were to be excited independently, there would be a mismatch loss at that port. Furthermore, if the two ports are excited simultaneously, but the relative drive voltages differ from the ratio resulting from the splitter scenario, each of the two ports will suffer a mismatch loss.

The subtleties of the use of a three-port as a combiner are most easily understood, and analyzed, using linear superposition. For example, to analyze the three-port as a coupler, we can first determine the reflected power at the first coupled input, setting the voltage source at the second coupled port to zero. Next, setting the voltage at the first coupled port to zero and that at the second coupled port to its appropriate value, the power coupled (or leaking) to the first port can be determined. There will be a perfect match (i.e., no reflected power) at the first port if the sum of the reflected and coupled power at the first coupled port exactly cancel. This is similar for the match at the second coupled port.

2.9.1 Input Impedance and Voltage Division, Driving Port 1 Alone

$$Z_{1\text{in}} = \left(\frac{2}{1 - S_{11} - \dfrac{\Gamma_2 S_{12}\left(S_{21} + \Gamma_3\left(S_{23}S_{31} - S_{21}S_{33}\right)\right) + \Gamma_3 S_{13}\left(S_{31} + \Gamma_2\left(S_{32}S_{21} - S_{22}S_{31}\right)\right)}{\left(1 - \Gamma_2 S_{22}\right)\left(1 - \Gamma_3 S_{33}\right) - \Gamma_2\Gamma_3 S_{23}S_{32}}} - 1 \right) Z_1 \tag{2.148}$$

$$\frac{V_2}{V_{S1}} = \frac{\left(1 + \Gamma_2\right)\left(S_{21} + \Gamma_3\left(S_{23}S_{31} - S_{21}S_{33}\right)\right)}{\left(1 - \Gamma_2 S_{22}\right)\left(1 - \Gamma_3 S_{33}\right) - \Gamma_2\Gamma_3 S_{23}S_{32}} \frac{Z_1 + Z_{1\_\text{in}}}{2\left(Z_1 + Z_{1\_\text{in}}\right)} \tag{2.149}$$

$$\frac{V_3}{V_{S1}} = \frac{\left(1 + \Gamma_3\right)\left(S_{31} + \Gamma_2\left(S_{32}S_{21} - S_{22}S_{31}\right)\right)}{\left(1 - \Gamma_2 S_{22}\right)\left(1 - \Gamma_3 S_{33}\right) - \Gamma_2\Gamma_3 S_{32}S_{23}} \frac{Z_1 + Z_{1\_\text{in}}}{2\left(Z_1 + Z_{1\_\text{in}}\right)} \tag{2.150}$$

These equations may be used to calculate power transfer.

2.9.2 Input Impedance and Voltage Division, Driving Port 2 Alone

$$Z_{2\text{in}} = \left(\frac{2}{1 - S_{22} - \dfrac{\Gamma_1 S_{21}\left(S_{12} + \Gamma_3\left(S_{13}S_{32} - S_{12}S_{33}\right)\right) + \Gamma_3 S_{23}\left(S_{32} + \Gamma_1\left(S_{12}S_{31} - S_{11}S_{32}\right)\right)}{\left(1 - \Gamma_1 S_{11}\right)\left(1 - \Gamma_3 S_{33}\right) - \Gamma_1\Gamma_3 S_{13}S_{31}}} - 1 \right) Z_2 \tag{2.151}$$

$$\frac{V_1}{V_{S2}} = \frac{(1+\Gamma_1)(S_{12}+\Gamma_3(S_{13}S_{32}-S_{12}S_{33}))}{(1-\Gamma_1 S_{11})(1-\Gamma_3 S_{33})-\Gamma_1\Gamma_3 S_{13}S_{31}} \frac{\underline{Z}_2+Z_{2\_in}}{2(Z_2+Z_{2\_in})} \tag{2.152}$$

$$\frac{V_3}{V_{S2}} = \frac{(1+\Gamma_3)(S_{32}+\Gamma_1(S_{12}S_{31}-S_{11}S_{32}))}{(1-\Gamma_1 S_{11})(1-\Gamma_3 S_{33})-\Gamma_1\Gamma_3 S_{13}S_{31}} \frac{\underline{Z}_2+Z_{2\_in}}{2(Z_2+Z_{2\_in})} \tag{2.153}$$

These equations may be used to calculate power transfer.

2.9.3 Input Impedance and Voltage Division, Driving Port 3 Alone

$$Z_{3in} = \left(\frac{2}{1-S_{33}-\dfrac{\Gamma_1 S_{31}(\Gamma_2 S_{12}S_{23}+S_{13}(1-\Gamma_2 S_{22}))+\Gamma_2 S_{32}(\Gamma_1 S_{13}S_{21}+S_{23}(1-\Gamma_1 S_{11}))}{(1-\Gamma_1 S_{11})(1-\Gamma_2 S_{22})-\Gamma_1\Gamma_2 S_{12}S_{21}}} - 1 \right) \underline{Z}_3 \tag{2.154}$$

$$\frac{V_1}{V_{S3}} = \frac{(1+\Gamma_1)(S_{13}+\Gamma_2(S_{12}S_{23}-S_{13}S_{22}))}{(1-\Gamma_1 S_{11})(1-\Gamma_2 S_{22})-\Gamma_1\Gamma_2 S_{12}S_{21}} \frac{\underline{Z}_3+Z_{3\_in}}{2(Z_3+Z_{3\_in})} \tag{2.155}$$

$$\frac{V_3}{V_{S3}} = \frac{(1+\Gamma_2)(S_{23}+\Gamma_1(S_{21}S_{13}-S_{11}S_{23}))}{(1-\Gamma_1 S_{11})(1-\Gamma_2 S_{22})-\Gamma_1\Gamma_2 S_{21}S_{12}} \frac{\underline{Z}_3+Z_{3\_in}}{2(Z_3+Z_{3\_in})} \tag{2.156}$$

These equations may be used to calculate power transfer.

2.9.4 Input Impedances, Driving Ports 1 and 2 Simultaneously

Frequently, when a three-port network is used to combine two input powers to sum as an output at the third port, the magnitudes of the two input voltages are equal. In this scenario, two of the most common cases are: (1) the voltages are in phase, or (2) the voltages are out-of-phase. When the two voltages are in-phase, this is referred to as an even-mode drive; when they are out-of-phase, this is referred to as an odd-mode drive. Alternatively, more commonly, the latter mode is also known as differential. When a three-port is used in this manner to combine two differential input powers into a single output, it is referred to as a balun (abbreviation of balanced-to-unbalanced). A balun can equally function in reverse to split a single input signal into two equal differential outputs. Some basic examples of three-port networks are shown in Figure 2.24.

Figure 2.25 shows the defining circuit variables for a three-port network under even-mode and odd-mode drives.

The impedances at the two input ports (1 and 2) are given by

$$Z_{1in} = \frac{((1+S_{11})(1-\Gamma_3 S_{33})+\Gamma_3 S_{13}S_{31})a_1+(S_{12}(1-\Gamma_3 S_{33})+\Gamma_3 S_{13}S_{32})a_2}{((1-S_{11})(1-\Gamma_3 S_{33})-\Gamma_3 S_{13}S_{31})a_1-(S_{12}(1-\Gamma_3 S_{33})+\Gamma_3 S_{13}S_{32})a_2} \underline{Z}_1 \tag{2.157}$$

and

$$Z_{2\text{in}} = \frac{\left(S_{21}\left(1-\Gamma_3 S_{33}\right)+\Gamma_3 S_{23} S_{31}\right)a_1+\left(\left(1+S_{22}\right)\left(1-\Gamma_3 S_{33}\right)+\Gamma_3 S_{23} S_{32}\right)a_2}{-\left(S_{21}\left(1-\Gamma_3 S_{33}\right)+\Gamma_3 S_{23} S_{31}\right)a_1+\left(\left(1-S_{22}\right)\left(1-\Gamma_3 S_{33}\right)-\Gamma_3 S_{23} S_{32}\right)a_2} Z_2 \quad (2.158)$$

where a_1 and a_2 are the incident voltage amplitudes at each port.

For the even-mode drive

$$a_2 = a_1 \quad (2.159)$$

and for the odd-mode drive

$$a_2 = -a_1 \quad (2.160)$$

thus, for the even-mode drive, the input port impedances are

$$Z_{1\text{even}} = \left(\frac{2\left(1-\Gamma_3 S_{33}\right)}{\left(1-S_{11}-S_{12}\right)\left(1-\Gamma_3 S_{33}\right)-\Gamma_3 S_{13}\left(S_{31}+S_{32}\right)}-1\right) Z_1 \quad (2.161)$$

$$Z_{2\text{even}} = \left(\frac{2\left(1-\Gamma_3 S_{33}\right)}{\left(1-S_{21}-S_{22}\right)\left(1-\Gamma_3 S_{33}\right)-\Gamma_3 S_{23}\left(S_{31}+S_{32}\right)}-1\right) Z_2 \quad (2.162)$$

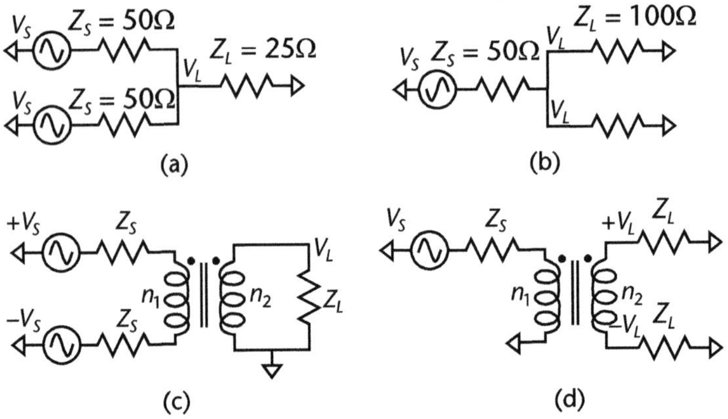

Figure 2.24 Basic three-port network examples: (a) in-phase discrete combiner, (b) in-phase discrete splitter, (c) differential transformer combiner, and (d) differential transformer splitter.

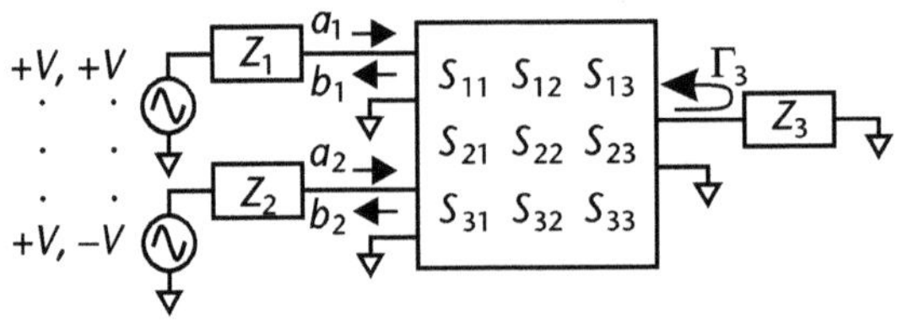

Figure 2.25 Three-port with even-mode and odd-mode drive.

and for the odd-mode drive

$$Z_{1\text{odd}} = \left(\frac{2(1-\Gamma_3 S_{33})}{(1-S_{11}+S_{12})(1-\Gamma_3 S_{33}) - \Gamma_3 S_{13}(S_{31}-S_{32})} - 1 \right) Z_1 \tag{2.163}$$

$$Z_{2\text{odd}} = \left(\frac{2(1-\Gamma_3 S_{33})}{(1+S_{21}-S_{22})(1-\Gamma_3 S_{33}) + \Gamma_3 S_{23}(S_{31}-S_{32})} - 1 \right) Z_2 \tag{2.164}$$

2.9.5 *S*-Parameters from Terminal Voltages and Currents

On occasion, having determined the port voltage-current relations for such a network, from, say, circuit analysis, it is desired to use these values to derive an S-parameter representation for the network.

Consider the three-port network shown in Figure 2.23, with the voltages and currents shown at each port. Suppose the following voltage-current relationships have been determined for the network

$$\begin{aligned} V_1 &= Z_{11}I_1 + Z_{12}I_2 + Z_{13}I_3 \\ V_2 &= Z_{21}I_1 + Z_{22}I_2 + Z_{23}I_3 \\ V_3 &= Z_{31}I_1 + Z_{32}I_2 + Z_{33}I_3 \end{aligned} \tag{2.165}$$

Assuming the same normalization impedance Z_0, on all three ports, the S-parameters for the network may be derived from the following formulae, with

$$\begin{aligned} D = {} & \alpha_{23}\alpha_{32}(\alpha_{11}+Z_0) + \alpha_{13}\alpha_{31}(\alpha_{22}+Z_0) + \alpha_{12}\alpha_{21}(\alpha_{33}+Z_0) \\ & -\alpha_{12}\alpha_{23}\alpha_{31} - \alpha_{13}\alpha_{21}\alpha_{32} - (\alpha_{11}+Z_0)(\alpha_{22}+Z_0)(\alpha_{33}+Z_0) \end{aligned} \tag{2.166}$$

The S-parameters are

$$S_{11} = 1 - \frac{2(\alpha_{23}\alpha_{32} - (\alpha_{22}+Z_0)(\alpha_{33}+Z_0))Z_0}{D} \tag{2.167}$$

$$S_{12} = \frac{2(\alpha_{13}\alpha_{32} - \alpha_{12}(\alpha_{33}+Z_0))Z_0}{D} \tag{2.168}$$

$$S_{13} = \frac{2(\alpha_{12}\alpha_{23} - \alpha_{13}(\alpha_{22}+Z_0))Z_0}{D} \tag{2.169}$$

$$S_{21} = \frac{2(\alpha_{23}\alpha_{31} - \alpha_{21}(\alpha_{33}+Z_0))Z_0}{D} \tag{2.170}$$

$$S_{22} = 1 - \frac{2(\alpha_{13}\alpha_{31} - (\alpha_{11}+Z_0)(\alpha_{33}+Z_0))Z_0}{D} \tag{2.171}$$

$$S_{23} = \frac{2\left(\alpha_{13}\alpha_{21} - \alpha_{23}\left(\alpha_{11} + \underline{Z}_0\right)\right)\underline{Z}_0}{D} \tag{2.172}$$

$$S_{31} = \frac{2\left(\alpha_{21}\alpha_{32} - \alpha_{31}\left(\alpha_{22} + \underline{Z}_0\right)\right)\underline{Z}_0}{D} \tag{2.173}$$

$$S_{32} = \frac{2\left(\alpha_{12}\alpha_{31} - \alpha_{32}\left(\alpha_{11} + \underline{Z}_0\right)\right)\underline{Z}_0}{D} \tag{2.174}$$

$$S_{33} = 1 - \frac{2\left(\alpha_{12}\alpha_{21} - \left(\alpha_{11} + \underline{Z}_0\right)\left(\alpha_{22} + \underline{Z}_0\right)\right)\underline{Z}_0}{D} \tag{2.175}$$

2.9.6 Derivation of Terminal Voltages and Currents from S-Parameters

As an inverse to the previous section, sometimes it is desired to determine the voltages and currents on the ports of a three-port network, given its S-parameters and nature of the externally applied sources.

Consider the three-port network shown in Figure 2.26, with the voltages, currents, and external driving sources shown at each port.

Assuming the same normalization impedance $\underline{Z}_0$, on all three ports, the voltages and currents at the network ports may be derived from the following formulae.

With

$$E_1 = \left(V_{S1} - \frac{\left(\underline{Z}_0 - Z_{S1}\right)S_{13}V_{S3}}{\underline{Z}_0 + Z_{S3} + \left(\underline{Z}_0 - Z_{S3}\right)S_{33}}\right)\underline{Z}_0 \tag{2.176}$$

$$E_2 = \underline{Z}_0 + Z_{S1} + \left(\underline{Z}_0 - Z_{S1}\right)S_{11} - \frac{\left(\underline{Z}_0 - Z_{S1}\right)\left(\underline{Z}_0 - Z_{S3}\right)S_{13}S_{31}}{\underline{Z}_0 + Z_{S3} + \left(\underline{Z}_0 - Z_{S3}\right)S_{33}} \tag{2.177}$$

$$E_3 = \left(\underline{Z}_0 - Z_{S1}\right)S_{12} - \frac{\left(\underline{Z}_0 - Z_{S1}\right)\left(\underline{Z}_0 - Z_{S2}\right)S_{13}S_{32}}{\underline{Z}_0 + Z_{S3} + \left(\underline{Z}_0 - Z_{S3}\right)S_{33}} \tag{2.178}$$

$$E_4 = \left(V_{S2} - \frac{\left(\underline{Z}_0 - Z_{S2}\right)S_{23}V_{S3}}{\underline{Z}_0 + Z_{S3} + \left(\underline{Z}_0 - Z_{S3}\right)S_{33}}\right)\underline{Z}_0 \tag{2.179}$$

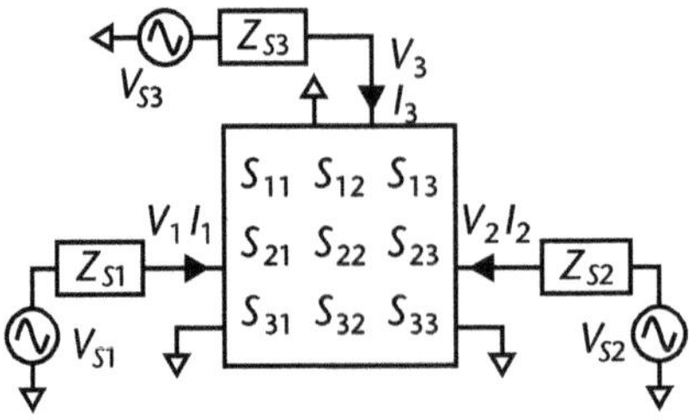

Figure 2.26 Three-port network with voltage and current variables.

$$E_5 = (\underline{Z}_0 - Z_{S2})S_{21} - \frac{(\underline{Z}_0 - Z_{S2})(\underline{Z}_0 - Z_{S3})S_{23}S_{31}}{\underline{Z}_0 + Z_{S3} + (\underline{Z}_0 - Z_{S3})S_{33}} \tag{2.180}$$

$$E_6 = \underline{Z}_0 + Z_{S2} + (\underline{Z}_0 - Z_{S2})S_{22} - \frac{(\underline{Z}_0 - Z_{S2})(\underline{Z}_0 - Z_{S3})S_{23}S_{32}}{\underline{Z}_0 + Z_{S3} + (\underline{Z}_0 - Z_{S3})S_{33}} \tag{2.181}$$

we have

$$a_1 = \frac{E_1E_6 - E_3E_4}{E_2E_6 - E_3E_5} \tag{2.182}$$

$$a_2 = \frac{E_2E_4 - E_1E_5}{E_2E_6 - E_3E_5} \tag{2.183}$$

$$a_3 = \frac{V_{S3}\underline{Z}_0 - (\underline{Z}_0 - Z_{S3})(S_{31}a_1 + S_{32}a_2)}{\underline{Z}_0 + Z_{S3} + (\underline{Z}_0 - Z_{S3})S_{33}} \tag{2.184}$$

The voltages are

$$V_1 = (1 + S_{11})a_1 + S_{12}a_2 + S_{13}a_3 \tag{2.185}$$

$$V_2 = S_{21}a_1 + (1 + S_{22})a_2 + S_{23}a_3 \tag{2.186}$$

$$V_3 = S_{31}a_1 + S_{32}a_2 + (1 + S_{33})a_3 \tag{2.187}$$

The currents are

$$I_1 = \frac{((1 - S_{11})a_1 - S_{12}a_2 - S_{13}a_3)}{\underline{Z}_0} \tag{2.188}$$

$$I_2 = \frac{(-S_{21}a_1 + (1 - S_{22})a_2 - S_{23}a_3)}{\underline{Z}_0} \tag{2.189}$$

$$I_3 = \frac{(-S_{31}a_1 - S_{32}a_2 + (1 - S_{33})a_3)}{\underline{Z}_0} \tag{2.190}$$

CHAPTER 3

Y and Z-Parameters: Key Relationships

For certain networks and interconnections, S-parameters may not always provide the most convenient matrix representation for circuit modeling and analysis. Some basic properties of two alternate sets of matrix parameters, namely, Y and Z, are presented in this chapter.

3.1 Y-Parameters

Y-parameters provide an alternative to S-parameters, for characterizing a multiport network with an unlimited number of ports. They are particularly advantageous when networks are to be connected in parallel. Y-parameters are admittance parameters and thus are current-based, rather than power-based, as are S-parameters. This makes them particularly suitable for analyzing solid-state designs comprising current-driven active components such as transistors.

A network with n ports, the Y-parameter voltage, and current variables and matrix parameters is shown schematically in Figure 3.1(a). In Figure 3.1(b), an example schematic is shown of a parallel connection of two Y-parameters represented two-port networks.

The Y-parameter matrix network equations are:

$$\begin{bmatrix} I_1 \\ \cdot \\ \cdot \\ I_n \end{bmatrix} = \begin{bmatrix} Y_{11} & \cdot & \cdot & Y_{1n} \\ \cdot & \cdot & \cdot & \cdot \\ \cdot & \cdot & \cdot & \cdot \\ Y_{n1} & \cdot & \cdot & Y_{nn} \end{bmatrix} \begin{bmatrix} V_1 \\ \cdot \\ \cdot \\ V_n \end{bmatrix} \tag{3.1}$$

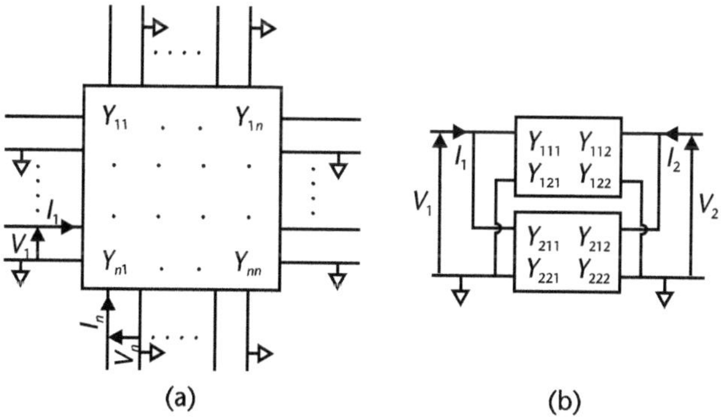

Figure 3.1 Multiport Y-parameter network schematics: (a) n-port Y network elements, and (b) two-port parallel Y-parameters.

that is,

$$I_1 = Y_{11}V_1 \cdots\cdots + Y_{1n}V_n$$
$$\vdots \qquad \vdots \qquad \vdots \tag{3.2}$$
$$I_n = Y_{n1}V_1 \cdots\cdots + Y_{nn}V_n$$

thus,

$$Y_{mn} = \left.\frac{I_m}{V_n}\right|_{V_i=0,\text{for } i\neq n} \tag{3.3}$$

Hence, the Y-parameters for each port are determined by the ratio of the terminal input current to an applied independent voltage source, while all other ports are simultaneously short-circuited. This can prove very challenging to achieve in a bench measurement. In light of (3.3), Y-parameters are often referred to as short-circuit admittance parameters.

Port admittances
$$Y_m = \frac{\left(\sum_n Y_{mn}V_n\right)}{V_m} \tag{3.4}$$

Thus, for the parallel-connected two-ports in Figure 3.1(b), the Y-parameters of the combined network are simply the sum of the corresponding individual Y-parameters on each port.

3.2 Some Useful Two-Port Y-Parameter Relationships

Complex RF networks are typically characterized by Y-parameter representations comprising multiple ports. However, it is frequently the case that, for final analysis, the network is reduced to a two-port representation by terminating the remaining ports with impedance networks. The characteristics of the network for transforming an input signal on one of the ports to an output signal at the second port can then be evaluated. Among the key characteristics of interest are the network input and output impedances, gain, and phase or group delay through the network. Figure 3.2 shows the key circuit parameters for analysis of a two-port network characterized by Y-parameters.

The Y-parameter matrix for a two-port is reduced to

$$[Y] = \begin{bmatrix} Y_{11} & Y_{12} \\ Y_{21} & Y_{22} \end{bmatrix} \tag{3.5}$$

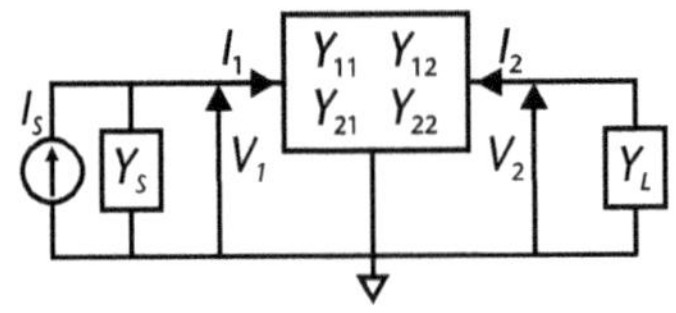

Figure 3.2 Two-port Y-parameter network variables.

$$I_1 = Y_{11}V_1 + Y_{12}V_2 \tag{3.6}$$

$$I_2 = Y_{21}V_1 + Y_{22}V_2 \tag{3.7}$$

with

$$Y_{11} = \left.\frac{I_1}{V_1}\right|_{V_2=0} \qquad Y_{12} = \left.\frac{I_1}{V_2}\right|_{V_1=0} \tag{3.8}$$

$$Y_{21} = \left.\frac{I_2}{V_1}\right|_{V_2=0} \qquad Y_{22} = \left.\frac{I_2}{V_2}\right|_{V_1=0} \tag{3.9}$$

Input admittance $\quad Y_1 = \dfrac{I_1}{V_1} = Y_{11} - \dfrac{Y_{12}Y_{21}}{Y_{22} + Y_L}$

Input impedance $\quad Z_1 = \dfrac{V_1}{I_1} = \dfrac{Y_{22} + Y_L}{Y_{11}(Y_{22} + Y_L) - Y_{12}Y_{21}} \qquad (3.10)$

Output admittance $\quad Y_2 = \dfrac{I_2}{V_2} = Y_{22} - \dfrac{Y_{12}Y_{21}}{Y_S + Y_{11}} \qquad (3.11)$

Output impedance $\quad Z_2 = \dfrac{V_2}{I_2} = \dfrac{Y_S + Y_{11}}{Y_{22}(Y_S + Y_{11}) - Y_{12}Y_{21}} \qquad (3.12)$

3.3 Common Two-Port Network Y-Parameters

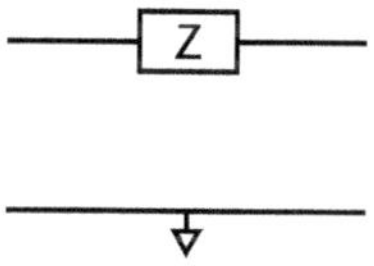

$$[Y] = \begin{bmatrix} \frac{1}{Z} & -\frac{1}{Z} \\ -\frac{1}{Z} & \frac{1}{Z} \end{bmatrix} \tag{3.13}$$

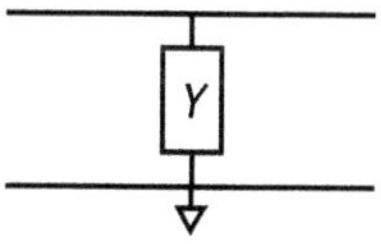

$$[Y] = \begin{bmatrix} \infty & \infty \\ \infty & \infty \end{bmatrix} \tag{3.14}$$

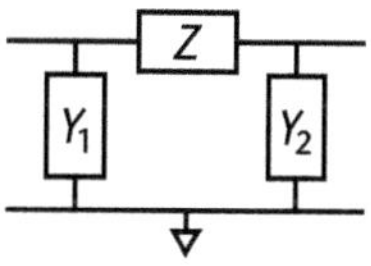

$$[Y] = \begin{bmatrix} \frac{Y_1 + 1}{Z} & -\frac{1}{Z} \\ -\frac{1}{Z} & \frac{Y_2 + 1}{Z} \end{bmatrix} \tag{3.15}$$

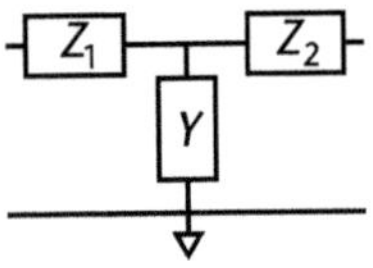

$$[Y] = \begin{bmatrix} \dfrac{1+YZ_2}{Z_1+Z_2+YZ_1Z_2} & \dfrac{-1}{Z_1+Z_2+YZ_1Z_2} \\ \dfrac{-1}{Z_1+Z_2+YZ_1Z_2} & \dfrac{1+YZ_1}{Z_1+Z_2+YZ_1Z_2} \end{bmatrix} \tag{3.16}$$

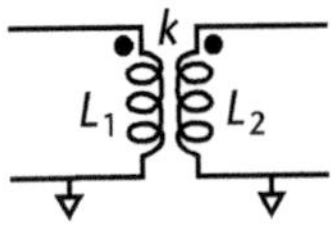

$$[Y] = \begin{bmatrix} \dfrac{-j}{\omega\left(1-k^2\right)L_1} & \dfrac{jk}{\omega\left(1-k^2\right)\sqrt{L_1L_2}} \\ \dfrac{jk}{\omega\left(1-k^2\right)\sqrt{L_1L_2}} & \dfrac{-j}{\omega\left(1-k^2\right)L_2} \end{bmatrix} \tag{3.17}$$

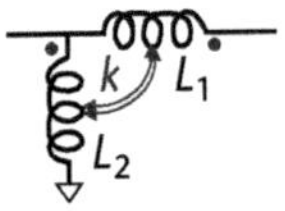

$$[Y] = \begin{bmatrix} \dfrac{L_1+L_2+2k\sqrt{L_1L_2}}{j\omega\left(1-k^2\right)L_1L_2} & \dfrac{L_2+k\sqrt{L_1L_2}}{j\omega\left(1-k^2\right)L_1L_2} \\ \dfrac{L_1+k\sqrt{L_1L_2}}{j\omega\left(1-k^2\right)L_1L_2} & \dfrac{L_2}{j\omega\left(1-k^2\right)L_1L_2} \end{bmatrix} \tag{3.18}$$

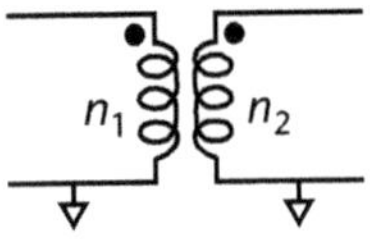

$$[Y] = \begin{bmatrix} \infty & \infty \\ \infty & \infty \end{bmatrix} \tag{3.19}$$

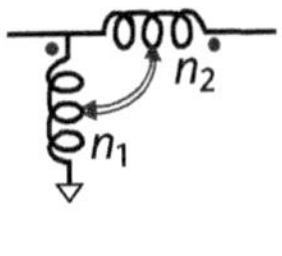

$$[Y] = \begin{bmatrix} \infty & \infty \\ \infty & \infty \end{bmatrix} \tag{3.20}$$

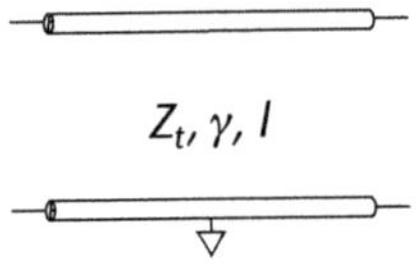

$$[Y] = \begin{bmatrix} \left(\dfrac{1}{Z_t}\right)\coth(\gamma l) & \dfrac{-1}{(Z_t \sinh(\gamma l))} \\ \dfrac{-1}{(Z_t \sinh(\gamma l))} & \left(\dfrac{1}{Z_t}\right)\coth(\gamma l) \end{bmatrix} \tag{3.21}$$

where Z_t = transmission-line impedance. The complex propagation constant $\gamma = \alpha + j(2\pi/\lambda)$, where α = dissipative coefficient, λ = wavelength, and l = length of transmission line in wavelengths.

Note that finite Y-parameters do not exist for several of the above networks. In those cases, Y-parameters cannot be used to characterize such networks.

3.4 Z-Parameters

Z-parameters provide another alternative set of variables for characterizing a multiport network with an unlimited number of ports. They are particularly advantageous when networks are to be connected in series. Z-parameters are impedance parameters and thus are voltage-based.

A network with n ports, the Z-parameter voltage, and current variables and matrix parameters is shown schematically in Figure 3.3(a). In Figure 3.3(b), an example schematic is shown of a series connection of two Z-parameters represented two-port networks.

The network is represented by the Z-parameter matrix equations:

$$\begin{bmatrix} V_1 \\ \cdot \\ \cdot \\ V_n \end{bmatrix} = \begin{bmatrix} Z_{11} & \cdot & \cdot & Z_{1n} \\ \cdot & \cdot & \cdot & \cdot \\ \cdot & \cdot & \cdot & \cdot \\ Z_{n1} & \cdot & \cdot & Z_{nn} \end{bmatrix} \begin{bmatrix} I_1 \\ \cdot \\ \cdot \\ I_n \end{bmatrix} \tag{3.22}$$

that is,

$$\begin{aligned} V_1 &= Z_{11}I_1 \cdots\cdots + Z_{1n}I_n \\ \cdot \quad & \quad \cdot \qquad \cdot \\ \cdot \quad & \quad \cdot \qquad \cdot \\ V_n &= Z_{n1}I_1 \cdots\cdots + Z_{nn}I_n \end{aligned} \tag{3.23}$$

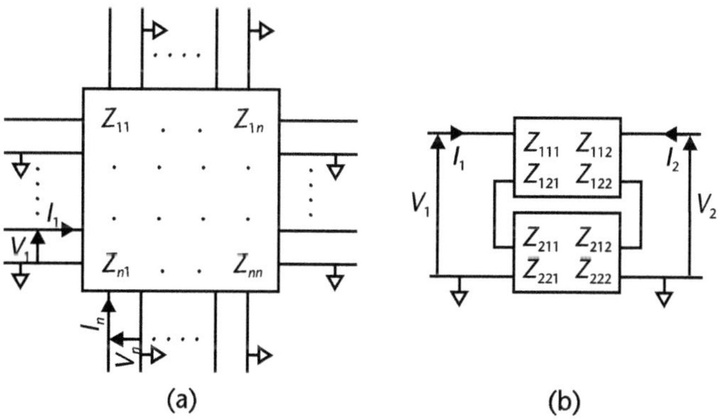

Figure 3.3 Definition of multiport Z-parameters: (a) n-port Z network elements, and (b) two-port series Z-parameters.

thus

$$Z_{mn} = \left.\frac{V_m}{I_n}\right|_{I_i=0,\text{for } i\neq n} \tag{3.24}$$

Hence, the Z-parameters for each port are determined by the ratio of the terminal voltage to an applied independent current source, while all other ports are simultaneously open-circuited. This can prove very challenging to achieve in a bench measurement.

Port impedances
$$Z_m = \frac{\left(\sum_n Z_{mn} I_n\right)}{I_m} \tag{3.25}$$

3.5 Some Useful Two-Port Z-Parameter Relationships

Complex RF networks are typically characterized by Z-parameter representations comprising multiple ports. However, it is frequently the case that, for final analysis, the network is reduced to a two-port representation by terminating the remaining ports with impedance networks. The characteristics of the network for transforming an input signal on one of the ports to an output signal at the second port can then be evaluated. Among the key characteristics of interest are the network input and output impedances, gain, and phase or group delay through the network. Figure 3.4 shows the key circuit parameters for analysis of a two-port network characterized by Z-parameters.

The Z-parameter matrix for a two-port is reduced to

$$[Z] = \begin{bmatrix} Z_{11} & Z_{12} \\ Z_{21} & Z_{22} \end{bmatrix} \tag{3.26}$$

that is,

$$V_1 = Z_{11}I_1 + Z_{12}I_2 \tag{3.27}$$

$$V_2 = Z_{21}I_1 + Z_{22}I_2 \tag{3.28}$$

with

$$Z_{11} = \left.\frac{V_1}{I_1}\right|_{I_2=0} \qquad Z_{12} = \left.\frac{V_1}{I_2}\right|_{I_1=0} \tag{3.29}$$

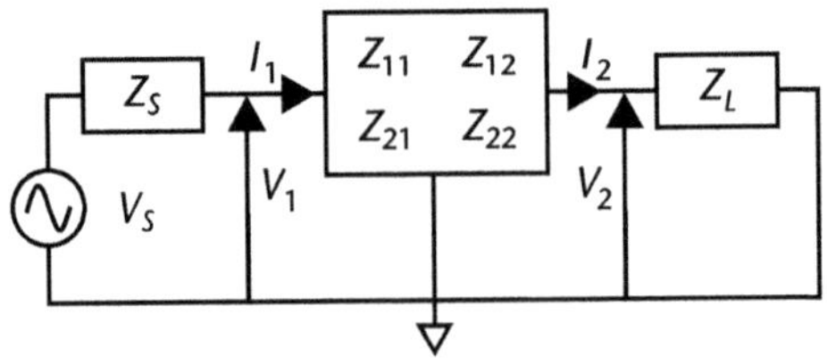

Figure 3.4 Two-port Z-parameter network variables.

$$Z_{21} = \left.\frac{V_2}{I_1}\right|_{I_2=0} \qquad Z_{22} = \left.\frac{V_2}{I_2}\right|_{I_1=0} \tag{3.30}$$

Thus, the Z-parameters on port 1 are obtained by open-circuiting port 2 (i.e., $V_2 = 0$) and vice versa.

Input admittance
$$Y_1 = \frac{I_1}{V_1} = \frac{Z_{22} + Z_L}{Z_{11}(Z_{22} + Z_L) - Z_{12}Z_{21}} \tag{3.31}$$

Input impedance
$$Z_1 = \frac{V_1}{I_1} = Z_{11} - \frac{Z_{12}Z_{21}}{Z_{22} + Z_L} \tag{3.32}$$

Output admittance
$$Y_2 = \frac{I_2}{V_2} = \frac{Z_S + Z_{11}}{Z_{22}(Z_S + Z_{11}) - Z_{12}Z_{21}} \tag{3.33}$$

Output impedance
$$Z_2 = \frac{V_2}{I_2} = Z_{22} - \frac{Z_{12}Z_{21}}{Z_S + Z_{11}} \tag{3.34}$$

3.6 Common Two-Port Network *Z*-Parameters

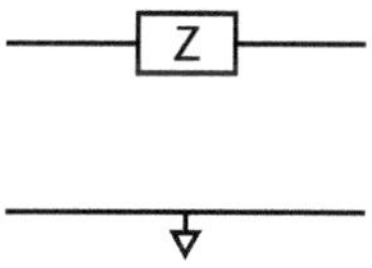

$$[Z] = \begin{bmatrix} \infty & \infty \\ \infty & \infty \end{bmatrix} \tag{3.35}$$

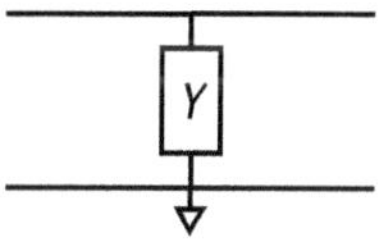

$$[Z] = \begin{bmatrix} \frac{1}{Y} & \frac{1}{Y} \\ \frac{1}{Y} & \frac{1}{Y} \end{bmatrix} \tag{3.36}$$

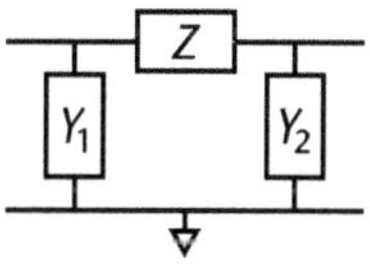

$$[Z] = \begin{bmatrix} \frac{1 + Y_2 Z}{Y_1 + Y_2 + Y_1 Y_2 Z} & \frac{1}{Y_1 + Y_2 + Y_1 Y_2 Z} \\ \frac{1}{Y_1 + Y_2 + Y_1 Y_2 Z} & \frac{1 + Y_1 Z}{Y_1 + Y_2 + Y_1 Y_2 Z} \end{bmatrix} \tag{3.37}$$

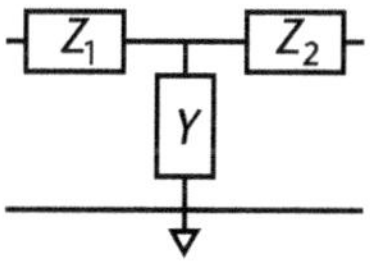

$$[Z] = \begin{bmatrix} Z_1 + \frac{1}{Y} & \frac{1}{Y} \\ \frac{1}{Y} & Z_2 + \frac{1}{Y} \end{bmatrix} \tag{3.38}$$

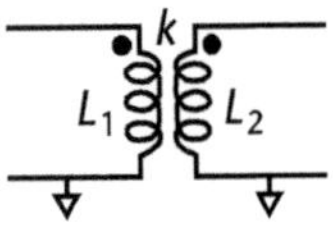

$$[Z] = \begin{bmatrix} j\omega L_1 & j\omega k\sqrt{L_1 L_2} \\ j\omega k\sqrt{L_1 L_2} & j\omega L_2 \end{bmatrix} \tag{3.39}$$

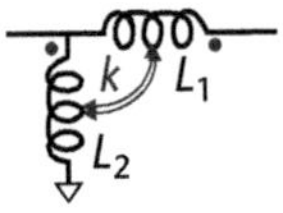

$$[Z] = \begin{bmatrix} j\omega L_2 & j\omega\left(L_2 + k\sqrt{L_1 L_2}\right) \\ j\omega\left(L_2 + k\sqrt{L_1 L_2}\right) & j\omega\left(L_1 + L_2 + 2k\sqrt{L_1 L_2}\right) \end{bmatrix} \tag{3.40}$$

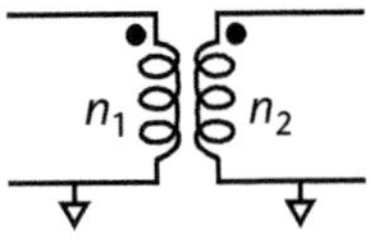

$$[Z] = \begin{bmatrix} \infty & \infty \\ \infty & \infty \end{bmatrix} \tag{3.41}$$

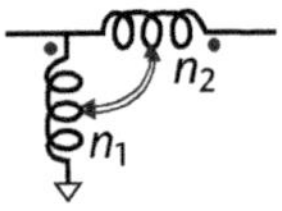

$$[Z] = \begin{bmatrix} \infty & \infty \\ \infty & \infty \end{bmatrix} \tag{3.42}$$

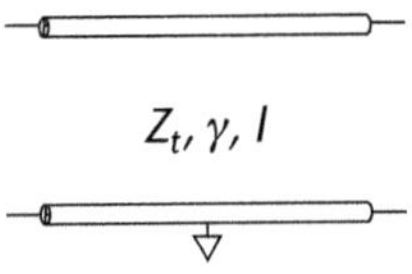

$$[Z] = \begin{bmatrix} Z_t \coth(\gamma l) & \dfrac{Z_t}{\sinh(\gamma l)} \\ \dfrac{Z_t}{\sinh(\gamma l)} & Z_t \coth(\gamma l) \end{bmatrix} \tag{3.43}$$

where Z_t = transmission-line impedance. The complex propagation constant $\gamma = \alpha + j(2\pi/\lambda)$, where α = dissipative coefficient, λ = wavelength, and l = length of the transmission line in wavelengths.

In Appendix C, formulae are given for converting four-port network S parameters to equivalent Y parameters.

CHAPTER 4

Power Relationships

Once a preferred set of parameters capable of accurately modeling the performance of a network has been determined, either experimentally or analytically, invariably, the designer will seek to use them to determine how efficiently the network performs an assigned task. For this, the power transfer relationships for the network are required.

There are two distinct measures of power-transfer efficiency (i.e., insertion loss) that can be used to characterize a network. While for many networks, in the band of interest, they may be very similar, there is a subtle but important difference between the two. Taken together, these two transfer parameters can tell us a lot about the network.

1. P_{del}/P_{in}: This ratio, referred to as the operating power gain, describes the power delivered to the load(s) relative to the input power (i.e., entering the network). This power loss characterizes any dissipative losses in the network.
2. $P_{del}/P_{avail.}$: This ratio, referred to as the transducer power gain, describes the power delivered to the load(s) relative to the maximum power available from the source. This power ratio accounts for power loss to the load arising from any mismatch loss on the input and power dissipated in the network.

In all cases, $P_{del}/P_{avail.} \leq P_{del}/P_{in}$. It is generally very instructive to plot these two insertion gain characteristics when analyzing any RF network. Over any frequency range in which the input to the network is well matched to the source, they will be essentially the same. Any discrepancy between the two is indicative that the network input impedance is not well-matched to the source impedance at that frequency. The parameter P_{del}/P_{in} is helpful in revealing the best power transfer efficiency that could possibly be achieved with the network, for the given source and load impedances, given the dissipative nature of the network.

4.1 Fundamental Power Relations

Power available from the source

$$P_{avail} = \Re\left(VI^*\right) = \frac{V_S^2}{4\Re\left(Z_S\right)} \tag{4.1}$$

Power into the network

$$P_{\text{in}} = \frac{\Re(Z_{\text{in}})V_S^2}{|Z_S + Z_{\text{in}}|^2} \tag{4.2}$$

Power delivered to the load

$$P_{\text{del}} = \Re\left(\frac{V_L V_L^*}{Z_L^*}\right) = \frac{|V_L|^2}{|Z_L|^2}\Re(Z_L) \tag{4.3}$$

where $\Re(\,)$ denotes the real part and * denotes the complex conjugate.

Hence, the transmission gain in decibels is

$$\begin{aligned}\frac{P_{\text{del}}}{P_{\text{avail}}} &= 10\log\left(\left|\frac{V_L/V_S}{Z_L}\right|^2 4\Re(Z_S)\Re(Z_L)\right) \\ &\equiv 20\log\left(2\left|\frac{V_L/V_S}{Z_L}\right|\sqrt{\Re(Z_S)\Re(Z_L)}\right)\end{aligned} \tag{4.4}$$

$$\begin{aligned}\frac{P_{\text{del}}}{P_{\text{in}}} &= 10\log\left(\left|\frac{(Z_S + Z_{\text{in}})V_L/V_S}{Z_L}\right|^2 \frac{\Re(Z_L)}{\Re(Z_{\text{in}})}\right) \\ &= 20\log\left(\left|\frac{(Z_S + Z_{\text{in}})V_L/V_S}{Z_L}\right|\sqrt{\frac{\Re(Z_L)}{\Re(Z_{\text{in}})}}\right)\end{aligned} \tag{4.5}$$

and, if Z_S, Z_L are real,

$$\frac{P_{\text{del}}}{P_{\text{Avail}}} = 10\log\left(\left|\frac{V_L}{V_S}\right|^2 \frac{4R_S}{R_L}\right) \equiv 20\log\left(2\left|\frac{V_L}{V_S}\right|\sqrt{\frac{R_S}{R_L}}\right) \tag{4.6}$$

4.2 [*ABCD*] Power Relations

Power into the network is

$$P_{\text{in}} = \Re\left(\frac{V_{\text{in}} V_{\text{in}}^*}{Z_{\text{in}}^*}\right) = \frac{\Re\left((AZ_L + B)(CZ_L + D)^*\right)V_S^2}{|AZ_L + B + Z_S(CZ_L + D)|^2} \tag{4.7}$$

Power delivered to the load is

$$P_{\text{del}} = \Re\left(\frac{V_L V_L^*}{Z_L^*}\right) = \frac{\Re(Z_L)}{|AZ_L + B + Z_S(CZ_L + D)|^2}V_S^2 \tag{4.8}$$

Hence,

$$\frac{P_{del}}{P_{in}} = 10\log\left(\frac{\Re(Z_L)}{\Re\left((AZ_L + B)(CZ_L + D)^*\right)}\right)\cdot \text{dB} \tag{4.9}$$

This expression provides a measure of the power dissipation in the network.

$$\frac{P_{del}}{P_{avail}} = 10\log\left(\frac{4\Re(Z_S)\Re(Z_L)}{\left|AZ_L + B + Z_S(CZ_L + D)\right|^2}\right)\cdot \text{dB} \tag{4.10}$$

This expression incorporates transmission losses due to both reflection on the input port and dissipation in the network.

4.3 Two-Port S-Parameter Power Relations

With

$$Z_{in} = \left(\frac{2}{1 - S_{11} - \Gamma_L S_{12}S_{21}/(1 - \Gamma_L S_{22})}\right)Z_1 \tag{4.11}$$

and

$$\frac{V_{in}}{V_S} = \frac{\left((1 + S_{11})(1 - \Gamma_L S_{22}) + \Gamma_L S_{12}S_{21}\right)Z_1}{(Z_1 + Z_S + S_{11}(Z_1 - Z_S))(1 - \Gamma_L S_{22}) + \Gamma_L S_{12}S_{21}(Z_1 - Z_S)} \tag{4.12}$$

The power into network is

$$P_{in} = \Re\left(V_{in}I_{in}^*\right) = \Re\left(V_{in}\frac{V_{in}^*}{Z_{in}^*}\right) = V_S^2\left|\frac{V_{in}}{V_S}\right|^2\Re\left(\frac{1}{Z_{in}}\right) \tag{4.13}$$

or

$$P_{in} = \frac{\Im\left(\left((1 + S_{11})(1 - \Gamma_L S_{22}) + \Gamma_L S_{12}S_{21}\right)Z_1\right)\Re\left((1 - S_{11})(1 - \Gamma_L S_{22}) - \Gamma_L S_{12}S_{21}\right)}{\left|(Z_1 + Z_S + S_{11}(Z_1 - Z_S))(1 - \Gamma_L S_{22}) + \Gamma_L S_{12}S_{21}(Z_1 - Z_S)\right|^2} \tag{4.14}$$

where $\Re(\ )$ and $\Im(\ )$ represent real and imaginary components.

The power delivered to the load is

$$\begin{aligned} P_{del} &= \left|\frac{V_L}{Z_L}\right|^2\Re(Z_L) \\ &= V_S^2\left|\frac{S_{21}(1 - \Gamma_S)(1 + \Gamma_L)}{2\left((1 - \Gamma_S S_{11})(1 - \Gamma_L S_{22}) - \Gamma_S\Gamma_L S_{12}S_{21}\right)Z_L}\right|^2\Re(Z_L) \end{aligned} \tag{4.15}$$

and

$$\frac{P_{\text{del}}}{P_{\text{in}}} = \left|\frac{S_{21}(1-\Gamma_S)(1+\Gamma_L)(Z_S+Z_{\text{in}})}{2\left((1-\Gamma_S S_{11})(1-\Gamma_L S_{22})-\Gamma_S\Gamma_L S_{12}S_{21}\right)Z_L}\right|^2 \frac{\Re(Z_L)}{\Re(Z_{\text{in}})} \tag{4.16}$$

This expression provides a measure of the power dissipation in the network.

$$\frac{P_{\text{del}}}{P_{\text{avail}}} = 20\log\left|\frac{S_{21}(1-\Gamma_S)(1+\Gamma_L)\sqrt{\Re(Z_S)\Re(Z_L)}}{\left((1-\Gamma_S S_{11})(1-\Gamma_L S_{22})-\Gamma_S\Gamma_L S_{12}S_{21}\right)Z_L}\right| \tag{4.17}$$

This expression incorporates transmission losses due to both reflection on the input port and dissipation in the network.

4.4 Two-Port *Y*-Parameter Power Relations

The power into the network is

$$P_{\text{in}} = \Re\left(\frac{V_{\text{in}}V_{\text{in}}^*}{Z_{\text{in}}^*}\right) = \frac{\Re\left((1+Z_LY_{22})\left(Y_{11}-Z_L(Y_{12}Y_{21}-Y_{11}Y_{22})\right)^*\right)V_S^2}{\left|1+Z_LY_{22}+Z_S\left(Y_{11}-Z_L(Y_{12}Y_{21}-Y_{11}Y_{22})\right)\right|^2} \tag{4.18}$$

The power delivered to the load is

$$P_{\text{del}} = \Re\left(\frac{V_LV_L^*}{Z_L^*}\right) = \frac{\Re(Z_L)|Y_{21}|^2}{\left|1+Z_LY_{22}+Z_S\left(Y_{11}-Z_L(Y_{12}Y_{21}-Y_{11}Y_{22})\right)\right|^2}V_S^2 \tag{4.19}$$

and

$$\frac{P_{\text{del}}}{P_{\text{in}}} = 10\log\left(\frac{\Re(Z_L)|Y_{21}|^2}{\Re\left((1+Z_LY_{22})\left(Y_{11}-Z_L(Y_{12}Y_{21}-Y_{11}Y_{22})\right)\right)^*}\right)\cdot\text{dB} \tag{4.20}$$

This expression provides a measure of the power dissipation in the network.

$$\frac{P_{\text{del}}}{P_{\text{avail}}} = 10\log\left(\frac{4\Re(Z_S)\Re(Z_L)|Y_{21}|^2}{\left|1+Z_LY_{22}+Z_S\left(Y_{11}-Z_L(Y_{12}Y_{21}-Y_{11}Y_{22})\right)\right|^2}\right)\cdot\text{dB} \tag{4.21}$$

4.5 Two-Port *Z*-Parameter Power Relations

The power into the network is

$$P_{in} = \Re\left(\frac{V_{in}V_{in}^*}{Z_{in}^*}\right) = \frac{\Re\left((Z_L Z_{11} + Z_{11}Z_{22} - Z_{12}Z_{21})(Z_L + Z_{22})^*\right)V_S^2}{\left|Z_L Z_{11} + Z_{11}Z_{22} - Z_{12}Z_{21} + Z_S(Z_L + Z_{22})\right|^2} \quad (4.22)$$

The power delivered to the load is

$$P_{del} = \Re\left(\frac{V_L V_L^*}{Z_L^*}\right) = \frac{\Re(Z_L)|Z_{21}|^2}{\left|Z_L Z_{11} + Z_{11}Z_{22} - Z_{12}Z_{21} + Z_S(Z_L + Z_{22})\right|^2} V_S^2 \quad (4.23)$$

and

$$\frac{P_{del}}{P_{in}} = 10\log\left(\frac{\Re(Z_L)|Z_{21}|^2}{\Re\left((Z_L Z_{11} + Z_{11}Z_{22} - Z_{12}Z_{21})(Z_L + Z_{22})^*\right)}\right) \cdot \text{dB} \quad (4.24)$$

This expression provides a measure of the power dissipation in the network.

$$\frac{P_{del}}{P_{avail}} = 10\log\left(\frac{4\Re(Z_S)\Re(Z_L)|Z_{21}|^2}{\left|Z_L Z_{11} + Z_{11}Z_{22} - Z_{12}Z_{21} + Z_S(Z_L + Z_{22})\right|^2}\right) \cdot \text{dB} \quad (4.25)$$

This expression incorporates transmission losses due to both reflection on the input port and dissipation in the network.

4.6 Some Useful Power Relationships

Due to their compact nature, mobile devices are invariably powered by low-voltage batteries, typically 5V or less. If they must drive a high-impedance output load (e.g., 50Ω), this substantially limits the power that can be delivered directly into the load. If more power is required to the load, an output matching network must be inserted between the network and the load to adapt the output impedance of the network to that of the load. Output matching networks are invariably required for RF power amplifiers (PAs) in a cellphone, for example. Regrettably, any output matching network will invariably reduce the system performance somewhat from that of the network alone. The matching network will result in additional power losses, reducing overall efficiency, and also reduce the system bandwidth. A principal goal of the designer must be to minimize both these downsides. Some formulae useful to these efforts are given below.

If η_0 = inherent network efficiency, into an ideal conjugately-matched load and dBL = post-network insertion gain, the resulting network efficiency (network + output circuitry) is

$$\eta = \eta_0 \cdot 10^{-dBL/10} \quad (4.26)$$

or inverting, if the measured efficiency of a network plus output circuitry is η, then the inherent transfer efficiency of the network is

$$\eta_0 = \eta \cdot 10^{dBL/10} \tag{4.27}$$

Mobile RF PAs invariably have some form of output power control. In the majority of cases, the power control is implemented in the active devices and associated with a change in efficiency; the output matching network usually remains fixed.

For a system delivering power into a load impedance R_L, the power is determined from

$$P_{\text{dBm}} = 30 + 10\log\left(\frac{V_{rms}^2}{R_L}\right) \tag{4.28}$$

or

$$P_{\text{dBm}} = 30 + 10\log\left(\frac{V_{\text{peak}}^2}{2R_L}\right) \tag{4.29}$$

thus

$$V_{\text{peak}} = \sqrt{2R_L}\,10^{(P_{\text{dBm}}/10-3)/2} \tag{4.30}$$

In a special case in which $R_L = 50\Omega$,

$$V_{\text{peak}} = 10^{(P_{\text{dBm}}/10-1)/2} \tag{4.31}$$

4.7 Maximum Available Gain: Optimum Conjugate Matching of a Passive Two-Port

As mentioned in the previous chapter, how to optimize the performance of a given network to provide maximum power transfer between a given source and load is typically of prime importance to the practicing RF engineer. In general, for a prescribed network, this may require adding additional elements to both the input and output of the network, to achieve optimum power transfer over a desired frequency range, as shown in Figure 4.1.

In seeking to understand what the given RF network is capable of, the designer often is interested in knowing the best possible performance that could be achieved

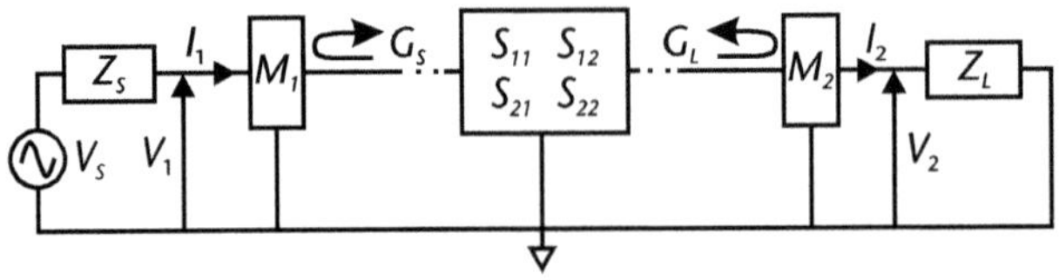

Figure 4.1 Two-port S-parameter conjugate matching.

from the network with optimum external matching networks. The term "optimum" here means those matching networks that would result in the network having the lowest power loss when driven by a specified source impedance on the input to a specified load impedance on the output. To achieve this minimum insertion loss state, the intermediate matching networks (M_1 and M_2 in Figure 4.1) must provide simultaneous complex matches Γ_S and Γ_L on the two ports of the network.

Achieving the simultaneous match condition is nontrivial as, in general, the two ports of the network interact. Thus, changing the external impedance on one port will modify the impedance seen at the second port. In consequence, for example, if an arbitrary load is applied to the output, an input matching network M_1 can be determined that provides for a complex conjugate match on the input port of the network. However, despite this input match, network analysis will generally show that the complex conjugate match condition is not met on the output. Due to the interactive complexities, to ensure the minimum possible insertion, or dissipative loss in the network, the conjugate matching networks are best determined analytically.

For a simultaneous-match analysis, the matching networks M_1 and M_2 are generally assumed to be lossless and continuously adaptable to maintain the simultaneous conjugate matching conditions at every frequency. As such, these matching networks may not be physically realizable. Typically, they can only be approximated by physical networks over a restricted frequency range. Nonetheless, the analysis can be helpful to the designer in understanding the circuit limitations and its capabilities.

There are a few important aspects to be aware of when performing a simultaneous conjugate match analysis of a passive two-port network:

1. The network must be linear.
2. Unique solutions for the matching networks only exist providing:

Stability factor
$$K = \frac{1 - |S_{11}|^2 - |S_{22}|^2 + |S_{11}S_{22} - S_{12}S_{21}|^2}{2|S_{12}S_{21}|} > 1 \tag{4.32}$$

3. For a lossless reciprocal network, $K = 1$. In this limiting case, there will exist an infinite possible set of solutions for the conjugate matching requirements. If the source or load impedances are chosen arbitrarily, a conjugate matching impedance at the alternate port may always be determined that results in 0-dB insertion loss through the system.

The simultaneous conjugate matching impedances are determined as follows. Defining

$$A = S_{11}\left(1 - |S_{22}|^2\right) + S_{12}S_{21}S_{22}^* \tag{4.33}$$

and

$$B = 1 - |S_{22}|^2 + \frac{|A|^2 - |S_{12}S_{21}|^2}{1 - |S_{22}|^2} \tag{4.34}$$

the required simultaneous conjugate matching reflection coefficients on the input and output ports are

$$\Gamma_S = \frac{B - \sqrt{B^2 - 4|A|^2}}{2A} \qquad \Gamma_L = \left(S_{22} + \frac{\Gamma_S S_{12} S_{21}}{1 - \Gamma_S S_{11}}\right)^* \tag{4.35}$$

The corresponding external impedances are then

$$Z_S = \frac{1 + \Gamma_S}{1 - \Gamma_S} Z_1 \qquad Z_L = \frac{1 + \Gamma_L}{1 - \Gamma_L} Z_2 \tag{4.36}$$

where Z_1 and Z_2 are the normalization impedances on ports 1 and 2, respectively.
From the formulae presented with simultaneous conjugate matches on the network, the minimum insertion loss through the network is

$$\left(\frac{P_{\text{del}}}{P_{\text{avail}}}\right)_{\max} = 20\log\left|\frac{S_{21}(1 - \Gamma_S)(1 + \Gamma_L)\sqrt{\Re(Z_S)\Re(Z_L)}}{((1 - \Gamma_S S_{11})(1 - \Gamma_L S_{22}) - \Gamma_S \Gamma_L S_{12} S_{21}) Z_L}\right| \cdot \text{dB} \tag{4.37}$$

where $\Re(\,)$ denotes the real component.

CHAPTER 5

Lumped-Element Basics

Typically, RF circuits comprise both active and passive components. These must be modeled accurately over frequency in any analytic model to be used for device simulation and design. The modeling of passive elements is especially critical, as active device models invariably include passive elements in their circuit representations. This chapter is thus focused on presenting circuit models, which can represent passive elements over an extended frequency range, and describing how the characterizing parametric values can be extracted from measurement.

Individual circuit elements can be divided up into three distinct groups, those whose function is primarily inductive, capacitive, or resistive in nature. The inclusion of the qualifying word "primarily" is especially critical for reactive elements, as their characteristics will typically vary significantly over frequency. This variation is due to undesired parasitics associated with any physical element. As a result, an element that is capacitive over one frequency range may appear inductive over another. Likewise, an element that is inductive over one frequency range may appear capacitive over another. For high-frequency RF circuits, such parasitics are critical to predicting device behavior over an extended frequency range and thus must be included in any analytic model if it is to be useful.

A common initial step to developing a valid equivalent circuit model for an RF component is to obtain its key electrical parameters by measurement on the bench. Alternatively, these parameters may also be obtained from complex, typically time-consuming and expensive, electromagnetic modeling software tools. Either way, the goal is to use the data to develop a lumped-element model of the minimum complexity to match the characteristics over frequency.

5.1 Parametric Model Extraction

Lumped-element component models are only useful if the parametric values for the models can be extracted from either measured or modeled data. Scattering parameters, commonly referred to as *S*-parameters, are typically used to characterize the electrical characteristics of an RF component. Commonly, they are normalized to a characteristic impedance $Z_0 = 50\Omega$.

5.1.1 Extracting Immittances from *S*-Parameters

Capacitors and inductors are one-port devices. To fit model parameters to measured component data, first the impedance Z, or admittance $Y = 1/Z$, of the element must be derived from the S-parameters. The S-parameters of a one-port element may be measured in one of three configurations, as shown in Figure 5.1.

With the circuit topology shown in Figure 5.1(a), only input S-parameter data S_{11} is taken. In that case, the elemental impedance and admittance are determined by Figure 5.1(a):

$$Z = Z_0 \frac{1+S_{11}}{1-S_{11}} \quad \text{and} \quad Y = Y_0 \frac{1-S_{11}}{1+S_{11}} \tag{5.1}$$

where $Y_0 = 1/Z_0$.

With the circuit topology shown in Figure 5.1(b), full two-port S-parameters may be taken. However, due to the circuit symmetry, there are only two independent parameters to consider, S_{11} and S_{21}. The elemental immittance values may be derived from either of the latter.

In Figure 5.1(b), from S_{11},

$$Z = -Z_0 \frac{1+S_{11}}{2S_{11}} \quad \text{or} \quad Y = -Y_0 \frac{2S_{11}}{1+S_{11}} \tag{5.2}$$

In Figure 5.1(b), from S_{21},

$$Z = Z_0 \left(\frac{S_{21}}{2(1-S_{21})} \right) \quad \text{or} \quad Y = 2Y_0 \left(\frac{1-S_{21}}{S_{21}} \right) \tag{5.3}$$

With the circuit topology shown in Figure 5.1(c), again full two-port S-parameters may be taken. As before, due to the circuit symmetry, there are only two independent parameters to consider S_{11} and S_{21}. The elemental immittance values may be derived from either of the latter.

In Figure 5.1(c), from S_{11},

$$Z = Z_0 \frac{2S_{11}}{1-S_{11}} \quad \text{or} \quad Y = Y_0 \frac{1-S_{11}}{2S_{11}} \tag{5.4}$$

In Figure 5.1(c), from S_{21},

$$Z = -2Z_0 \left(\frac{1-S_{21}}{2-S_{21}} \right) \quad \text{or} \quad Y = -Y_0 \left(\frac{2-S_{21}}{2(1-S_{21})} \right) \tag{5.5}$$

In the two-port measurement configurations (Figure 5.1(b, c)), the component immittances can be determined in two ways. This can be a useful check to verify the device is actually behaving as a true one-port. If the immittances obtained from

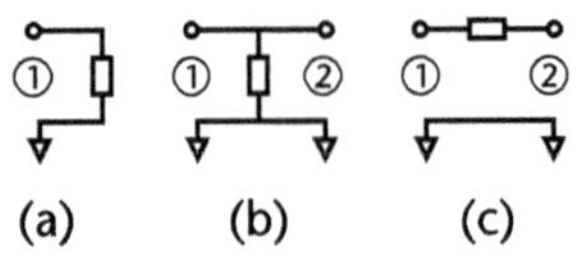

Figure 5.1 Alternate topologies for component S-parameter characterization: (a) termination, (b) shunt element, and (c) series element.

the dual calculations are significantly different, this implies that the device is likely coupling to a neighboring device, or ground plane, and that it cannot be reliably represented in the circuit as a simple one-port component.

Once the immittances of a physical component have been obtained from measurement, as described above, the next step in creating a lumped-element equivalent circuit model of the components is to determine the value of the elements over frequency. The formulations for doing this for both capacitors and inductors follow. Both two and three-element models are considered and their usefulness contrasted.

5.2 Capacitor Lumped-Element Models

An ideal capacitor is a passive circuit element that presents a negative series reactance at the frequency of interest, which is inversely proportional to frequency, that is,

$$Z = -j/\omega C \tag{5.6}$$

where $\omega = 2\pi f$, f = frequency, and C = capacitance, typically expressed in picofarads (pF) or nanofarads (nF).

Unfortunately, due to parasitics, (5.6) is generally insufficient for describing the behavior of a capacitor in any analytic model that is to be useful for reliable RF design.

The first aspect lacking in (5.6) is that any practical capacitor will have loss, or dissipation, associated with it. Thus, its impedance Z will have a real component to it.

Such dissipation can easily be accommodated in an equivalent circuit model by the addition of a series resistor, as in Figure 5.2(b). In this lumped-element model for a capacitor, C_{eff} = effective capacitance of the element at a given frequency, and R_{eff} = resistance (Ω), included to account for losses in the element. This model is commonly used both in the literature and for design.

A prime motivation for employing a two-element model for a capacitor is that the elemental values can be readily determined from its characterization data. As discussed above, this is the impedance, or admittance, of the device. Because these complex parameters comprise only two discrete values (i.e., real and imaginary), only two equations can be directly written to yield two parameters for a model. Despite the great advantage of this simplicity, unfortunately, the two-element model has severe limitations.

For any lumped-element device model to be useful over a wide frequency range, it is highly desirable that the elements of the model do not have a strong frequency dependence. For the model of Figure 5.2(b), however, this is frequently not the case. The reason is that the reactance of a physical capacitor does not generally behave exactly inversely proportional to frequency, as predicted by (5.6). It may even change sign. In consequence, C_{eff} in Figure 5.2(b) will often be determined to have values

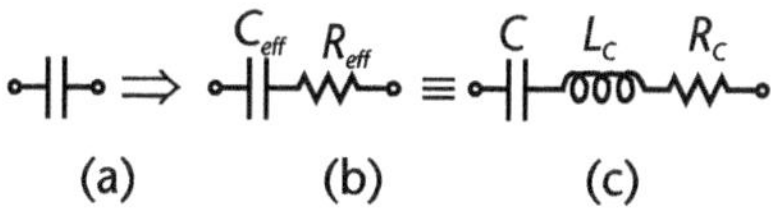

Figure 5.2 Capacitor lumped-element representations: (a) ideal, (b) added dissipative elements, and (c) added dissipative and inductive elements.

that vary greatly with frequency and may even go negative, when extracted from physical data.

To address this issue, a greatly improved lumped-element capacitor model may be achieved by the addition of a series inductor to Figure 5.2(b), as shown in Figure 5.2(c). In the latter, C = capacitance, with parasitic elements, L_C = inductance (typically, nH or μH), and R_C = resistance. As a result of the inclusion of this additional parasitic inductance, the elements C, L_C, and R_C, generally have a much-reduced frequency dependence over those in Figure 5.2(b), when fitting measured data. In many cases, over even wide bandwidths, the elements can frequently be assumed to be relatively constant.

The implications and significant differences between the models of Figure 5.2(b) and Figure 5.2(c) are now examined.

5.2.1 Capacitor: Extracting Two-Element Model Values

The two-element model for a capacitor is shown in Figure 5.3.

Defining

$$Z = Z_r + jZ_i \qquad \text{and} \qquad Y = Y_r + jY_i \tag{5.7}$$

where the subscripts denote real and imaginary parts. Then

$$C_{\text{eff}} = \frac{-1}{\omega Z_i} \qquad \text{and} \qquad R_{\text{eff}} = Z_r \tag{5.8}$$

or

$$C_{\text{eff}} = \frac{|Y|^2}{\omega Y_i} \qquad \text{and} \qquad R_{\text{eff}} = \frac{Y_r}{|Y|^2} \tag{5.9}$$

5.2.2 Capacitor: Extracting Three-Element Model Values

The three-element model for a capacitor is shown in Figure 5.4.

Because this model has three values to be determined, and impedance, or admittance, data from measurement have only real and imaginary values, the fitting problem appears to be overconstrained. Additional boundary conditions must be introduced to determine the most appropriate values for the three elements.

Figure 5.3 Capacitor two-element model.

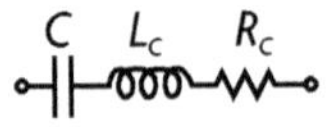

Figure 5.4 Capacitor three-element model.

One possible approach would be to assume that the element values remain fixed between two frequencies and determine the element values to satisfy the boundary conditions at the two frequencies. However, such an approach requires an arbitrary choice of frequencies and does not yield a continuous representation as to how the element values might vary over frequency. Size and performance are dual critical design requirements for passive on-die RF circuit design, invariably involving a compromise between the two. To enable optimal layout for a capacitor, or inductor, the designer needs to carefully evaluate how the component characteristics, and its associated parasitics, vary over frequency with layout changes. Otherwise, incorrect conclusions may be drawn. For example, if a smaller capacitor footprint leads to a higher associated parasitic series inductance, the equality factor may appear to degrade. However, it may well be that this is simply an artifact of the latter and of no consequence if the capacitor is to be series-resonated in the circuit.

Rather than the dual-frequency boundary condition approach mentioned above, another approach was developed for determining the values of three-element capacitor representation of Figure 5.4. This approach was found to be extremely simple and effective to use in the design process. The element values are extracted over frequency directly from capacitor impedance data with no designer inputs. The necessary third equation for determining the three element values is obtained by equating the derivative of the capacitive reactance, at each frequency, of the model and the measured data.

First, we define

$$Z_i' = dZ_i/d\omega \tag{5.10}$$

A prime motivation of using the three-element model in place of the two-element model is to obtain a model whose element values vary only slowly with frequency. Consistent with the latter, we make the following assumptions in deriving the element values using the reactive derivative of the component, namely,

$$\omega C' << C \qquad \text{and} \qquad \omega L_C' << L_C \tag{5.11}$$

From analysis, we then determine the lumped-element values as:

$$C \approx \frac{2}{\omega(\omega Z_i' - Z_i)} \quad \text{and} \quad L_C \approx \frac{(Z_i/\omega + Z_i')}{2} \quad \text{and} \quad R_C = Z_r \tag{5.12}$$

5.2.3 Capacitor: Quality Factor

The unloaded quality factor (Q) of a reactive element is a measure of how much energy is stored to how much energy is dissipated in the element, per RF cycle. The latter is typically due to resistive or radiative losses. It is a critical factor in determining insertion losses in RF circuits employing the component. In general, the higher the Q factor, the less the loss in the circuit. Q-factor is determined by

$$Q = |Z_i| / Z_r \tag{5.13}$$

For a capacitor having the equivalent circuit of Figure 5.4, we have

$$Z = \frac{\omega C R_C - j\left(1 - \omega^2 L_C C\right)}{\omega C} \tag{5.14}$$

Thus,

$$Q = \frac{\left|1 - \omega^2 L_C C\right|}{\omega C R_C} \tag{5.15}$$

Hence, if $L_C = 0$,

$$Q_{L0} = \frac{1}{\omega C R_C} \tag{5.16}$$

From (5.16), we can see that the Q-factor of a pure capacitor (i.e., one with $L_C = 0$) is inversely proportional to frequency. However, for a practical capacitor, with some associated series inductance, the Q value decreases faster than the inverse dependence, according to (5.14), and approaches zero as the frequency approaches a resonance condition where

$$\omega_r^2 L_C C = 1 \tag{5.17}$$

For frequencies above the resonance, the reactance changes sign, implying that, for higher frequencies, the capacitor reactance is inductive.

High-quality capacitors used employed in RF circuits typically have Q values in the range of 50 to 100, provided that they are operated significantly below self-resonance. The Q value of a capacitor is inversely proportional to both frequency and capacitor value; thus, it is more difficult to achieve a high-quality factor for a capacitor: (1) at higher frequencies, and (2) with higher capacitance values.

5.2.4 Capacitor: Comparing Two and Three-Element Models

We shall now compare the major characteristics of the two and three-element equivalent circuit models for a capacitor, with a view to their use in circuit design.

All physical capacitors have a self-resonance frequency, above which they become inductive. In practice, the type and form factor of a capacitor are specifically selected to ensure that this frequency is well above the desired operating frequency to neglect any resonance effects. However, sometimes this may not be possible. In other cases, it may not even be necessary if the capacitor is to be used in a series resonant circuit, in which case, it is only necessary to choose a capacitor that has its self-resonance frequency above that of the series resonant frequency circuit.

To compare the major differences in the two equivalent-circuit models for a capacitor, we shall consider a capacitor with the following characteristics:

Capacitance:	$C = 2.8$ pF	
Series resistance:	$R_C = 0.4\Omega$	(5.18)
Series inductance:	$L_C = 0.4$ nH	

With these values, the capacitor has a self-resonant frequency ~3,052 MHz.

For the two-element model, if we extract the element values according to the formulae in Section 5.2.1, we find the values of the elements over frequency to have the dependencies shown in Figure 5.5(a).

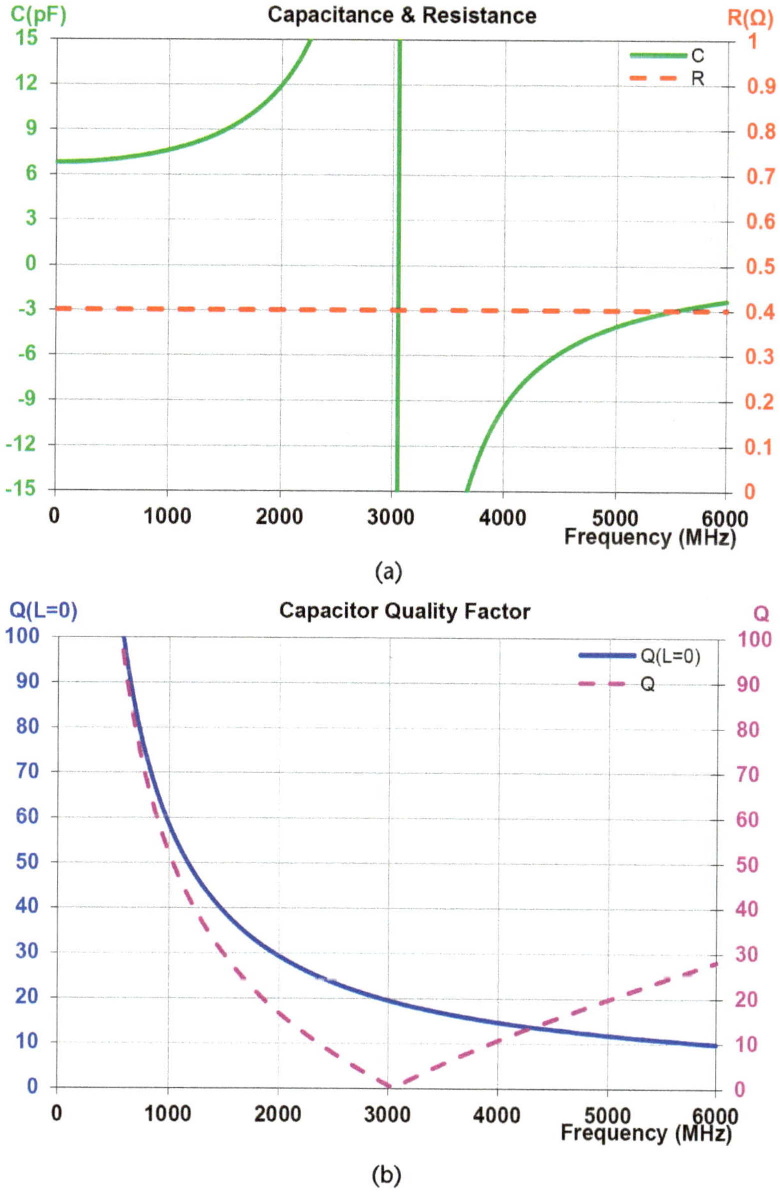

Figure 5.5 Two-element capacitor parametric plots.

The resistance, extracted from the data, shows a constant series resistance of 0.4Ω, consistent with the model. However, the capacitance value, determined from the data, varies greatly over frequency. As expected, it shows a strong resonance around 3 GHz. However, below this, even far below the self-resonance, it has a strong frequency dependence, rising continuously with increasing frequency. A lumped-element model whose elements are strongly frequency-dependent is highly undesirable for use in analytic design approaches. To be useful, the elements of the model should remain relatively constant at least across the desired frequency band of operation.

Figure 5.5(b) shows the dependence of the capacitor quality factor, Q, over frequency. The trace on the left shows the Q of a capacitor having the elemental values of (5.18), but with $L_C = 0$. This is the inverse frequency dependence expected of "pure" capacitor with finite dissipation. In contrast, the trace on the right shows the Q of a capacitor, having the element values of (5.18), calculated for the two-element model from the equations of Section 5.2.1. The Q approaches zero as the capacitor approaches its self-resonant frequency. At all frequencies below resonance, the Q value of the capacitor is visibly lowered by the presence of the series inductance, from what would be expected of the lossy capacitor alone. This can obfuscate a clear understanding of the key physical parameters of the capacitor.

Now we turn to examining the characteristics of the three-element model. If we extract the element values according to the three-element formulae in Section 5.2.2, we find the values of the various parameters of the model over frequency to have the dependencies shown in Figure 5.6(a). Except at very low frequencies, the values of the capacitance and resistance determined from the equations are in excellent agreement with the model values in (5.18) used to generate the data.

The plot referred to in the left axis of Figure 5.6(b) shows the calculated series inductance of the capacitor. Again, except at very low frequencies, this is in excellent agreement with the constant value expected from (5.18).

As long as the electrical characteristics of a capacitor are slowly varying with frequency, that is, the inequalities in (5.11) are valid, the three-element formulae in (5.12) will be appropriate for extracting the lumped-element values.

To illustrate this, we consider a capacitor having the same constant reactive element values of (5.18), but with a resistance that is frequency-dependent. The dependence on frequency, shown in Figure 5.7(a), is $\propto 1/\sqrt{f}$ that is classically associated with skin resistance.

Figure 5.7(b) shows the extracted reactive parameters for the capacitor having the frequency-dependent resistance. The capacitance and inductance values are in close alignment with the model and essentially identical with those in Figure 5.6(a).

Observe that the parametric values of the elements extracted with the three-element capacitor model are well-behaved with no singularities over frequency. This is dramatically different from the capacitance value derived for the two-element model, which showed a strong frequency dependence and a singularity at resonance.

The three-element equivalent-circuit model is thus much superior to the two-element model for characterizing a capacitor. The values are much less frequency-dependent; they do not exhibit any singularity at self-resonance; and they give a much better physical insight into the key physical properties of the capacitor.

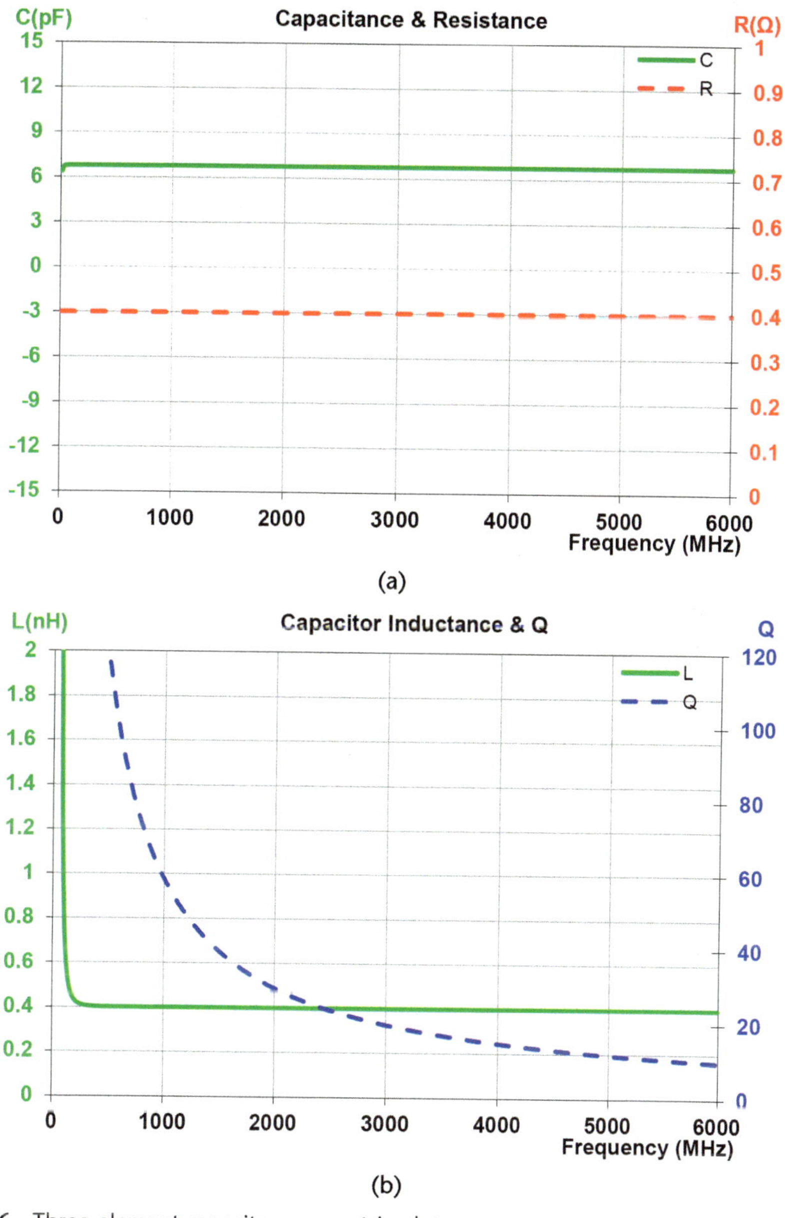

Figure 5.6 Three-element capacitor parametric plots.

5.3 Inductor Lumped-Element Models

An ideal inductor is a passive circuit element that presents a positive series reactance that is directly proportional to frequency, that is,

$$Z = j\omega L \tag{5.19}$$

where $\omega = 2f$, f = frequency, and L = inductance, typically expressed in microHenries (μH) or nanoHenries (nH).

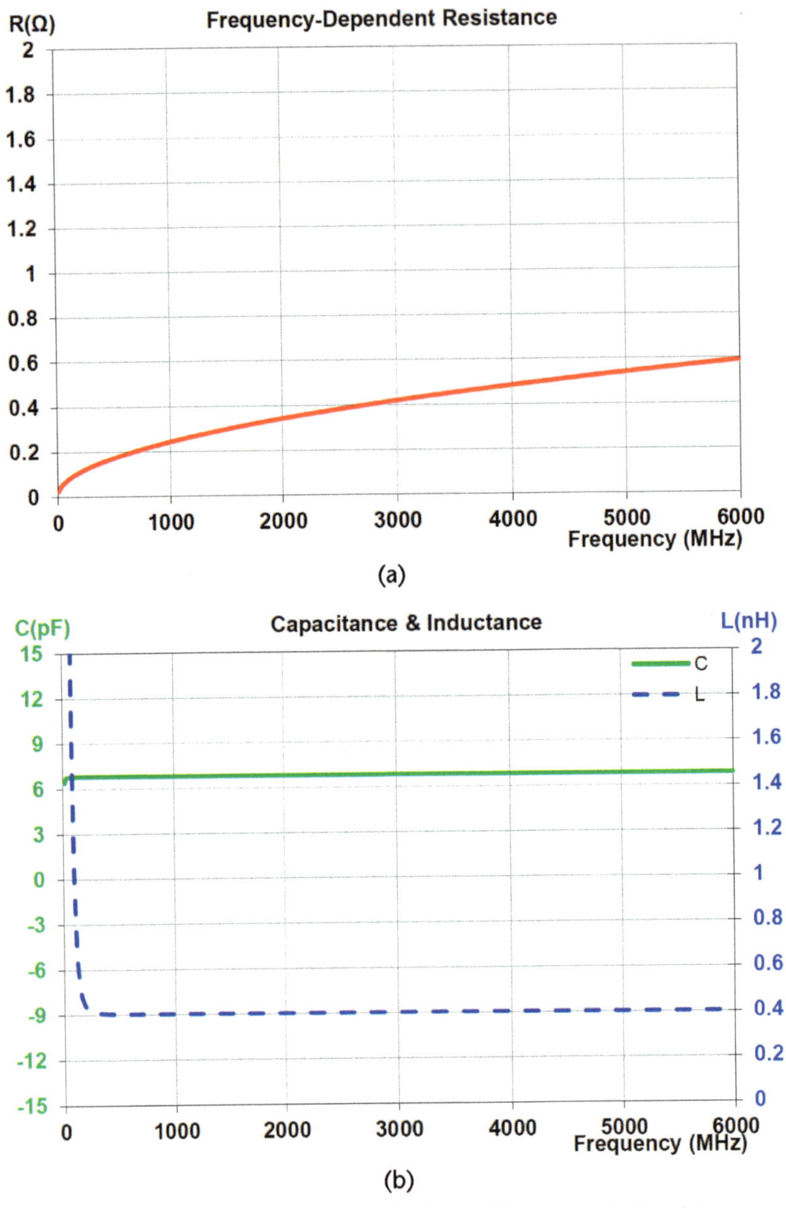

Figure 5.7 Three-element capacitor parametric plots with nonconstant resistance.

As with a capacitor, due to parasitics, (5.19) is generally insufficient for describing the behavior of an inductor in an analytic model that is to be used for RF design. Again, the first aspect lacking in (5.19) is that any practical inductor will have loss or dissipation associated with it. Thus, its impedance Z will have a real component to it.

Such dissipation can easily be accommodated in an equivalent circuit model by the addition of a series resistor, as in Figure 5.8(b). In this lumped-element model for an inductor, L_{eff} = effective inductance of the element at a given frequency, and R_{eff} = resistance (Ω), included to account for losses in the element. This model is commonly used both in the literature and design.

As with the capacitor, a prime motivation for employing a two-element model for an inductor is that the elemental values can be readily determined from its

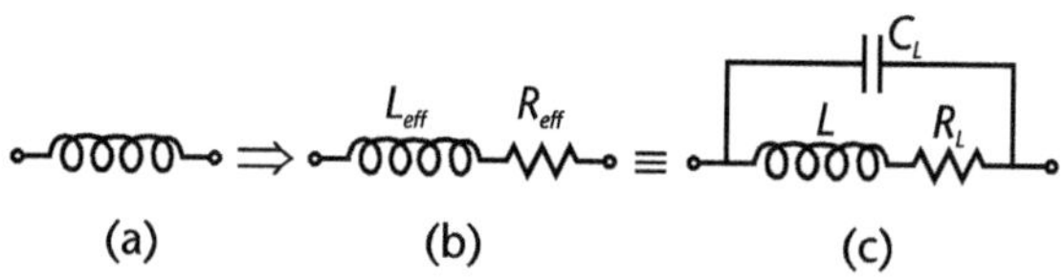

Figure 5.8 Inductor lumped-element representations: (a) ideal, (b) added dissipative elements, and (c) added dissipative and capacitive elements.

characterization data, which again is a single parameter measurement of the impedance, or admittance, of the element. Because a complex parameter comprises only two discrete values (i.e., real and imaginary), only two equations can be directly written to yield two parameters for a model. Despite the great advantage of this simplicity, unfortunately, the two-element model for an inductor has severe limitations.

As with a capacitor, for any lumped-element device model to be useful over a wide frequency range, it is highly desirable that the elements of the model do not have a strong frequency dependence. For the model of Figure 5.8(b), however, this is frequently not the case. The reason is that the reactance of a physical capacitor does not generally behave directly proportional to frequency, as predicted by (5.19). In fact, it may even change sign. In consequence, L_{eff} in Figure 5.8(b) will often be determined to have values that vary greatly with frequency and may even go negative when extracted from physical data.

To address this issue, a greatly improved lumped-element capacitor model may be achieved by the addition of a parallel inductor to Figure 5.8(b), as shown in Figure 5.8(c). In the latter, L = inductance, with parasitic elements, C_L = capacitance, and R_L = resistance. As a result of the inclusion of this additional parasitic inductance, the elements L, C_L, and R_L generally have a much-reduced frequency dependence over those in Figure 5.8(b), when fitting measured data. In many cases, over even wide bandwidths, the elements can frequently be assumed to be constant.

Note that, in Figure 5.8(c), the parasitic capacitor C_L is added around the series elements L and R_L, not directly across L. This ensures that the model predicts dissipative losses even when the inductor and capacitor are close to parallel resonance. This more closely reflects the electrical behavior of a physical inductor.

The implications and significant differences between the models in Figure 5.8(b) and Figure 5.8(c) are now examined.

5.3.1 Inductor: Extracting Two-Element Model Values

The two-element model for an inductor is shown in Figure 5.9.

$$Z = Z_r + jZ_i \qquad \text{and} \qquad Y = Y_r + jY_i \tag{5.20}$$

where the subscripts denote real and imaginary parts. Then

$$L_{\text{eff}} = \frac{Z_i}{\omega} \qquad \text{and} \qquad R_{\text{eff}} = Z_r \tag{5.21}$$

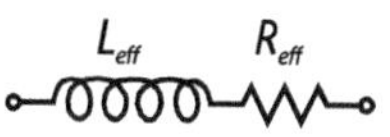

Figure 5.9 Inductor two-element model.

or

$$L_{\text{eff}} = -\frac{Y_i}{\omega |Y|^2} \quad \text{and} \quad R_{\text{eff}} = \frac{Y_r}{|Y|^2} \tag{5.22}$$

5.3.2 Inductor: Extracting Three-Element Model Values

The three-element model for an inductor is shown in Figure 5.10. As with the capacitor, because there are three values to be determined, and impedance, or admittance, data from measurement, have only real and imaginary values, the fitting problem appears to be overconstrained. A necessary third equation is thus again required for determining the three element values. Similar to the capacitor, the necessary third equation can be obtained by equating the derivative of the inductive reactance, at each frequency, of the model and the measured data.

First, we define

$$Y_i' = dY_i / d\omega \tag{5.23}$$

Again, we shall assume that the three element values vary only slowly with frequency. Consistent with this, we make the following assumptions in deriving the element values using the reactive derivative of the component, namely,

$$\omega L' << L \quad \text{and} \quad \omega C_L' << C_L \quad \text{and} \quad R_L' << \frac{\omega L^2}{R_L} \tag{5.24}$$

Derivation of the element values, under these assumptions, is significantly more complex than was the case for the three-element model for the capacitor. This is because, in the inductor, lumped elements are interconnected in a series and parallel configuration. However, after some analysis, we can obtain the solutions in the following form.

Defining

$$\alpha = \frac{\omega Y_i' - Y_i}{Y_r^2} \tag{5.25}$$

$$\omega L \approx \frac{\sqrt[3]{\alpha\left(54 + \alpha^2 Y_r^2 + 18\sqrt{9 + \alpha^2 Y_r^2 / 3}\right)} + \sqrt[3]{\alpha\left(54 + \alpha^2 Y_r^2 - 18\sqrt{9 + \alpha^2 Y_r^2 / 3}\right)}}{6 \cdot \sqrt[3]{Y_r^2}} - \frac{\alpha}{3} \tag{5.26}$$

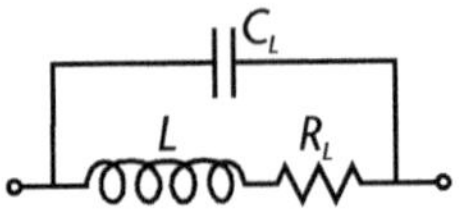

Figure 5.10 Inductor three-element model.

$$R_L \approx \omega L Y_r \sqrt{\frac{2\omega L}{\omega Y_i' - Y_i}} = \omega L \sqrt{\frac{2\omega L}{\alpha}} \tag{5.27}$$

$$C_L \approx Y_i' + \frac{\left(R_L^2 - (\omega L)^2\right)L}{\left(R_L^2 + (\omega L)^2\right)^2} \tag{5.28}$$

The solution for the inductor L simplifies significantly, if the following additional assumption can be made

$$R_L << \omega L \qquad \text{that is,} \qquad Y_r \approx 0 \tag{5.29}$$

Then

$$\omega L \approx \frac{2(\omega Y_i' - Y_i)^3}{\left(4Y_r^2 + (\omega Y_i' - Y_i)^2\right)^2} \tag{5.30}$$

5.3.3 Inductor: Quality Factor

As discussed in the capacitor section, the unloaded quality factor (Q) of a reactive element is determined by

$$Q = \frac{|Z_i|}{Z_r} \tag{5.31}$$

For an inductor having the equivalent circuit of Figure 5.10, we have

$$Z = \frac{R_L + j\omega\left(L - \left((\omega L)^2 + R_L^2\right)C_L\right)}{\left(1 - \omega^2 L C_L\right)^2 + (\omega C_L R_L)^2} \tag{5.32}$$

Thus,

$$Q = \frac{\left|\omega\left(L - \left((\omega L)^2 + R_L^2\right)C_L\right)\right|}{R_L} \tag{5.33}$$

Hence, if $C_L = 0$,

$$Q_{CO} = \frac{|\omega L|}{R_L} \tag{5.34}$$

From (5.34), we can see that the Q-factor of a pure inductor (i.e., one with $C_L = 0$) is directly proportional to frequency. However, for a practical inductor with some associated series capacitance, (5.33) predicts that, beginning from zero at low frequency, the Q value will grow increasingly more slowly with frequency. Eventually, it will achieve a stationary maximum value before decreasing to zero at a higher frequency. At the frequency corresponding to zero Q, the inductor is self-resonant.

From (5.33), the frequency at which the inductor is self-resonant, with $Q = 0$, occurs when

$$1 - \omega_r^2 L C_L = R_L^2 C_L / L \approx 0 \tag{5.35}$$

For frequencies above the resonance at ω_r, the inductor has a negative reactance, characteristic of a capacitor.

Let us now determine where the maximum Q value occurs, relative to the self-resonant frequency, and how this relates to the inherent Q of the inductor (i.e., with $C_L = 0$).

From (5.35),

$$C_L = \frac{1}{\omega_r^2 L + R_L^2 / L} \tag{5.36}$$

thus, from (5.33),

$$Q = \frac{\omega L^3}{R_L} \frac{\left|\omega_r^2 - \omega^2\right|}{(\omega_r L)^2 + R_L^2} \tag{5.37}$$

The maximum value occurs at

$$\omega_{\max} = \omega_r / \sqrt{3} \approx 0.577 \omega_r \tag{5.38}$$

and the maximum value is

$$Q_{\max} = \frac{2(\omega_{\max} L)^3}{\left(3(\omega_{\max} L)^2 + R_L^2\right) R_L} \tag{5.39}$$

Also, if $R_L << \omega_{\max} L$,

$$Q_{\max} \approx \frac{2}{3(\omega_{\max} L / R_L)} \tag{5.40}$$

These expressions give great insight into the physical characteristics of the inductor. We see that the self-resonant frequency of an inductor is predicted to occur at ~$\sqrt{3}$ × the maximum Q frequency. Also, at the frequency at which the inductor Q is a maximum, the apparent Q is reduced by 30% from what it would be if the inductor had no parallel capacitance.

It is common for designers to seek to determine the Q of various inductors from bench measurements so they can determine which configurations may be the most advantageous for their designs. As we see from the above equations, however, the Q of an inductor is multifaceted. Simply comparing which inductor exhibits the maximum Q value may be inappropriate, depending on how the inductor is to be used in the circuit.

If the inductor is to be used as a series element, then the apparent (i.e., measured) Q value is appropriate. However, if the inductor is to be used in a parallel (e.g., tank) circuit, then the inherent inductor Q, ignoring the inductor's parallel capacitance, is more appropriate. The inductor's parallel capacitance can be absorbed into the parallel capacitance required in the circuit. This design approach is valid so long as the inductor's self-resonant frequency is less or equal to that required for the parallel combination. In the limit that the inductor's self-resonant frequency is equal to that of the required parallel combination, no additional capacitor will even be required across the inductor.

In general, RF inductors cannot achieve the high Q values possible for inductors. They are the primary loss element in passive RF circuits. Typically, good RF inductors have Q values in the range of 30 to 70, limited by skin-effect resistance, eddy-current, and radiative loss mechanisms. Inductor Qs can be increased by allowing an increased volume for the inductor (i.e., a larger footprint) and increasing the distance from neighboring conductors or ground planes. Unfortunately, the ultra-compact size requirements of modern RF circuits for mobile applications make large-volume inductors impractical. Apart from this, at gigahertz frequencies, even with no size constraint, it is still an overwhelming challenge to achieve inductor Qs approaching 100.

The Q value of a pure inductor is directly proportional to both frequency and inductor value; thus, it is more difficult to achieve a high-quality factor for an inductor at lower frequencies and with smaller inductance values.

5.3.4 Inductor: Comparing Two and Three-Element Models

We shall now compare the major characteristics of the two-element and three-element equivalent circuit models for an inductor, with a view to their use in circuit design.

All physical inductors have a self-resonance frequency above which they become capacitive. As discussed in the previous section, if the inductor is to be used as a series element, its design must ensure that the self-resonant frequency occurs significantly above the frequencies of interest. If not, the network Q will be degraded and exhibit undesirable frequency dependence. If the inductor is to be used in a parallel (i.e., tank network), it is only necessary that the self-resonant frequency of the inductor be at or above the nominal resonant frequency of the tank circuit.

To compare the major differences in the two equivalent-circuit models for an inductor, we shall consider an inductor with the following characteristics:

$$\begin{aligned}
&\text{Inductance:} && L = 3.0\ \text{nH} \\
&\text{Series resistance:} && R_L = 0.4\Omega \\
&\text{Series capacitance:} && C_L = 1.0\ \text{pF}
\end{aligned} \tag{5.41}$$

With these values, the capacitor has a self-resonant frequency ~2,906 MHz.

For the two-element model, if we extract the element values according to the formulae in Section 5.3.1, we find the values of the elements over frequency to have the dependencies shown in Figure 5.11(a).

For the inductor, we see that both the inductance value and resistance, as determined from the data, vary greatly over frequency. Both are singular around 3 GHz, which corresponds to the self-resonant frequency of the inductor. Even far below this self-resonance, both parameters continue to have a significant frequency

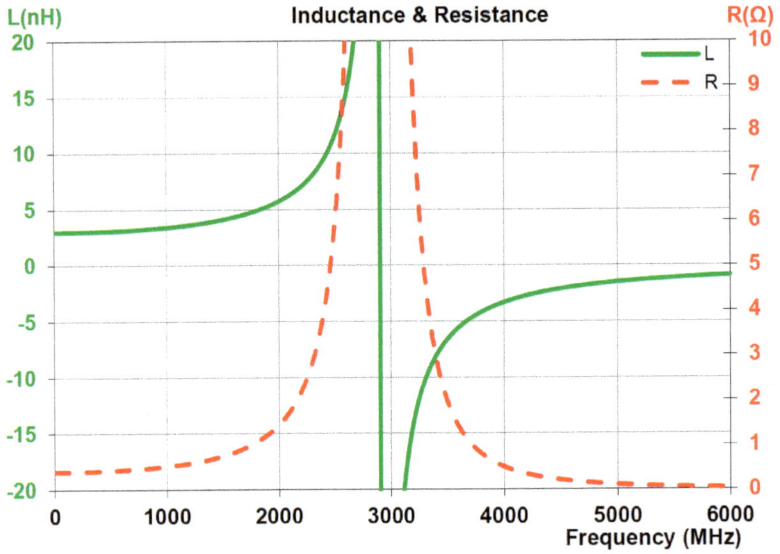

(a)

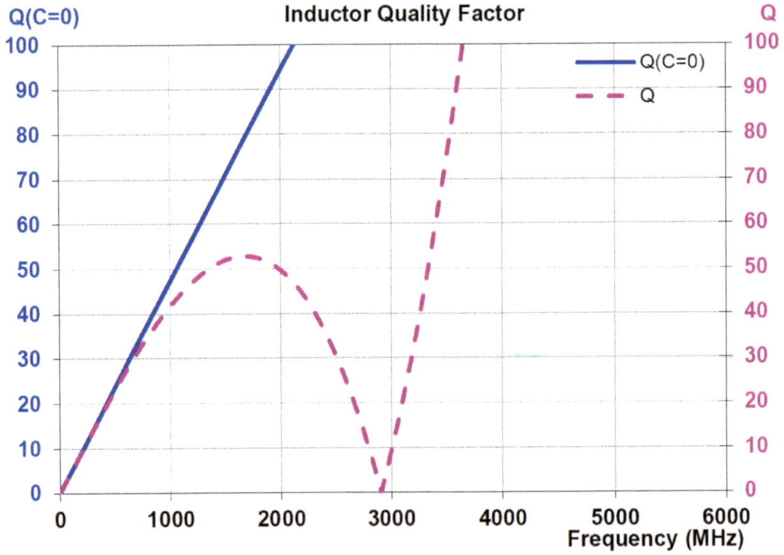

(b)

Figure 5.11 Two-element inductor parametric plots.

dependence, both rising continuously with increasing frequency. A lumped-element model whose elements are strongly frequency-dependent is highly undesirable for use in analytic design approaches. To be useful for the latter, the elements of the model should remain relatively constant at least across the desired frequency band of operation.

Figure 5.11(b) shows the dependence of the capacitor quality factor Q over frequency. The trace on the left shows the Q of a capacitor having the elemental values of (5.41), but with $C_L = 0$. This is the frequency-proportional dependence expected of a "pure" inductor with finite dissipation. In contrast, the trace on the right shows the Q of an inductor, having the element values of (5.41), calculated for the two-element model from the equations of Section 5.3.1. As predicted in Section 5.3.3, the Q grows from DC, although slower than proportional, and then reaches a maximum value before decreasing with increasing frequency until $Q = 0$ around the inductor self-resonance frequency of ~2.9 GHz. A maximum effective inductor value Q_{max} = ~52.7 occurs ~1.68 GHz, compared to the inherent inductor Q (i.e., $C_L = 0$) ~79.2 at this frequency. The ratio 52.7/79.2 = ~0.66, in line with (5.40). Also, the ratio of the frequency at Q_{max} to the self-resonant frequency, 1.68/2.91 = ~0.577, is as predicted by (5.38).

The singular nature of both the inductance and resistance values of the two-element equivalent circuit models for an inductor make this model of limited utility in circuit analysis and synthesis. Such rapidly changing values obfuscate any clear understanding of the key physical characteristics of the element. In order to optimize the layout of an inductor in design, it is critical to have a clear picture of how the key elements of the inductor respond to design changes.

Having seen disadvantages of the two-element inductor model, we now turn to examining the characteristics of the three-element model. If we extract the element values according to the three-element formulae in Section 5.3.2, we find the values of the various parameters of the model over frequency to have the dependencies shown in Figure 5.12(a). The values of the inductance and resistance determined from the equations are in excellent agreement with the model values in (5.41), used to generate the data.

The plot referred to the left axis of Figure 5.12(b) shows the calculated series inductance of the capacitor. Again, except at very low frequencies, this is in excellent agreement with the constant value expected from (5.41).

As long as the electrical characteristics of an inductor are slowly varying with frequency, that is, the inequalities in (5.24) are valid, the three-element formulae in (5.25) to (5.28) will be appropriate for extracting the lumped-element values.

To illustrate this, we consider a capacitor having the same constant reactive element values of (5.41), but with a resistance that is frequency-dependent. The dependence on frequency, shown in Figure 5.13(a), is $\propto 1/\sqrt{f}$ that is classically associated with skin resistance.

Figure 5.13(b) shows the extracted reactive parameters for the inductor with the frequency-dependent resistance. The inductance and capacitance values are in close alignment with the model and essentially identical with those in Figure 5.12.

Observe that the parametric values of the elements extracted with the three-element inductor model are well-behaved with no singularities over frequency. This

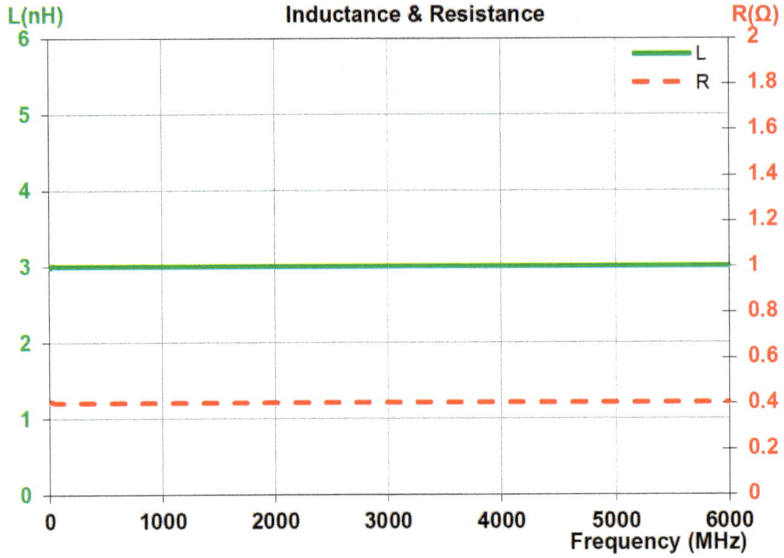

(a)

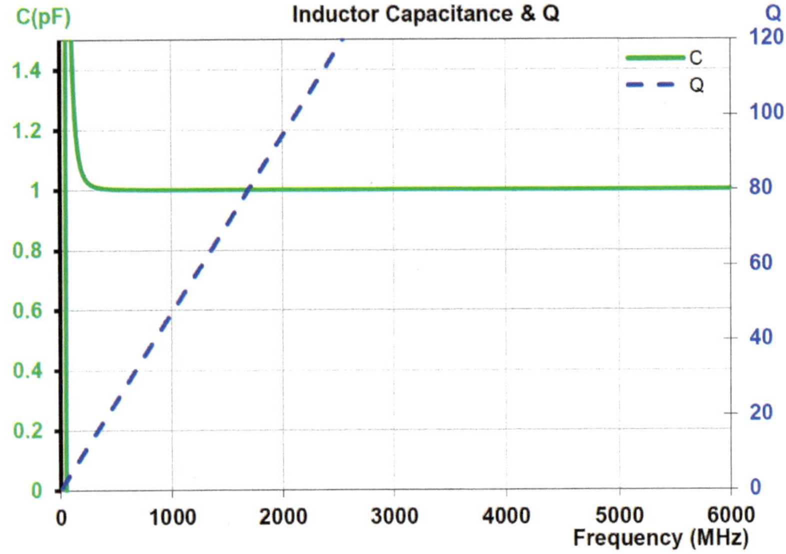

(b)

Figure 5.12 Three-element inductor parametric plots.

is dramatically different from the inductance value derived for the two-element model which showed a strong frequency dependence and a singularity at resonance.

The three-element equivalent-circuit model is thus much superior to the two-element model for characterizing an inductor. The values are much less frequency-dependent; they do not exhibit any singularity at self-resonance; and they give a much better physical insight into the key physical properties of the inductor.

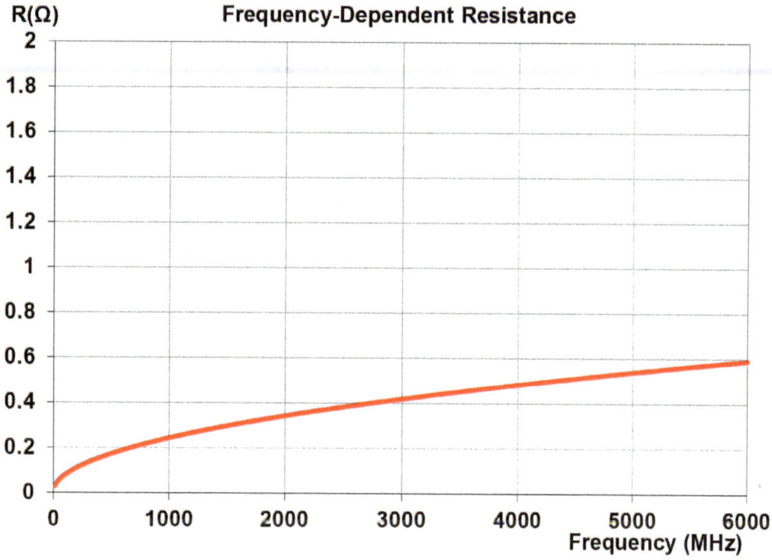

(a)

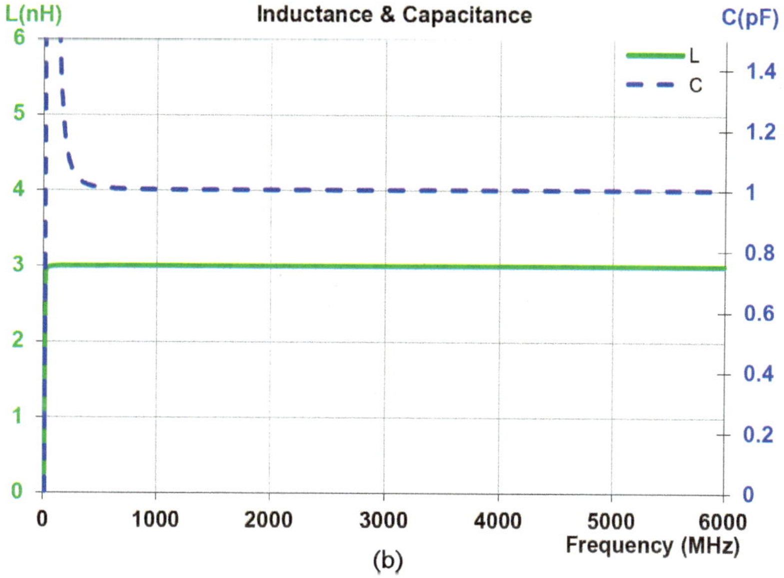

(b)

Figure 5.13 Three-element inductor parametric plots with nonconstant resistance.

5.4 Quadratic Interpolation for $dY/d\omega$ and $dZ/d\omega$

For derivation of the three-element capacitor and inductor equivalent circuit models, the derivatives of impedance or inductance must be determined on a frequency-by-frequency basis. This is best done by fitting a quadratic polynomial through the frequency of interest and the frequency data above and below it. Assuming, for admittance,

$$Y_n = a_0 + a_1 f_n + a_2 f_n^2 \tag{5.42}$$

and for impedance

$$Z_n = b_0 + b_1 f_n + b_2 f_n^2 \tag{5.43}$$

based on expressions in Appendix E, after some manipulation, we determine

$$\left[\frac{dY}{df}\right]_{f2} = \frac{-(f_3 - f_2)^2 Y_1 + (f_3 - f_1)(f_1 - 2f_2 + f_3) Y_2 + (f_2 - f_1)^2 Y_3}{(f_3 - f_2) f_1^2 - (f_3 - f_1) f_2^2 + (f_2 - f_1) f_3^2} \tag{5.44}$$

$$\left[\frac{dZ}{df}\right]_{f2} = \frac{-(f_3 - f_2)^2 Z_1 + (f_3 - f_1)(f_1 - 2f_2 + f_3) Z_2 + (f_2 - f_1)^2 Z_3}{(f_3 - f_2) f_1^2 - (f_3 - f_1) f_2^2 + (f_2 - f_1) f_3^2} \tag{5.45}$$

5.5 Integration of RF Inductors into a Compact Module Design

Inductors in RF modules are invariably planar in construction. They may be implemented in one or more layers in a laminate carrier, on an active die, or on a passive die. For optimum performance characteristics and highest yield, critical RF inductors are invariably implemented on die, rather than in a laminate. The reason for this is that on die the inductor traces are controlled lithographically and thus subject to less variation in the manufacturing process. Smaller trace widths and gaps between the traces can also be achieved.

Figure 5.14 shows examples of some common planar on-die RF inductor layouts. To minimize the footprint area required to achieve a given inductance value, they typically comprise metal traces wrapped tightly around one another. If both ends of the inductor need to be in close physical proximity, then some type of via or underpass metallization is required on the wafer. Such compact planar inductors may be implemented on an active die, such as a PA. Alternatively, a separate passive die may be designed for inclusion in a module. This approach typically has the advantage of lower cost.

One particularly advantageous approach for implementing a low-cost compact PA module utilizes flip-chip attach technology. Such a module is typically realized with two or more dies mounted on a laminate carrier. Passive dies for implementing multiple planar RF inductors required for biasing, matching, and filtering functions, have the dual advantages of low cost and low production variability. Active die carry the transistors and associated circuitry required for signal processing and amplification. Flip-chip bumps are plated on each die. Figure 5.15 shows a profile of a representative on-die stud bump.

The active and passive die are flip-chip-attached to a laminate carrier by reflowing the Sn solder cap on top of the stud bumps. After reflow, the module is overmolded with a resin and the individual modules are then singulated. This construction technique is shown schematically in Figure 5.16.

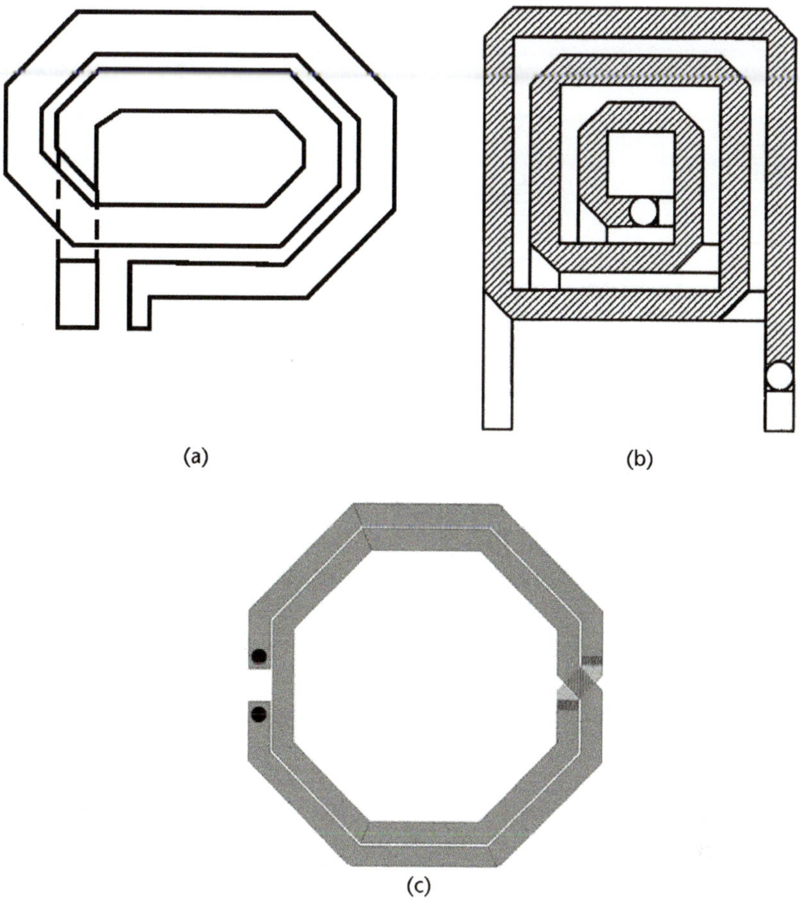

Figure 5.14 Examples of on-die critical RF inductor layouts.

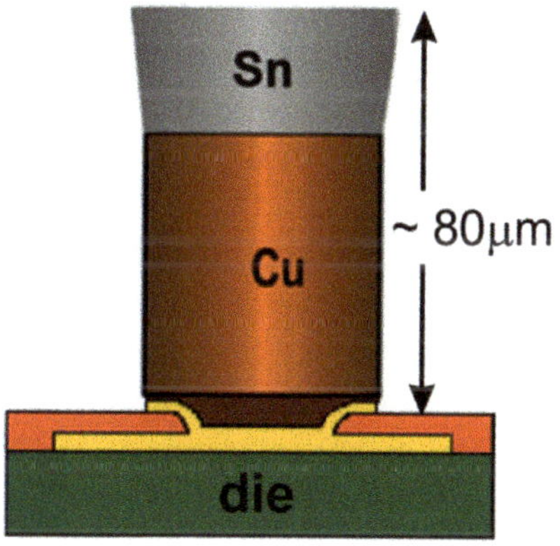

Figure 5.15 Multiple inductive passive die and Cu stud bump profile.

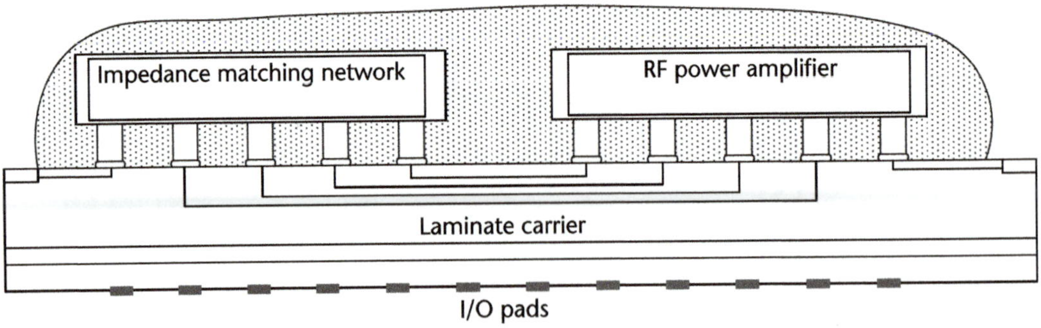

Figure 5.16 Flip chip PA module construction.

This flip-chip module construction technique has several very significant advantages over conventional die wire bonding.

1. The Cu bumps have very low resistance.
2. The Cu bumps have very low inductance.
3. There is minimal RF magnetic coupling between the bumps and other circuit elements.
4. Because the bumps are lithographically defined, they have minimal variability in production.
5. The bumps permit a very low-profile module.

5.6 Summary

Reactive elements (i.e., capacitors and inductors) are critical elements in all RF circuits and circuit models, whether active or passive. Because they are essentially one-port devices, they are characterized by only a single complex parameter, namely impedance or admittance, Z or Y. Consequently, lumped-element models of reactive elements generally comprise only single reactive and resistive elements.

In order to develop useful analytic synthesis approaches for the RF circuit, lumped-element models for inductors and capacitors, with relatively constant values over frequency, are essential. In practice, all reactive elements have an associated self-resonance, above which their reactance changes sign. If the self-resonance is sufficiently above the frequencies of interest for the design, it may be of little consequence and may be ignored. However, for high-frequency RF circuits, for example, hundreds of megahertz or gigahertz, this is frequently not the case and becomes ever truer as designers are continuously pressured to reduce the size of their designs.

When self-resonances of the reactive components cannot be ignored, simple two lumped-element representations are not useful for analytic circuit design. The reason for this is that the element values become increasingly frequency-dependent as the resonance is approached and change sign above resonance. With such behavior, it is nearly impossible to derive analytic formulae useful for circuit design. It also masks any clear understanding of how the key design parameters affect the element's performance and how it might be improved by redesign.

Formulae are presented in this chapter for deriving three-element, equivalent-circuit models for reactive elements from the one-port data. These formulae do not rely on a best fit of the models to the data over a moving range of frequencies. Instead, the element values are derived in a closed form on a frequency-by-frequency basis. The additional parameter required for these solutions is the derivative of the reactance at each frequency.

The three-element, equivalent-circuit models for reactive elements are greatly advantageous compared to the two-element models. The three-element component values are much better behaved and typically vary only slowly and smoothly over frequency. This makes them eminently suitable for deriving analytic formulae for RF circuit design. It also greatly aids in understanding how design trade-offs impact the key RF characteristics of the reactive element.

When referencing the quality factor Q of a reactive element, there are two alternate values that are relevant, depending on how the element will be used in the circuit. The first we shall call the inherent Q factor of the element. For a capacitor, this would be its Q factor, ignoring any series inductance of the element; for an inductor, this would be its Q factor, ignoring any parallel capacitance of the element. The second Q factor would be the apparent Q factor of the reactive as determined by measurement.

The inherent Q factor of a capacitor (i.e., ignoring any self-inductance) is the appropriate parameter to use when a capacitor is to be used in a series configuration. In this case, any series inductance can be absorbed in the series inductance required in the circuit, thereby reducing the required value of the inductance. This design approach is applicable, providing only that the self-resonant frequency of the capacitor is at or above the desired self-resonant frequency of the series configuration.

The inherent Q of an inductor (i.e., ignoring any self-capacitance) is the appropriate parameter to use when an inductor is to be used in a parallel (i.e., tank) configuration. In this case, any parallel capacitance can be absorbed in the parallel capacitance required in the circuit, thereby reducing the required value of the capacitance. This design approach is applicable, providing only that the self-resonant frequency of the inductor is at or above the desired self-resonant frequency of the parallel configuration.

CHAPTER 6

Efficient Analytic Optimization Approach

The starting point for any RF design is the set of electrical requirements for the circuit supplied by the end user. The choice of circuit topology to best meet these requirements is generally left to the designer. Frequently, there can be multiple possible design approaches with the potential to meet the sought-after design goals. It is up to the designer to select the best circuit for the application in terms of size, cost, performance, and production yield.

When faced with a new set of required performance specifications for an RF module, the designer must decide as quickly as possible the topology to be used that will have the best chance to meet these requirements. The choices are typically to modify an existing design or to start afresh with a new approach that may offer the chance for improved performance, smaller size, or lower cost. While the latter may be advantageous in the long run, in the short run, it requires significantly more time and effort. To minimize such an effort, the approach as detailed in this book is to develop a set of easily applied test benches that can rapidly provide the designer with all the key design trade-offs for each approach. These test benches will allow the optimum design architecture to be selected as expeditiously as possible and at the same time give the designer a comprehensive appreciation of what each circuit option is, and is not, capable of.

Like a set of colored paints for an artist, a set of test benches can be developed for each of the common basic circuit functions that a designer may face. These may include input and output amplifier impedance matching in various forms, filtering, diplexing, power splitting and combining, and couplers. The test benches to be described are based on analytic circuit analysis combined with a novel approach to circuit optimization that makes them extremely time-efficient.

To illustrate the fundamentals of this approach, consider the schematic of an archetypal two-section amplifier matching architecture shown in Figure 6.1. The two cascaded lowpass π-sections comprise a total of 10 elements. The amplifier, on

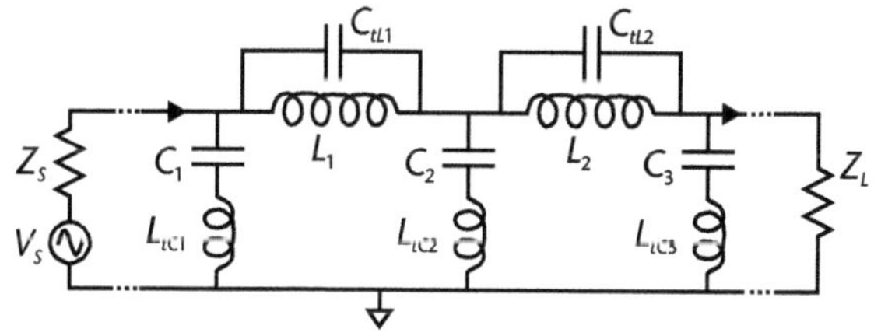

Figure 6.1 Example of two-section amplifier output match with frequency traps.

the left, has complex source impedance Z_S, and the load impedance, on the right, has a complex impedance Z_L. This circuit is representative of those commonly used for matching the low output impedance of an RF PA to a higher impedance load, typically the antenna.

Optional reactive elements $C_{t\#\#}$ and $L_{t\#\#}$ are included in the matching circuit to implement frequency traps at user-specified frequencies. Commonly, these may be harmonic frequencies of the passband, for example, second and third harmonics, where the network is required to have desired impedances or rejection characteristics. Because the circuit has 10 elements, determining unique values for the elements would require 10 equations with associated boundary conditions (i.e., input specifications). Given that both the complex source and load impedances are specified, for a low insertion loss through the network, we require the input impedance to the network $Z_{in} = Z_S^*$, where "*" denotes a complex conjugate. Combined with this impedance requirement, we can write the following equations for the network:

1. $(Z_{in})_{real} = (Z_S)_{real}$.
2. $(Z_{in})_{imag} = -(Z_S)_{imag}$.
3. $L_{tC1} = 1/\omega_{C1}^2 C_1$, where ω_{C1} = trap frequency for C_1.
4. $L_{tC2} = 1/\omega_{C2}^2 C_2$, where ω_{C2} = trap frequency for C_2.
5. $L_{tC3} = 1/\omega_{C3}^2 C_3$, where ω_{C3} = trap frequency for C_3.
6. $C_{tL1} = 1/\omega_{L1}^2 L_1$, where ω_{L1} = trap frequency for L_1.
7. $C_{tL2} = 1/\omega_{L2}^2 L_2$, where ω_{L2} = trap frequency for L_2.

This comprises only seven equations, yet the network has a total of 10 element values to be determined. Consequently, the network synthesis problem is underconstrained, implying that there is a limitless set of solutions that is possible for the element values. While this might at first appear to be disadvantageous, presenting an insurmountable challenge for the optimization of the network via synthesis, the opposite is actually true. The fact that a limitless set of network solutions is possible, each with differing performance characteristics, presents the designer with the freedom to select an optimum configuration, for the application, from an infinite spectrum of possibilities. No single "optimum" solution exists. Among the possible configuration choices, each may be optimum for one parameter, for example, insertion loss, input impedance, and harmonic rejection. Invariably, none will be optimum for each of the desired characteristics simultaneously.

Because there is no one optimum solution for the network, what is the best and most time-efficient way for the designer to find the best compromise for the application? A common approach, typically used in powerful RF computer-aided design (CAD) software suites, is to begin with a starting solution for the network and arrive at an improved solution by means of an optimizing algorithm. However, this approach typically has the following drawbacks:

1. The designer must set design goals for key performance characteristics.
2. Only a single design configuration is returned.
3. The optimization can be very time-consuming.
4. There is a risk that only a local optimum will be found.
5. Optimization must be rerun if minor design parameters are changed.

For an unfamiliar network, it can be a challenge for the designer to set goals for key network characteristics a priori. The designer first needs to have a good feel (i.e., understanding) how individual goals might impact each other. For example, it might be that requiring too good an impedance match across a passband limits the minimum insertion loss that can be achieved.

Optimization algorithms, if they are successful, generally stop when they find the first set of component values, which result in the circuit performance being just within the design goals. They do not continue to improve the circuit performance beyond that point, because a computer algorithm cannot decide what is better without further user input. With only a single design solution, the designer has little insight into the limitations, capabilities, and sensitivities of the circuit. This can only be gained by running repeated circuit optimizations with different goal-setting objectives. Such an approach can be very challenging and time-consuming.

Optimization algorithms are notoriously challenged by being trapped by a local optimum. If the initial circuit values going into the optimizer are not well-chosen, the best performing circuit design returned may not be truly the best solution possible. Wider ranging, randomized optimization algorithms do exist, but in practice the results are frequently disappointing. Conservative optimization algorithms applied to an already well-conceived design typically yield the best results.

Finally, once circuit values have been selected for a design, using some type of optimization process, designers will frequently want to slightly modify its performance or make minor modifications to the circuit. In Figure 6.1, for example, the designer might want to replace one or more of the ideal elements by a three-element equivalent circuit, as discussed in Chapter 1, to account for parasitics. Alternatively, there might be a need to improve harmonic rejection. To maintain optimum performance, such small changes would inevitably result in the need to modify all the circuit elements, even if only slightly. Consequently, using the software optimizer approach, every time the designer wants to make any changes to the circuit, the optimizer must be rerun. The designer does not get instantaneous feedback to circuit modifications, which can slow down the design process.

The analytic optimization approach developed overcomes the drawbacks listed above. It allows circuit prototyping to be performed very rapidly, at minimal cost, on computers with only moderate processor capability. It is very flexible and allows rapid comparison of different design approaches by clearly providing the performance trade-offs of each architecture.

Consider again the amplifier output matching circuit in Figure 6.1. It is impractical to consider all possibilities for the network as, even if only 10 values for each element were to be considered, there would be a total of 10^{10} solutions for the network to be analyzed, which would take considerable computational time. For this reason, software optimization algorithms generally begin with a set of "seed" network element values and proceed by methodically making progressive and incremental changes to each of the elements in turn, and then repeating the process, in an attempt to bring the network performance closer to the design goals. This approach is very time-consuming, requires considerable computation, and has no guarantee of finding a global optimum.

If we apply circuit analysis, as detailed above, we can write only seven circuit equations for the network, which has 10 unknowns. We thus need three additional boundary conditions (i.e., equations) to determine unique solutions for the element values of the network. One possible approach might be to set arbitrary values for three of the elements and then determine values for the remaining elements from the circuit equations. If the values of these three elements are then stepped, through a predetermined range, a spectrum of solutions for the network could be obtained. However, such an approach is highly problematic for the following reasons:

1. The key circuit performance parameters will have varying sensitivities to each of the individual circuit elements (in this case, 10). Consequently, the three elements chosen for the analysis will be critical to the outcome.
2. With no a priori knowledge of the appropriate values for the elements chosen, it is extremely difficult to set well-chosen values for the span and increments for the elements being varied.
3. Circuit solutions may not even exist for some of the three-element value combinations being evaluated.

Thus, choosing element values for the additional circuit boundary conditions to generate a spectrum of circuit solutions is not a recommended approach. The better choice, and the one that forms the basis of the approaches in this book, is to identify key circuit characteristics whose limits can be identified a priori and then step though the ranges for each of them to provide the spectrum of circuit solutions. These alternate analytic solutions can then be compared and the one that provides the best compromise of electrical characteristics selected.

By way of example, the circuit of Figure 6.1 can be viewed as a concatenation of two π-networks. Each of these networks will have a transmission phase shift associated with it. In addition, at the common node of the networks, the impedance looking to the right and left will be complex conjugates of each other if the network is to provide the required input match. The phase shifts of the two sections plus the real impedance at the common node can then be chosen to complete the three variables to be used for the network scan. Limits can be placed on these three parameters relatively easily without preassigning any of the element values. This approach has been found to be very effective at providing a rapid and comprehensive knowledge of a network's capabilities and trade-offs, as is illustrated in the following chapter.

It should be emphasized that the analytic optimization approaches described here are not proposed as an alternative to a full, typically three-dimensional (3D) CAD analysis. Quite the opposite, they are meant to complement it. On being given a list of desired device specifications, a designer must first decide upon the best circuit architecture to employ for achieving the goals and to what extent they are achievable. In practice, there are only a relatively few number of passive circuit architectures employed for PA output matching. Flexible analytic models for these can be extremely helpful to the designer to decide which architecture is the best starting point for a full CAD simulation and optimization.

CHAPTER 7

Excel: A Powerful Interactive RF Design Tool

One of the most important aspects of any circuit analysis and optimization software tool is its visualization capability. The design equations and optimization algorithms presented in this book may be implemented by a competent designer using any number of programming choices. However, in this book, we shall be illustrating the RF circuit design approaches using Microsoft Excel. This may strike many designers, who have likely never contemplated Excel as being capable of complex RF analysis, to be a surprising choice. However, the latest versions of Excel have many overlooked capabilities that can be exploited for efficient RF design. We shall explain how these capabilities can be used to build powerful RF design and analysis software tools. The built-in conditional statements functionality in Excel is particularly powerful in enabling the creation of dynamic circuit visualizations, which are more flexible and fluid than those available in even the most expensive RF software design suites.

The principal advantages for RF circuit analysis in Excel are:

1. Dynamic circuit visualization.
2. Low cost, much lower than any commercial RF design software suite.
3. Widely available, being part of the Microsoft Office suite.
4. No additional licensing fee required.
5. Design tools can be run anywhere, without the need to access a license server.
6. Readily customizable.
7. Easy to exchange data with other programs.
8. Programming help and tips widely available, for free, on the internet.

Excel can be used as RF analysis tool as it has complex functional and calculation capabilities built in. It is a very flexible tool used by millions around the world for a multitude of disparate purposes. To satisfy the varying needs of its wide user base, its data manipulation capabilities have been dramatically expanded on a continual basis to the point where the typical user today is only cognizant of a very small range of its capabilities. We shall attempt, in this book, to demonstrate how some of these capabilities may be applied to efficient and user-friendly RF analysis.

While Excel has the capacity for use as an RF design tool, it does not have any built-in RF functions. It is, in essence, a programming platform with an excellent visual interface, which can be used for building RF applications.

7.1 Visualizations in Excel

As listed above, a major incentive for building RF applications in Excel is its visual display capabilities. There are three classes of visual interaction tools available to the designer:

1. Data display plots. The data may be displayed in a variety of plot formats, for example, Cartesian, polar, bar, and π. Excel incorporates a wide variety of built-in display formats. They provide the designer with near unrestricted freedom to display in the best format for visualization.
2. Dynamic circuit schematic layout. This is made possible by the "conditional format" capability of Excel. To arrive at an optimum design, the RF designer needs to begin by considering a layout that encompasses as many variants as possible. For example, if the design is for a filter, the order of the filter should be a variable; if the design is for a matching network, the number of sections and whether they should be highpass or lowpass networks should be investigated. Depending on the analysis/optimization, the circuit schematics could look very different. Excel has the unique ability to cleanly present a schematic for the chosen solution on the spreadsheet, with no crossed-out elements or disconnects. Even the most expensive RF design tools are not capable of this.
3. Contour plots. In addition to the dynamic schematic layout, the "conditional format" capability also makes a form of contour plotting relatively easy to implement. This is achieved by populating a grid of cells with colors that correspond to a discrete range of element values. Multiple contour plots of this kind may be linked together to how important network parameters can be traded off with one another in the design and linked to the schematic layout. These contour plots are elements to design optimization.

7.1.1 Display Plots

The majority of RF network performance data is typically presented in a rectangular Cartesian plot format. Such plots may have left and right axes to display, for example, displaying various characteristics of the network. Most commonly, the abscissa, or horizontal axis, is frequency; however, it may be another network variable such as the value of one of the elements.

Figure 7.1 shows a Cartesian plot example from an Excel spreadsheet for a coupler. In this example, the coupler was assumed to have a coupling coefficient of 22 dB, and 20 dB directivity. The load reflectivity for the plot was VSWR = 2.5. The coupler parameters are plotted versus the phase of the mismatch on the output of the coupler. Note that multiple parameters may appear on one plot and each may be referred to either of the vertical axes. In addition, parameters can be turned on or off in the display window at will.

In RF design, reflection data and input and output impedances are invariably displayed on a Smith chart. Fortunately, Smith charts can easily be created on an Excel spreadsheet. Complex data is displayed using the polar chart option in Excel. Hence, the parametric data to be plotted need simply to be converted to its polar

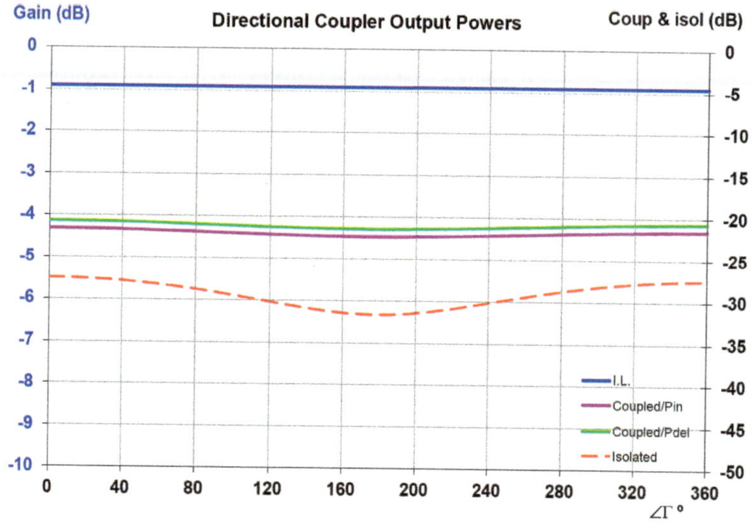

Figure 7.1 Cartesian plot example for a coupler in Excel.

format (i.e., magnitude and phase). If a Smith chart graphic is then used as the fill option for the chart, a classic Smith chart representation of the data may be obtained.

Figure 7.2 shows examples of two different Smith chart fill options displaying the same data. The first has an immittance background, showing both impedance and admittance contours. The second shows only the impedance contours. Multiple contours may be included on the same chart. Note also in Figure 7.2 that the titles capability may be used to display dynamic information related to the data displayed in the chart. In the charts shown, the frequency range of the data is displayed on the left at the top and bottom of the charts. A description of the data being plotted is shown on the top right, which is also dynamically linked to the system variables.

On the Cartesian plots, the ranges on the abscissa and ordinate axes can be modified interactively and the resulting ranges are displayed on each of the axes. Thus, for example, if the abscissa displays frequency data, the user can change the frequency range for the plot and the modified frequency range of the data will be instantly and clearly displayed on the axis. However, for the Smith chart plots, varying the range of the data being displayed is a little more complicated. The normal extent of the data on both axes is −1 to + 1, the nominal range for a reflection coefficient. However, what we are commonly interested in seeing on a Smith chart is data over a limited frequency range. For example, we might only want to see the input impedance of a filter over the passband frequency range. How do we accomplish this? The workaround is relatively simple. In two cells, close to the Smith chart, allow the user to enter the range to be displayed on the chart. For example, if the independent variable is frequency, the two cells could look like Table 7.1. On the Excel sheets, for convenience, data values in cells that the user is free to modify are always assigned to the color blue.

These values can then be used in conditional statements in each of the data cells where the values to be displayed are calculated. For example, if the entries specify a frequency range and the frequency associated with a data value is within that range, the cell value is left untouched. However, if the associated frequency is outside the specified range, the cell value is set to NA(), the null value. Cells with

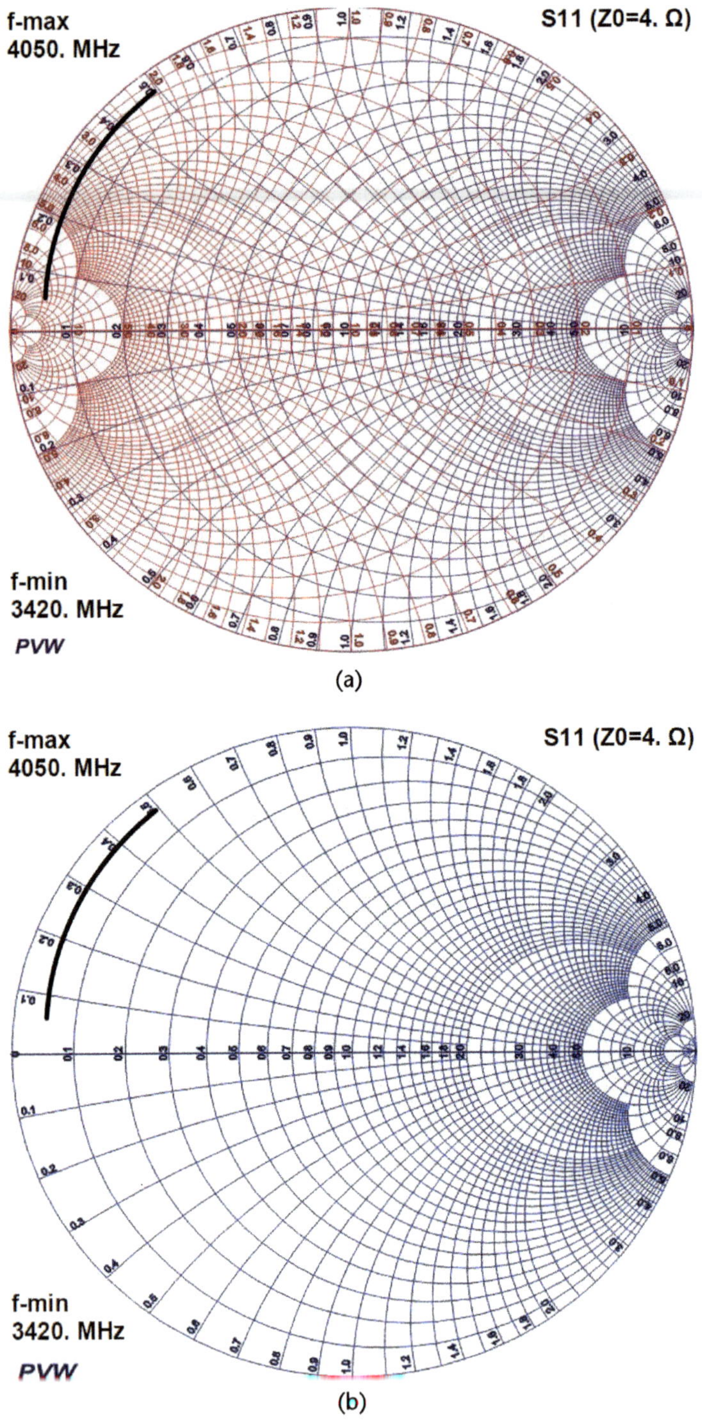

Figure 7.2 Smith chart examples in Excel.

Table 7.1 Smith Chart Display Range Entry on Spreadsheet

Smith range (MHz)	1700	2025

this value (#N/A) will not be displayed on the chart. Hence, even if the data range in the Smith chart plot window is set to include the full extent of the data, only data within the specified range will be displayed.

By way of example, Figure 7.3 illustrates how the input impedance of a filter may be displayed over differing frequency ranges, depending on the user inputs,

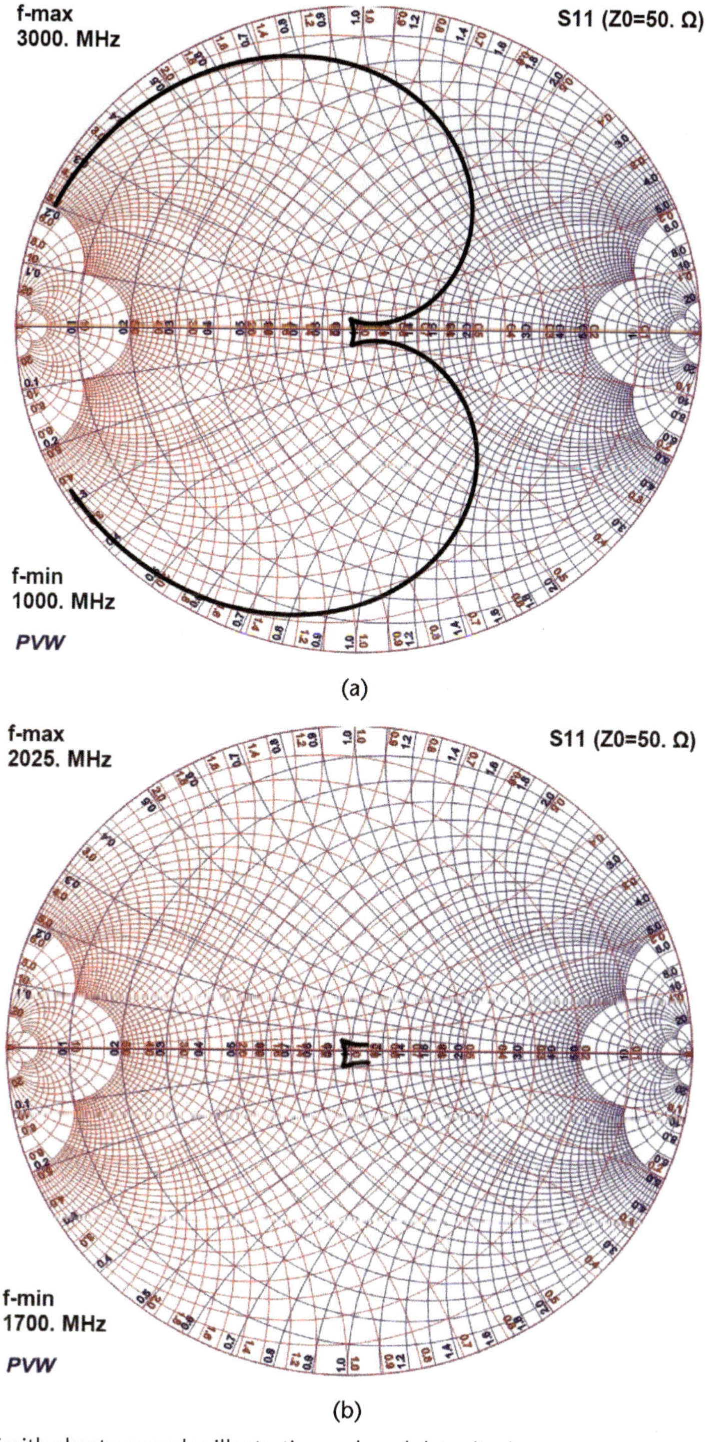

Figure 7.3 Smith chart examples illustrating reduced data display range.

using this technique. On a spreadsheet in the workbook, frequently hidden from the user, the filter characteristics are calculated over an extended frequency range, say, 10 to 12,000 MHz. The range of the input impedance data to be displayed on the chart is set to encompass this full range. However, in the case of the plot on the left, the user specifies a desired frequency display range for the data from 1,000 to 3,000 MHz by means of entries such as those in Table 7.1. For the plot on the right, the user specifies a reduced frequency display range, corresponding to the filter passband, from 1,700 to 2,025 MHz by changing the entries in cells such as those in Table 7.1.

This facility in Excel of conditionally equating a data value to the null value is a very useful tool that can be exploited in many ways beyond the example described for the Smith chart. When displaying multiple data sets on the same chart, it can be used to dynamically turn data sets on and off. If the designer is trying to meet customer-defined performance specifications, it can be used to include dynamically variable template limits on the chart.

7.1.2 Dynamic Circuit Schematic Layout

The RF design challenge begins with a set of electrical performance specifications that the circuit will be required to meet. In most cases, there will be multiple alternate circuit approaches to consider that could potentially meet the specifications. The challenge for the designer is to narrow down these choices, with a minimal expenditure of time and resources, to arrive at an "optimum" solution based on a set of critical criteria.

Alternate circuit design solutions that an engineer may want to consider in the initial stage of the design will usually comprise significantly different circuit topologies. Thus, the circuit schematics for each approach will be different. Not only that, but even for a single approach, there may be a choice between highpass and lowpass elements or the number of sections required in the circuit. Therefore, as the engineer progresses through the early stages of design, many circuit schematics will typically have to be drawn and redrawn many times. These schematics can also become inevitably cluttered with many elements that are open-circuited or short-circuited in the schematic that can lead to confusion.

Even the most advanced RF design tools have static circuit schematic visualizations. Thus, if a matching network has three sections, even after optimization, it will still have three sections albeit if some of the elements are unneeded. Likewise, a lowpass section or filter in a schematic will remain lowpass in nature, regardless of any circuit optimization applied to it.

In contrast to the commercial RF design tools, the "conditional format" capability of Excel enables the creation of dynamic circuit schematics on a spreadsheet. This means that if, as a result of optimization, for example, a highpass network is required in place of a lowpass one, the schematic can immediately be redrawn as the former with no superfluous elements in the schematic. Similarly, if as a result of optimization or user inputs, a two-section matching network is preferred to a three-section network, the schematic can immediately morph from the latter to the former.

The dynamic schematic morphing capability, combined with the associated optimization and analysis capabilities that can be incorporated into an Excel workbook, makes it a powerful tool for RF design. Lossy and reactive parasitics can be included in the circuit analysis, as described in Chapter 1. While these parasitics are relatively basic, and do not incorporate higher-order parasitic element coupling, they are generally sufficient to give the designer a solid understanding of the circuit behavior and capabilities. Because schematic changes can be evaluated easily and essentially instantaneously, the designer can rapidly compare different circuit topologies to determine the optimum for the application.

Figure 7.4 shows an example of the dynamic schematic capability on an Excel spreadsheet. Both circuit topologies are from the same spreadsheet, but with differing user inputs. As can be seen, inductors are represented as rectangles, but, other than that, other components take their conventional form.

Figure 7.4(a) shows a two-section lowpass PA matching network, with a bias feed network for the amplifier. In contrast, Figure 7.4(b) shows a two-section highpass PA matching network, with the same bias feed, but with a frequency trap on one of the series inductors, a series blocking capacitor, and a two-port network (e.g., a filter) on the output of the network. These significantly different circuit schematics are displayed in the identical area on the spreadsheet. The morphing

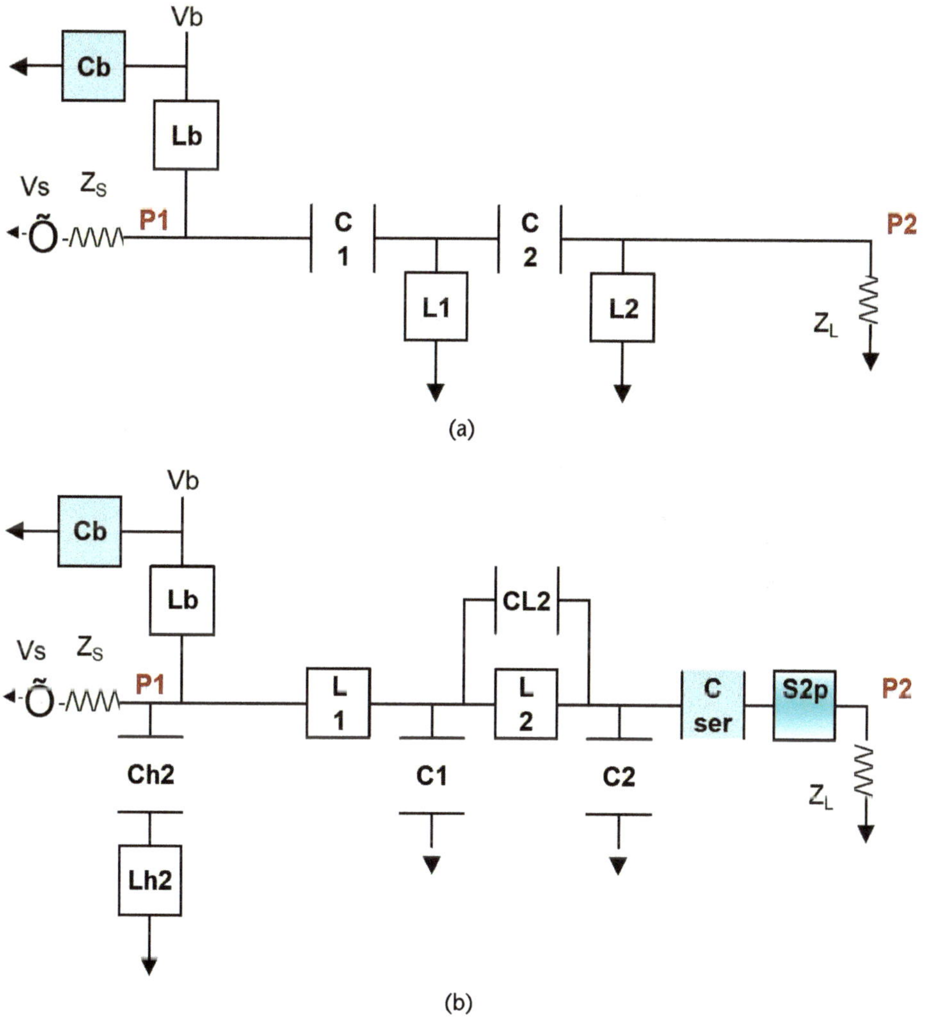

Figure 7.4 Dynamic output matching network schematics: (a) basic two-section lowpass network, and (b) network with added elements.

is accomplished using the "conditional format" capability of Excel in response to changes in user inputs.

A more striking example of the dynamic schematic capability of Excel is shown in Figure 7.5. In that figure, four different schematics are shown corresponding to the filter type selected by the user. Again, all four schematics shown occupy the

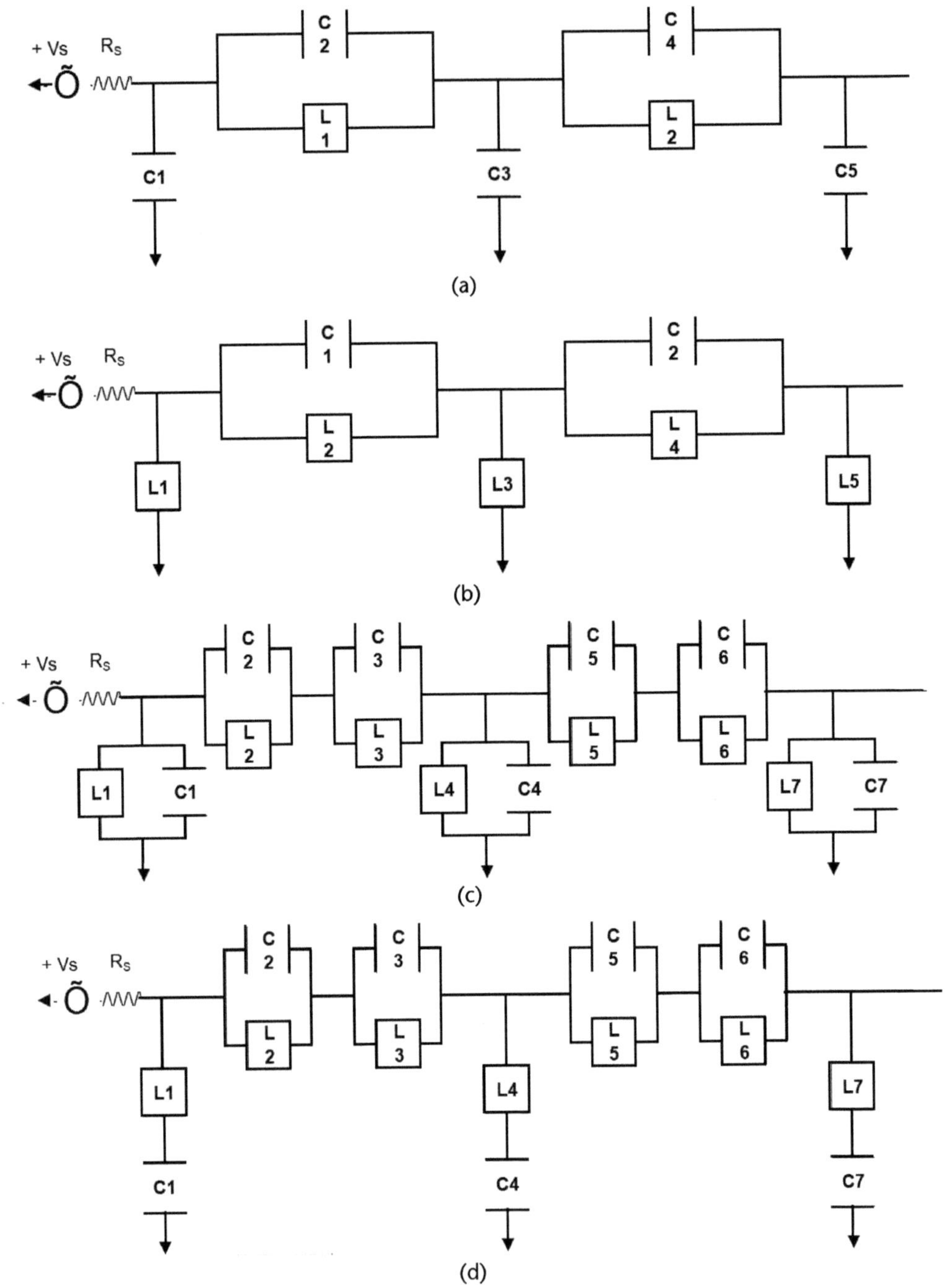

Figure 7.5 Dynamic filter network schematics: (a) lowpass, (b) highpass, (c) passband, and (d) stopband.

same area on the spreadsheet. This spreadsheet allows the filter designer to rapidly change the filter type, with all other parameters being equal, to compare the attributes of the varying filter type for the application in question. In addition, while in Figure 7.5 all the filters comprise two sections, by changing only one entry on the spreadsheet, the designs can be immediately reconfigured to have anywhere from one to five cavities. Each time, the schematics are redrawn accordingly and instantly. This greatly facilitates the RF design challenge by significantly reducing design cycle time and permits the designer to consider more options that can hopefully result in a more "optimum" design.

An example of how conditional formatting can be used to cause two groups of cells to display as either a vertically orientated capacitor or an inductor is shown in Table 7.2.

7.1.3 Contour Plots

As discussed earlier, there is invariably never a single design whose performance is optimal in all key aspects of design. A design that is best for, say, insertion loss may not be best for input impedance or harmonic rejection. Faced with a spectrum of possible design solutions, the engineer must choose the one that offers the best compromise between the critical performance parameters for the RF functionality of the module. A visual representation of the compromises available to the designer for the various circuit options available can greatly facilitate this choice. Colored contour plots are an effective option for doing this.

To generate the contour plots in Excel, the designer needs to first identify two key independent circuit variables that can best be used to characterize the network. This approach was described in Chapter 6. By sweeping these variables in turn and determining the associated analytic solutions for the network, the values of key performance parameters can be entered on a 2 × 2 grid on the spreadsheet. Once such a scan is complete, colors can be assigned to each of the cells, based on the cell value relative to the extreme values in the grid, using the "conditional format" capability in Excel.

Figure 7.6 shows an example of two such grids for a scan of a two-section amplifier output match, such as that shown in Figure 7.4(a). The two independent variables $\delta\varphi_{\pi 1}$ and $\delta\varphi_{\pi 2}$ are the phase shifts through the two lowpass π matching sections, as will be explained more fully in Section 11.8. The first contour plot, Figure 7.6(a), shows the value of network insertion gain as a function of $\delta\varphi_{\pi 1}$ and $\delta\varphi_{\pi 2}$. The second contour plot, Figure 7.6(b), shows the magnitude of the complex input impedance mismatch of the network also as a function of $\delta\varphi_{\pi 1}$ and $\delta\varphi_{\pi 2}$. In each chart, the green cells have the most desirable values for the corresponding

Table 7.2 *LC*-Series Resonator Design Parameters

Formula	*Border Format*	*Applies To*
=IF($E$5=1,TRUE,FALSE)	AaBbCcYyZz	=$F$17:$G$18,$V$17:$W$18
=IF(OR($E$5=2,$E$5=4),TRUE,FALSE)	AaBbCcYyZz	=$F$17:$G$18,$V$17:$W$18

dependent variable, while the dark red cells have the least desirable values. The cells in each chart that are dark green are the most favorable value in each chart. The cells in each chart that are yellow represent the current design selection, which can be simply changed by selecting a different cell in one of the charts. Each time the active cell is changed, the color of the corresponding cells in each of the charts is changed to yellow to highlight where the current design choice lies on each of the charts. The schematic and all network performance data and plots on the spreadsheet are determined based on the active cell selection. Thus, by simply moving the mouse around on the contour plot, the schematic and performance parameters displayed are instantly updated to reflect the design changes.

Figure 7.6 shows only two contour plots for the matching network, yet, in practice, there would typically be more. The designer can create as many of the coupled contour plots as are necessary to display all the key parameters of interest for the

Selecting a cell in table below will automatically load phases in δΦπ1 & δΦπ2 11 8

Insertion Gain (n_type=low-low,opt.=ins. loss)

$\delta\Phi_{\pi 2}$ \ $\delta\Phi_{\pi 1}$	50	52	53	55	56	58	59	61	63	64	66	67	69	71	72	74	75	77	78	80
80	-0.63	-0.63	-0.63	-0.62	-0.62	-0.62	-0.62	-0.62	-0.62	-0.62	-0.62	-0.62	-0.62	-0.62	-0.63	-0.63	-0.63	-0.64	-0.64	-0.65
78	-0.61	-0.61	-0.61	-0.6	-0.6	-0.6	-0.6	-0.6	-0.6	-0.6	-0.6	-0.6	-0.6	-0.61	-0.61	-0.61	-0.62	-0.62	-0.62	-0.63
77	-0.6	-0.6	-0.6	-0.6	-0.59	-0.59	-0.59	-0.59	-0.59	-0.59	-0.59	-0.59	-0.6	-0.6	-0.6	-0.6	-0.61	-0.61	-0.62	-0.62
75	-0.59	-0.58	-0.58	-0.58	-0.58	-0.58	-0.58	-0.57	-0.58	-0.58	-0.58	-0.58	-0.58	-0.58	-0.59	-0.59	-0.59	-0.6	-0.6	-0.61
74	-0.58	-0.57	-0.57	-0.57	-0.57	-0.57	-0.57	-0.57	-0.57	-0.57	-0.57	-0.57	-0.57	-0.58	-0.58	-0.58	-0.59	-0.59	-0.59	-0.6
72	-0.56	-0.56	-0.56	-0.56	-0.55	-0.55	-0.55	-0.55	-0.55	-0.56	-0.56	-0.56	-0.56	-0.56	-0.57	-0.57	-0.57	-0.58	-0.58	-0.59
71	-0.55	-0.55	-0.55	-0.55	-0.55	-0.55	-0.55	-0.55	-0.55	-0.55	-0.55	-0.55	-0.55	-0.56	-0.56	-0.57	-0.57	-0.57	-0.58	-0.58
69	-0.54	-0.54	-0.54	-0.54	-0.54	-0.53	-0.53	-0.54	-0.54	-0.54	-0.54	-0.54	-0.54	-0.55	-0.55	-0.55	-0.56	-0.56	-0.57	-0.57
67	-0.53	-0.53	-0.53	-0.52	-0.52	-0.52	-0.52	-0.52	-0.53	-0.53	-0.53	-0.53	-0.53	-0.54	-0.54	-0.55	-0.55	-0.55	-0.56	-0.56
66	-0.52	-0.52	-0.52	-0.52	-0.52	-0.52	-0.52	-0.52	-0.52	-0.52	-0.52	-0.53	-0.53	-0.53	-0.54	-0.54	-0.54	-0.55	-0.55	-0.56
64	-0.51	-0.51	-0.51	-0.51	-0.51	-0.51	-0.51	-0.51	-0.51	-0.51	-0.52	-0.52	-0.52	-0.53	-0.53	-0.53	-0.54	-0.54	-0.55	-0.55
63	-0.51	-0.51	-0.5	-0.5	-0.5	-0.5	-0.5	-0.51	-0.51	-0.51	-0.51	-0.51	-0.52	-0.52	-0.52	-0.53	-0.53	-0.54	-0.54	-0.55
61	-0.5	-0.5	-0.5	-0.5	-0.5	-0.5	-0.5	-0.5	-0.5	-0.5	-0.5	-0.51	-0.51	-0.52	-0.52	-0.52	-0.53	-0.53	-0.54	-0.54
59	-0.49	-0.49	-0.49	-0.49	-0.49	-0.49	-0.49	-0.49	-0.49	-0.5	-0.5	-0.5	-0.5	-0.51	-0.51	-0.52	-0.52	-0.53	-0.53	-0.54
58	-0.49	-0.48	-0.48	-0.48	-0.48	-0.49	-0.49	-0.49	-0.49	-0.49	-0.5	-0.5	-0.5	-0.51	-0.51	-0.52	-0.52	-0.53	-0.53	-0.54
56	-0.48	-0.48	-0.48	-0.48	-0.48	-0.48	-0.48	-0.48	-0.49	-0.49	-0.49	-0.49	-0.5	-0.5	-0.51	-0.51	-0.51	-0.52	-0.53	-0.53
55	-0.48	-0.48	-0.48	-0.48	-0.48	-0.48	-0.48	-0.48	-0.48	-0.48	-0.49	-0.49	-0.49	-0.5	-0.5	-0.51	-0.51	-0.52	-0.52	-0.53
53	-0.47	-0.47	-0.47	-0.47	-0.47	-0.47	-0.47	-0.48	-0.48	-0.48	-0.48	-0.49	-0.49	-0.5	-0.5	-0.51	-0.51	-0.52	-0.52	-0.53
52	-0.47	-0.47	-0.47	-0.47	-0.47	-0.47	-0.47	-0.47	-0.48	-0.48	-0.48	-0.48	-0.49	-0.5	-0.5	-0.5	-0.51	-0.52	-0.52	-0.53
50	-0.46	-0.46	-0.46	-0.46	-0.47	-0.47	-0.47	-0.47	-0.47	-0.48	-0.48	-0.48	-0.49	-0.49	-0.5	-0.5	-0.51	-0.51	-0.52	-0.53

(a)

|Z-Zs| (n_type=low-low,opt.=ins. loss)

$\delta\Phi_{\pi 2}$ \ $\delta\Phi_{\pi 1}$	50	52	53	55	56	58	59	61	63	64	66	67	69	71	72	74	75	77	78	80
80	0.5	0.5	0.5	0.4	0.4	0.4	0.4	0.3	0.3	0.3	0.3	0.3	0.3	0.3	0.3	0.4	0.4	0.4	0.4	0.5
78	0.5	0.4	0.4	0.4	0.4	0.3	0.3	0.3	0.2	0.2	0.3	0.3	0.3	0.4	0.4	0.4	0.4	0.5	0.5	0.5
77	0.4	0.4	0.4	0.3	0.3	0.3	0.3	0.2	0.3	0.3	0.3	0.3	0.3	0.4	0.4	0.4	0.5	0.5	0.5	0.5
75	0.4	0.3	0.3	0.3	0.3	0.2	0.2	0.3	0.3	0.3	0.3	0.4	0.4	0.4	0.4	0.5	0.5	0.5	0.6	0.6
74	0.3	0.3	0.3	0.3	0.2	0.2	0.3	0.3	0.3	0.3	0.4	0.4	0.4	0.4	0.5	0.5	0.5	0.6	0.6	0.6
72	0.3	0.2	0.2	0.3	0.3	0.3	0.3	0.3	0.4	0.4	0.4	0.4	0.5	0.5	0.5	0.5	0.6	0.6	0.6	0.6
71	0.2	0.3	0.3	0.3	0.3	0.3	0.3	0.4	0.4	0.4	0.4	0.4	0.5	0.5	0.5	0.6	0.6	0.6	0.6	0.7
69	0.3	0.3	0.3	0.3	0.3	0.4	0.4	0.4	0.4	0.4	0.5	0.5	0.5	0.6	0.6	0.6	0.6	0.7	0.7	0.7
67	0.4	0.4	0.4	0.4	0.4	0.4	0.4	0.4	0.5	0.5	0.5	0.5	0.6	0.6	0.6	0.6	0.7	0.7	0.7	0.7
66	0.4	0.4	0.4	0.4	0.4	0.4	0.4	0.5	0.5	0.5	0.5	0.6	0.6	0.6	0.6	0.7	0.7	0.7	0.7	0.8
64	0.4	0.4	0.4	0.5	0.5	0.5	0.5	0.5	0.5	0.6	0.6	0.6	0.6	0.7	0.7	0.7	0.7	0.8	0.8	0.8
63	0.5	0.5	0.5	0.5	0.5	0.5	0.5	0.5	0.6	0.6	0.6	0.6	0.6	0.7	0.7	0.7	0.7	0.8	0.8	0.8
61	0.5	0.5	0.5	0.5	0.5	0.6	0.6	0.6	0.6	0.6	0.6	0.7	0.7	0.7	0.7	0.8	0.8	0.8	0.8	0.9
59	0.6	0.6	0.6	0.6	0.6	0.6	0.6	0.6	0.6	0.7	0.7	0.7	0.7	0.8	0.8	0.8	0.8	0.8	0.9	0.9
58	0.6	0.6	0.6	0.6	0.6	0.6	0.6	0.6	0.7	0.7	0.7	0.7	0.7	0.8	0.8	0.8	0.8	0.9	0.9	0.9
56	0.6	0.6	0.6	0.7	0.7	0.7	0.7	0.7	0.7	0.7	0.7	0.8	0.8	0.8	0.8	0.8	0.9	0.9	0.9	0.9
55	0.7	0.7	0.7	0.7	0.7	0.7	0.7	0.7	0.7	0.7	0.8	0.8	0.8	0.8	0.8	0.9	0.9	0.9	0.9	1
53	0.7	0.7	0.7	0.7	0.7	0.7	0.7	0.8	0.8	0.8	0.8	0.8	0.8	0.9	0.9	0.9	0.9	0.9	1	1
52	0.8	0.7	0.7	0.7	0.7	0.8	0.8	0.8	0.8	0.8	0.8	0.8	0.9	0.9	0.9	0.9	0.9	1	1	1
50	0.8	0.8	0.8	0.8	0.8	0.8	0.8	0.8	0.8	0.8	0.9	0.9	0.9	0.9	0.9	1	1	1	1	1

(b)

Figure 7.6 Contour plots for (a) insertion gain, and (b) mismatch.

design. Such contour plots can greatly aid the design process. The designer can clearly choose the parametric area that offers the best compromise in performance characteristics. They also give a clear visual indication of areas to avoid, where one or more of the parameters is too close to a design region where performance degrades rapidly. The fact that the schematic and all circuit parametric plots also change immediately in response to a change in the selected active design cell, reinforces the feedback to the designer.

The ability to include contour plots on the design spreadsheet is highly advantageous both in letting the designer quickly explore design concepts and in clearly conveying the design compromises of which the network is capable.

7.1.4 Bar Plots

The engineer is under constant pressure to minimize both the size and cost of an RF module destined for use in a modern mobile RF device. Passive components in these modules, inductors and capacitors, typically consume a significant area of the footprint and comprise a significant portion of the module bill of materials. For capacitors, higher-quality factors typically require higher-cost components. For printed RF inductors, a higher-quality factor generally requires a larger footprint. In both cases, therefore, there is an inherent cost for higher Q elements. A key challenge in coming up with a cost-effective design is to understand how the Q factor of the individual components impacts the overall module performance. This can be visualized easily in Excel by use of bar chart plots.

Using an Excel macro, a bar chart can be generated showing the sensitivity of a module performance parameter to variations in parasitic parameters of the passive elements.

Figure 7.7 shows the variation in module insertion gain for bracketed changes in the *ESR* (equivalent series resistance) of the capacitors in the circuit.

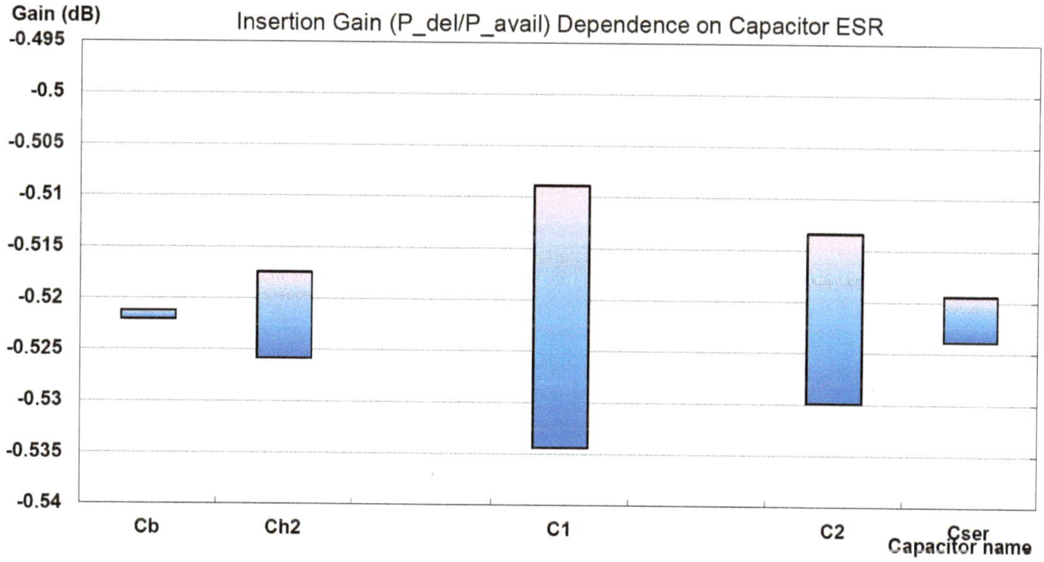

Figure 7.7 Bar plot showing insertion gain dependence on capacitor *ESR*.

Figure 7.8 shows the variation in module insertion gain for bracketed changes in the Q factors of the inductors in the circuit.

It is clear from both plots that improved module performance can be achieved by focusing on the design of only a few key components.

7.2 Complex Expressions in Excel

The ability to evaluate complex expressions is an essential requirement of any RF software design tool. Fortunately, Excel contains all the basic mathematical complex functions for fulfilling this requirement. Some of the principal complex functions are:

To represent a complex number, for example, a + jb	"= complex(a, b)"
To sum complex numbers, for example, $Z_1 + Z_2$	"= imsum(Z_1, Z_2)"
To subtract complex numbers, for example, $Z_1 - Z_2$	"= imsub(Z_1, Z_2)"
To multiply two complex numbers, for example, $Z_1 \times Z_2$	"= improduct(Z_1, Z_2)"
To divide two complex numbers, for example, Z_1/Z_2	"= imdiv(Z_1, Z_2)"
For real part of complex number, for example, real(Z)	"= imreal(Z)"
For imaginary part of complex number, for example, imag.(Z)	"= imaginary(Z)"
For conjugate part of complex number, for example, imag.(Z)	"= imconjugate(Z)"
For magnitude of complex number, for example, imag.(Z)	"= imabs(Z)"
For square root of complex number, for example, $\sqrt{Z}$	"= imsqrt(Z)"
For exponent of complex number, for example, e^Z	"= imexp (Z)"
For power of complex number, for example, Z^n	"= impower (Z, n)"

There are many more. With such built-in functions, essentially any complex expression may be evaluated within a cell. Evaluating expressions with many terms can be a little cumbersome due to the verbose nature of some of the functions. However, where this is problematic, the expression may be broken down in parts and evaluated over more than one cell.

To keep the user interface uncluttered and unconfusing, it is recommended that all the calculation cells be assigned to a separate spreadsheet from the one that accepts user inputs and displays the results. It is further recommended that the calculation spreadsheets be even hidden from the user to prevent inadvertent corruption of the embedded formulae.

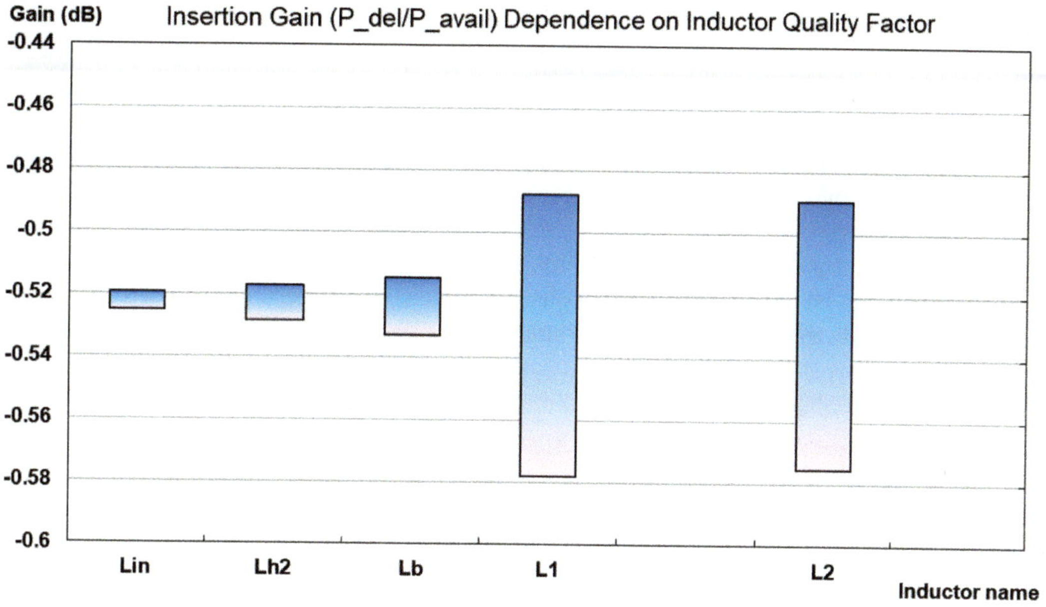

Figure 7.8 Bar plot showing insertion gain dependence on inductor *Q*.

On the user interface spreadsheet, there will inevitably be some cells assigned for user inputs, while others contain the results from calculations on other pages of the workbook. It is important that the user can easily distinguish these two types of cells so only the user-input cells are modified. A simple way to do this is to assign a specific color to those cells that the user is free to modify. In the examples provided in this book, data in those cells are always assigned the color blue.

7.3 Use of Macros in Excel

Excel incorporates a Visual Basic programming capability. Subroutines written in Visual Basic may be saved and run as macros by selecting a control box on one of the spreadsheets. While macros can be very powerful, but they have one potential drawback that the user should guard against. Namely, if cell values are dependent upon being updated by a macro after, say, the user makes input changes, the values will remain unchanged until the macro is executed. Since there is no way to tell if a macro has been executed, after changes, cell values may be inconsistent with the data inputs on the spreadsheet.

Macros are also inconsistent with instantaneous user feedback. Each time an input is modified, ideally the network characteristics should be immediately updated in response to the change. If equations are written directly in cells on a calculation spreadsheet, the cells will be immediately recalculated as desired. The user will therefore see the results of the changes immediately. However, if the calculations are implemented in a Visual Basic subroutine, the user will only see the results of any changes if the macro is executed, which thus requires an extra step that might even be forgotten.

Where macros are best used is in stepping through a series of changes. This is where they excel. For the contour plots discussed earlier, macros provide the means to step through a sequence of input changes and tabulate the results. Another very useful utility they have is in being able to change chart parameters on the spreadsheet. If there are multiple charts, or plots, on a spreadsheet and we wish to change the frequency span of the data on multiple charts, it is time-consuming to have to do this manually and interactively on each of the charts in turn. A macro can relieve this task. By clicking on a single macro button, the *x* or *y* ranges on one or more charts can be changed instantaneously. Below is an example of a set of Visual Basic commands to change the *x*-axis range on one chart:

```
ActiveSheet.Shapes("chart_name").Select
Name = ActiveChart.ChartTitle.Caption
ActiveChart.Axes(xlCategory).MinimumScale = x1
ActiveChart.Axes(xlCategory).MaximumScale = x2
```

CHAPTER 8

LC Resonator Basics

LC resonators, at their simplest, comprise a single inductor (L) and capacitor (C). The two elements may be connected either in series, or in parallel, depending on the application. Figure 8.1 shows the basic configurations of an LC-series and an LC-parallel resonator incorporating parasitic resistances R_S and R_P, respectively, to account for dissipative losses. In both cases, the two elements will resonate at a specific frequency f_r, where the reactive impedances of the two elements are equal to each other, but of opposite sign, that is,

$$j\omega_r L - \frac{j}{\omega_r C} = 0 \tag{8.1}$$

Thus,

$$\omega_r^2 LC = 1 \tag{8.2}$$

At resonance, energy in the elements rapidly exchanges between magnetic and electric storage.

RF systems are invariably optimized to process signals within a finite specified range of frequencies. If strong enough, frequencies outside this range can often result in negative effects, such as saturation of an amplifier or nonlinear mixing resulting in the generation of unwanted spurious emissions. To attenuate unwanted signal energy relative to the desired RF signal, LC resonators are frequently employed by suitable placement in the RF circuit. Depending how they are incorporated into the circuit, they may be used to effect either bandpass or bandstop filtering. When employed as bandstop elements, they are usually referred to as "traps."

Both the series and parallel resonator types can be designed to have very similar bandpass or bandstop characteristics in the neighborhood of the design (i.e., resonant) frequency. The major distinguishing feature between the two is the impedance that they present in the circuit.

If the resonating elements are in series, below resonance, they exhibit a capacitive reactance. Above resonance, they exhibit an inductive reactance. At the resonant

Figure 8.1 *LC* resonator configurations: (a) *LC*-series, and (b) *LC*-parallel.

frequency, ideally, they have zero reactance and thus present no inhibition to the flow of RF energy.

If the resonating elements are in parallel, below resonance, they exhibit an inductive reactance. Above resonance, they exhibit a capacitive reactance. At the resonant frequency, ideally, they have infinite reactance and thus completely restrict the flow of RF energy.

The impedance of a frequency trap on the output of an RF PA can be critical to its efficiency. To understand why, it is helpful to consider their interaction with the energy in the desired system passband. Due to inevitable nonlinearities in an RF PA, mixing products (i.e., harmonic products of the desired signal inputs) appear on the output.

A high impedance mismatch on the output of an amplifier serves to attenuate this energy by reflecting it back into the amplifier. Depending on the phase of the reflected energy, it may remix with the desired signal energy to reduce distortion of the waveform, thereby increasing efficiency, or increase the distortion leading to reduced efficiency. Using a mixer analogy, we think of the efficiency being maximized when the harmonic energy is reflected back into the amplifier with the appropriate phase to cancel the harmonic energy being generated.

Whether used in a bandpass or bandstop function, *LC* resonators have two defining characteristics for their design:

1. The center frequency f_r, that is, the frequency at which the resonator provides minimum or maximum attenuation.
2. Its quality factor (Q). This incorporates aspects of both insertion loss and bandwidth. In general, the higher the resonator Q, the lower will be its dissipative losses at resonance. However, the higher the Q, the narrower will be the bandwidth of the resonance.

For an *LC*-series resonator, with series parasitic resistance R_S, and in terms of ω_r from (11.2):

$$Q_S = \frac{1}{R_S}\sqrt{\frac{L_S}{C_S}} = \frac{\omega_r L_S}{R_S} \tag{8.3}$$

For an *LC*-parallel resonator, with parallel parasitic resistance R_P, and in terms of ω_r from (11.2):

$$Q_P = R_P\sqrt{\frac{C_P}{L_P}} = \omega_r C_P R_P \tag{8.4}$$

8.1 Formulae for Equivalency Between *LC*-Series and Parallel Resonators

As mentioned above, in theory, *LC* resonators implemented with either the series or parallel configurations can be designed to have near-identical gain characteristics in the neighborhood of resonance. Given one form, the element values for the alternate form, with similar characteristics, can be obtained from the following mappings.

Mapping from parallel to series resonators is:

$$C_S = L_P / Z_0^2 \tag{8.5}$$

$$L_S = C_P Z_0^2 \tag{8.6}$$

$$R_S = Z_0^2 / R_P \tag{8.7}$$

and from series to parallel resonators is:

$$C_P = L_S / Z_0^2 \tag{8.8}$$

$$L_P = C_S Z_0^2 \tag{8.9}$$

$$R_P = Z_0^2 / R_S \tag{8.10}$$

where Z_0 = line impedance.

By way of example, employing the above formulae, consider the equivalent *LC*-parallel and *LC*-series implemented traps, resonant at 3 GHz, with the following parameters:

LC-series resonator 1	*LC-parallel resonator 1*	
$R_S = 2.36\Omega$	$R_P = 1061\Omega$	
$L_S = 5$ nH	$L_P = 1.41$ nH	
$C_S = 0.56$ pF	$C_P = 2$ pF	(8.11)
$Q_S = 40$	$Q_P = 40$	

For the same resonant frequency, if the inductance and resistance are both doubled for the *LC*-series resonator, the quality factor will remain unchanged (see (8.3)). The equivalent *LC*-resonator element values are then

LC-series resonator 2	*LC-parallel resonator 2*	
$R_S = 4.71\Omega$	$R_P = 531\Omega$	
$L_S = 10$ nH	$L_P = 0.7$ nH	
$C_S = 0.28$ pF	$C_P = 4$ pF	(8.12)
$Q_S = 40$	$Q_P = 40$	

These solutions are used in the sections below to illustrate some basic *LC* resonator design concepts.

8.2 Design of *LC* Resonators for Passband Filtering

In this section, we consider the use of LC-series and LC-parallel resonators in RF circuits, primarily targeted as implementing a bandpass function. The resonant frequency f_r is the center of the desired signal bandwidth and the resonator's purpose is to provide increasing attenuation of RF energy away from the passband. The two resonator types are incorporated into the RF circuit as shown in Figure 8.2.

Figure 8.3 shows the bandpass characteristics of the two resonator types with the values listed in Section 8.1. For reference, the dashed line shows the passband gain of series and parallel LC resonators with the element values of (8.11), but with no dissipative loss.

The series and parallel LC equivalent bandpass implementations have near identical gain characteristics, as expected. Despite the fact that the type 1 and 2 resonator designs all have an identical Q of 40, however, note that the type 2 designs have a much narrower resonant bandwidth.

This illustrates that the major design parameter for controlling bandwidth is actually the ratio of the two reactive elements used in the resonators. For the LC-series resonator, the bandwidth is reduced with increasing inductor value L_S. In contrast, for the LC-parallel resonator, the bandwidth is reduced with the increasing capacitor C_P. These observations are consistent with (8.3) and (8.4).

In Figure 8.3, comparing the type 1 resonator designs, with and without loss, we observe only a very minor difference in the resonant bandwidths. Changing the resonator Qs from 40 to ideally infinity has only a minor impact on the resonant bandwidth. The major difference, again as would be expected, is in the insertion loss, especially close to the resonance.

Although the passband gain characteristics for the LC-series and LC-parallel equivalent resonators are essentially identical, their impedance behavior in the neighborhood of resonance is significantly different, as shown in Figure 8.4. The series LC resonator impedance transitions from capacitive to inductive as the frequency moves up through resonance. In comparison, the parallel LC resonator impedance transitions from inductive to capacitive as the frequency moves up through resonance.

Also of note is that the type 2 LC resonators exhibit a wider swing in reactance across resonance than do the type 1 designs. It is this that is responsible for their narrower passband width.

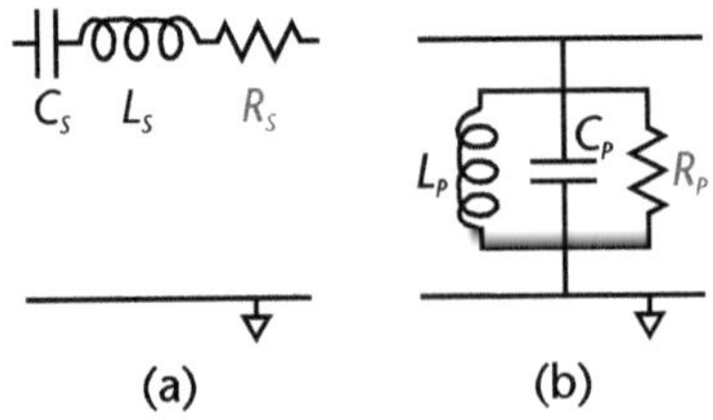

Figure 8.2 *LC*-bandpass resonator configurations: (a) *LC*-series, and (b) *LC*-parallel.

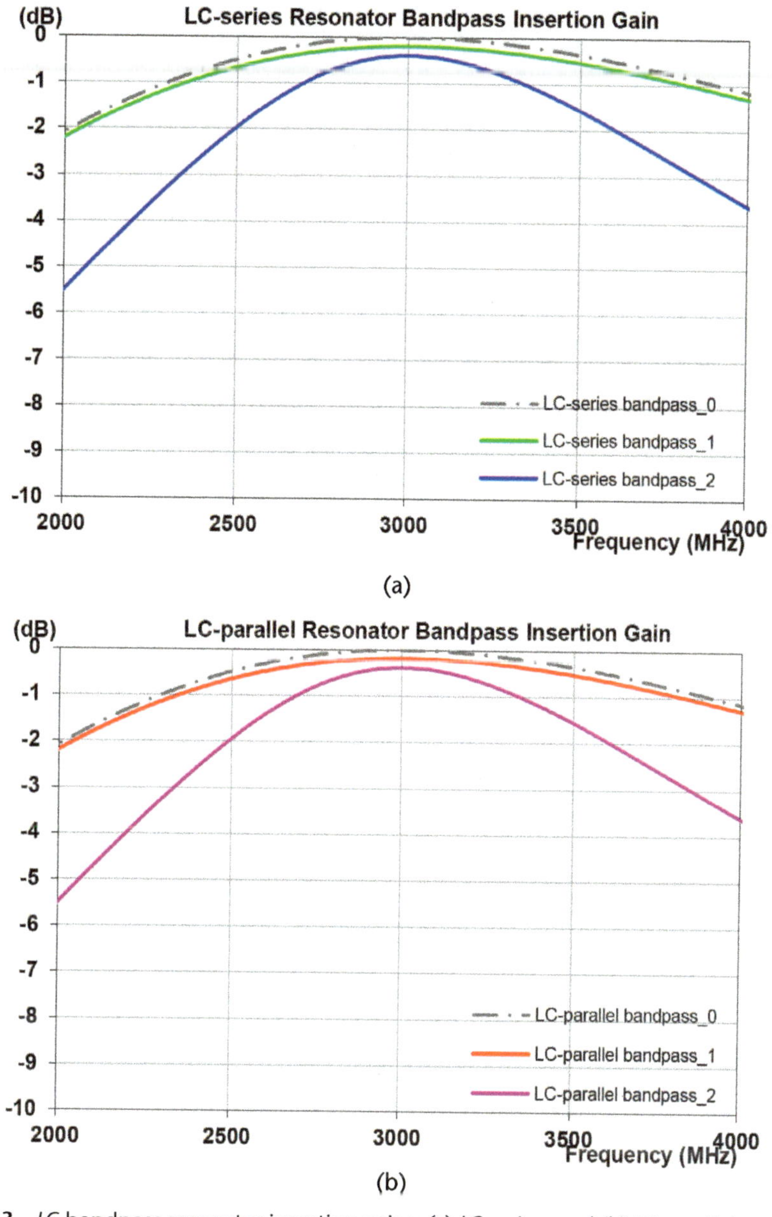

Figure 8.3 *LC*-bandpass resonator insertion gains: (a) *LC*-series, and (b) *LC*-parallel.

8.3 Design of *LC* Resonators for Stopband Rejection

In this section, we consider the use of LC-series and LC-parallel resonators in RF circuits, primarily targeted as implementing a stopband rejection function. In such traps, the resonant frequency f_r is the center of the desired rejection bandwidth and the resonators' purpose is to provide high levels of attenuation of the unwanted RF energy. The two resonator types are incorporated into the RF circuit as shown in Figure 8.5.

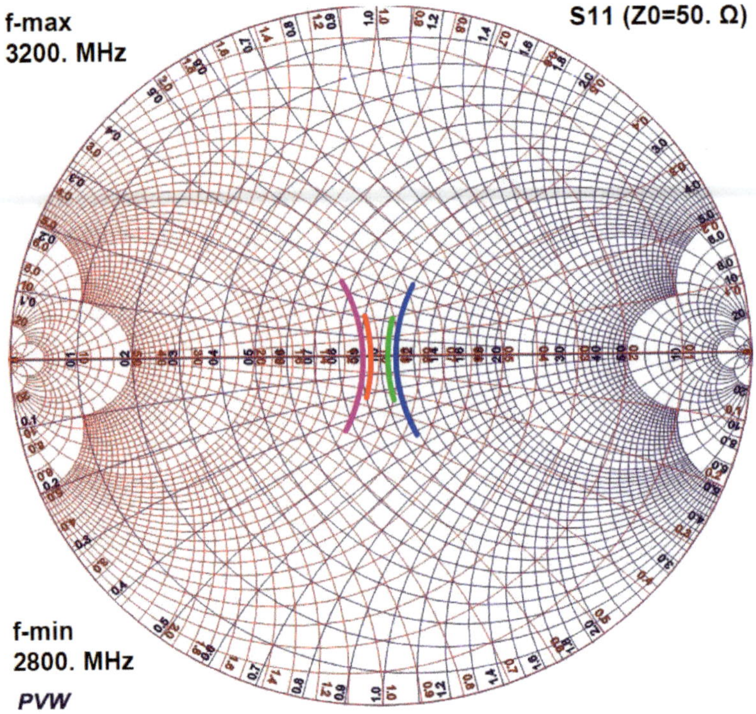

Figure 8.4 Smith chart impedances of bandpass *LC*-series and *LC*-parallel resonators.

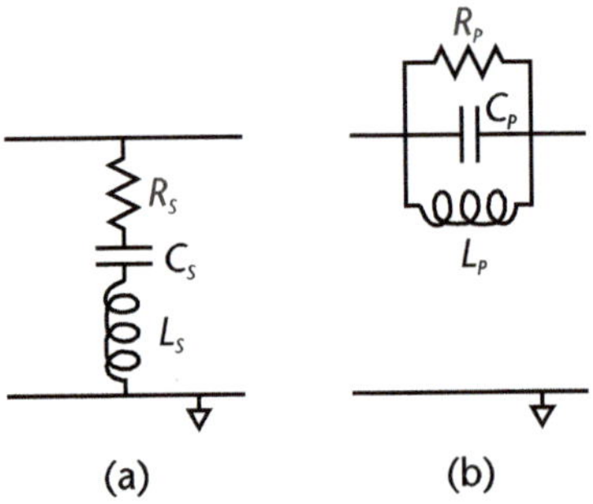

Figure 8.5 *LC*-resonator trap architectures: (a) *LC*-series, and (b) *LC*-parallel.

Figure 8.6 shows the bandpass characteristics of the two resonator types with the values listed in Section 8.1. For reference, the dashed gray line shows the passband gain of series and parallel LC resonators with the element values of (8.11), but with no dissipative loss.

The series and parallel LC equivalent bandpass implementations have near-identical gain characteristics, as expected. Despite the fact that the type 1 and 2 resonator designs all have an identical Q of 40, however, note that the type 2 designs have a much narrower resonant bandwidth.

This illustrated that the major design parameter for controlling bandwidth is actually the ratio of the two reactive elements used in the resonators. For the LC-series resonator, the bandwidth is reduced with an increasing inductor value L_S. In

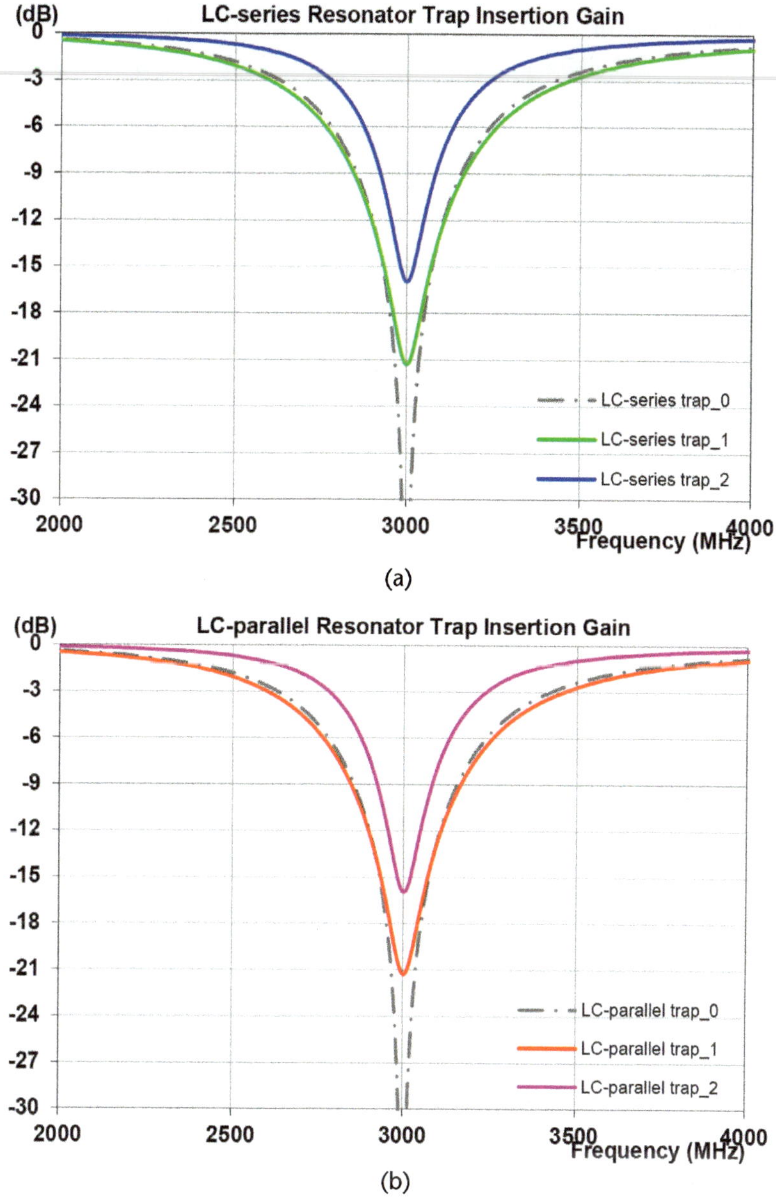

Figure 8.6 *LC*-bandstop resonator insertion gains: (a) *LC*-series, and (b) *LC*-parallel.

contrast, for the *LC*-parallel resonator, the bandwidth is reduced with increasing capacitor C_P. These observations are consistent with (8.3) and (8.4).

In Figure 8.6, comparing the type 1 resonator designs, with and without loss, we observe only a very minor difference in the resonant bandwidths. Changing the resonator Qs from 40 to ideally infinity has only a minor impact on the rejection bandwidth. The major difference is in the depth of the notch. Thus, dissipation in the bandstop configurations primarily limits the maximum attenuation achievable and has little influence on the rejection bandwidth.

Although the passband gain characteristics for the *LC*-series and *LC*-parallel equivalent resonators are essentially identical, their impedance behavior in the neighborhood of resonance are dramatically different, as shown in Figure 8.7. The series LC resonator impedance transitions from capacitive to inductive as the frequency moves up through resonance. In comparison, the parallel LC resonator impedance transitions from inductive to capacitive as the frequency moves up through resonance. However, in the neighborhood of resonance, the series LC resonator presents close to a short circuit to the source. In contrast, the parallel LC resonator presents close to an open circuit.

Also, of note is that the type 2 LC resonators exhibit a wider swing in reactance across resonance than do the type 1 designs. It is this that is responsible for their narrower passband width.

8.4 Design of *LC*-Series Resonators with Desired In-Band Capacitance and a High-Side Resonance

In an RF circuit where a capacitor is needed in the passband, for matching or other purposes, it can be implemented with an LC-series resonator, rather than a simple capacitor. The advantage is in addition to the passband capacitance; the resonator can be designed to have its self-resonance at a desired frequency above the passband. This can be useful for achieving attenuation of unwanted high-frequency signal components.

For such a resonator design, there are two requirements for the circuit designer: (1) achieve the desired capacitance at the passband frequency f_0, and (2) effect a series resonance at a frequency f_r above the passband. Because we only have two elemental values to determine (i.e., L and C), there is a unique solution to this problem.

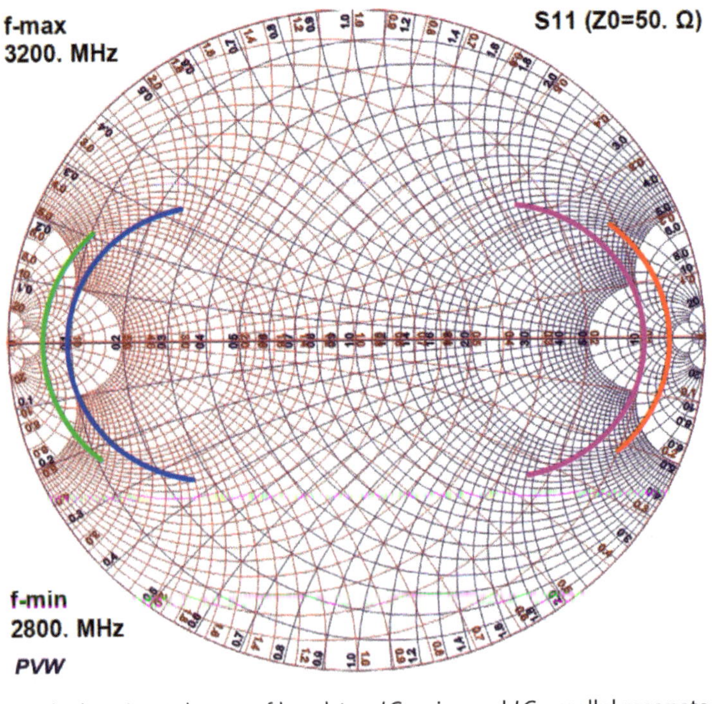

Figure 8.7 Smith chart impedances of bandstop *LC*-series and *LC*-parallel resonators.

It is important to note here that the capacitance in the passband will be affected by the high-side resonance. While the desired capacitance can be designed for f_0, the effective capacitance of the resonator will vary across the passband. The closer f_r is to f_0, the greater will be the capacitance variation across the passband. Fortunately, in many cases, the frequency band for rejection is at a harmonic frequency and thus significantly removed from the passband.

As discussed in Chapter 5, the majority of components used in practical RF circuits have significant associated dissipative and reactive parasitics. For any useful synthesis, these parasitic elements must be taken into account in the design equations.

Figure 8.8 shows an LC-series resonator with a composite equivalent incorporating the parasitic elements described in Chapter 5. Rather than determine the values of L and C directly from the design constraints, we seek to determine the values of L' and C', which incorporate the effects of the parasitics on the overall resonator response. Because the reactive parasitics have the dominant influence on the resonator characteristics, the resistive elements can typically be ignored in such a synthesis. The parasitic resistive elements contribute little to frequency shifts of the resonator and predominantly result in dissipative losses that are unavoidable.

The required capacitance is obtained from the solution to a quadratic equation, with

$$a \approx \omega_r^2\left(\left(1/C + \omega_0^2 L_C\right) L_C C_L - 1/\left(\omega_r^2 - \omega_0^2\right)C\right)$$

$$b \approx 1 - C_L/C - \left(\omega_r^2 + \omega_0^2\right) L_C C_L$$

$$c \approx C_L$$

The solution is

$$C' \approx \frac{-b - \sqrt{b^2 - 4ac}}{2a} \tag{8.13}$$

and

$$L' \approx \frac{1}{\omega_r^2\left(C_L + 1/\left(1/C' - \omega_r^2 L_C\right)\right)} \tag{8.14}$$

By way of example, consider the LC-series resonator parameters in Table 8.1. As in Chapter 7, the entries in blue are user inputs. The parasitic element values associated with both the inductor and capacitor are typically obtained from manufacturers' data sheets, bench measurement, or sophisticated RF electromagnetic simulation software. The element values in the two right columns of the table are

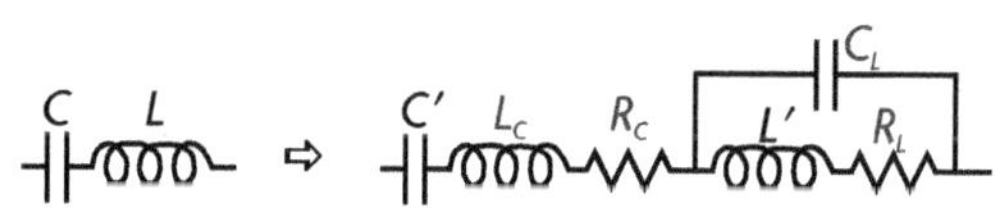

Figure 8.8 *LC*-series resonator with parasitics.

Table 8.1 *LC*-Series Resonator Design Parameters

Series (Type 1)							
f_0	2000 MHz	*Parasitics*				*Elements*	
f_r	4000 MHz	$R_C(\Omega)$	L_C(nH)	Q_L	C_L(pF)	L' (nH)	C' (pF)
C_S (at f_0)	0.6 pF	0.2	0.1	40	0.15	2.46633	0.47708

the required element values for the resonator obtained by applying the formulae above which incorporate any detuning due to the parasitics.

In this case, the *LC*-series resonator is desired to have a capacitance of 0.6 pF at the center of the band f_0 = 2,000 MHz and to have a series resonance at f_r = 4,000 MHz.

Figure 8.9 shows the key performance characteristics of the *LC*-series resonator obtained when implemented with the values for L' and C' given in Table 8.1. Note that both the key goals for the resonator have been achieved.

8.5 Design of *LC*-Series Resonators with Desired In-Band Inductance and a Low-Side Resonance

In an RF circuit where an inductor is needed in the passband, for matching or other purposes, it can be implemented with an *LC*-series resonator, rather than a simple

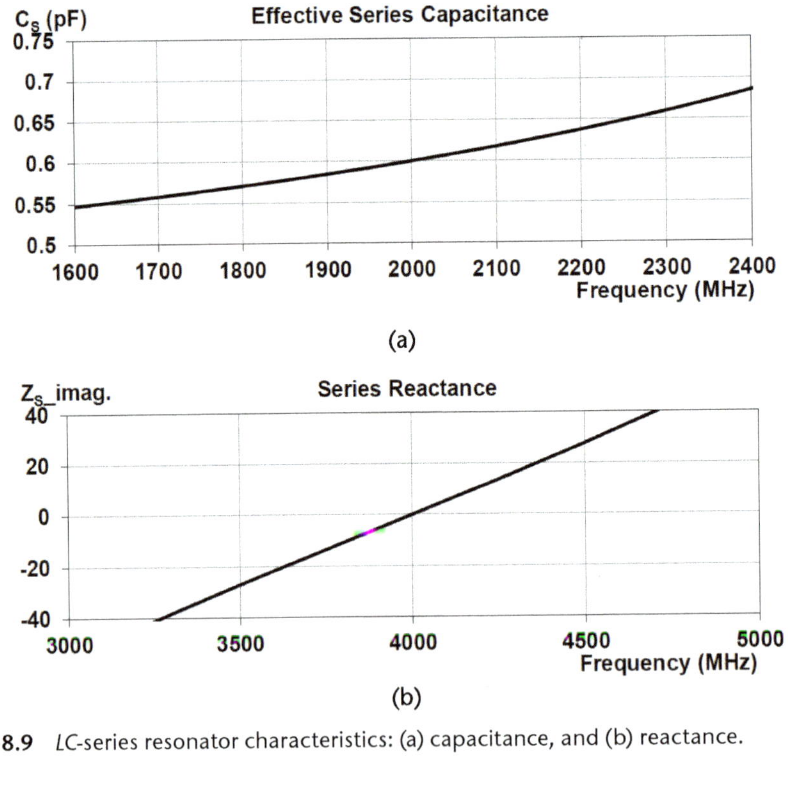

Figure 8.9 *LC*-series resonator characteristics: (a) capacitance, and (b) reactance.

inductor. The advantage is in addition to the passband inductance, the resonator can be designed to have its self-resonance at a desired frequency below the passband. This can be useful for achieving attenuation of unwanted low-frequency signal components.

For such a resonator design, there are two requirements for the circuit designer: (1) achieve the desired inductance at the passband frequency f_0, and (2) effect a series resonance at a frequency f_r below the passband. Because we only have two elemental values to determine (i.e., L and C), there is a unique solution to this problem.

It is important to note here that the inductance in the passband will be affected by the low-side resonance. While the desired inductance can be designed for f_0, the effective inductance of the resonator will vary across the passband. The closer f_r is to f_0, the greater will be the inductance variation across the passband.

As discussed in Chapter 5, the majority of components used in practical RF circuits have significant associated dissipative and reactive parasitics. For any useful synthesis, these parasitic elements must be taken into account in the design equations.

As in the previous section, Figure 8.8 shows an LC-series resonator with a composite equivalent incorporating the parasitic elements described in Chapter 5. Rather than determine the values of L and C directly from the design constraints, we seek to determine the values of L' and C', which incorporate the effects of the parasitics on the overall resonator response. Because the reactive parasitics have the dominant influence on the resonator characteristics, the resistive elements can typically be ignored in such a synthesis. The parasitic resistive elements contribute little to frequency shifts of the resonator and predominantly result in dissipative losses that are unavoidable.

The required capacitance is obtained from the solution to a quadratic equation, with

$$a \approx \omega_0^2\omega_r^2\left((L - L_C)L_C C_L + L/\left(\omega_0^2 - \omega_r^2\right)\right)$$

$$b \approx -1 - \omega_0^2 L C_L + \left(\omega_0^2 + \omega_r^2\right)L_C C_L$$

$$c \approx -C_L$$

The solution is

$$C' \approx \frac{-b + \sqrt{b^2 - 4ac}}{2a} \tag{8.15}$$

and

$$L' \approx \frac{1}{\omega_r^2\left(C_L + 1/\left(1/C' - \omega_r^2 L_C\right)\right)} \tag{8.16}$$

By way of example, consider the LC-series resonator parameters in Table 8.2. As in Chapter 7, the entries in blue are user inputs. The parasitic element values associated with both the inductor and capacitor are typically obtained from manufacturers' data sheets, bench measurement, or sophisticated RF electromagnetic

Table 8.2 *LC*-Series Resonator Design Parameters

Series (Type 2)							
f_0	2000 MHz	*Parasitics*				*Elements*	
f_r	1000 MHz	R_C (Ω)	L_C (nH)	Q_L	C_L (pF)	L' (nH)	C' (pF)
L_S (at f_0)	2 nH	0.2	0.2	45	0.3	2.15791	10.4909

simulation software. The element values in the two right columns of the table are the required element values for the resonator obtained by applying the formulae above which incorporate any detuning due to the parasitics.

In this case, the *LC*-series resonator is desired to have an inductance of 2.0 nH at the center of the band f_0 = 2,000 MHz, and to have a series resonance at f_r = 1,000 MHz.

Figure 8.10 shows the key performance characteristics of the *LC*-series resonator obtained when implemented with the values for L' and C' given in Table 8.2. Note that both the key goals for the resonator have been achieved.

8.6 Design of *LC*-Parallel Resonators with Desired In-Band Inductance and a High-Side Resonance

In an RF circuit where an inductor is needed in the passband, for matching or other purposes, it can be implemented with an *LC*-parallel resonator, rather than

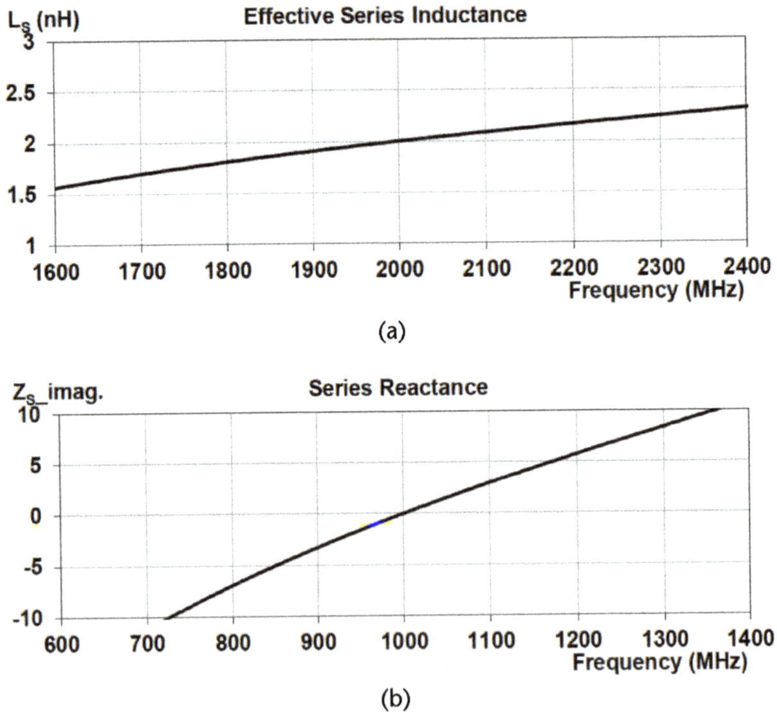

Figure 8.10 *LC*-series resonator characteristics: (a) capacitance, and (b) reactance.

a simple inductor. The advantage is, in addition to the passband inductance, the resonator can be designed to have its self-resonance at a desired frequency above the passband. This can be useful for achieving attenuation of unwanted high-frequency signal components.

For such a resonator design, there are two requirements for the circuit designer: (1) achieve the desired inductance at the passband frequency f_0, and (2) effect a parallel resonance at a frequency f_r above the passband. Because we only have two elemental values to determine (i.e., L and C), there is a unique solution to this problem.

It is important to note here that the inductance in the passband will be affected by the high-side resonance. While the desired inductance can be designed for f_0, the effective inductance of the resonator will vary across the passband. The closer f_r is to f_0, the greater will be the inductance variation across the passband. Fortunately, in many cases, the frequency band for rejection is at a harmonic frequency and thus significantly removed from the passband.

As discussed in Chapter 5, the majority of components used in practical RF circuits have significant associated dissipative and reactive parasitics. For any useful synthesis, these parasitic elements must be taken into account in the design equations.

Figure 8.11 shows an LC-parallel resonator with a composite equivalent incorporating the parasitic elements described in Chapter 5. Rather than determine the values of L and C directly from the design constraints, we seek to determine the values of L' and C', which incorporate the effects of the parasitics on the overall resonator response. Because the reactive parasitics have the dominant influence on the resonator characteristics, the resistive elements can typically be ignored in such a synthesis. The parasitic resistive elements contribute little to frequency shifts of the resonator and predominantly result in dissipative losses that are unavoidable.

The required capacitance is obtained from the solution to a quadratic equation, with

$$a \approx \omega_r^2\left(\left(1/L + \omega_0^2 C_L\right) L_C C_L - 1/\left(\omega_r^2 - \omega_0^2\right) L\right)$$
$$b \approx 1 - L_C/L - \left(\omega_r^2 + \omega_0^2\right) L_C C_L$$
$$c \approx L_C$$

The solution is

$$L' \approx \frac{-b - \sqrt{b^2 - 4ac}}{2a} \tag{8.17}$$

and

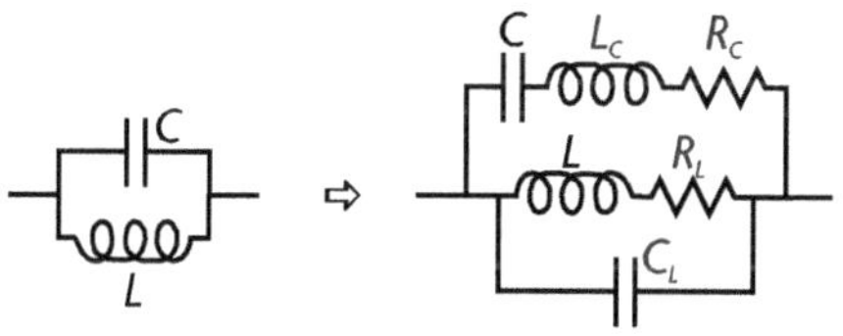

Figure 8.11 *LC*-parallel resonator with parasitics.

$$C' \approx \frac{1}{\omega_r^2\left(L_C + 1/\left(1/L' - \omega_r^2 C_L\right)\right)} \quad (8.18)$$

By way of example, consider the LC-parallel resonator parameters in Table 8.3. As in Chapter 7, the entries in blue are user inputs. The parasitic element values associated with both the inductor and capacitor are typically obtained from manufacturers' data sheets, bench measurement, or sophisticated RF electromagnetic simulation software. The element values in the two right columns of the table are the required element values for the resonator obtained by applying the formulae above, which incorporate any detuning due to the parasitics.

In this case, the LC-parallel resonator is desired to have an inductance of 2.0 nH at the center of the band f_0 = 2,000 MHz and to have a parallel resonance at f_r = 4,000 MHz.

Figure 8.12 shows the key performance characteristics of the LC-series resonator obtained when implemented with the values for L' and C' given in Table 8.3. Note that both the key goals for the resonator have been achieved.

Table 8.3 *LC*-Parallel Resonator Design Parameters

Parallel (Type 1)							
f_0	2000 MHz	*Parasitics*				*Elements*	
f_r	4000 MHz	$R_C(\Omega)$	L_C(nH)	Q_L	C_L(pF)	L' (nH)	C' (pF)
L_P (at f_0)	2 nH	0.2	0.3	45	0.3	1.53344	0.64315

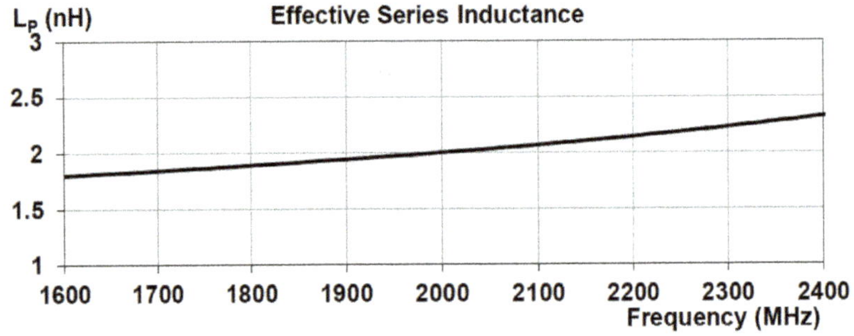

(a)

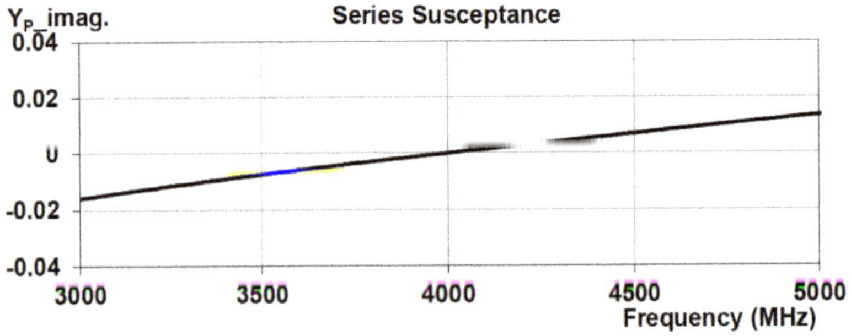

(b)

Figure 8.12 *LC*-parallel resonator characteristics: (a) inductance, and (b) susceptance.

8.7 Design of *LC*-Parallel Resonators with Desired In-Band Capacitance and a Low-Side Resonance

In an RF circuit where a capacitor is needed in the passband, for matching or other purposes, it can be implemented with an *LC*-parallel resonator, rather than a simple capacitor. The advantage is in addition to the passband inductance, the resonator can be designed to have its self-resonance at a desired frequency below the passband. This can be useful for achieving attenuation of unwanted low-frequency signal components.

For such a resonator design, there are two requirements for the circuit designer: (1) achieve the desired inductance at the passband frequency f_0, and (2) effect a parallel resonance at a frequency f_r above the passband. Because we only have two elemental values to determine (i.e., L and C), there is a unique solution to this problem.

It is important to note here that the inductance in the passband will be affected by the low-side resonance. While the desired inductance can be designed for f_0, the effective inductance of the resonator will vary across the passband. The closer f_r is to f_0, the greater will be the inductance variation across the passband.

As discussed in Chapter 5, the majority of components used in practical RF circuits have significant associated dissipative and reactive parasitics. For any useful synthesis, these parasitic elements must be taken into account in the design equations.

As in the previous section, Figure 8.11 shows an *LC*-parallel resonator with a composite equivalent incorporating the parasitic elements described in Chapter 5. Rather than determine the values of L and C directly from the design constraints, we seek to determine the values of L' and C', which incorporate the effects of the parasitics on the overall resonator response. Because the reactive parasitics have the dominant influence on the resonator characteristics, the resistive elements can typically be ignored in such a synthesis. The parasitic resistive elements contribute little to frequency shifts of the resonator and predominantly result in dissipative losses that are unavoidable.

The required capacitance is obtained from the solution to a quadratic equation, with

$$a \approx \omega_0^2\omega_r^2\left((C - C_L)L_C C_L + C/\left(\omega_0^2 - \omega_r^2\right)\right)$$

$$b \approx -1 - \omega_0^2 L_C C + \left(\omega_0^2 + \omega_r^2\right)L_C C_L$$

$$c \approx -L_C$$

The solution is

$$L' \approx \frac{-b + \sqrt{b^2 - 4ac}}{2a} \tag{8.19}$$

and

$$C' \approx \frac{1/L' - \omega_r^2 C_L}{\omega_r^2\left(1 + \left(1/L' - \omega_r^2 C_L\right)L_C\right)} \tag{8.20}$$

By way of example, consider the LC-parallel resonator parameters in Table 8.4. As in Chapter 7, the entries in blue are user inputs. The parasitic element values associated with both the inductor and capacitor are typically obtained from manufacturers' data sheets, bench measurement, or sophisticated RF electromagnetic simulation software. The element values in the two right columns of the table are the required element values for the resonator obtained by applying the formulae above, which incorporate any detuning due to the parasitics.

In this case, the LC-parallel resonator is desired to have a capacitance of 3.0 pF at the center of the band f_0 = 2,000 MHz and to have a parallel resonance at f_r = 1,000 MHz.

Figure 8.13 shows the key performance characteristics of the LC-parallel resonator obtained when implemented with the values for L' and C' given in Table 8.4. Note that both the key goals for the resonator have been achieved.

Table 8.4 *LC*-Parallel Resonator Design Parameters

Parallel (Type 2)							
f_0	2000 MHz	*Parasitics*				*Elements*	
f	1000 MHz	$R_C(\Omega)$	L_C(nH)	Q_L	C_L(pF)	L' (nH)	C' (pF)
C_P (at f_0)	3 pF	0.2	0.5	45	0.5	7.73831	2.62942

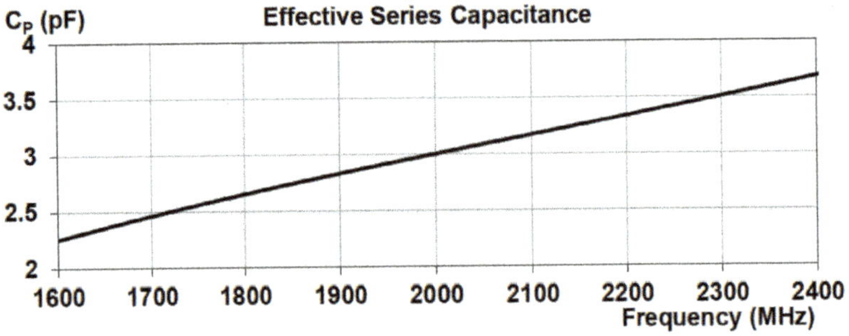

(a)

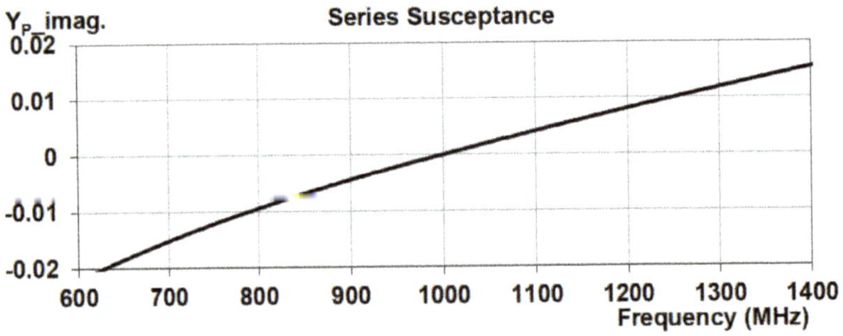

(b)

Figure 8.13 *LC*-parallel resonator characteristics: (a) inductance, and (b) susceptance.

8.8 Practical Bandpass Filter Design

LC-series and -parallel resonator elements are critical components in many RF designs. For optimum performance, it is critical that the resonator quality factor be as high as possible. This improves the fidelity of the desired frequency characteristics and reduces parasitic losses.

Figure 8.14 shows the schematic of a Cauer elliptic bandpass filter designed to meet stringent passband characteristics. Implementing the filter required seven parallel *LC* resonator sections. Each of the resonators was designed for a unique individual resonant frequency. This necessitated a different capacitor and inductor design for each of the resonators. The filter was implemented on a 200-mm high-resistivity Si IPD wafer with two thick Cu layers available, which were critical to maximizing inductor quality factor. Even so, there were several important aspects of the physical realization that were critical to achieving optimum performance characteristics for the filter.

The physical die layout for the filter is shown in Figure 8.15.

1. In the layout, each of the inductors was designed to maximize its inherent *Q* (see Section 5.3.3), while ensuring that its self-resonant frequency was inferior to the desired tank frequency for the resonator. This absorbed the parasitic capacitance of the inductor into the design while eschewing a common error in assuming that the latter implies low inductor *Q*.
2. In a high-*Q* parallel *LC* resonator, the circulating currents in the resonator are substantially higher than the currents entering and leaving the resonator. In each RF cycle, the energy stored in the resonator flips back and forth between the magnetic and capacitive components. Hence, it is critical that any resistance in the path between the two components of the resonator be minimized as much as possible. To this end, as seen in Figure 8.15, each of the tank capacitors was placed directly across the terminals of the inductors. This greatly enhanced the *Q* factor of the resonators.

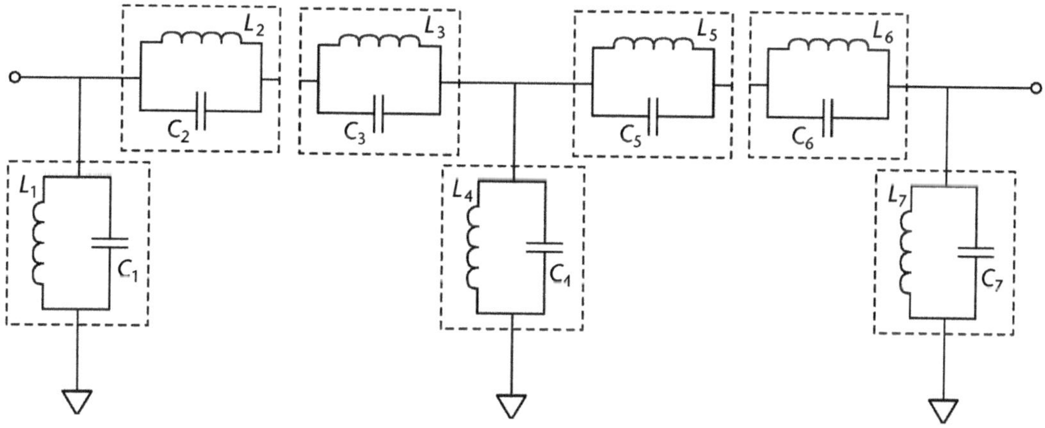

Figure 8.14 Bandpass elliptic filter circuit schematic.

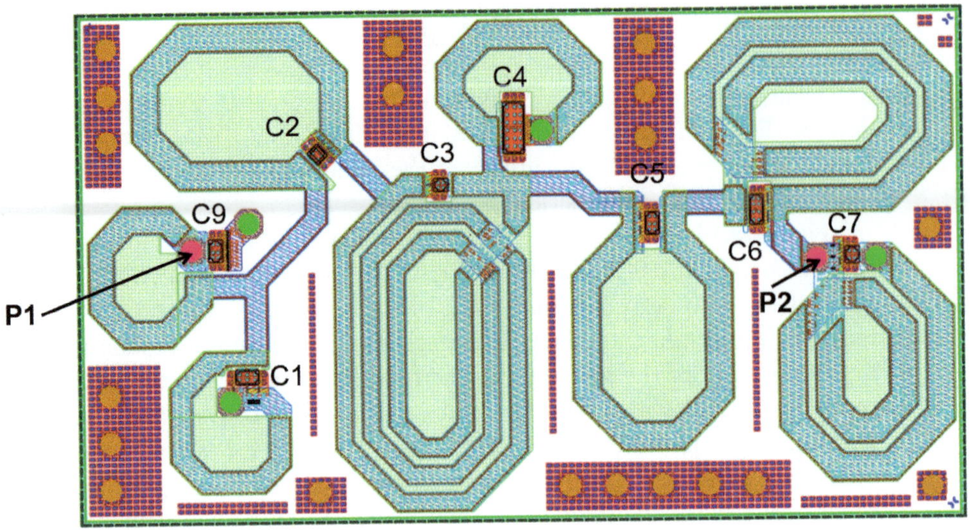

Figure 8.15 Bandpass elliptic filter circuit IPD layout.

3. As is evident in Figure 8.14, the filter required ground connections to certain of the resonators. To minimize resistance in the ground connections, the filter was designed for a flip-chip attach to a laminate carrier, as described in Section 5.5. The Cu flip-chip bumps, as shown in Figure 5.15(b), provided minimal series resistance to ground.

Figure 8.16 shows the measured response for the filter. At the center of the passband, the insertion gain is ~ −1.3 dB. This is extremely low considering the wide passband and steep rejection shirts. (For details of this development, see Wright,

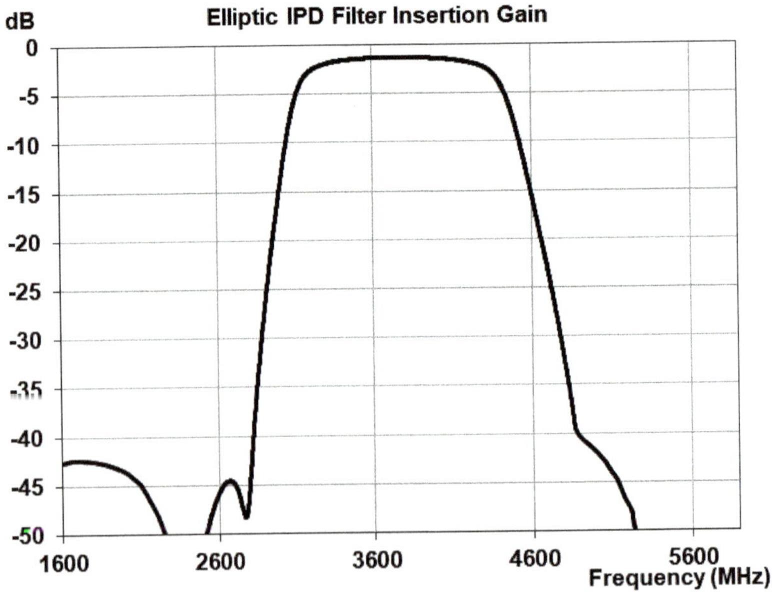

Figure 8.16 Bandpass elliptic filter measured insertion characteristics.

P. V., "Radio Frequency Filtering Circuitry on Integrated Passive Die," US Patent 11,489,506, November 1, 2022.)

8.9 Novel Resonator Pairing for Bandpass Shaping

The design and basic properties of LC-series and LC-parallel resonators were presented in some detail in at the beginning of this chapter. If the former is employed in shunt with the line, or the latter in series with the line, as shown in Figure 8.17, significant attenuation can be achieved for frequencies close to the resonances. However, in the desired passband, which occurs away from the resonance, either higher or lower in frequency, the traps will exhibit some residual reactance. The extent of this reactance, and whether it significantly impacts the insertion loss of the desired signal energy, depends on the frequency separation between the passband and the traps. This residual in-band reactance can also be negated by matching network, subsequent to the trap, as discussed in Chapter 9.

In some cases, one frequency trap may be required above the passband while another is required below it. In this case, we may take advantage of the requirement by code-signing the trap pair such that there is no residual reactance at the center of the passband to disrupt the signal path. The result is a bandpass-type response for the desired signal.

8.10 Novel *LC*-Series Resonator Pairing for a Passband-Type Response

Figure 8.18 shows an LC-series resonator pair connected in shunt across the signal line. Let the critical system frequencies be defined as:

f_1 ~ lower trapband frequency.
f_0 ~ center passband frequency.
f_2 ~ upper trapband frequency.

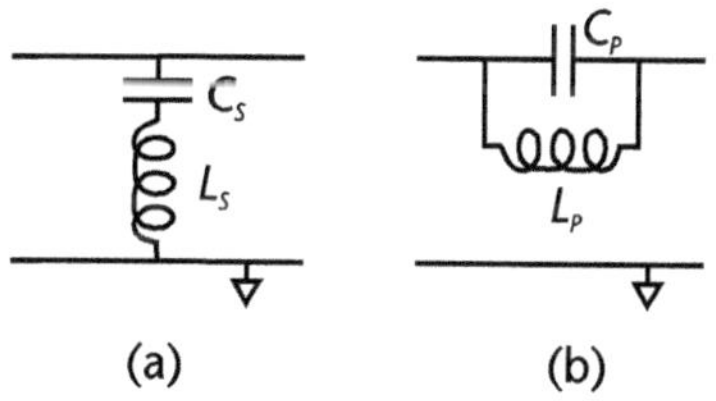

Figure 8.17 *LC*-resonator traps employed for filtering: (a) *LC*-series, and (b) *LC*-parallel.

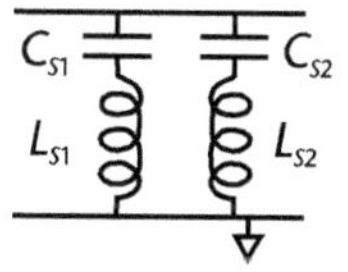

Figure 8.18 *LC*-series resonator traps employed for bandpass filtering.

Suppose the first resonator L_{S1}/C_{S1} resonates at f_1, while the second resonator L_{S2}/C_{S2} resonates at f_2. At the center of the passband f_0, resonator 1, which is above resonance, will be inductive. In contrast, resonator 2, which is below resonance, will remain capacitive. Because these two loading reactances are of opposite sign, we may design them to cancel as parallel elements at f_0, thereby negating any net loading reactance on the signal line.

If the capacitance of resonator 2 at f_0 is C_0, the solutions for the critical element values are

$$C_{S1} = \left((\omega_0/\omega_1)^2 - 1\right)C_0 \tag{8.21}$$

$$L_{S1} = \frac{1}{\omega_1^2 C_{S1}} \tag{8.22}$$

$$C_{S2} = \left(1 - (\omega_0/\omega_2)^2\right)C_0 \tag{8.23}$$

$$L_{S2} = \frac{1}{\omega_2^2 C_{S2}} \tag{8.24}$$

The value of C_0 is a design parameter that the designer can use to compromise the depth of the notches for the passband insertion loss. Smaller values for C_0 are associated with lower notch depths but a wider passband. This can be seen in Figure 8.19.

Two factors contribute to the passband loss: (1) reactive mismatch away from f_0, and (2) dissipative losses due to nonideal components. The responses shown in Figure 8.19 assume each inductor has a $Q = 35$, while each capacitor has an Equivalent Series Resistance (ESR) = 0.2Ω.

8.11 Novel *LC*-Parallel Resonator Pairing for a Passband-Type Response

Figure 8.20 shows an *LC*-parallel resonator pair connected in series with the signal line. Let the critical system frequencies be defined as:

f_1 ~ lower trapband frequency.
f_0 ~ center passband frequency.
f_2 ~ upper trapband frequency.

Suppose that the first resonator L_{P1}/C_{P1} resonates at f_1, while the second resonator L_{P2}/C_{P2} resonates at f_2. At the center of the passband f_0, resonator 1, which is above resonance, will be capacitive. In contrast, resonator 2, which is below resonance, will remain inductive. Because these two series reactances are of the opposite sign, we may design them to cancel at f_0, thereby negating any net series reactance in the signal line.

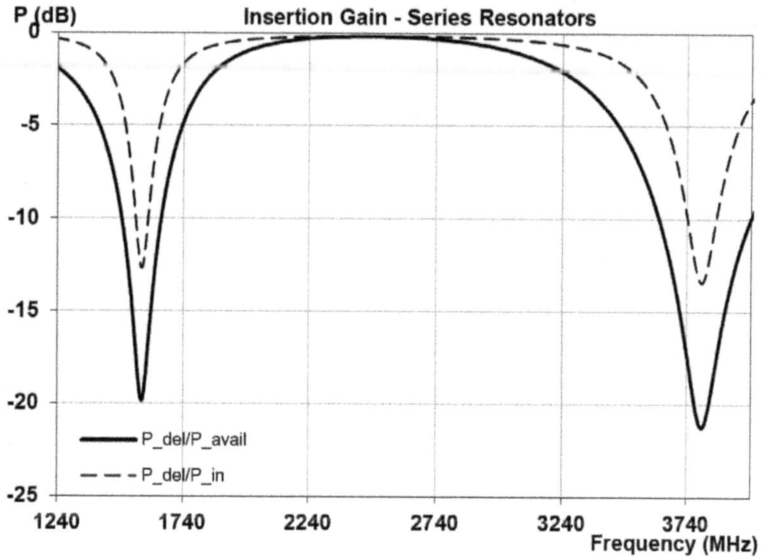

(a)

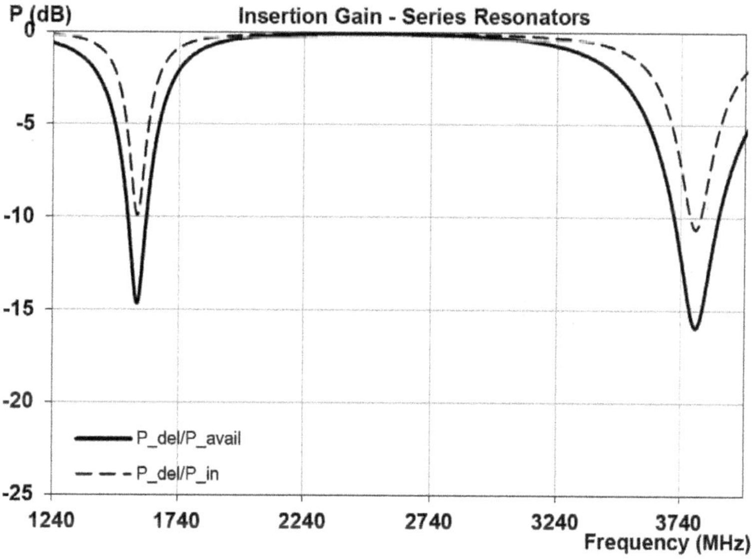

(b)

Figure 8.19 *LC*-series resonator pairing passband-type response: (a) C_0 = 0.8 pF, and (b) C_0 = 0.4 pF.

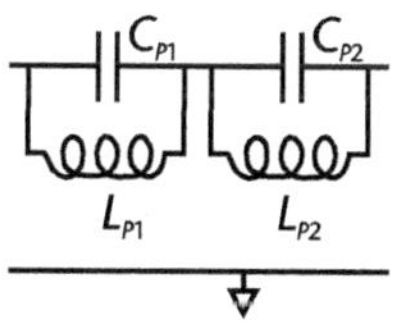

Figure 8.20 *LC*-parallel resonator traps employed for bandpass filtering.

If the capacitance of resonator 1 at f_0 is C_0, the solutions for the critical element values are

$$C_{P1} = \frac{C_0}{1-(\omega_1/\omega_0)^2} \tag{8.25}$$

$$L_{P1} = \frac{1}{\omega_1^2 C_{P1}} \tag{8.26}$$

$$C_{P2} = \frac{C_0}{(\omega_2/\omega_0)^2 - 1} \tag{8.27}$$

$$L_{P2} = \frac{1}{\omega_2^2 C_{P2}} \tag{8.28}$$

The value of C_0 is a design parameter that the designer can use to compromise the depth of the notches for the passband insertion loss. Larger values for C_0 are associated with lower notch depths but a wider passband. This can be seen in Figure 8.21.

Two factors contribute to the passband loss: (1) reactive mismatch away from f_0, and (2) dissipative losses due to nonideal components. The responses shown in Figure 8.21 assume that each inductor has a $Q = 35$, while each capacitor has an Equivalent Series Resistance (ESR) = 0.2Ω.

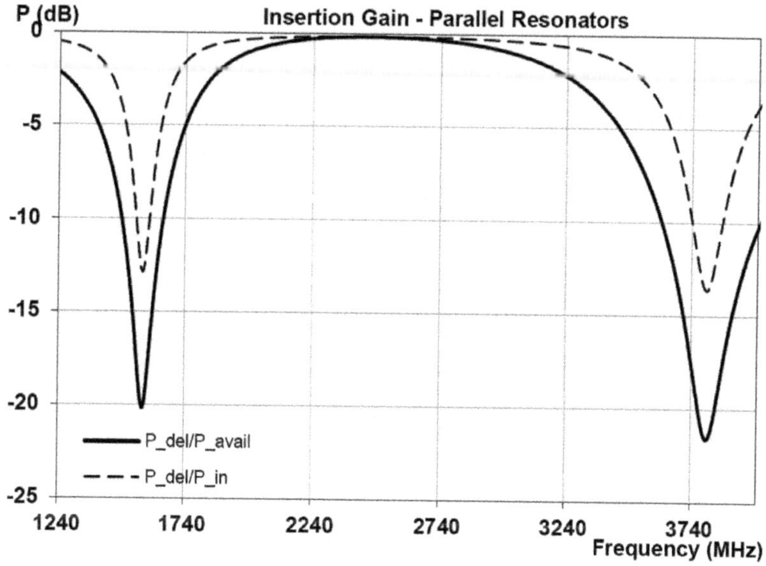

(a)

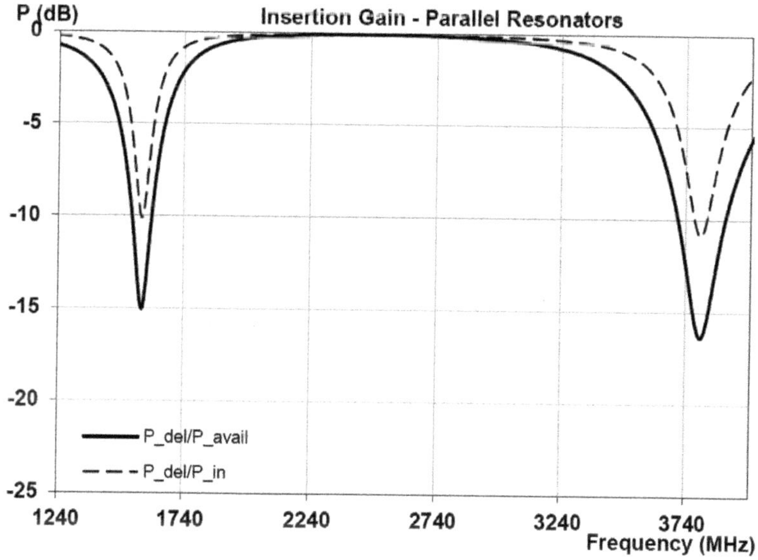

(b)

Figure 8.21 *LC*-parallel resonator pairing a passband-type response: (a) $C_0 = 2$ pF, and (b) $C_0 = 4$ pF.

CHAPTER 9

Fundamentals of Amplifier Output Matching

Power amplifiers (PAs) are critical components in all of today's mobile communications devices. They are essential to raise the power level of an incoming modulated RF signal, containing information, to a sufficiently high-power level such that when it is transmitted it can be demodulated by a receiver at some distance away.

In a mobile device, the supply voltage is invariably low (~3V), since it is supplied by a battery, usually Li-ion. A typical transmit power for a handheld device is ~1W (30 dBm). Since power $\propto V^2/R$, the amplifier output impedance (R) must necessarily be also low, ~2–5Ω. However, the amplifier must typically drive a load with a much higher impedance, ~50Ω. Common examples of amplifier loads are filters, couplers, and antennas.

A network is thus required between the output of the PA and its load to avoid power loss due to the high impedance mismatch between the two. This matching network must transform the low PA impedance to that of the load with as little loss as possible and across the full bandwidth of the signal. However, it also must also be designed to meet other stringent requirements. In addition to achieving the required passband loading impedance, desired loading impedances across the harmonic frequencies are invariably specified. These are important for waveform shaping in the PA and maximizing PA efficiency. Desired harmonic impedances, typically close to open or short, are generally dependent on the modulation waveform. Yet another requirement generally imposed on the output matching network is for it to provide an out-of-band frequency filtering function. Rejection requirements for the network may be both below and above the passband. As a result, the design of such networks can be very challenging, invariably requiring compromises between the various requirements. A number of varying circuit topologies is employed for PA matching networks. The one chosen for a particular application typically being dependent on the designer's previous experience, ease of implementation within the module, and the overall system performance specifications.

In addition to a PA requiring a matching network following the PA, it must be also be biased from the DC power source in the mobile device. This is typically accomplished via a bias inductor, L_b, connected to the supply line, V_{cc}, and the output of the PA. To reduce induced RF ripple on the supply line, which might interfere with the performance of other components within the module, the bias inductor must be bypassed by a capacitor C_b to ground.

Further to the bias inductor, as mentioned in Chapter 6, a shunt trap is generally included across the output of the amplifier. The trap serves two purposes:

1. It provides a low-impedance termination on the amplifier output over a harmonic frequency band, most frequently the second harmonic, for waveform shaping, which serves to maximize the amplifier efficiency.
2. It provides rejection of unwanted harmonic output, again commonly the second harmonic.

Figure 9.1 shows a simplified block diagram of a typical single-ended mobile PA matching network.

It should be noted that the resonant frequency of the shunt series L_h/C_h trap is not typically set to correspond exactly to the second-harmonic frequency. This design freedom allows the harmonic impedance on the output of the amplifier to be adjusted from slightly inductive, or slightly capacitive, as required by the PA for optimal in-band efficiency.

The PA architecture in Figure 9.1 shows a single PA and signal line. That architecture is referred to as "single-ended." Another PA architecture, in common use, is typically referred to as "differential" or sometimes as "push-pull." In the differential PA architecture two amplifiers and signal lines are used with the signal energy on the two lines of equal amplitude but 180° out of phase with each other.

Figure 9.2 shows a simplified block diagram of a differential PA architecture.

In Figure 9.2, coupled inductors (often referred to as transformers) are shown for implementing both the differential drive to the two amplifier chains, and combining the outputs for delivery to the load. In practice, lumped elements may also be used for either the differential phase splitting or combining instead of the coupled inductors.

While single-ended and differential PA architectures are most commonly employed, other dual or multiple PA architectures also find application. Each has its own distinct advantages and drawbacks. In general, the single-ended architecture offers a size advantage over the multiple PA architectures. A comparison of single-ended versus differential architectures is presented in Chapter 13.

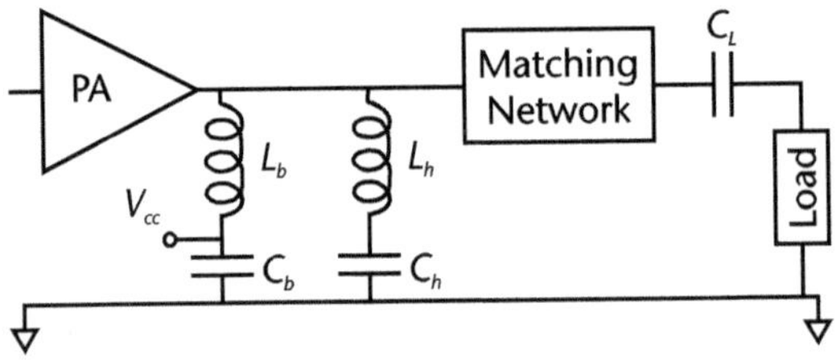

Figure 9.1 Simplified block diagram of single-ended PA.

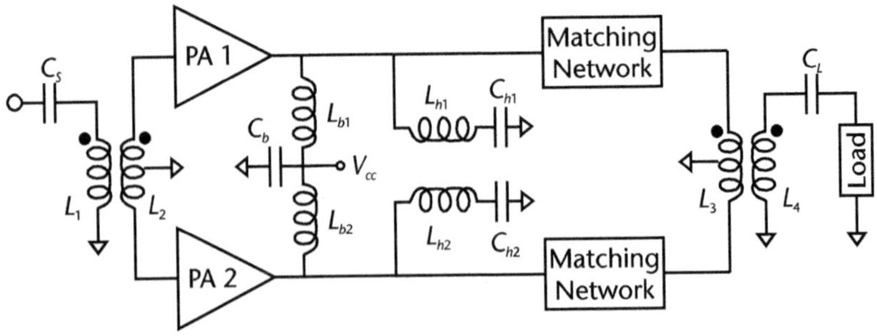

Figure 9.2 Simplified block diagram of a differential PA.

Whether a single or multiple PA architecture, there is a commonality of matching and termination characteristics for each of the amplifiers. We may thus consider only a single PA in isolation to illustrate some of these fundamentals. Consider the simplified schematic of a single PA output stage such as that shown in Figure 9.3. More complex and multisection LC-matching networks for single-ended PAs are considered in Chapter 11. The source voltage V_S represents the PA, with output impedance Z_S. A simple lowpass matching network comprising L_m and C_m is employed to match the low output impedance of the PA to that of the load. Note that, even in this very basic circuit, there are six components whose values must be chosen to optimize the performance of the amplifier. To achieve optimum performance, it is essential that the values for these elements be chosen simultaneously and not individually.

We will now present a rational approach for doing this. First, consider the bias inductor L_b and its bypass capacitor C_b. The latter is required to shunt any RF energy "leaking" through L_b on to the supply line V_{CC}. For this to be the case, we require

$$|Z(C_b)| << |Z(L_b)| \tag{9.1}$$

which implies

$$C_b << 1/\omega_0^2 L_b \tag{9.2}$$

This inverse relationship between C_b and L_b signifies that lower values for L_b will require larger values for C_b to effectively bypass it. While small values for the bias inductor might be advantageous in terms of size, the associated requirement for a large bypass capacitor can be problematic, for reasons other than size. A common technique in today's mobile power amplifiers, often referred to as Envelope Tracking (ET), is based on being able to modulate the supply line with the envelope of the signal. This can significantly boost the overall efficiency of the amplifier. However, for such a technique to be practical, there must not be too much capacitance across the supply line or the modulator will have difficulty driving the line. Given that there are usually multiple amplifiers connected to the supply line, the restrictions on supply-line capacitance for each amplifier can be extremely restrictive. A typical limit on C_b for any individual PA is ~50 pF. Given (9.2), this puts a practical lower limit on the value of the bias inductor L_b.

For an initial exploration of some key aspects of PA matching, let us assume that C_b is sufficiently large (and lossless), such that all RF energy on the supply line V_{CC} is fully shorted to ground. This is represented in Figure 9.3 by the dashed short

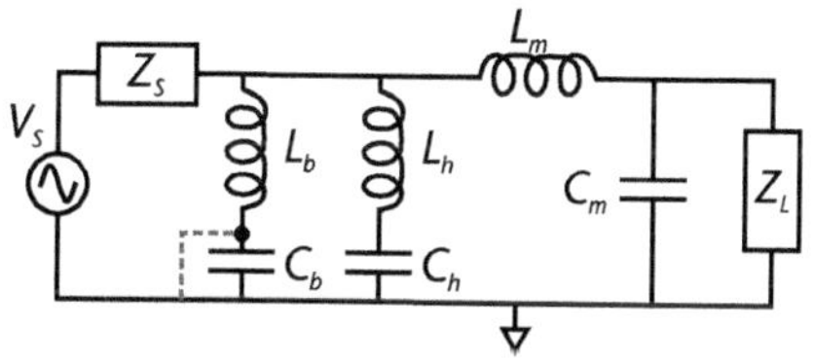

Figure 9.3 Simplified single PA lowpass matching network.

circuit shown across C_b. For an illustrative analysis, assume the following values for the circuit parameters in Figure 9.3.

$$\begin{array}{ll} f_{\text{lo}} = 3{,}300\ \text{MHz} & f_{\text{hi}} = 4{,}200\ \text{MHz} \\ f2_{\text{lo}} = 6{,}600\ \text{MHz} & f2_{\text{hi}} = 8{,}400\ \text{MHz} \\ Z_S = 5\Omega & Z_L = 50\Omega \\ Q_L = 40,\ \text{for all inductors} & ESR = 0.15\Omega,\ \text{for all capacitors} \end{array} \tag{9.3}$$

Thus, the fundamental band of interest (i.e., signal passband) is from 3,300 to 4,200 MHz. The second-harmonic band limits are double the latter.

In the simplified circuit of Figure 9.3, L_h and C_h are determined such that their series resonant frequency is close to the center of the second-harmonic bandwidth. This places a low impedance across the amplifier in this band and also results in high attenuation across the band. The lower the Q of this resonator (i.e., the lower the value of L_h) the lower the impedance and the higher the harmonic rejection that can be achieved. However, in the passband, at approximately half the frequency, the L_h/C_h resonator presents a positive (i.e., capacitive) susceptance across the line. For low loss in the passband, this susceptance must be compensated by an inductive susceptive loading on the line, which can come from two places: (1) the bias inductor L_b, or (2) the lowpass matching network, comprising L_m and C_m.

We shall consider first the situation in which the harmonic resonator susceptance is fully compensated by the bias inductor alone. In this case, the matching network L_m/C_m is only required to match the real source impedance (5Ω) to the real load impedance (50Ω). The latter requirement uniquely determines the values for L_m and C_m.

9.1 Passband Harmonic Susceptance Compensated Uniquely by Bias Inductor

Because the harmonic resonator must resonate at or close to the center of the second-harmonic band, C_h, will be dependent on L_h. However, L_b and L_h are interdependent because we require no residual susceptance across the signal line at the center of the passband. This leaves us with only a single independent variable L_b or L_h.

Although the bias inductor can be chosen to completely cancel the residual susceptance of the harmonic resonator at a single frequency, typically at the center of the passband, this cancellation becomes less complete moving away from the center of the band. This is especially true on the high side. The smaller the resonator inductor L_h, the greater is the susceptance variation across the passband, due to the resonance of L_h/C_h close to the second harmonic. This results in an increased insertion loss towards the edges of the passband due to reactive mismatch.

In enumerating the design options for the circuit of Figure 9.3, rather than taking the value of either of the inductors as the independent variable, it is generally more instructive to take the maximum in-band insertion loss as the independent variable, as insertion loss is a prime PA system performance parameter for the PA designer. The maximum insertion loss, for the case we are considering here, is in the passband 3,300 to 4,200 MHz, with the maximum loss invariably occurring at 4,200 MHz. Higher insertion losses will be associated with lower values for L_h, greater second-harmonic rejection, and less impedance variation across the second-harmonic bandwidth. This can be seen in Figure 9.4.

In Figure 9.4, note that L_b and L_h decrease in tandem with increasing system insertion loss. This is because the bias inductor must provide an increasing susceptive loading on the signal line to compensate for the increasingly large residual susceptance of the harmonic resonator in the passband.

Figures 9.5 and 9.6 show the in-band and second-harmonic reflection coefficients for 1-dB and 2-dB maximum system insertion loss, respectively. As expected,

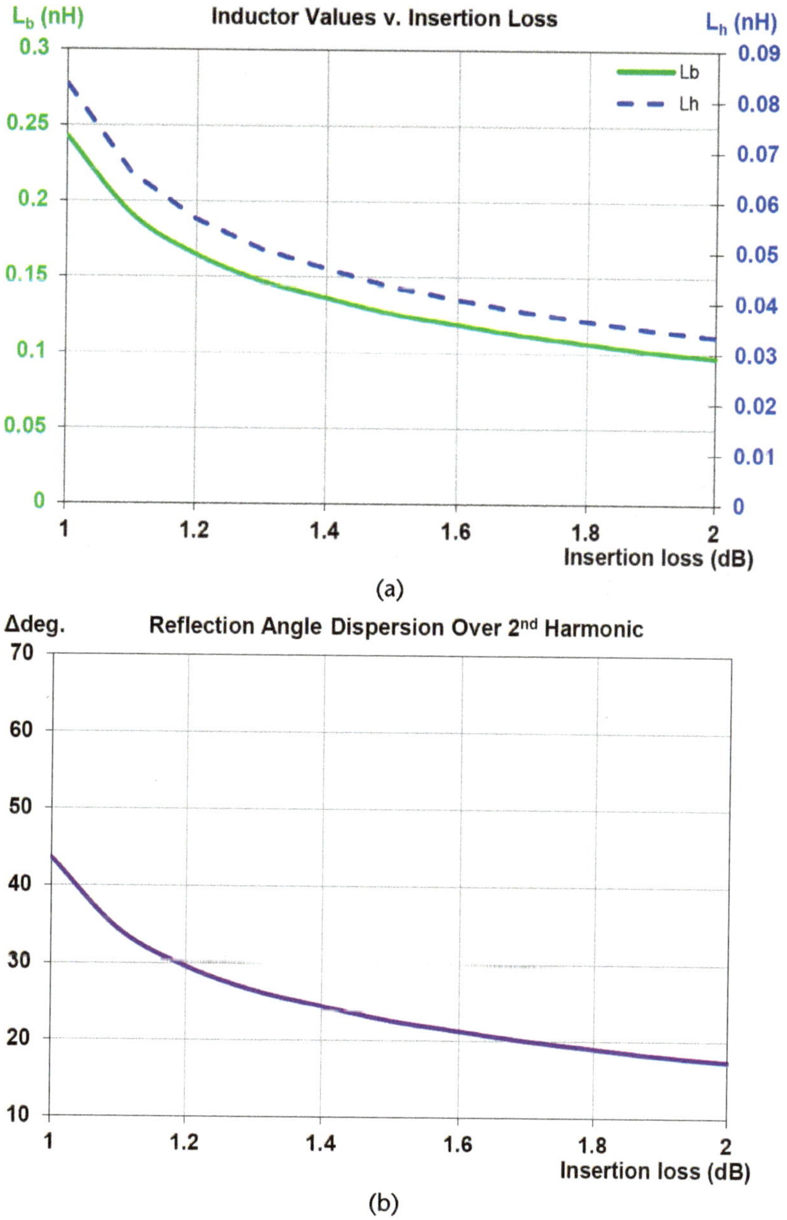

Figure 9.4 Key circuit variables versus insertion loss: (a) inductor variation with loss, and (b) second-harmonic reflection angle dispersion.

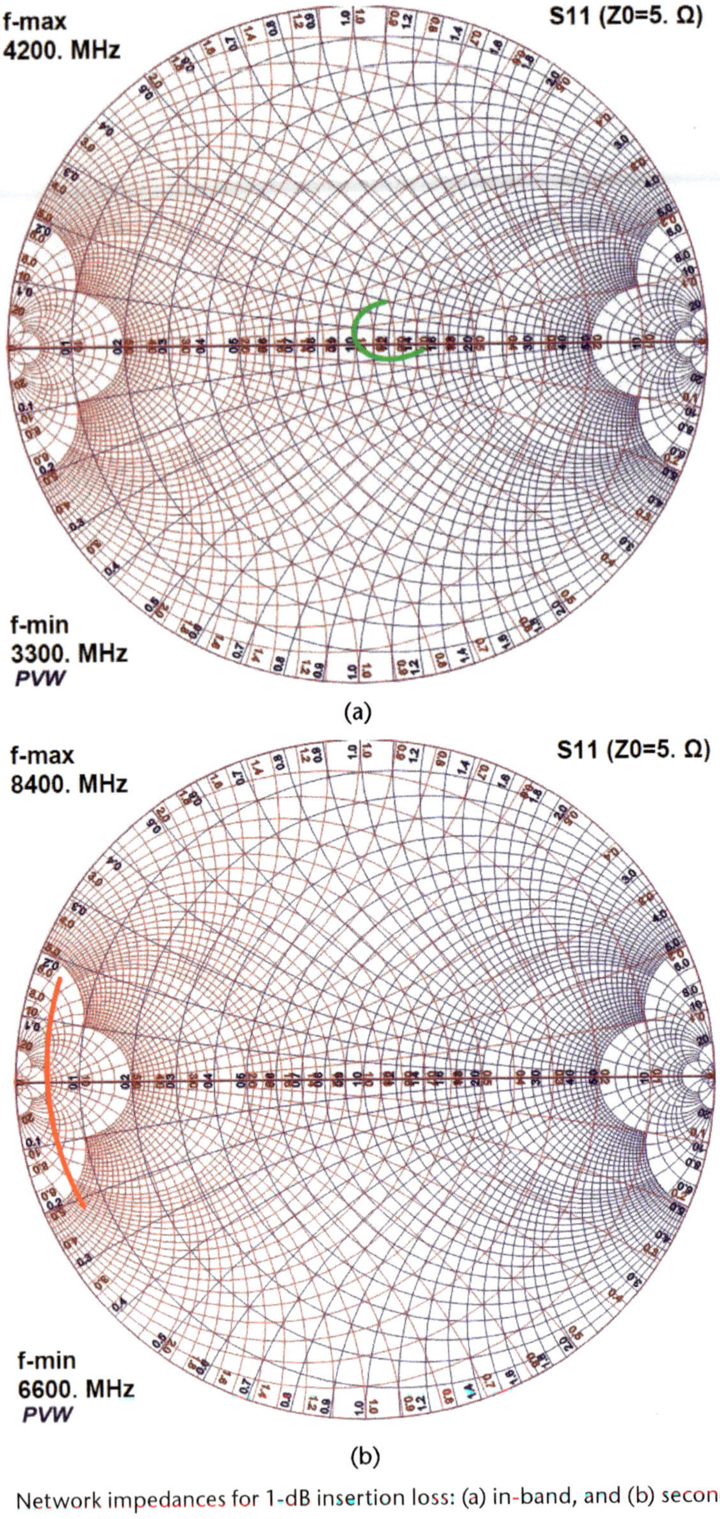

Figure 9.5 Network impedances for 1-dB insertion loss: (a) in-band, and (b) second-harmonic band.

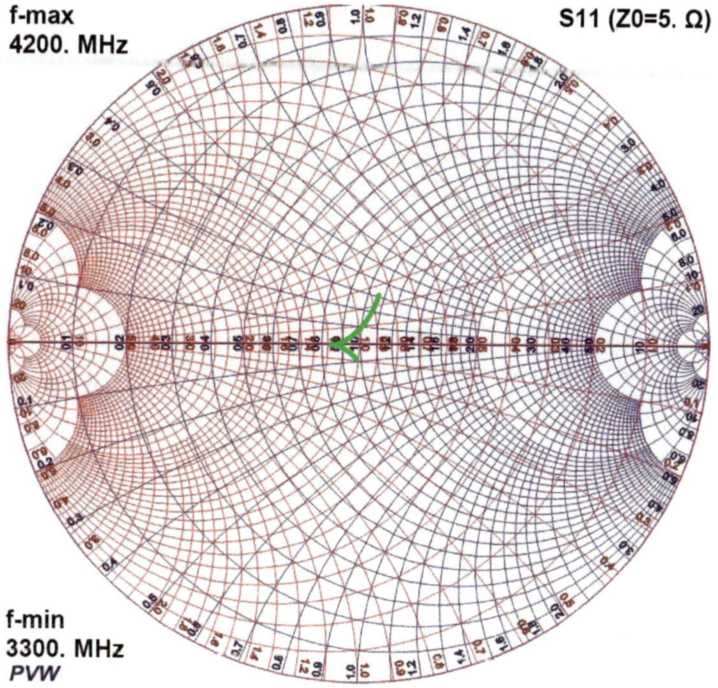

(a)

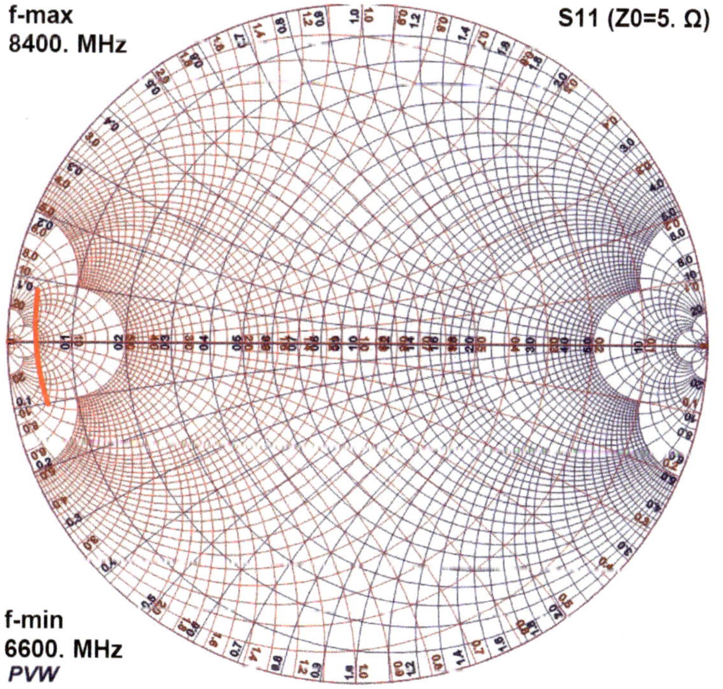

(b)

Figure 9.6 Network impedances for 2-dB insertion loss: (a) in-band, and (b) second-harmonic band.

there is more variation in the match across the passband, but the phase variation in the second harmonic reflection coefficient is significantly reduced at the higher insertion loss. Corresponding insertion losses across the two bands are shown in Figures 9.7 and 9.8.

The data clearly illustrates that the greater the second-harmonic rejection required or the less variation in the second-harmonic impedance, the higher the associated system insertion loss. The designer must trade off these two parameters against each other and decide on an acceptable compromise.

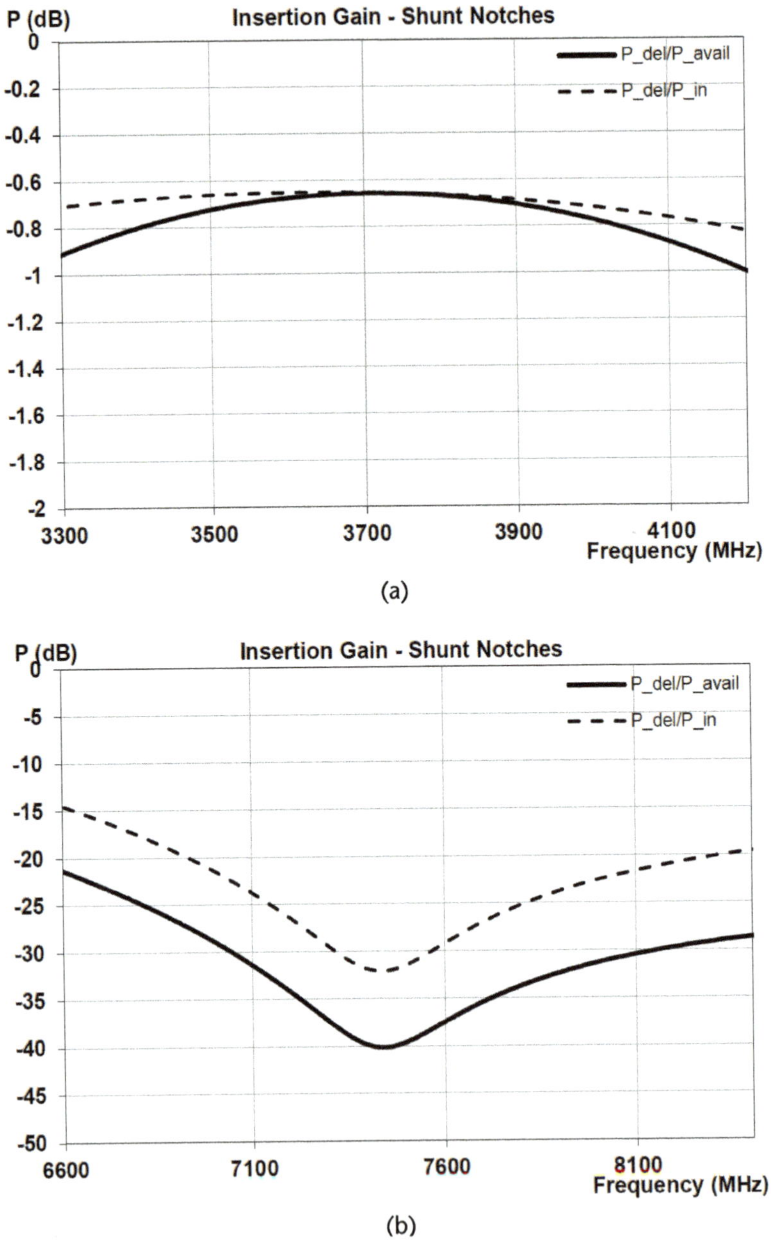

Figure 9.7 Network losses for 1-dB insertion loss: (a) passband insertion gain, and (b) second-harmonic insertion gain.

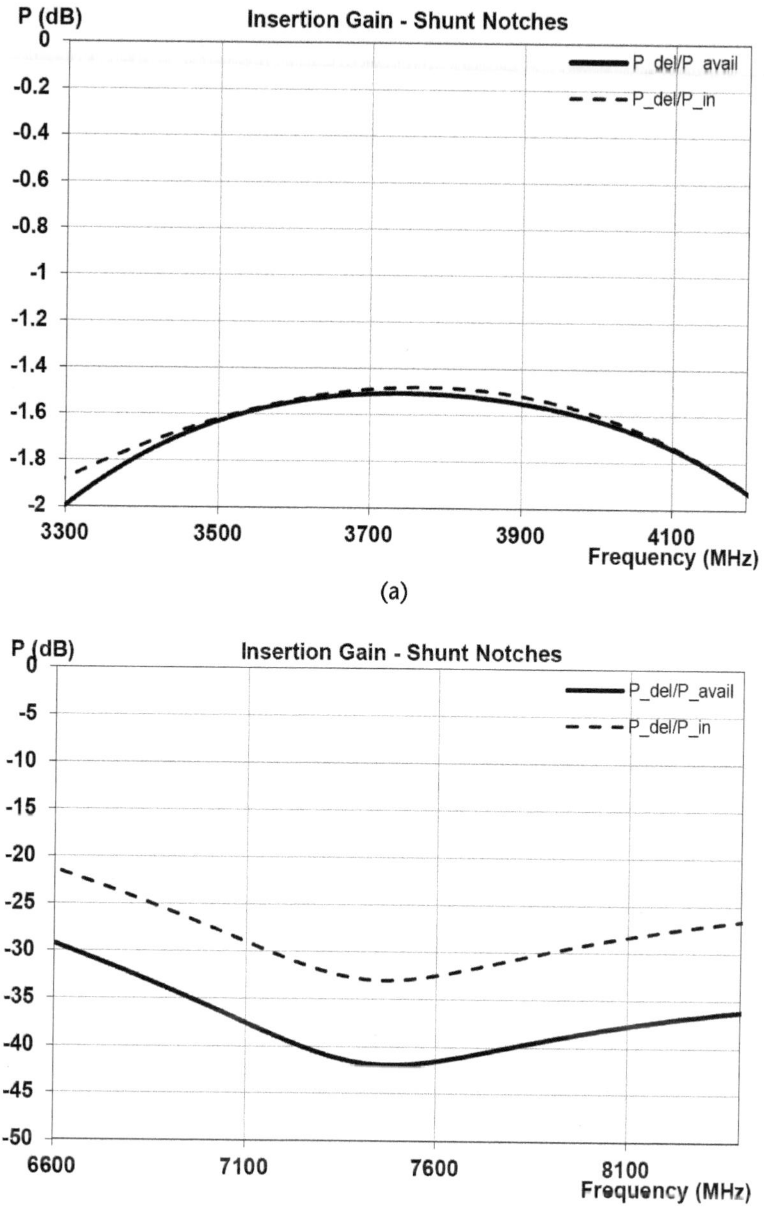

(b)

Figure 9.8 Network losses for 2-dB insertion loss: (a) passband insertion gain, and (b) second-harmonic insertion gain.

In the previous analysis, the harmonic resonator susceptance across the passband is compensated entirely by the bias inductor. However, this requires increasingly lower values for the bias inductor for reduced second-harmonic variation. As discussed above, this would, in turn, require larger values for the bypass capacitor C_b beyond what is acceptable. This is illustrated in Figure 9.9.

Figure 9.9 shows key in-band performance characteristics for the circuit parameters above having a 2-dB insertion loss, but C_b = 100 pF in place of the perfect

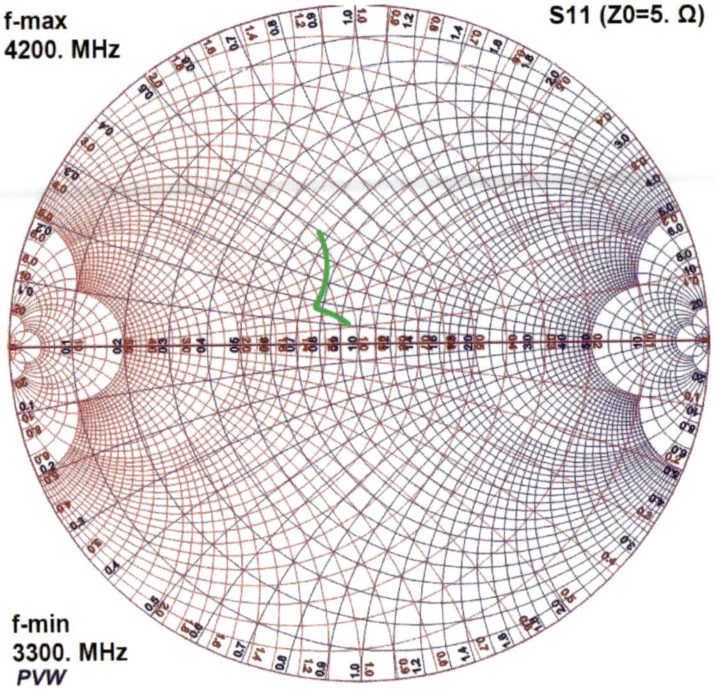

(a)

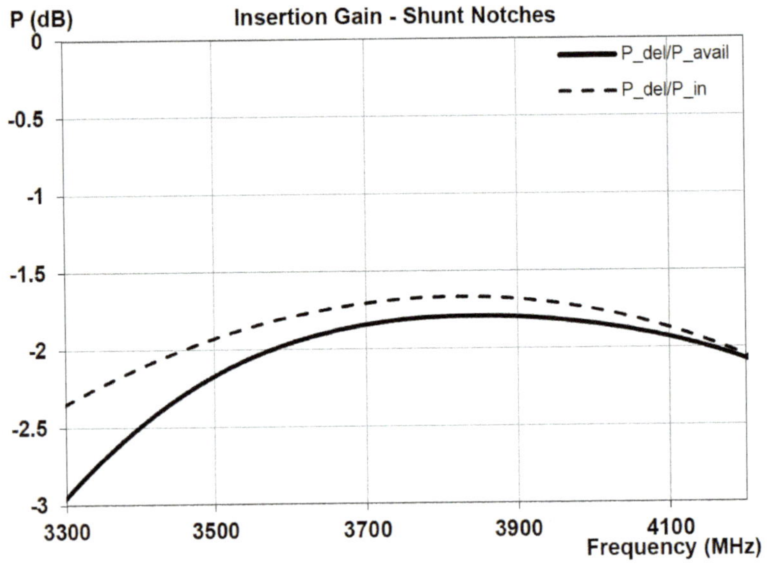

(b)

Figure 9.9 In-band characteristics with 100-pF bypass capacitor: (a) reflection coefficient, and (b) insertion loss.

short previously assumed. There is a clear degradation in the passband reflection coefficient seen in Figure 9.9(a) compared with that in Figure 9.6(a). An even more significant degradation is observed in insertion gain, comparing Figure 9.8(a) with Figure 9.9(b). Primarily, these degradations are caused by a change in the effective shunt susceptance of the bias inductor L_b across the RF line. This indicates that the 100-pF bypass capacitor is inadequate for effectively grounding the V_{CC} supply end of L_b. Such a change in susceptance could easily be compensated for, by a change in the value of the bypass capacitor. However, that would not address the remaining problem of inadequate grounding the V_{CC} supply line. Permitting RF energy to leak on to the supply line, via the bias inductor, degrades the Power-Supply Rejection (*PSR*), a critical parameter in any mobile device. Thus, the data indicates that a 100-pF bypass capacitor is insufficient for a bias inductor $L_b \approx 0.1$ nH.

As a practicality, therefore, in most cases, the bias inductor alone cannot be used to fully compensate for the residual susceptance of the harmonic resonator across the passband. Partial reactive compensation must also be provided by the matching network.

For example, this would require, in the illustrative case here, that the matching network match a complex impedance at the output of the amplifier, to the 50Ω load, rather than a real 5Ω impedance. By requiring the matching network to compensate for any residual susceptance left over from the bias inductor and harmonic resonator at the center of the passband, the value of the bias inductor can be set independent of network insertion loss.

9.2 Passband Harmonic Susceptance Compensated by Bias Inductor and Matching Network

Defining a residual susceptance to be absorbed by the matching network on the output of a PA, such as that shown in Figure 9.3, gives the designer the freedom to set the bias inductor L_b to a preferred value. This, in turn, can minimize the value required for the bypass capacitor C_b. Unfortunately, there is a price to pay for this design flexibility; in general, the greater the residual susceptance the matching network must absorb, the greater the network insertion loss. The second-harmonic impedance variation for a given insertion loss will also typically be higher. This is illustrated in the example data in Figure 9.10.

For each insertion loss data point, the residual susceptance B_1, required of the matching network for each insertion loss, was selected to maintain a constant value for the bias inductor of 0.5 nH. The variation required for B_1 is shown in Figure 9.10(b), and the values for L_b and L_h are shown in Figure 9.10(a). Also shown in Figure 9.10(b) is the angular dispersion of the second-harmonic reflection coefficient.

Comparing Figure 9.4(b) and Figure 9.10(b), we can see significantly more dispersion in the second-harmonic reflection coefficient, at a given insertion loss, in the latter. This is due to an increase in both resistive dissipation and mismatch loss in the matching network when it must adapt the complex source impedance to the load, rather than when the source impedance is pure real. The amount of performance degradation will depend on how much residual susceptance must be absorbed and the nature of the matching network.

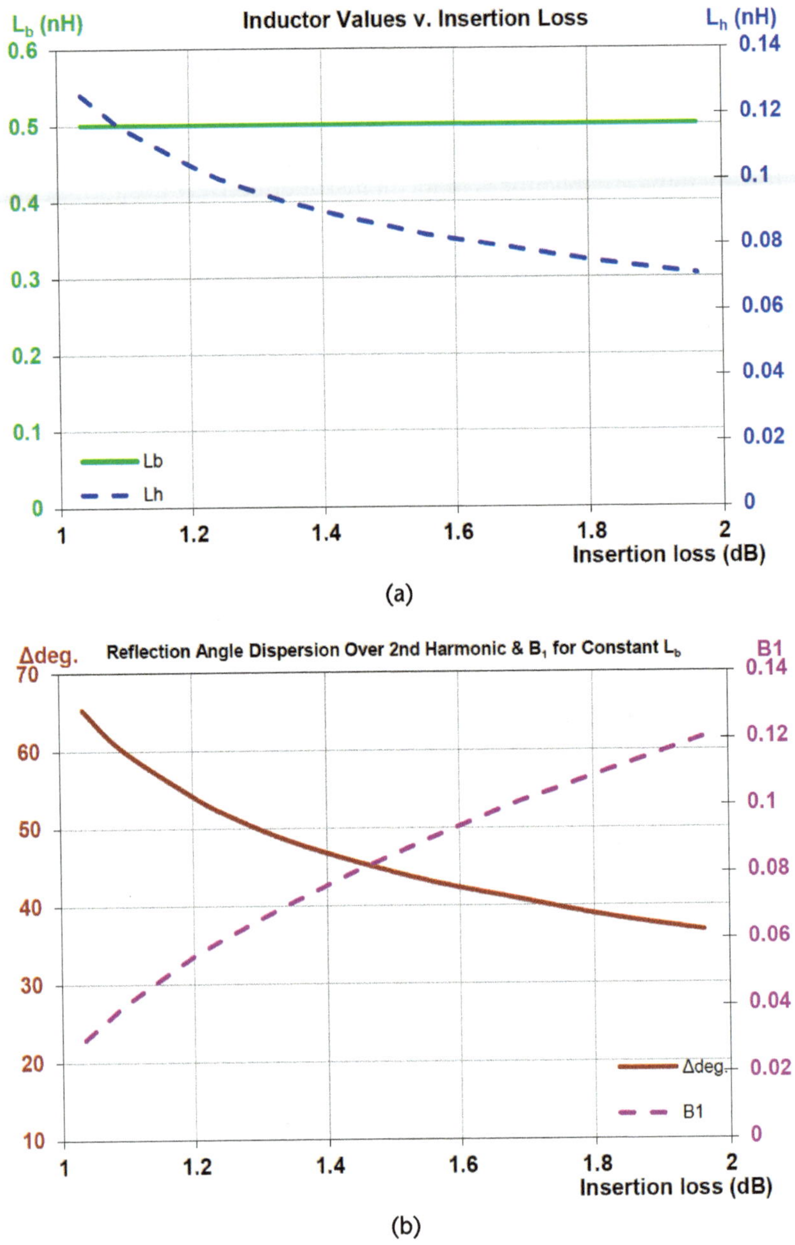

Figure 9.10 Key circuit variables versus insertion loss for constant L_b: (a) inductor variation with loss, and (b) second-harmonic reflection angle dispersion and residual susceptance.

CHAPTER 10

Basic RF PA Bias and Harmonic Trap Networks

In the previous chapter, some of the basic concepts and trade-offs associated with matching the output of an RF PA were introduced. Principal among these are:

1. Series LC traps in shunt on the output of a PA output are effective in providing a low impedance and rejection across harmonic bands.
2. The wider or deeper the harmonic trap (i.e., smaller L_S), the greater will be its susceptive loading in the passband leading to increased insertion loss.
3. The susceptance of harmonic traps in the passband must be canceled by a combination of the bias inductor and matching network.
4. An appropriate value for the bias inductor may be chosen by adjusting the "residual" susceptance in the passband of the matching network. This residual susceptance negates that of the bias inductor plus traps in the passband.
5. The greater the residual susceptance for which a matching network is required to compensate, the higher will be its insertion loss.

In this chapter, the design and analysis of some basic PA bias and harmonic termination networks are presented.

10.1 PA Shunting Inductance and Single Harmonic Trap

Figure 10.1 shows a schematic of single shunt inductance in parallel with a series LC trap. Such an arrangement is commonly placed directly at the output of an RF PA. The inductance L_{eff} represents the combined inductive susceptance of the bias inductor and any residual inductance of the matching network at the PA. The inductance L_{eff} is required to negate the unwanted capacitive susceptance of the harmonic network (or networks) in the passband.

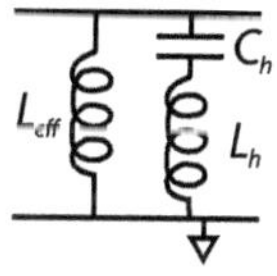

Figure 10.1 PA shunting inductance and single harmonic trap.

In the case of the simple network in Figure 10.1, taking L_b as the independent variable, the value of the effective inductance required is

$$L_{\text{eff}} = \left((\omega_r/\omega_0)^2 - 1\right)L_b \tag{10.1}$$

and

$$C_b = 1/\omega_r^2 L_b \tag{10.2}$$

where ω_r is the center frequency of the harmonic trap and ω_0 is the center frequency of the passband. If, as is common, the harmonic trap is placed close to the second-harmonic frequency (i.e., $\omega_r \approx 2\omega_0$), then

$$L_{\text{eff}} \approx 3L_b \tag{10.3}$$

For analysis, the combined admittance of the network is determined as

$$Y = \frac{1}{R_{L\text{eff}} + j\omega L_{\text{eff}}} + \frac{1}{R_{Lb} + R_C + j\omega L_b - \dfrac{j}{\omega C_b}} \tag{10.4}$$

where R_{element} = resistance of associated element.

To compare and contrast the performance characteristics of this simple single-trap architecture, with those that follow, we use the following basic parameters for all the data:

$$\begin{array}{ll} f_{\text{lo}} = 1{,}710 \text{ MHz} & f_{\text{hi}} = 2{,}025 \text{ MHz} \\ f2_{\text{lo}} = 3{,}420 \text{ MHz} & f2_{\text{hi}} = 4{,}050 \text{ MHz} \\ Z_S = 5\Omega & Z_L = 5\Omega \\ Q_L = 40, \text{ for all inductors} & Q_C \approx 100, \text{ for all capacitors} \end{array} \tag{10.5}$$

In the analyses, L_{eff} is assumed to be a single inductor, although in practice it would be a combination of the bias inductor L_b with a susceptive inductance generated by the matching network. In addition, it is assumed that L_{eff} shunts the signal line directly to ground, thereby neglecting the series reactance and resistance of the bypass capacitor C_b. These additional complexities are avoided here in an effort to more clearly illustrate the basic differences between the architectures being presented.

The data shown in Figure 10.2 corresponds to a circuit with L_b set to the minimum value consistent with a maximum in-band insertion loss ≤0.1 dB. For meaningful comparisons to subsequent architectures, this same insertion loss target will be used.

The reflection coefficient Γ_{in} in the passband is very tight, consistent with the low insertion loss requirement. However, the reflection coefficient across the second-harmonic band shows much more variation, with significant phase dispersion. The

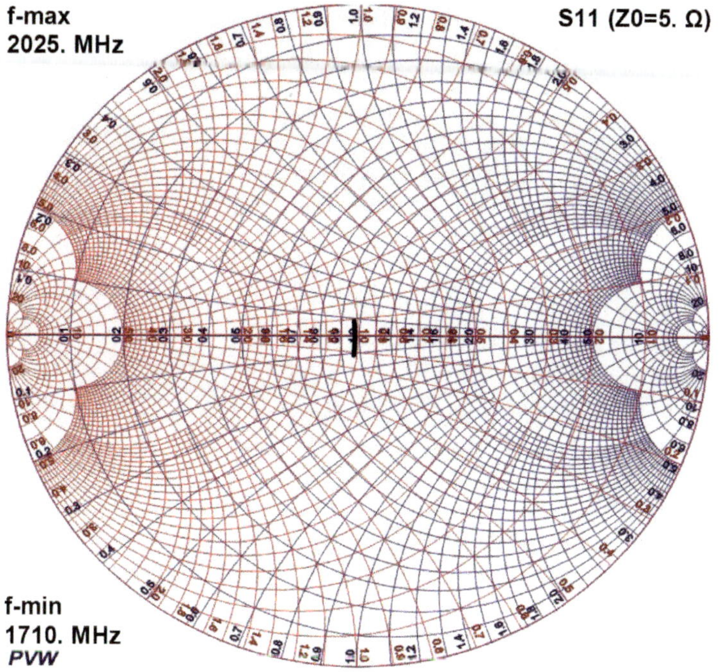

(a)

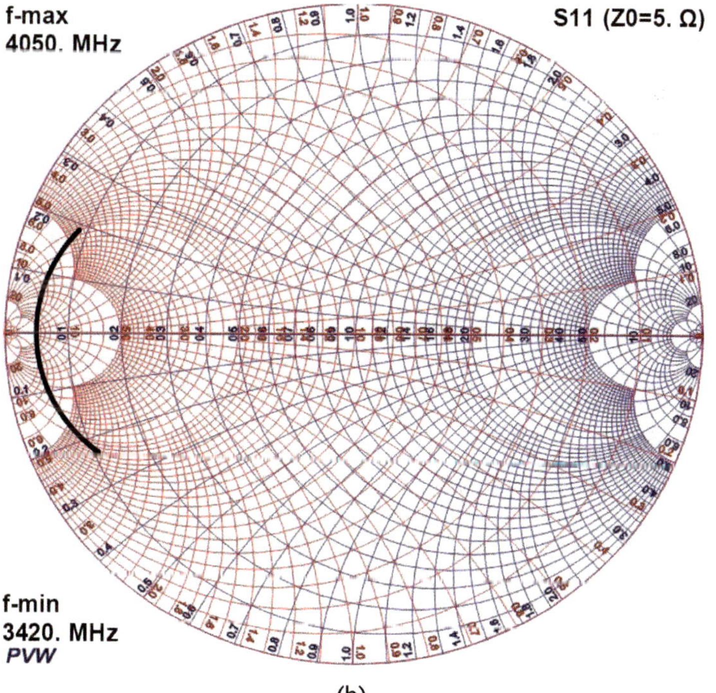

(b)

Figure 10.2 Single-trap characteristics: (a) Γ_{in} across passband, (b) Γ_{in} across second-harmonic band, (c) passband insertion loss, and (d) second-harmonic insertion loss.

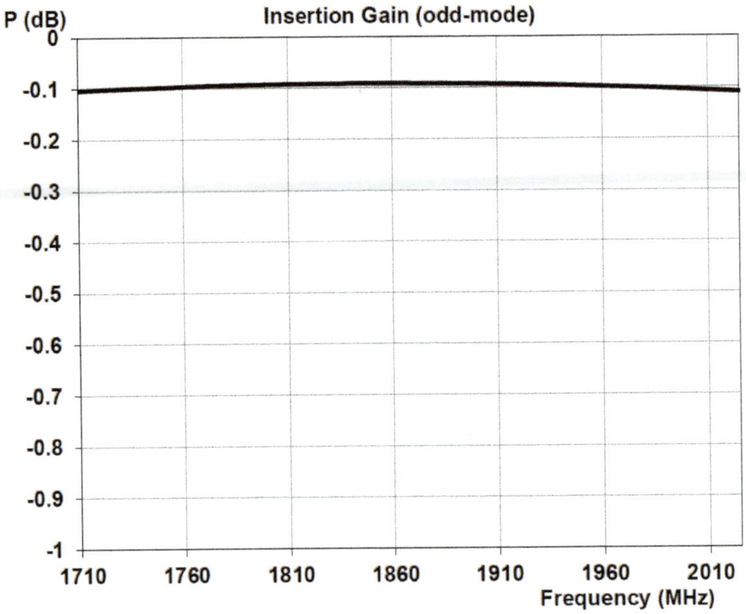

(c)

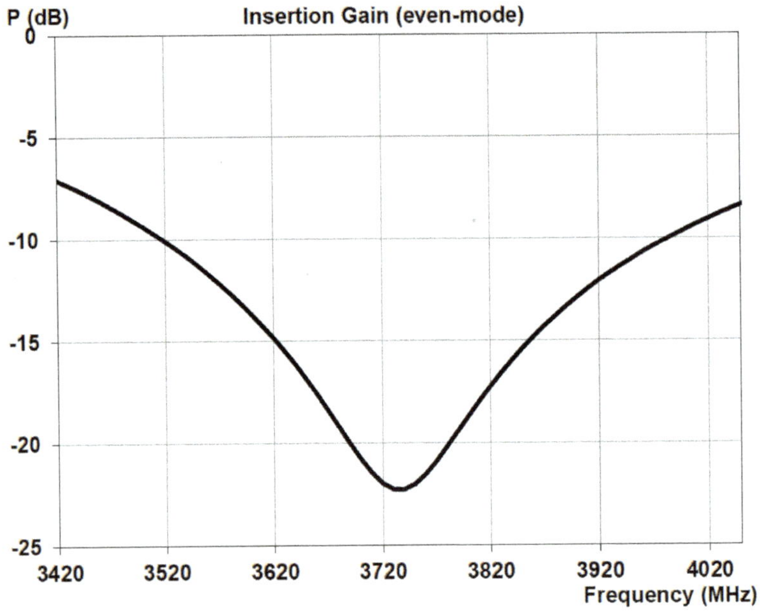

(d)

Figure 10.2 *(Continued)*

second-harmonic rejection also varies greatly across the band, with as low as 7-dB rejection at the low end of the band.

As discussed in Chapter 9, the second-harmonic response characteristics can be tightened by lowering the value of L_b; however, this will result in an increase in the in-band insertion loss. In an effort to increase the minimum rejection and lower the impedance variation in the harmonic band, without degrading the in-band characteristics, designers have considered numerous circuit approaches.

One such circuit approach that is a simple extension of the single trap is to add a second trap in the circuit. This approach is examined in the next section.

10.2 PA Shunting Inductance and Dual Harmonic Traps

Figure 10.3 shows a schematic of single shunt inductance L_{eff} in parallel with two series *LC* traps. Again, the single inductance L_{eff} represents the combined inductive susceptance of the bias inductor and any residual inductance of the matching network at the PA. The inductance L_{eff} is required to negate the unwanted capacitive susceptance of the harmonic networks in the passband.

The object of adding the additional trap is to broaden the second-harmonic characteristics without increasing the passband insertion loss. We shall now examine whether this technique can really be effective.

First, because the goal is to widen the harmonic passband and keep it flat, we require the relative bandwidth of the two series resonators to be equal. For this the resonant quality, factors must be equal, that is,

$$\frac{L_{h1}}{C_{h1}} = \frac{L_{h2}}{C_{h2}} \tag{10.6}$$

If lower trap frequency = f_{h1} and upper trap frequency = f_{h2}, then

$$C_{h1} = \frac{1}{\omega_{h1}^2 L_{h1}} \quad \text{and} \quad C_{h2} = \frac{1}{\omega_{h2}^2 L_{h2}} \tag{10.7}$$

and

$$L_{h2} = (\omega_{h1}/\omega_{h2})\, L_{h1} \tag{10.8}$$

Second, the central frequency of the combined trap responses should remain the center of the harmonic band. Thus, the two trap frequencies should be equally disposed about the center of the band.

Defining

$$f_{h1} = f_r - \alpha \cdot \Delta f_r/2 \tag{10.9}$$

$$f_{h2} = f_r + \alpha \cdot \Delta f_r/2 \tag{10.10}$$

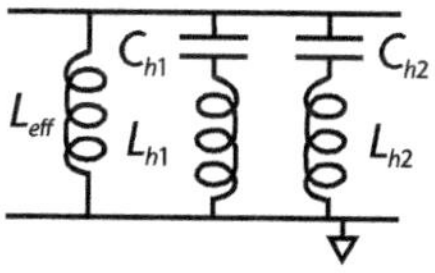

Figure 10.3 PA shunting inductance and dual harmonic traps.

where Δf_r = harmonic bandwidth, and α characterizes the separation of the trap frequencies.

For $\alpha = 0$,

$$f_{h1} = f_{h2} = f_r \tag{10.11}$$

For $\alpha = 1$,

$$f_{h1} = \text{lower end of harmonic band} \tag{10.12}$$

and

$$f_{h2} = \text{upper end of harmonic band} \tag{10.13}$$

For in-band susceptance cancelation, we require

$$L_{\text{eff}} = \frac{1}{\dfrac{\omega_0^2}{\left(\omega_{h1}^2 - \omega_0^2\right) L_{h1}} + \dfrac{\omega_0^2}{\left(\omega_{h2}^2 - \omega_0^2\right) L_{h2}}} \tag{10.14}$$

For analysis, the combined admittance of the network is determined as

$$Y = \frac{1}{R_{L\text{eff}} + j\omega L_{\text{eff}}} + \frac{1}{R_{Lh1} + R_{Ch1} + j\omega L_{h1} - \dfrac{j}{\omega C_{h1}}} + \frac{1}{R_{Lh2} + R_{Ch2} + j\omega L_{h2} - \dfrac{j}{\omega C_{h2}}} \tag{10.15}$$

where R_{element} = resistance of associated element.

For a valid comparison of the characteristics of the single and dual-trap architectures, in the dual-trap architecture, inductor L_b was set, as in the previous section, to ensure a minimum value consistent with a maximum in-band insertion loss of ≤0.1 dB. Key electrical response data for both the single and dual-trap architectures is shown in Figure 10.4. The frequency separation of the dual traps was consistent with $\alpha = 0.3$.

The design element values and parameters used for the data are provided in Table 10.1.

Comparing the passband characteristics of the single and dual-trap architectures, they are seen to be nearly identical. However, the characteristics of the second-harmonic band are significantly different. The goal of adding the second trap was to widen the second-harmonic response, in particular, to decrease the impedance, reduce its variation, and increase the rejection. None of these goals have been achieved.

The angular dispersion of the reflection coefficient across second-harmonic band of the single and dual traps is identical. Also, the impedance is not noticeably increased anywhere in the band. More troubling still is the rejection characteristics achieved with the two traps. Instead of a single maximum rejection point at center

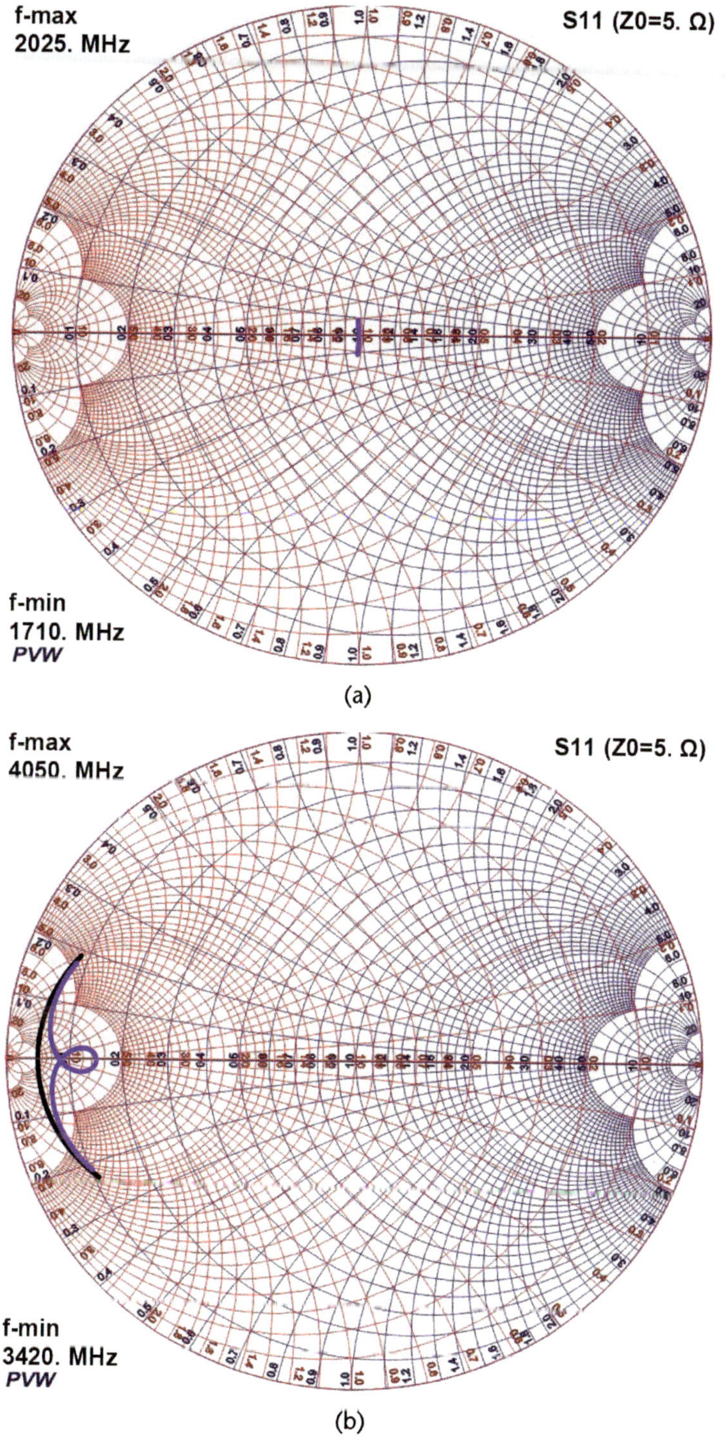

Figure 10.4 Single and dual-trap ($\alpha = 0.3$) characteristics: (a) Γ_{in} across passband, (b) Γ_{in} across second-harmonic band, (c) passband insertion losses, and (d) second-harmonic insertion losses.

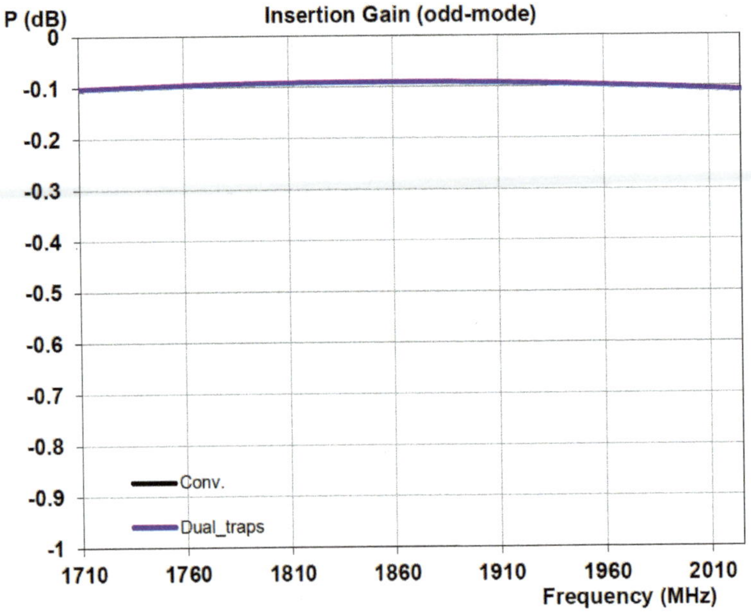

(c)

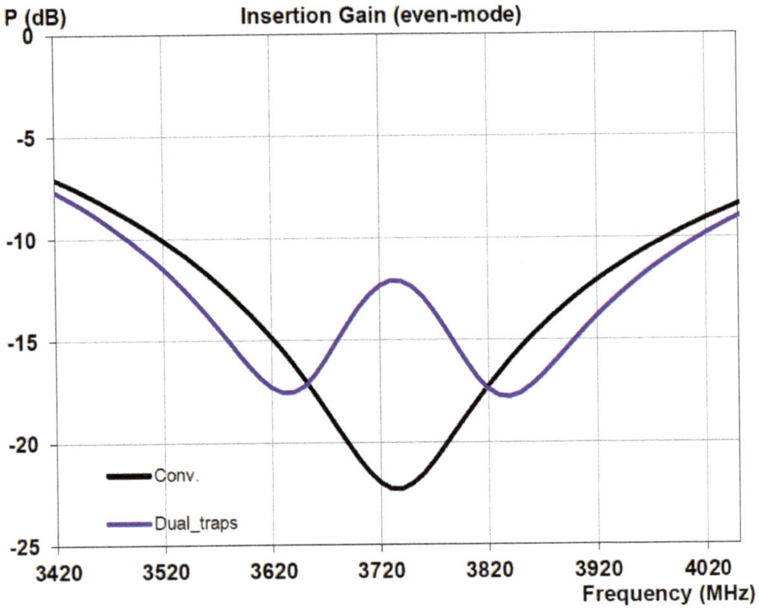

(d)

Figure 10.4 *(Continued)*

Table 10.1 Dual Trap ($\alpha = 0.3$, $k = 0$) Elements

Element Values	
$\alpha = 0.3$	$k = 0$
$Q_L = 40$	$Q_C = 100$
$L_{eff} = 0.905$ nH	
$L_{h1} = 0.62$ nH	$C_{h1} = 3.083$ pF
$L_{h2} = 0.589$ nH	$C_{h2} = 2.931$ pF

band, we have two maximums equally disposed about the center. At the center of the band, the rejection response actually has a local minimum value, corresponding to a significant increase in the impedance of the combined traps. In actuality, if ideal elements had been assumed for the analysis, at the center of the harmonic band, the traps would have presented no impedance loading and would have achieved zero rejection.

In Figure 10.4(d), it can also be noted that rejection at the band edges is little changed. Thus, in addition to the other deficiencies, adding a second trap does not even serve to widen the harmonic rejection response. Based on this data, therefore, in the majority of cases, there does not appear to be anything to be gained by replacing a single harmonic trap with two in parallel. However, there is one limited exception. If a PA is required only to operate at two ends of a frequency band, and not in the middle of the band, then the two-trap architecture may be beneficial. If the two maximum rejection frequencies, as in Figure 10.4(d), are centered on these two separated bands, then the rejection can be improved across both compared to employing just a single trap.

The reason that there is a pole (zero loss with ideal components) at the center of the band is because the two traps are in parallel with no intermediate series elements. All practical filter architectures employ both series and shunt elements. If there were, for example, a parallel LC resonator in the line between the two traps, this could be used to decouple the shunt traps at the center frequency. However, such an arrangement would not meet the other circuit goals for harmonic impedance and in-band loss.

In the dual-trap architecture, at the center frequency of the harmonic band, resonator 1 is above resonance, while resonator 2 is below resonance. This means that resonator 1 presents a negative (i.e., inductive) susceptance on the signal line, while resonator 2 presents a positive (i.e., capacitive) susceptance. These two mutually cancel, resulting in the response pole for the network.

In addition to the poor performance of the dual-trap architecture, save for one limiting case, there is yet another drawback to its practical implementation. Both traps need to be physically located close to the output of the PA. This requires that the two inductors in the traps also be located in close proximity to each other on the circuit board, which can result in mutual coupling between the two. Mutual coupling is also an aspect that might be considered to improve the performance of the two-trap approach. The effects of mutual coupling between the trap inductors are considered in the next section.

10.3 PA Shunting Inductance and Dual Coupled Harmonic Traps

Figure 10.5 shows a schematic of single shunt inductance L_{eff} in parallel with two series LC traps. This circuit is identical with that in the previous section with the exception that finite coupling is assumed between the two trap inductors. Such coupling may be intentional or parasitic. In practice, in a compact RF module design, if two traps were employed, the two inductors would inevitably have to be in close proximity, and this would inevitably lead to some coupling. It is instructive to examine the effects of such coupling on the circuit and whether it might be useful to improve the characteristics of the circuit.

The coupling constant k is defined in terms of mutual inductance M, by

$$M = k\sqrt{L_1 L_2} \tag{10.16}$$

where $0 \le k \le 1$.

As in the previous section, the single inductance L_{eff} represents the combined inductive susceptance of the bias inductor and any residual inductance of the matching network at the PA. The inductance L_{eff} is required to negate the unwanted capacitive susceptance of the harmonic networks in the passband.

For a flat harmonic passband, we again require the quality factors of the two series resonators to be equal, that is,

$$\frac{L_{h1}}{C_{h1}} = \frac{L_{h2}}{C_{h2}} \tag{10.17}$$

Also, the two trap frequencies are taken to be equally disposed about the center of the band. That is,

$$f_{h1} = f_r - \alpha \cdot \Delta f_r / 2 \tag{10.18}$$

$$f_{h2} = f_r + \alpha \cdot \Delta f_r / 2 \tag{10.19}$$

where Δf_r = harmonic bandwidth and α characterizes the separation of the trap frequencies.

For $\alpha = 0$,

$$f_{h1} = f_{h2} = f_r \tag{10.20}$$

For $\alpha = 1$,

$$f_{h1} = \text{lower end of harmonic band} \tag{10.21}$$

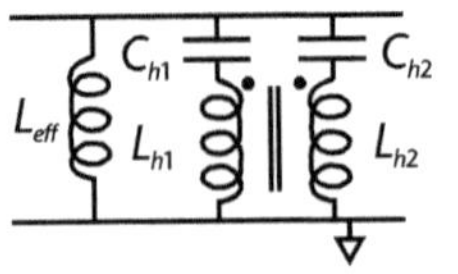

Figure 10.5 PA shunting inductance and dual-coupled harmonic traps.

and

$$f_{h2} = \text{ upper end of harmonic band} \tag{10.22}$$

From circuit analysis, seeking solutions for C_{h1} and C_{h2} consistent with the above conditions, we find that solutions only exist providing

$$|k| \leq \frac{2(\Delta\omega_h/\omega_r)}{1+(\Delta\omega_h/\omega_r)^2} \tag{10.23}$$

where

$$\Delta\omega_h = \alpha \cdot \Delta\omega_r/2 \tag{10.24}$$

Consistent with the previous section, the following basic parameters are to be used in the analysis

$$\begin{array}{ll} f_{\text{lo}} = 1{,}710 \text{ MHz} & f_{\text{hi}} = 2{,}025 \text{ MHz} \\ f2_{\text{lo}} = 3{,}420 \text{ MHz} & f2_{\text{hi}} = 4{,}050 \text{ MHz} \\ Z_S = 5\Omega & Z_L = 5\Omega \\ Q_L = 40, \text{ for all inductors} & Q_C \approx 100, \text{ for all capacitors} \\ \alpha = 0.3 & \end{array} \tag{10.25}$$

In this case, from (10.23), we find the limiting value for the mutual coupling coefficient is $|k| \leq 0.05$. This extremely small value indicates that, in the presence of finite mutual coupling between the inductors, it is not possible to achieve two equal and distinct minima in the insertion loss through the network.

Given that network solutions, with the above constraints, do not exist for reasonable values of mutual coupling, we use the same element solutions as in the previous section for $|k| = 0$, that is,

$$L_{h2} = (\omega_{h1}/\omega_{h2})L_{h1} \tag{10.26}$$

$$C_{h1} = \frac{1}{\omega_{h1}^2 L_{h1}} \tag{10.27}$$

$$C_{h2} = \frac{1}{\omega_{h2}^2 L_{h2}} \tag{10.28}$$

from uncoupled analysis

$$L_{\text{eff}} = \frac{1}{\dfrac{\omega_0^2}{\left(\omega_{h1}^2 - \omega_0^2\right)L_{h1}} + \dfrac{\omega_0^2}{\left(\omega_{h2}^2 - \omega_0^2\right)L_{h2}}} \tag{10.29}$$

For analysis, the admittance of the first trap is

$$Y_1 = \frac{R_{Lb2} + R_{Cb2} + j(\omega L_{b2} - 1/\omega C_{b2}) - j\omega k\sqrt{L_{b1}L_{b2}}}{(R_{Lb1} + R_{Cb1} + j(\omega L_{b1} - 1/\omega C_{b1}))(R_{Lb2} + R_{Cb2} + j(\omega L_{b2} - 1/\omega C_{b2})) + \omega^2 k^2 L_{b1}L_{b2}} \tag{10.30}$$

while the admittance of the second trap is

$$Y_2 = \frac{R_{Lb1} + R_{Cb1} + j(\omega L_{b1} - 1/\omega C_{b1}) - j\omega k\sqrt{L_{b1}L_{b2}}}{R_{Lb2} + R_{Cb2} + j(\omega L_{b2} - 1/\omega C_{b2}) - j\omega k\sqrt{L_{b1}L_{b2}}} Y_1 \tag{10.31}$$

where R_{element} = resistance of associated element. The combined admittance of the network is then

$$Y = \frac{1}{R_{\text{Leff}} + j\omega L_{\text{eff}}} + Y_1 + Y_2 \tag{10.32}$$

Figure 10.6 shows the key response characteristics of the dual coupled traps together with that of the conventional single trap of Section 10.1. Coupled trap responses are shown for $k = -0.05$ and $k = 0.05$. In both cases, the response characteristics are degraded from that of the single trap. While the passband responses are little changed, the second-harmonic impedances and rejection characteristics are much inferior. The maximum rejection harmonic frequency is shifted up or down in frequency, depending on the sign of the mutual coupling. The harmonic characteristics are also highly asymmetric, with one rejection minimum much reduced from the other. Increasing the magnitude of the coupling coefficient beyond 0.05, already very small, degrades the characteristics further.

The data in Figure 10.6 clearly illustrates one further disadvantage of trying to employ dual traps to widen the amplifier's harmonic characteristics without incurring the usual in-band insertion loss penalty. Namely, any coupling between the two trap inductors is extremely deleterious to the performance of the network and thus great efforts must be made to minimize it on the circuit board.

To summarize, dual *LC*-series harmonic traps in parallel on the output of a PA are ineffective in widening the second-harmonic rejection bandwidth. Any inductive coupling between the trap inductors further degrades the performance. Only in the restrictive case that a PA need operate on two close, but separate bands, could it be beneficial to employ the two-trap configuration.

10.4 Differential PA Shunting Inductances and Harmonic Traps

An amplifier or PA with a cascaded chain of amplification along a single signal line is referred to as "single-ended." In addition to single-ended architectures, there are many alternative PA circuit architectures employing multiple PA chains whose outputs are combined in a phasing network before the power is delivered to the load. There may be two or more amplifier chains in such PA architectures, the

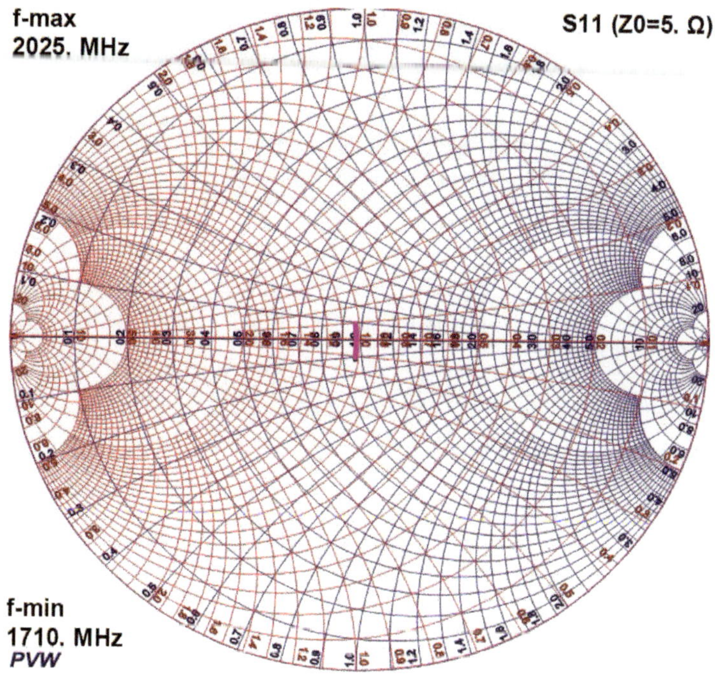

(a)

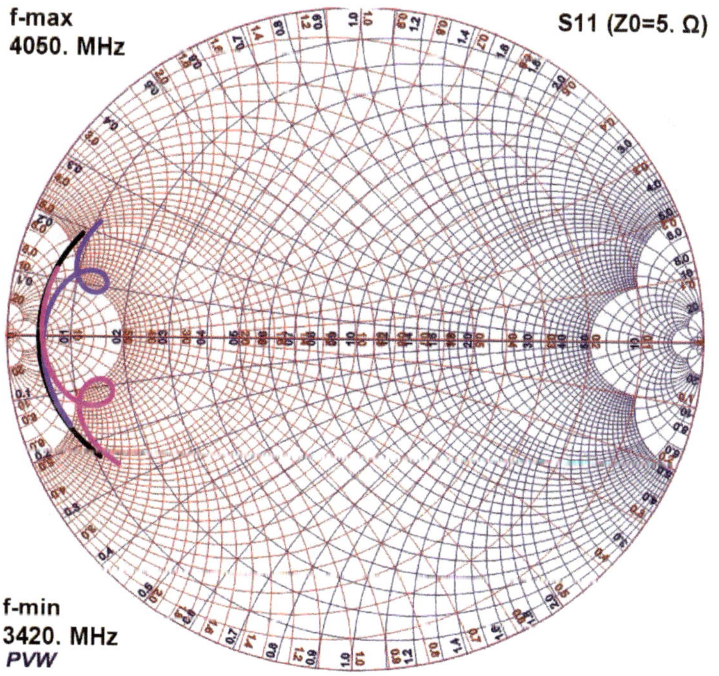

(b)

Figure 10.6 Single and dual coupled traps ($\alpha = 0.3$, $|k| = 0.05$) characteristics: (a) Γ_{in} across passband, (b) Γ_{in} across second-harmonic band, (c) passband insertion losses, and (d) second-harmonic insertion losses.

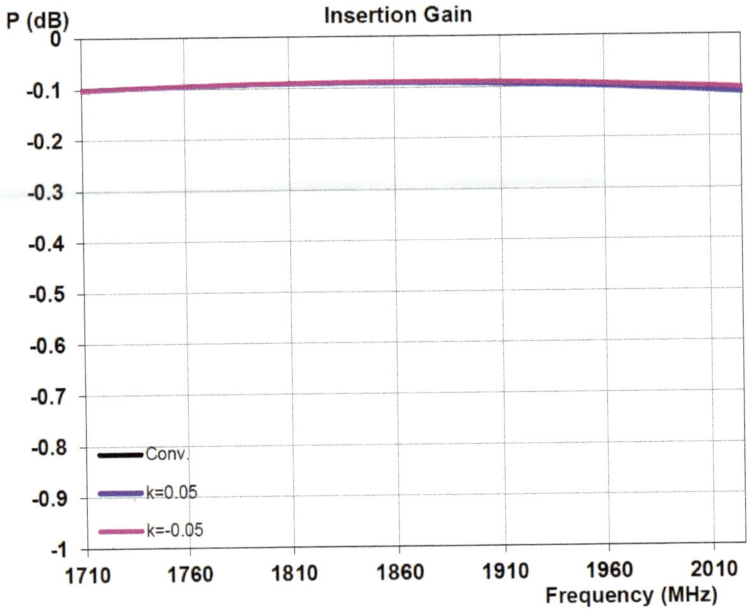

(c)

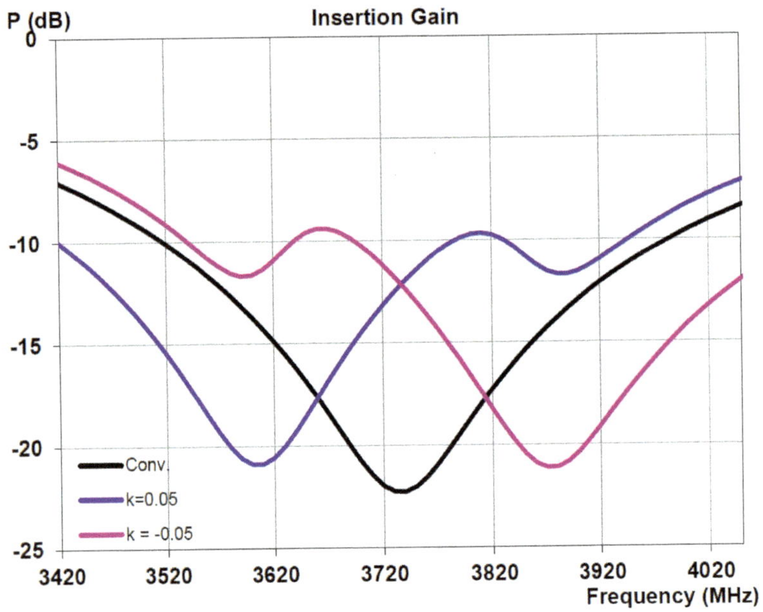

(d)

Figure 10.6 *(Continued)*

particular embodiment chosen depending on its merits for the application needed. Among the alternative PA architectures to the single-ended, the differential PA architecture employing two amplifier chains finds most widespread application. In this architecture, the desired signals on the two signal lines are of opposite phase. An example of a simplified schematic of a differential PA is shown in Figure 9.2.

Each of the signal lines in a differential PA can be regarded as two separate single-ended amplifiers up to the point where the signals are combined. Thus, all the discussions in the previous sections of this chapter apply equally to the individual signal paths in a differential PA. With this in mind, Figure 10.7 shows the two signal paths in a differential PA with the effective shunting inductance L_{eff} (a combination of the bias inductors with the matching networks in each chain) plus the LC-series traps to provide a low impedance and rejection across the second-harmonic bandwidth.

Since each of the amplifier chains behaves essentially as an independent single-ended PA chain, the passband and harmonic characteristics would be expected to be near identical. Such is the case.

Figure 10.8 shows the fundamental and second-harmonic characteristics of the differential architecture of Figure 10.7, together with the responses of the conventional single-ended architecture of Figure 10.1. For equivalency, the harmonic trap inductor L_h has again been set to maintain a maximum in-band insertion loss of 0.1 dB. In addition, to maintain power equivalency, the line impedance for the differential architecture is 10Ω, compared to the 5Ω impedance for the single-ended data.

For the differential configuration, the signal energy on the two lines is of equal magnitude but inverted in phase. This is represented in Figure 10.7 by the (+, –) drive notation, commonly referred to as odd-mode. However, the second-harmonic components on the signal lines, generated by nonlinearities in the two PAs, will be in-phase, as will all even harmonic components. In Figure 10.7, the even harmonic component phase relationships are represented by the (+, +) drive notation, commonly termed even-mode. In contrast, all generated odd harmonic components, like the fundamental, will be in anti-phase.

In Figure 10.8, for the differential architecture, the fundamental signal data is shown for odd-mode drive, while that for the second harmonic is shown for an even-mode drive. For the circuit of Figure 10.7, because the two paths are essentially independent, the even and odd-mode responses in all bands are identical. However, this is not the case if there is any coupling between the two signal paths, as is considered in the subsequent sections.

As expected, the fundamental and second-harmonic characteristics of the differential architecture are identical to those of the conventional single-ended, single LC resonator architecture. This confirms, as anticipated, that simply transitioning from a single-ended architecture to an equivalent differential architecture brings no benefit to the fundamental trade-off between the second-harmonic stop bandwidth and the passband insertion loss. Because the two differential amplifier chains can be viewed as two independent amplifiers chains, no change in fundamental characteristics could be expected.

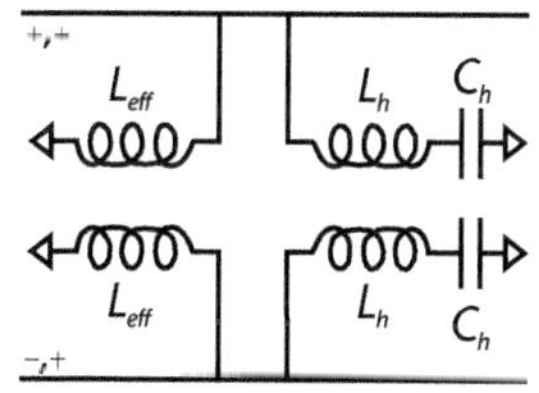

Figure 10.7 Differential PA shunting inductances and harmonic traps.

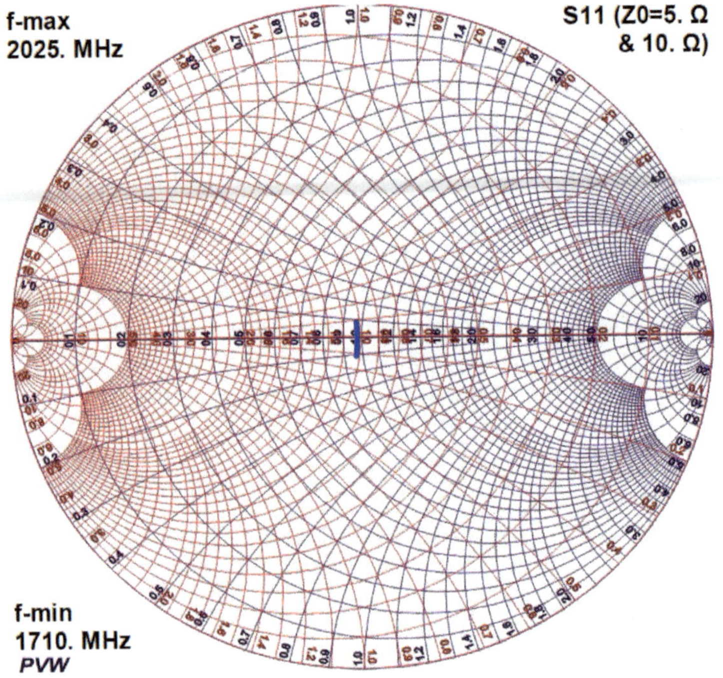

(a)

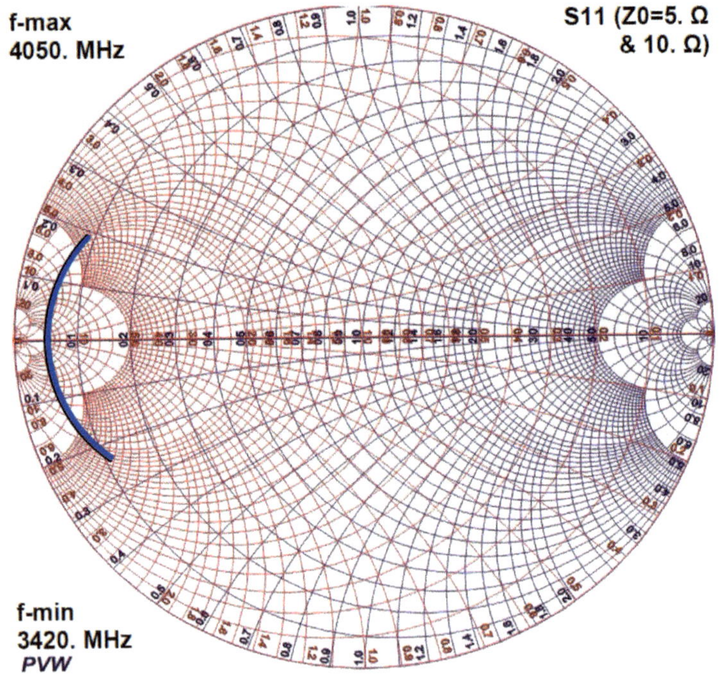

(b)

Figure 10.8 Single-ended and differential characteristics: (a) Γ_{in} across passband, (b) Γ_{in} across second-harmonic band, (c) passband insertion losses, and (d) second-harmonic insertion losses.

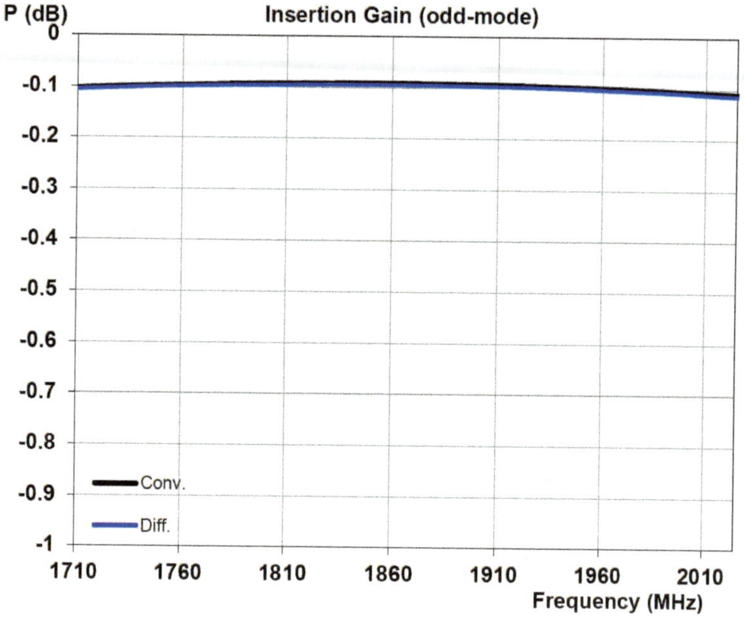

(c)

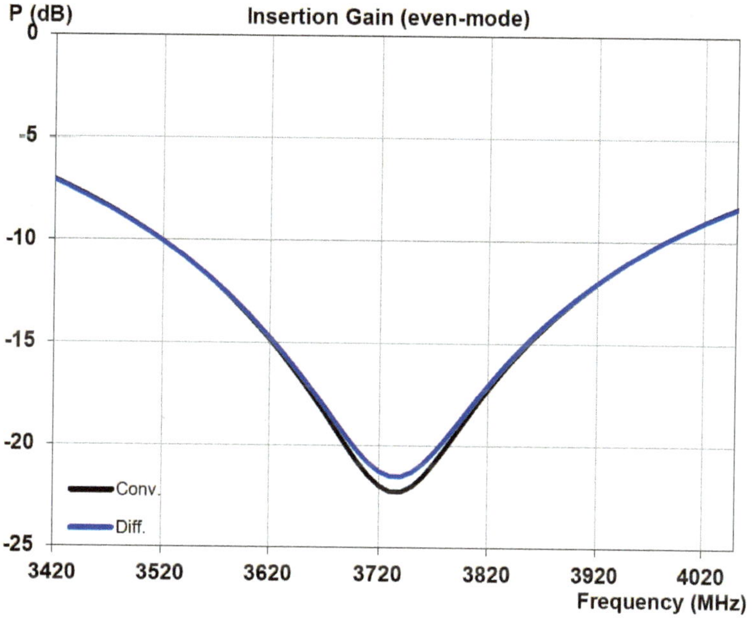

(d)

Figure 10.8 *(Continued)*

10.5 Differential PA Shunting Inductances and Coupled Harmonic Traps

For the single-ended PA architecture with dual coupled harmonic traps in Section 10.3, the coupling was shown to degrade the overall characteristics of the circuit. In that circuit, the two traps were at different frequencies and in parallel with each other on a single signal line. However, in the differential circuit of Figure 10.7, the two harmonic traps are separate signal lines and are centered at the same frequency. Thus, inductively coupling the two harmonic traps in the differential circuit may have benefits that did not occur in the single-ended case.

Figure 10.9 shows a similar differential architecture to that in the previous section, but with finite coupling assumed between the two trap inductors. In this circuit, the coupling constant k_h is defined in terms of mutual inductance M_h, by

$$M_h = k_h L_h \tag{10.33}$$

where $0 \le k_h \le 1$.

There are three potential benefits sought from seeking to couple the harmonic inductors.

1. Decreasing the effective susceptance of L_h in the passband.
2. Decreasing the physical value required for the harmonic inductors L_h.
3. Decreasing the Q of the harmonic traps.

Why should inductive coupling be a benefit for the differential circuit, when it degraded the performance of the single-ended dual resonator configuration? The key is the relative phase difference between the signals in the passband and harmonic bands in the differential circuit.

The even harmonic on the signal lines are products of nonlinear mixing within the PAs. Because they are products of even-ordering mixing, they are necessarily in phase on the two lines. In contrast, since all odd harmonic products are products of odd-ordering mixing, they are out-of-phase as are the incoming fundamental signals.

Because the second-harmonic energy on the two paths in Figure 10.9 are in phase, and of equal amplitude, currents flowing through the inductors L_h will likewise be in phase and serve to increase the net magnetic flux in the inductors. In consequence, the value of the inductors L_h, required to achieve a given effective inductance value, can be decreased. This reduction in the physical value of inductors L_h, by itself, would result in a reduced loading susceptance on the lines in the passband, resulting in a reduced insertion loss. However, there is yet another advantage to be gained from the coupling. In the passband, the signal energies are

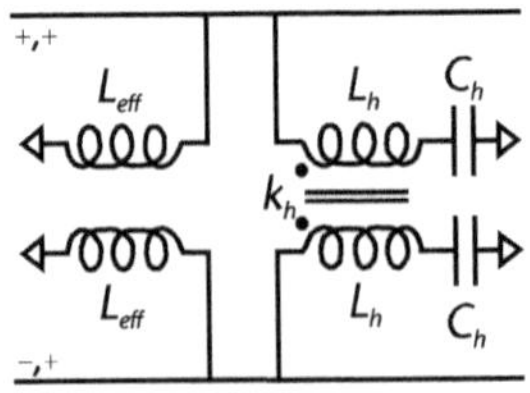

Figure 10.9 Differential PA shunting inductances and coupled harmonic traps.

in anti-phase. Thus, in contrast to the even-harmonic bands, the currents flowing in the two harmonic inductors L_h are also in anti-phase. This effectively reduces the susceptive loading of the harmonic traps still further.

Coupling between the two harmonic inductors L_h, in the differential case, is thus advantageous in breaking the unavoidable trade-off required between the second-harmonic characteristics and passband insertion loss in the single-ended architectures. A physical reduction can be achieved in the size of the harmonic inductors L_h required, allowing for a more compact design. In addition, because the effective susceptive loading of the inductors in the passband is reduced, the Q of the harmonic traps can be reduced leading to improved harmonic characteristics (i.e., less dispersion and higher rejection). These benefits are evident in the data in Figure 10.10.

Figure 10.10 shows the passband and second-harmonic characteristics of the differential architecture of Figure 10.9 without, and with, coupling between the harmonic inductors L_h. The characteristics of the differential architecture with no L_h coupling are identical to those for the single-ended architecture as was demonstrated in the previous section. For near-equivalent pass characteristics, the second-harmonic characteristics with coupling are notably improved. There is both reduced dispersion and increased rejection evident across the full harmonic bandwidth. For the data in Figure 10.10, a coupling coefficient $k_h = 0.7$ was assumed, which is easily achievable on a typical RF module laminate board.

A very significant reduction in the value of the harmonic inductors is also obtained by introducing the coupling as can be seen from the data in Table 10.2. There is an accompanying modest reduction in the effective bias inductance required L_{eff}.

10.6 Differential PA Shunting Inductances and Coupled Bias and Harmonic Traps

In the previous section, significant advantages were demonstrated for inductively coupling the two harmonic trap inductors. Additional advantages may be obtained for the differential circuit if the bias inductors L_{eff} are also inductively coupled, as shown in Figure 10.11.

In this circuit in Figure 10.11, the coupling constant k_b, between inductors L_{eff}, is defined in terms of mutual inductance M_b, by

$$M_b = k_b L_{eff} \tag{10.34}$$

where $0 < k_b \leq 1$.

Observe that opposite ends of the bias inductors L_{eff} are connected to their respective signal lines, in contrast to the connection of the harmonic inductors. For analysis, this is accounted for by a negative value for the coupling coefficient k_b. As a result, the currents are flowing in the same direction through the inductors L_{eff} in the passband but in opposite directions in the even-harmonic bands.

The motivations for inductively coupling the bias inductors are similar to those for coupling the harmonic inductors. In the passband, the magnetic flux in the bias inductors is enhanced, thereby increasing the effective inductance. This allows the values of the bias inductors required in the circuit to be reduced, thus helping to

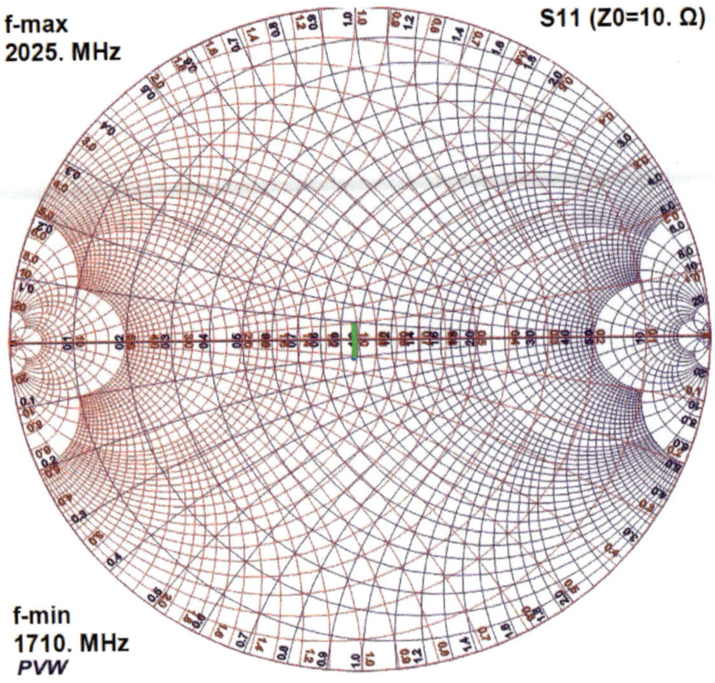

(a)

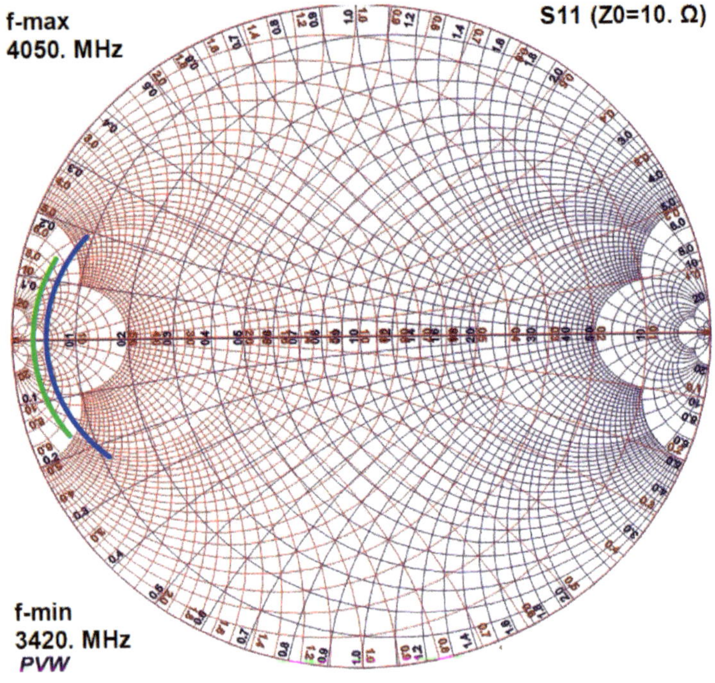

(b)

Figure 10.10 Differential characteristics, without and with harmonic coupling ($k_h = 0.7$): (a) Γ_{in} across passband, (b) Γ_{in} across second-harmonic band, (c) passband insertion loss, and (d) second-harmonic insertion losses.

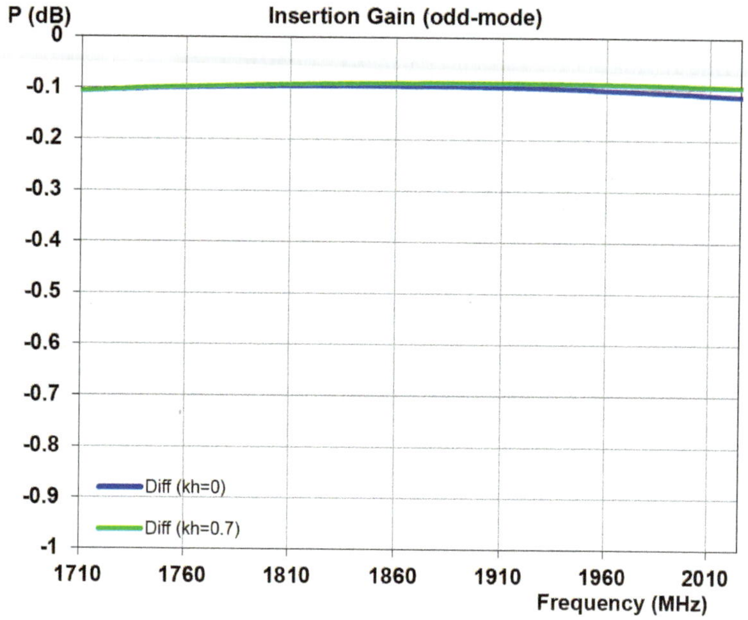

(c)

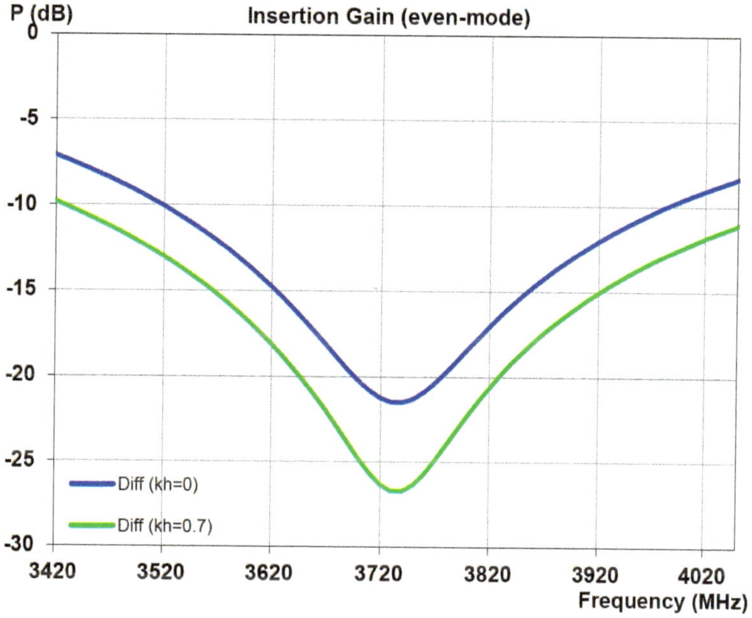

(d)

Figure 10.10 *(Continued)*

Table 10.2 Differential Coupling and Inductor Values

Coupling	L_{eff}	L_h
None	1.8 nH	0.6 n H
$k_h = 0.7$	1.56 nH	0.24 nH

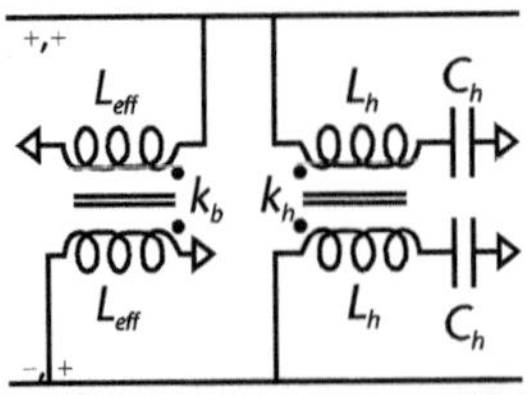

Figure 10.11 Differential PA with coupled bias and harmonic traps.

reduce the size of the module. In the even harmonic bands, because the currents flow in opposite directions through the bias inductors, their susceptive loading in these bands is reduced, which helps to further improve the even-harmonic characteristics.

Figure 10.12 shows the key characteristics of the differential circuit of Figure 10.11 with coupling coefficients $k_h = 0.7$ and $k_b = -0.7$. For reference, the responses of the same circuit with no coupling and $k_h = 0.7$, $k_b = 0$ are also shown.

In the figures, the passband characteristics are all essentially equivalent. In each case, the values of the inductors L_h were set to the minimum value consistent with a maximum in-band insertion loss of 0.1 dB. However, passband characteristics differ significantly. The harmonic response is improved by coupling only the harmonic inductors, as in the previous section. It is improved yet again by additionally coupling the bias inductors. The differential harmonic passband with both bias and harmonic inductors coupled, as in Figure 10.11, shows the lowest impedance, least dispersion, and greatest rejection of all the networks. In addition, a further benefit is that this implementation can be realized with the smallest footprint.

Table 10.3 shows the inductor element values for each of the three alternatives. The best performing architecture, with bias and harmonic inductor coupling, has significant inductor values that would thus require less area to implement.

The magnitude of the mutual coupling coefficient value of 0.7 used in the analysis is relatively arbitrary. The response characteristics of the circuit are not critically dependent on this value but steadily improve for higher coupling. The value of 0.7 is simply one that usually easy to achieve in practice.

For analysis, the admittance on the signal lines in odd mode (i.e., fundamental and odd harmonics) is

$$Y_O = \frac{1}{R_{\mathrm{Leff}} + j\omega(1-k_b)L_{\mathrm{eff}}} + \frac{1}{R_{Lh} + R_{Ch} + j\left(\omega(1-k_h)L_h - 1/\omega C_h\right)} \tag{10.35}$$

while the admittance on the signal lines in even mode (i.e., second and even harmonics) is

$$Y_E = \frac{1}{R_{\mathrm{Leff}} + j\omega(1+k_b)L_{\mathrm{eff}}} + \frac{1}{R_{Lh} + R_{Ch} + j\left(\omega(1+k_h)L_h - 1/\omega C_h\right)} \tag{10.36}$$

where R_{element} = resistance of the associated element.

The capacitance required for resonance at the harmonic frequency f_r is

$$C_h = \frac{1}{\omega_r^2(1+k_h)L_h} \tag{10.37}$$

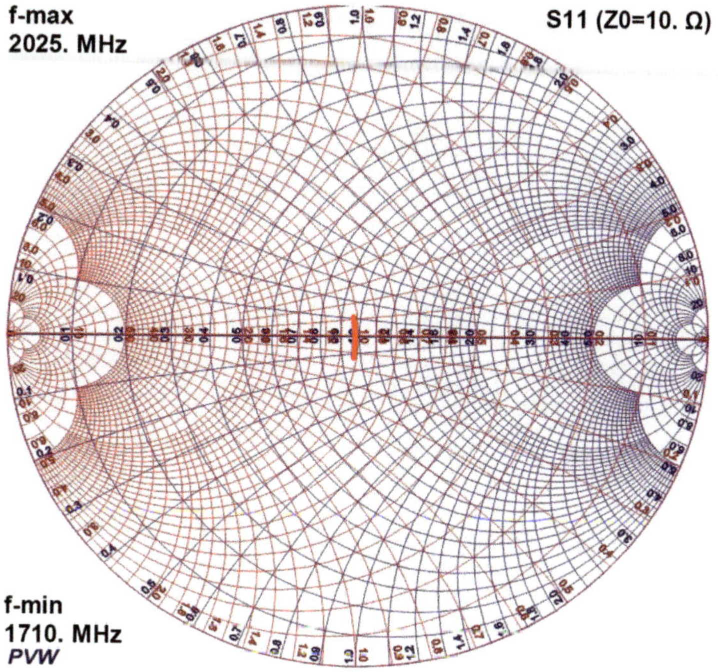

(a)

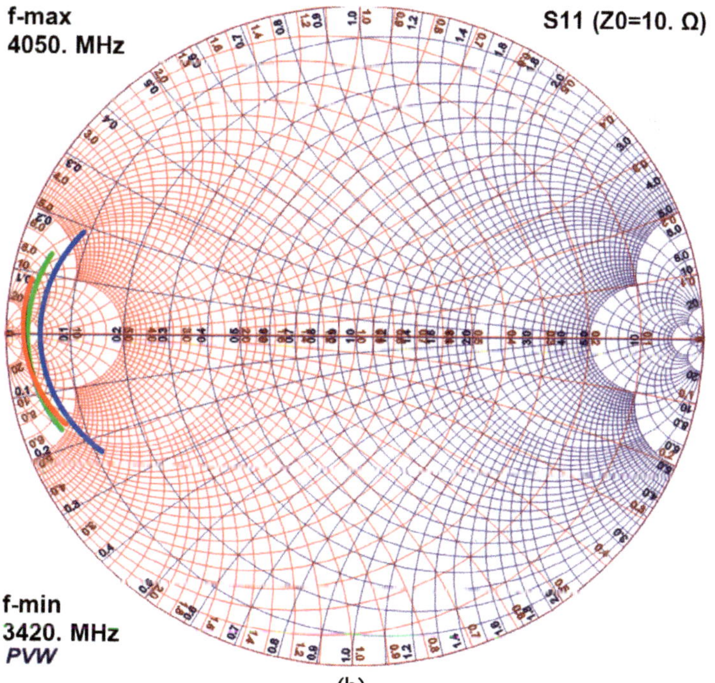

(b)

Figure 10.12 Differential characteristics with bias and harmonic coupling ($|k_h|$=0.7): (a) Γ_{in} across passband, (b) Γ_{in} across second-harmonic band, (c) passband insertion losses, and (d) second-harmonic insertion losses.

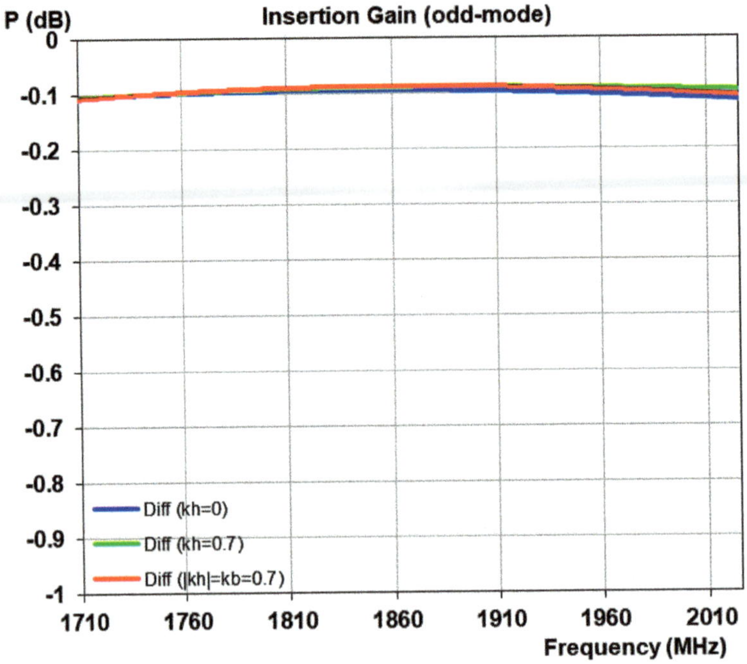

(c)

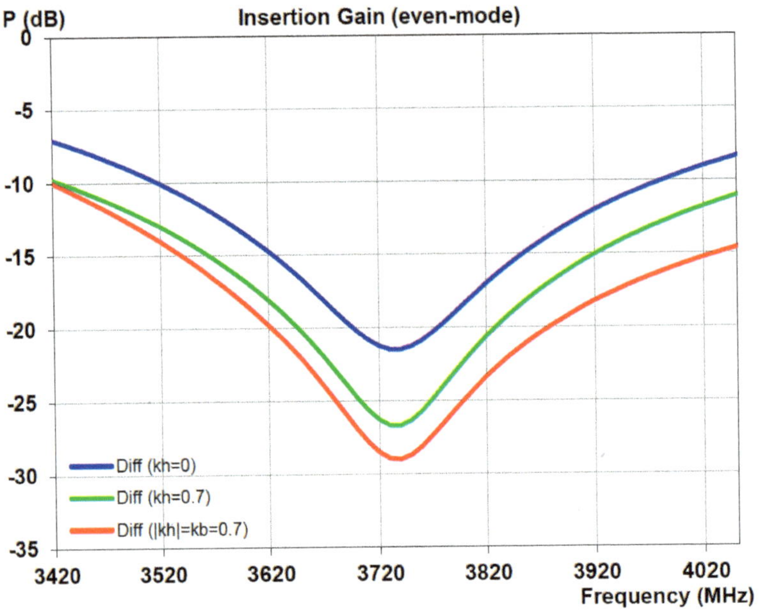

(d)

Figure 10.12 *(Continued)*

and the value required for the effective bias inductance, for the minimal passband loss at f_0, is

$$L_{\text{eff}} = \frac{(\omega_r/\omega_0)^2(1+k_b)-1+k_b}{1-k_b}L_b \tag{10.38}$$

Table 10.3 Differential Coupling and Inductor Values

Coupling	L_{eff}	L_h
None	1.8 nH	0.6 nH
$k_h = 0.7$	1.56 nH	0.24 nH
$k_h = 0.7$, $k_b = -0.7$	0.69 nH	0.18 nH

10.7 All-Pass Bridge-*T* Lowpass Differential Network

The all-pass bridge-*T* lowpass network is a circuit commonly employed in RF circuits for introducing a phase shift into the RF signal path. It is termed an "all-pass network" in light of the fact that, if implemented with ideal circuit elements, it introduces a phase delay into the signal, with no associated insertion loss across all frequency. In the ideal case, the input and output impedances of the network are equal to the characteristic impedance of the signal line and have no frequency dependence. The schematic for the basic all-pass bridge-*T* lowpass network is shown in Figure 10.13(a).

The bridge-*T* lowpass network comprises two equal-valued inductors and two unequal value capacitors. The network is left-right symmetric, consistent with presenting the same impedance on both sides. In a generalization of the network, some magnetic coupling may be assumed between the two inductors, as shown in Figure 10.13(b). This may be inadvertent, the result of unwanted coupling on the circuit board, or deliberate to gain more flexibility in the design.

In still another variant, shown in Figure 10.13(c), the nodal connections to one of the inductors may be reversed, resulting in a negative effective coupling coefficient between the inductors.

For these networks, the element values are determined by

$$\omega L = \frac{\left(\sin\phi + \sqrt{\sin^2\phi + 4(1-\cos\phi)^2(1+k)/(1-k)}\right)R_0}{2(1-\cos\phi)(1+k)} \tag{10.39}$$

$$C_1 = \frac{(1+k)L}{2R_0^2} \tag{10.40}$$

$$C_2 = \frac{2(1-k)L}{R_0^2} \tag{10.41}$$

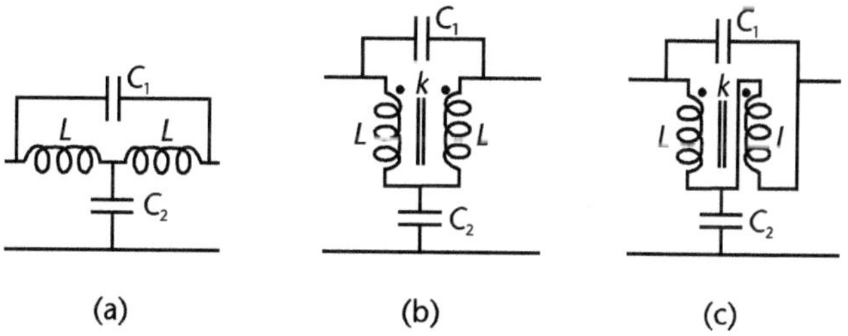

Figure 10.13 All-pass bridge-*T* lowpass networks: (a) conventional, (b) with positive inductor coupling, (c) with negative inductor coupling.

where Φ = desired phase delay and R_0 = characteristic line real impedance.

While the phase shift characteristics of the bridge-T lowpass network are not directly relevant to the biasing and harmonic termination of a differential PA, the circuit architecture itself may be exploited to great advantage in the differential PA architecture.

Figure 10.14(a) shows the schematic of a biasing and harmonic termination network for a differential amplifier that is essentially identical to the bridge-T network of Figure 10.13(c). To understand the functioning of this network, consider its properties under odd-mode and even-mode drives.

For the fundamental and odd-harmonic frequencies, the network is driven by an odd-mode excitation (+, −) on the two signal lines. The circuit has top-to-bottom symmetry, so under the odd-mode drive, the horizontal center line through the circuit is a virtual ground plane. As a result, the equivalent circuit under the odd-mode drive is that shown in Figure 10.14(b). If the parallel resonance of the coupled inductors L_h and capacitor $2C_t$ is set to the center of the passband, the insertion loss will be minimal.

For the second and even-harmonic frequencies, the network is driven by an even-mode excitation (+, +) on the two signal lines. Due to the top-to-bottom symmetry of the circuit, no current will flow through C_t. The equivalent circuit under the even-mode drive is thus that shown in Figure 10.14(c). In this figure, the loading of the bias inductor is shown in gray, as generally L_b can be made large enough to have negligible loading effect on the circuit. Unlike the previous single and differential circuits, in this circuit L_b can be chosen independently of the passband (odd-mode), as it is virtually grounded in the passband.

In Figure 10.14(c), it can be seen if the series resonance of the coupled inductors L_h and capacitors $C_h/2$ is set to coincide with the center of the second-harmonic band, a short-circuit impedance will be realized in the band. The bridge-T bias and harmonic trap architecture is more compact than the more conventional approaches described in the previous sections, as it uses only one bias inductor. It is also much more flexible as the harmonic and passband equivalent circuits are effectively decoupled. The decoupling of the passband and harmonic band characteristics gives this network significant performance advantages.

Figure 10.15 shows the key performance characteristics of the bridge-T network, compared to two of those considered previously. With equal in-band insertion loss for all the networks, the bridge-T circuit has much improved second-harmonic characteristics. The dispersion across the harmonic band is greatly reduced and the

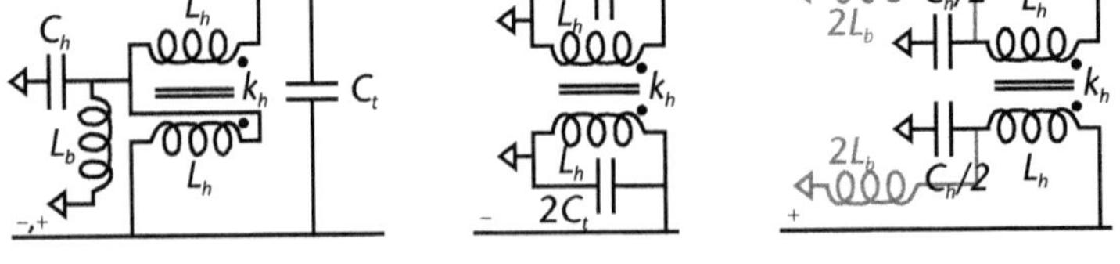

(a) (b) (c)

Figure 10.14 Differential bridge-T lowpass network: (a) schematic, (b) odd mode, and (c) even mode.

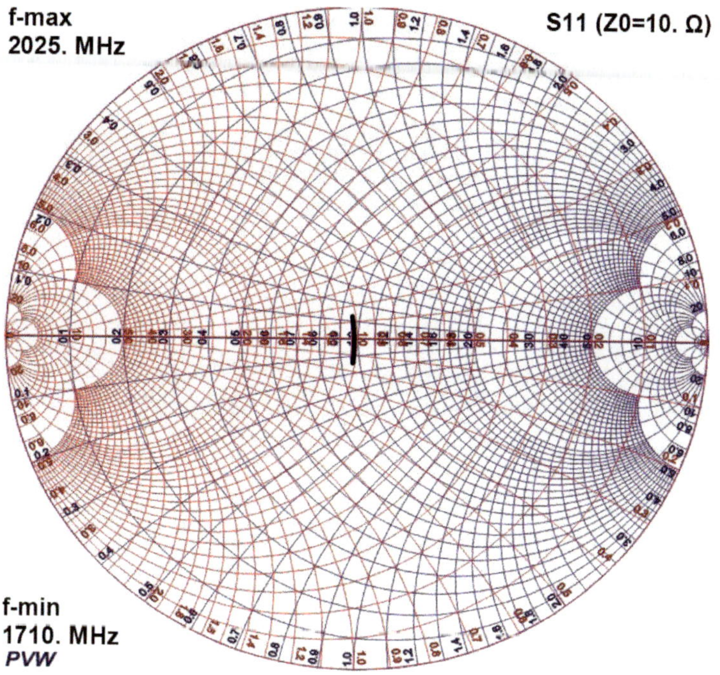

(a)

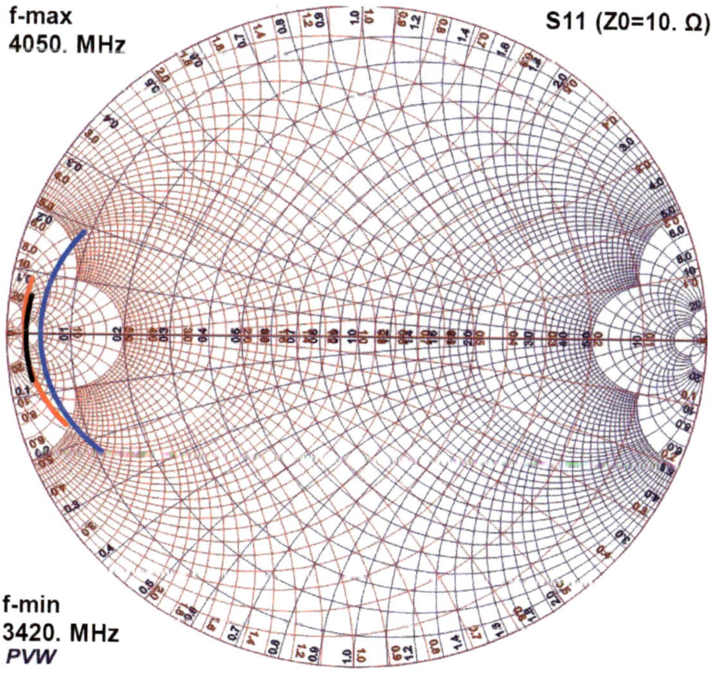

(b)

Figure 10.15 Differential characteristics of bridge-*T* lowpass with previous circuits: (a) Γ_{in} across passband, (b) Γ_{in} across second-harmonic band, (c) passband insertion losses, and (d) second-harmonic insertion losses.

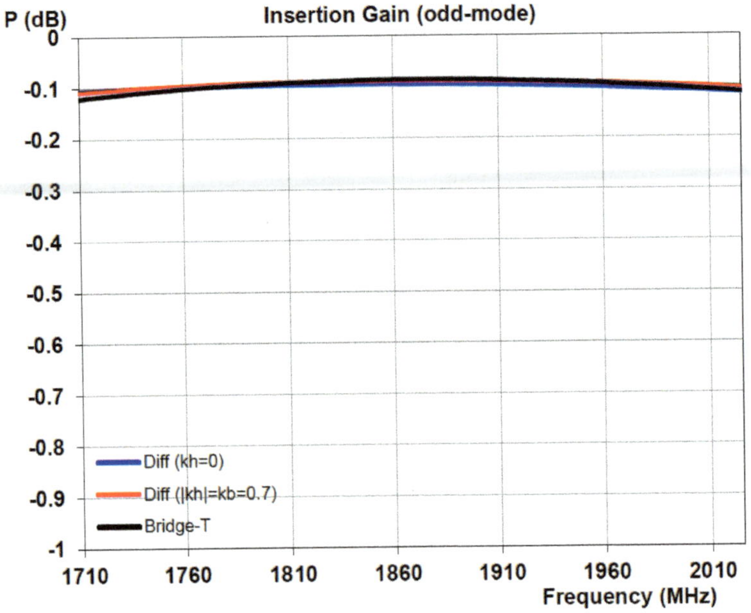

(c)

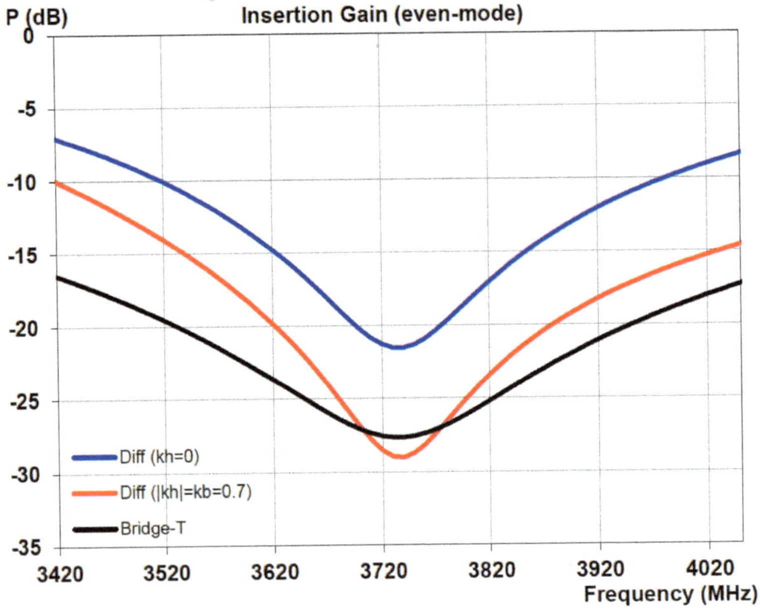

(d)

Figure 10.15 *(Continued)*

rejection across the band is much deeper and flatter than the previous alternatives. Compared to the differential architecture with no coupling, which is identical to that for the single-ended circuit, there is between 5 and 10-dB increase in rejection across the band.

In all responses shown in Figure 10.15, an arbitrary coupling coefficient $|k|$ = 0.7 was assumed.

If needed, the bridge-*T* circuit of Figure 10.14(a) can also be easily modified to additionally provide high third-harmonic rejection (odd-mode). This requires only one additional inductor L_3, as shown in Figure 10.16(a). The second-harmonic response is essentially unaffected by this addition, because no current flows through L_3/C_t under even-mode excitation. In the passband and third-harmonic band, which are odd-mode, the equivalent circuit of Figure 10.14(b) is transformed into that in Figure 10.16(b). To maintain minimum insertion loss in the passband L_3 and C_t are chosen to have an equivalent series capacitive reactance to that of $2C_t$ in Figure 10.14(b). For low impedance and high rejection across the third-harmonic band, L_3 and C_t are chosen to have a series resonance at the center of the band.

The very significant reduction in impedance across the third-harmonic band, achieved by the addition of inductor L_3 to the bridge-*T* circuit, can be seen in Figure 10.17. There is little dispersion across the entire band and a very low impedance is demonstrated. All other passband and second-harmonic responses are essentially

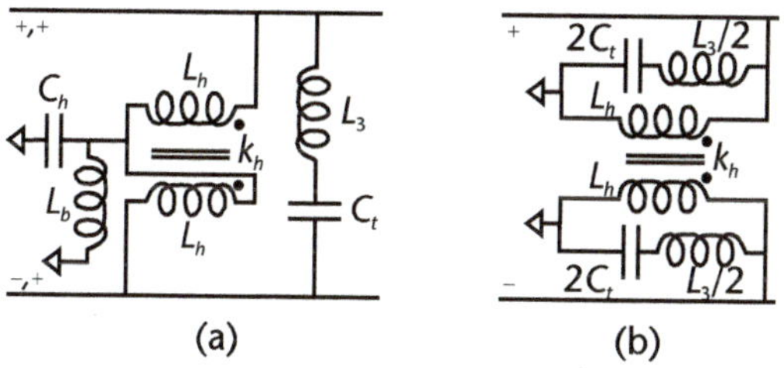

Figure 10.16 Differential bridge-*T* lowpass network with second and third-harmonic shorts: (a) schematic, and (b) odd mode.

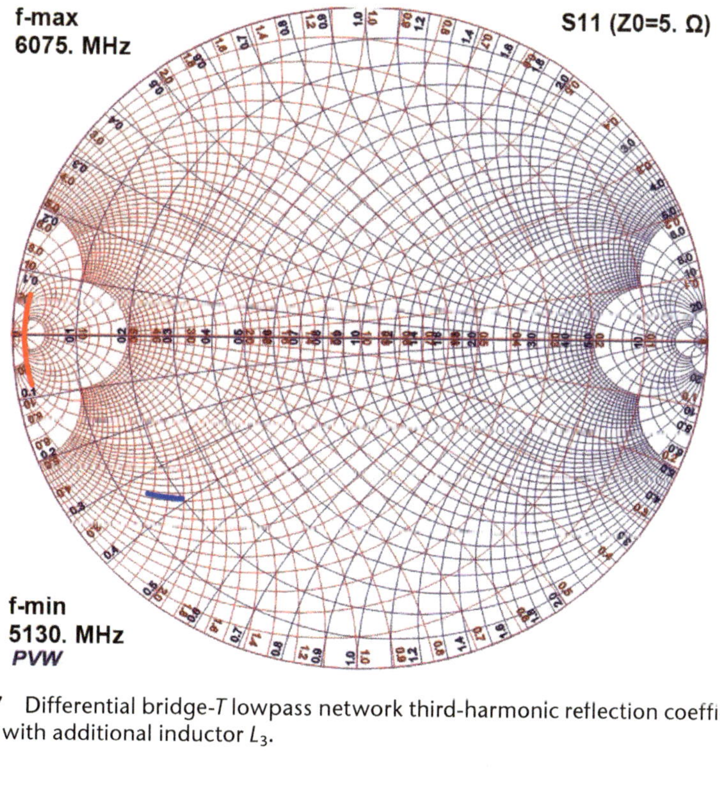

Figure 10.17 Differential bridge-*T* lowpass network third-harmonic reflection coefficients, without and with additional inductor L_3.

unchanged from those shown in Figure 10.15, save for a very minimal increase in passband insertion loss due to the nonideal nature assumed for the components.

The element values for the circuit are (accounting for L_b):

$$C_b = \frac{2}{\omega_{2h}^2 (1+k_b) L_b} + \frac{1}{\omega_{2h}^2 L_b} \tag{10.42}$$

For no third-harmonic short

$$C_t = \frac{1}{2(1-k_b)\omega_0^2 L_b} \tag{10.43}$$

with third-harmonic short

$$L_3 = \frac{1}{\omega_{3h}^2 C_t} \tag{10.44}$$

and

$$C_t = \frac{1/\omega_0^2 - 1/\omega_{3h}^2}{2(1-k_b) L_b} \tag{10.45}$$

The odd-mode loading impedance (i.e., fundamental and odd harmonics) on the two lines is

$$Z_O = \frac{(R_{Lb} + j\omega(1-k_b)L_b)(R_{Ct} + R_{L3} + j(\omega L_3 - 1/\omega C_t))/2}{R_{Lb} + j\omega(1-k_b)L_b + (R_{Ct} + R_{L3} + j(\omega L_3 - 1/\omega C_t))/2} \tag{10.46}$$

while the even-mode loading impedance (i.e., even harmonics) on the two lines is, accounting for L_b,

$$Z_E = R_{Lb} + j\omega(1+k_b)L_b + \frac{2}{\dfrac{1}{R_{Cb} - j/\omega C_b} + \dfrac{1}{R_{Lb} + j\omega L_b}} \tag{10.47}$$

where R_{element} = resistance of associated element.

CHAPTER 11

LC Single-Ended Matching Networks Overview

LC networks are the workhorses of passive RF networks. As building blocks, they are used for implementing matching networks, splitters, couplers, phase shifters, and filters. In general, such networks comprise multiple ports that interface with other network elements in an RF module. In this chapter, however, we shall focus on the most basic network form, the two-port. It is essential to have an intimate understanding of the capabilities and limitations of two-port networks to be able to be able to build upon them to implement networks of greater complexity, which are discussed later in this book.

Generally, two-port networks are divided into three classes: lowpass, highpass, and bandpass. Lowpass networks have insertion loss characteristics that increase with frequency, while highpass networks have insertion loss characteristics that decrease with frequency. To achieve a bandpass response, resonant elements, such as those discussed in Chapter 8, must be introduced into the networks. Typical frequency characteristics of the three network classes are shown in Figure 11.1.

For impedance matching the typically low output impedance (~5Ω) of a mobile power amplifier (PA) to a near universal 50Ω load, multiple cascaded matching sections are often required. The network class chosen for the sections may be the same or different. In addition, for optimum transmission characteristics, the intermediate impedances at the common nodes between sections must be allowed to be complex. Given the design complexity, achieving optimum overall network performance is dependent upon having a well-founded analytic description of each section.

The most basic form of a discrete matching network configuration comprises two elements, one element typically capacitive and the second element inductive. The possible configurations can take four different forms, as shown in Figure 11.2.

There are two lowpass options, with the appropriate choice dependent upon the relative impedances on the two ports. Likewise, there are two highpass options, again with the appropriate choice dependent upon the relative impedances on the two ports.

More complex discrete matching configurations comprise three elements and employ coupling between inductive elements or combinations thereof. General configurations for three of the most important two-port matching network architectures, used in RF design, are shown in Figure 11.3.

The π and T-network forms are widely used in PA matching applications. Here, the elements Y_n and Z_n represent the admittance and impedance, respectively, of

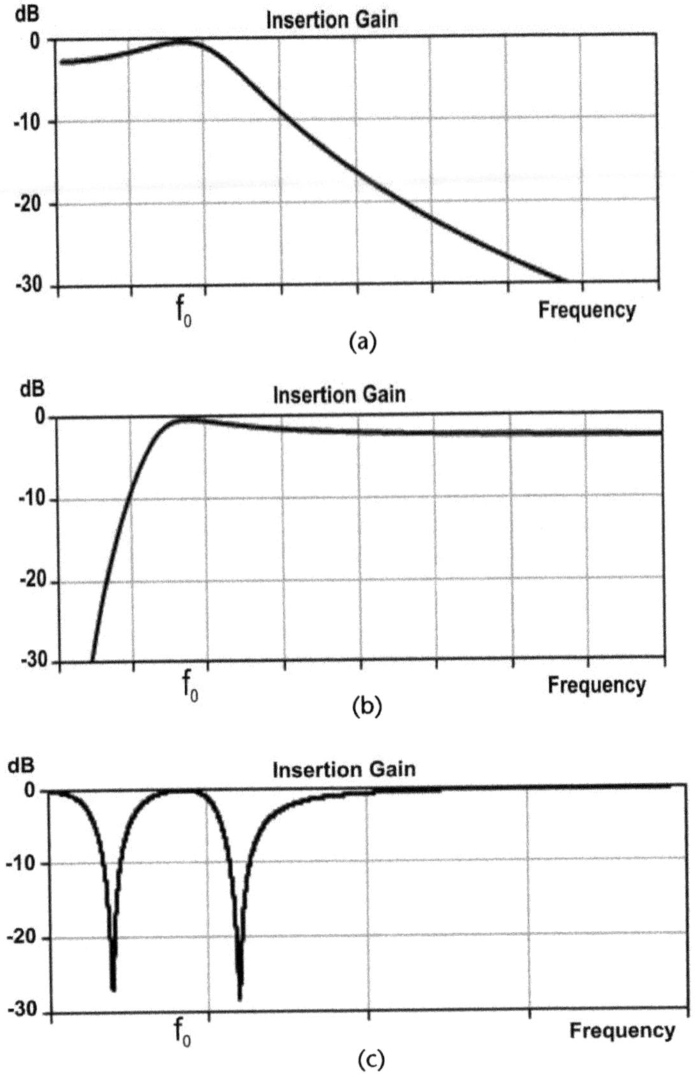

Figure 11.1 Typical transmission characteristics of two-port *LC* networks: (a) lowpass response, (b) highpass response, and (c) bandpass response.

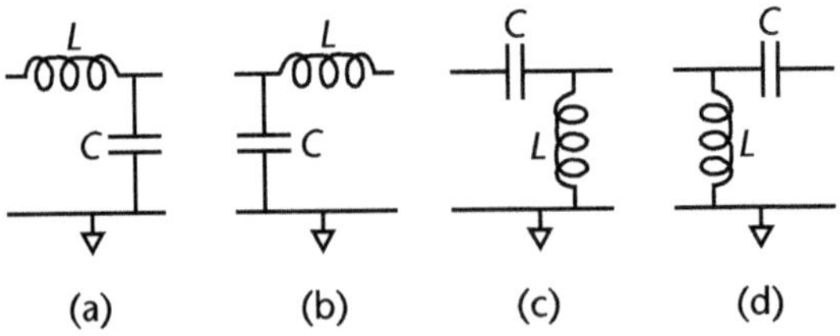

Figure 11.2 Basic two-port *LC* network architectures: (a) lowpass 1, (b) lowpass 2, (c) highpass 1, and (d) highpass 2.

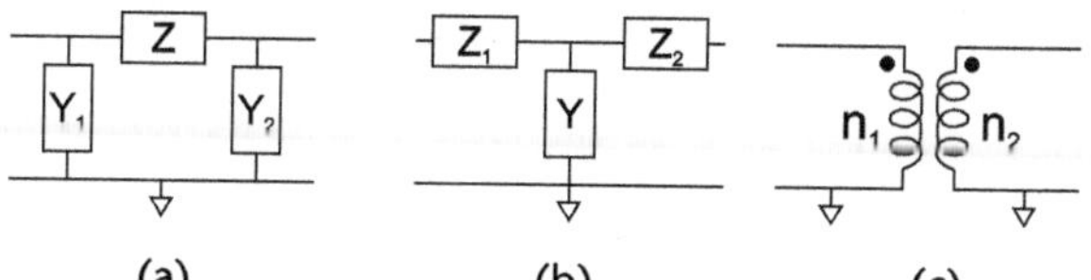

Figure 11.3 More complex two-port network architectures: (a) π-network, (b) T-network, and (c) transformer.

either inductive or capacitive circuit elements. The nature of the latter is dependent on the source and load impedances on the network.

A simplified two-port transformer schematic, as is commonly employed in low-frequency applications, is shown in Figure 11.3(c). Such a schematic is not physically realizable in high megahertz frequency ranges and above. This is due to the dual challenge of realizing large inductance values simultaneous with very tight coupling. Its function can be approximated by a pair of coupled inductors in RF circuits, although it is rarely used in single-ended impedance matching applications. However, coupled inductors do find widespread application in differential impedance matching networks. The application of coupled inductors for impedance matching for both single-ended and differential networks is examined in detail in subsequent chapters.

This chapter will focus on the design of π and T-networks and their two-element lowpass and highpass derivatives. The following parametric definitions are common for all the analyses:

Source impedance
$$Z_S = R_S + jX_S \tag{11.1}$$

Load impedance
$$Z_L = R_L + jX_L \tag{11.2}$$

Source admittance
$$Y_S = G_S + jB_S = \frac{R_S}{R_S^2 + X_S^2} - \frac{jX_S}{R_S^2 + X_S^2} \tag{11.3}$$

Load admittance
$$Y_L = G_L + jB_L = \frac{R_L}{R_L^2 + X_L^2} - \frac{jX_L}{R_L^2 + X_L^2} \tag{11.4}$$

11.1 Basic Two-Element Matches

The simple two-element network configurations shown in Figure 11.2 can be used for effecting impedance matching either in isolation or in a cascade arrangement. For a single section, as there are only two network variables to determine (i.e., L and C) and a complex impedance match is required, the elemental solutions are unique.

The choice of whether a lowpass or highpass match is most appropriate in a given design must be made by the designer in the light of system requirements. Once that choice is made, which of the alternate versions is required depends entirely on the source and load impedances to be matched.

For the lowpass match, the lowpass 1 type match is required, when

$$|Z_L| > \sqrt{R_S R_L} \tag{11.5}$$

or, if $X_L = 0$,

$$R_L > R_S \tag{11.6}$$

with required element values

$$\omega C = \frac{X_L \pm \sqrt{\left(R_L^2 + X_L^2\right) R_L / R_S - R_L^2}}{R_L^2 + X_L^2} \tag{11.7}$$

$$\omega L = \left(\omega C\left(R_L^2 + X_L^2\right) - X_L\right) R_S / R_L - X_S \tag{11.8}$$

For the lowpass match, the lowpass 2 type match is required, when

$$|Z_S| > \sqrt{R_S R_L} \tag{11.9}$$

or, if $X_S = 0$,

$$R_S > R_L \tag{11.10}$$

with required element values

$$\omega C = \frac{X_S \pm \sqrt{\left(R_S^2 + X_S^2\right) R_S / R_L - R_S^2}}{R_S^2 + X_S^2} \tag{11.11}$$

$$\omega L = \left(\omega C\left(R_S^2 + X_S^2\right) - X_S\right) R_L / R_S - X_L \tag{11.12}$$

For the highpass match, the highpass 1 type match is required, when

$$|Z_L| > \sqrt{R_S R_L} \tag{11.13}$$

or, if $X_L = 0$,

$$R_L > R_S \tag{11.14}$$

with required element values

$$\omega L = \frac{-X_L \pm \sqrt{\left(R_L^2 + X_L^2\right) R_L / R_S - R_L^2}}{1 - R_L / R_S} \tag{11.15}$$

$$C = \frac{L R_L / R_S}{\omega L\left(X_S R_L / R_S + X_L\right) + R_L^2 + X_L^2} \tag{11.16}$$

For the highpass match, the highpass 2 type match is required, when

$$|Z_S| > \sqrt{R_S R_L} \tag{11.17}$$

or, if $X_S = 0$,

$$R_S > R_L \tag{11.18}$$

with required element values

$$\omega L = \frac{-X_S \pm \sqrt{\left(R_S^2 + X_S^2\right) R_S / R_L - R_S^2}}{1 - R_S / R_L} \tag{11.19}$$

$$C = \frac{L R_S / R_L}{\omega L \left(X_L R_S / R_L + X_S\right) + R_S^2 + X_S^2} \tag{11.20}$$

11.2 Basic Two-Element Matching Networks Characteristics

Characteristics of the basic two-element matching networks of the previous section can be obtained by straightforward application of the relationships provided in Chapters 1 and 4. Some examples are given below.

First, see the design parameters given in Table 11.1.

With these external impedances, the lowpass and highpass matching networks take the form shown in Figure 11.4 with the parameters in Table 11.2.

Table 11.1 Design Parameters for Basic Two-Element *LC* Matches

Design Parameters	
f_{lo} 1,710 MHz	f_{hi} 2,025 MHz
$Z_S = 5\Omega$	$Z_L = 50\Omega$
Inductors, $Q_L = 60$	Capacitors, $Q_C \approx 100$

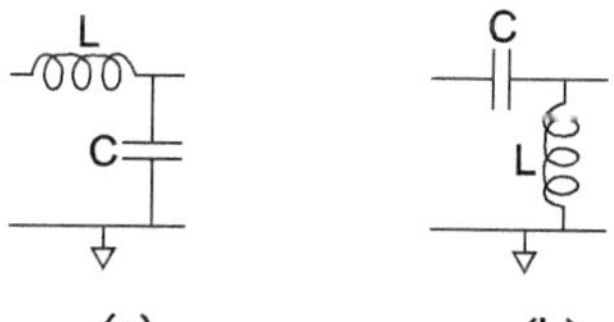

Figure 11.4 Basic two-element *LC* matching networks, with $R_S < R_L$: (a) lowpass, and (b) highpass.

Table 11.2 Design Parameters for Basic Two-Element *LC* Matches, with $R_S < R_L$

Lowpass Parameters	
$L = 1.278$ nH	$C = 5.11$ pF
Highpass parameters	
$C = 5.682$ pF	$L = 1.420$ nH

The impedance and gain characteristics of the lowpass network are shown in Figure 11.5.

With the elemental dissipation factors specified in Table 11.1, the minimum passband insertion gain, $P_{del}/P_{avail} \approx 0.71$ dB. Note that, in Figure 11.5(c), the two dependencies P_{del}/P_{avail} and P_{del}/P_{in} increasingly diverge at the band edges. The latter

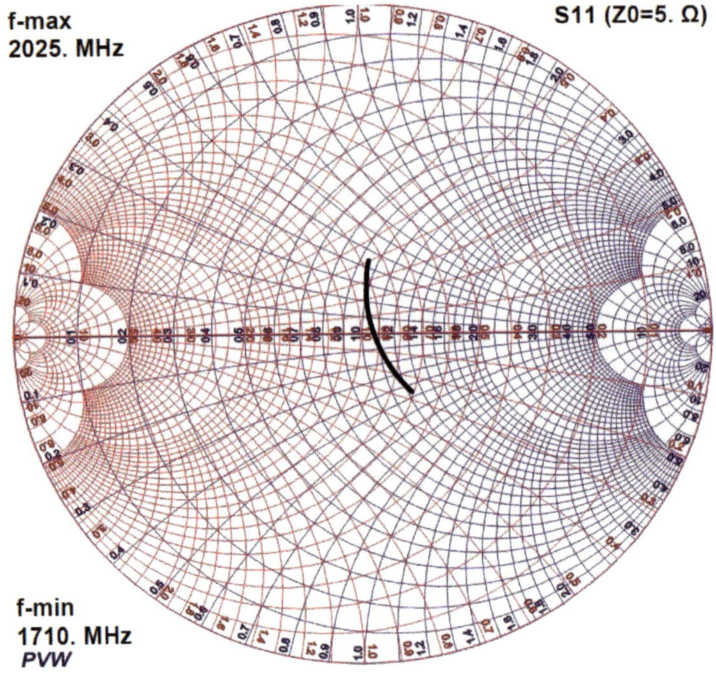

(a)

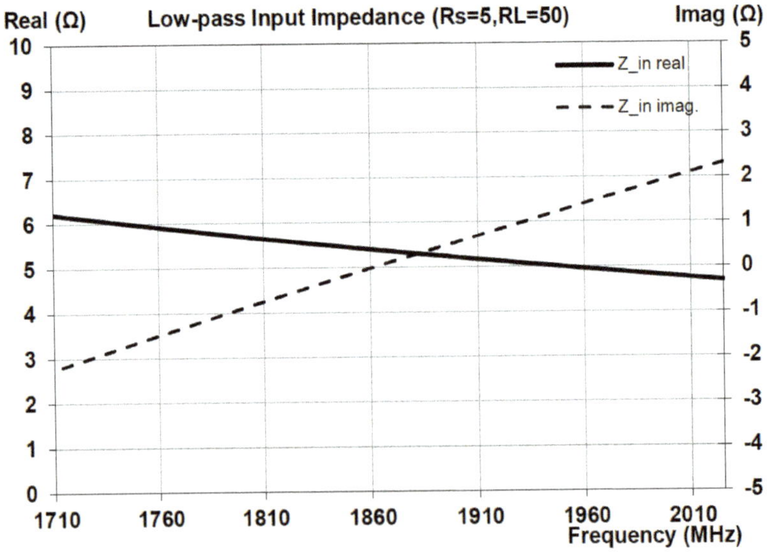

(b)

Figure 11.5 Lowpass *LC* characteristics, with $R_S < R_L$: (a) input reflection coefficient, (b) input impedance, and (c) insertion gain.

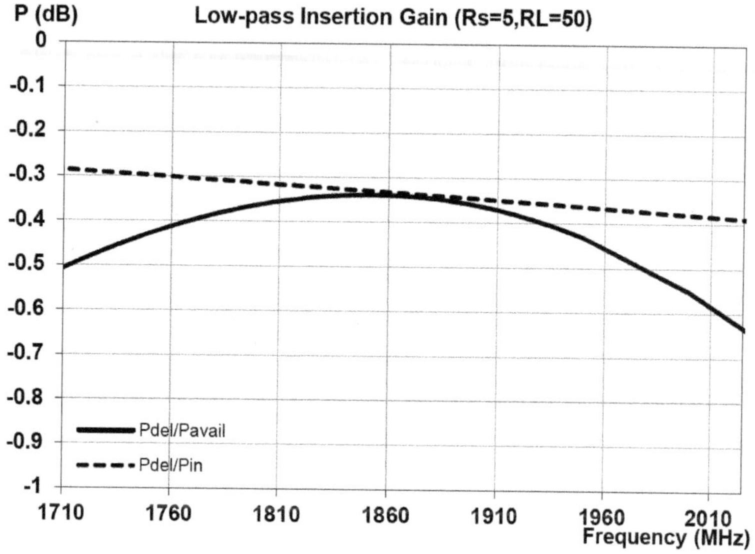

(c)

Figure 11.5 *(Continued)*

represents the power delivered to the load relative to the power entering the network. This insertion gain dependency is thus indicative of losses within the network due to dissipation in the elements. The insertion gain dependency P_{del}/P_{avail}, in addition to P_{del}/P_{in}, also includes reflective losses on the input to the network. Thus, the significant divergence between the two dependencies, particularly at the band edges, indicates that the simple two-element network is not capable of maintaining a good impedance match across the full bandwidth, with the 10:1 impedance mismatch. The variation of the input match over frequency can be seen in Figure 11.5(a, b). At the top end of the band, ~0.25-dB insertion loss can be seen to be attributable to mismatch.

The impedance and gain characteristics of the highpass network, for the design parameters of Table 11.1, are shown in Figure 11.6.

With the elemental dissipation factors specified in Table 11.1, the minimum passband insertion gain, $P_{del}/P_{avail} \approx 0.69$ dB, is almost identical to the lowpass configuration. Similar to the lowpass configuration, there is a significant divergence between the two insertion gain dependencies. An additional loss due to mismatch ~0.3 dB can be seen in Figure 11.6(c), this time at the low end of the band.

In conclusion, assuming the same elemental quality factors, the insertion gains across the passband for the two-element lowpass and highpass network configurations are approximately equivalent. Thus, the decision as which is most advantageous for a given application is generally decided by considerations of the transmission characteristics either above, or below, the passband.

11.3 Three-Element Network Dependency Options

The π and T-networks both comprise three independent elements, as is seen in Figure 11.3(a, b). However, to permit an impedance match between a complex source and

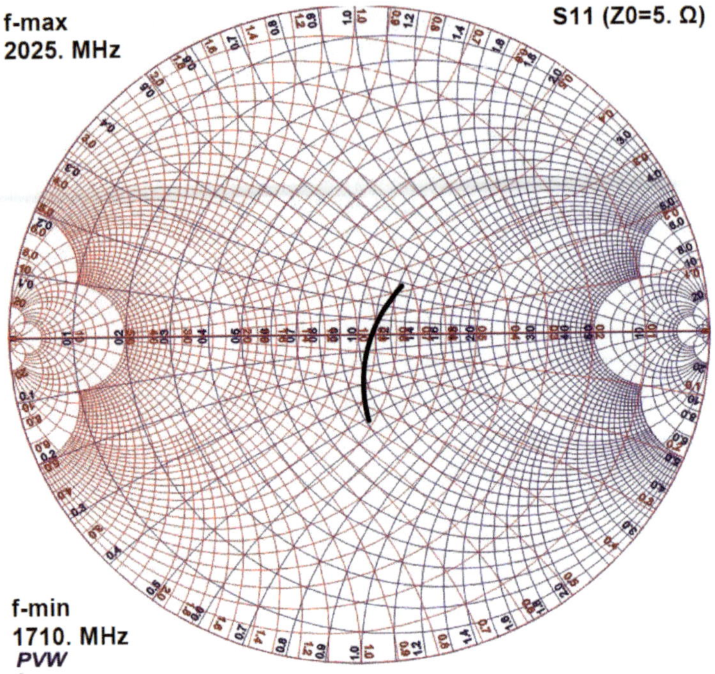

(a)

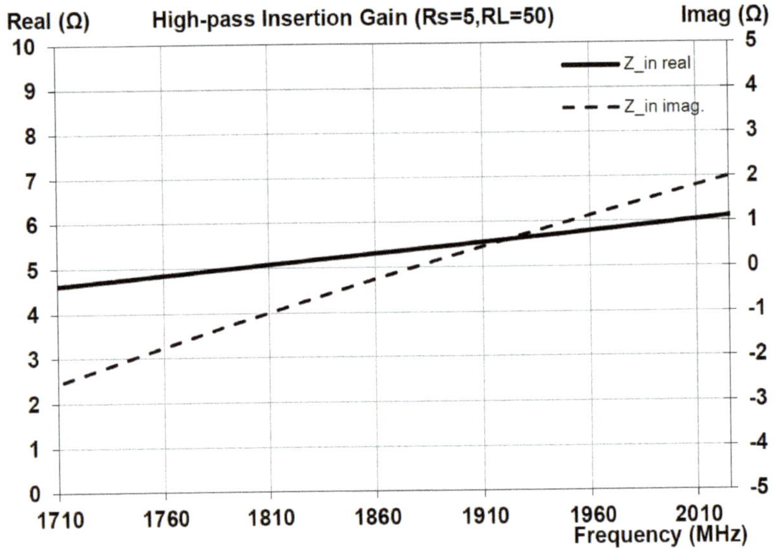

(b)

Figure 11.6 Highpass *LC* characteristics, with $R_S < R_L$: (a) input reflection coefficient, (b) input impedance, and (c) insertion gain.

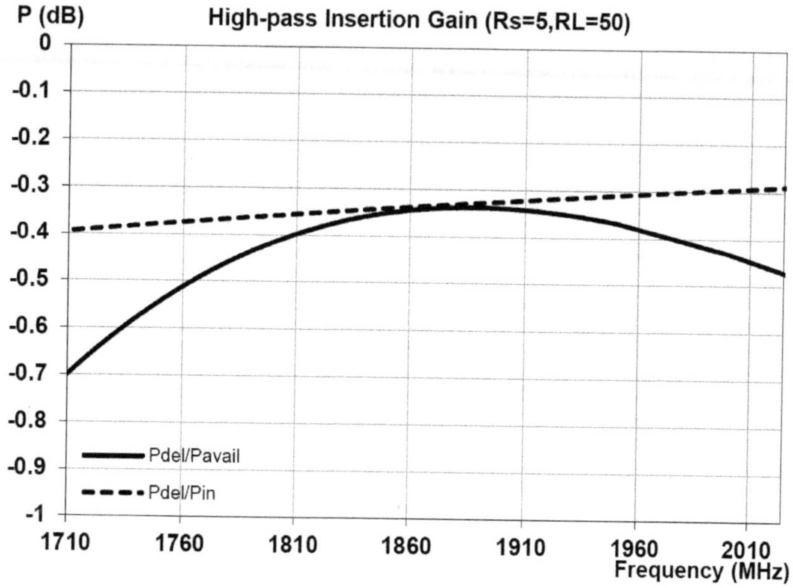

(c)

Figure 11.6 *(Continued)*

a complex load, only two equations must be satisfied, for the real and imaginary parts of either the terminated input or output impedance of the network. With three variables, the two equations are undetermined and capable of an infinite number of solutions. To address this issue, an additional constraint must be introduced to arrive at a unique set of solutions for the equations.

A first choice for an additional constraint is to treat one of the elements as an independent variable. Solutions for the remaining two elements will then be expressed in terms of this variable. Depending on which of the three circuit elements is selected as the independent variable, this would then yield three independent sets of solutions. Instead of one of the element values, a less obvious, but extremely useful, independent variable that can be chosen for each of the networks is the through-network phase shift. Choosing the network phase shift as the independent variable has the advantage of a frequency-independent quantity. This greatly facilitates the reuse of cascaded multisection amplifier matching network designs in differing frequency bands.

In the following sections for π and T networks, distinct sets of design solutions are developed for the alternate options of independent variables.

11.4 π-Network Design

For the synthesis solutions provided, all three network elements are assumed to be ideal (i.e., purely reactive). However, the solutions provided are for matching an

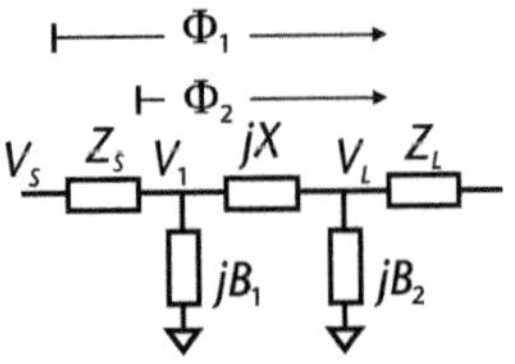

Figure 11.7 π-network reactive element definitions.

arbitrary complex source to an arbitrary complex load. The nomenclature for the reactive components of the network elements is detailed in Figure 11.7.

Case 11.1: Independent Variable = X

In terms of the parametric values in Figure 11.7, and those in (11.1) to (11.4), if the independent variable is taken to be the series reactance X, in order to match the complex source to the complex load impedance, require

$$B_1 = \left(1 \mp \sqrt{G_S\left(1/G_L - G_S X^2\right)}\right)/X - B_S \tag{11.21}$$

$$B_2 = \left(1 \mp \sqrt{G_L\left(1/G_S - G_L X^2\right)}\right)/X - B_L \tag{11.22}$$

Case 11.2: Independent Variable = B_1

In terms of the parametric values in Figure 11.7 and those in (11.1) to (11.4), if the independent variable is taken to be the left shunt susceptance B_1, in order to match the complex source to the complex load impedance, require

$$X = \frac{B_S + B_1 \pm G_S\sqrt{G_S^2 + (B_S + B_1)^2/G_S G_L - 1}}{G_S^2 + (B_S + B_1)^2} \tag{11.23}$$

$$B_2 = \left(1 \mp \sqrt{G_L\left(1/G_S - G_L X^2\right)}\right)/X - B_L \tag{11.24}$$

Case 11.3: Independent Variable = B_2

In terms of the parametric values in Figure 11.7 and those in (11.1) to (11.4), if the independent variable is taken to be the right shunt susceptance B_2, in order to match the complex source to the complex load impedance, we require

$$X = \frac{B_L + B_2 \pm G_L\sqrt{\left((B_L + B_2)^2 + G_L^2\right)/G_S G_L - 1}}{G_L^2 + (B_2 + B_L)^2} \tag{11.25}$$

$$B_1 = \left(1 \mp \sqrt{G_S\left(1/G_L - G_S X^2\right)}\right)/X - B_S \tag{11.26}$$

Case 11.4: Independent Variable = Network Phase Shift

Either the phase shift from the source to the load can be referenced, that is,

$$\Phi_1 = -\angle(V_L/V_S) = -\tan^{-1}(V_L/V_S) \tag{11.27}$$

in which case

$$X = \frac{\pm(R_S \sin\Phi_1 - X_S \cos\Phi_1)}{\sqrt{R_S G_L}} \tag{11.28}$$

or the phase shift through the network can be referenced, that is,

$$\Phi_2 = -\angle(V_L/V_1) = -\tan^{-1}(V_L/V_1) \tag{11.29}$$

in which case

$$X = \frac{\pm\sin\Phi_2}{\sqrt{R_S G_L}} \tag{11.30}$$

For both phase references from (11.21),

$$B_1 = \left(1 \mp \cos\Phi_2 \sqrt{G_S/G_L}\right)/X - B_S \tag{11.31}$$

and, from (11.22),

$$B_2 = \left(1 \mp \cos\Phi_2 \sqrt{G_L/G_S}\right)/X - B_L \tag{11.32}$$

For lowpass networks, the series reactance X must be inductive (i.e., $X > 0$). For highpass networks, the series reactance X must be capacitive (i.e., $X < 0$).

$B_1 = 0$, for

$$\Phi_2\left(B_1(0)\right) = \pm\cos^{-1}\left(\frac{(1 - B_S X)}{\sqrt{G_S/G_L}}\right) \tag{11.33}$$

$B_2 = 0$, for

$$\Phi_2\left(B_2(0)\right) = \pm\cos^{-1}\left(\frac{(1 - B_L X)}{\sqrt{G_L/G_S}}\right) \tag{11.34}$$

11.4.1 Dual Section π-Network Design

As mentioned above, multiple cascaded matching sections are frequently used to implement large impedance matching transformations. Employing two sections in the place of one invariably increases the bandwidth of the impedance match and significantly reduces insertion loss. However, these advantages are gained at the expense of an increase in the footprint of the network.

In the two cascaded sections, there are a total of five elements. As with the single-section match, only two equations need to be satisfied for the real and imaginary components of impedance. Thus, in this case, three independent variables must be defined to determine a unique set of element values.

A convenient choice for two of these variables is the phase shifts in each of the sections. Again, this choice is advantageous for ease of application to multiple frequency scenarios, as the phase shifts are independent of frequency.

For the third variable, an optimal choice, which is also frequency-independent, is the conductance G_N, at the common node between the two sections. This is not meant to imply any assumption that the impedance at this node is pure real. In general, for optimal network characteristics, particularly bandwidth and insertion loss, it will not be. However, requiring a conjugate match at the common combining node implies a common real conductance at this point, namely, G_N.

The network element definitions, for a dual-section π-network impedance match, are shown in Figure 11.8.

In this case, the phase shifts for the two sections are chosen as

$$\Phi_1 = -\angle(V_N/V_1) = -\tan^{-1}(V_N/V_1) \tag{11.35}$$

and

$$\Phi_2 = -\angle(V_L/V_N) = -\tan^{-1}(V_L/V_N) \tag{11.36}$$

In order for the cascaded sections to realize an impedance match between complex impedances Z_S to Z_L, we require

$$X_1 = \frac{\sin\Phi_1}{\sqrt{G_S G_N}} \tag{11.37}$$

$$X_2 = \frac{\sin\Phi_2}{\sqrt{G_N G_L}} \tag{11.38}$$

and

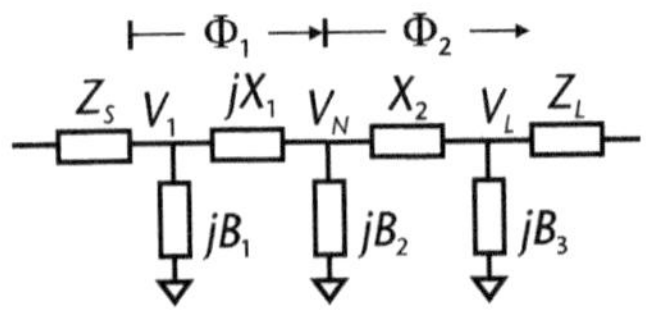

Figure 11.8 Dual-section π-network impedance matching network.

$$B_1 = \frac{\left(1 \mp \sqrt{G_S\left(1/G_N - G_S X_1^2\right)}\right)}{X_1 - B_S} \tag{11.39}$$

$$B_2 = \frac{\left(1 \mp \sqrt{G_N\left(1/G_S - G_N X_1^2\right)}\right)}{X_1} + \frac{\left(1 \mp \sqrt{G_N\left(1/G_L - G_N X_2^2\right)}\right)}{X_2} \tag{11.40}$$

$$B_3 = \frac{\left(1 \mp \sqrt{G_L\left(1/G_N - G_L X_2^2\right)}\right)}{X_2 - B_L} \tag{11.41}$$

Thus, in these design equations, there are three independent variables to define the match: Φ_1, Φ_2, and G_N.

11.5 *T*-Network Design

For the synthesis solutions provided, all three network elements are assumed to be ideal (i.e., purely reactive). However, the solutions provided are for matching an arbitrary complex source to an arbitrary complex load. The nomenclature for the reactive components of the network elements is detailed in Figure 11.9.

Case 11.5: Independent Variable = *B*

In terms of the parametric values in Figure 11.9, and those in (11.1) to (11.4), if the independent variable is taken to be the series reactance *X*, in order to match the complex source to the complex load impedance, we require

$$X_1 = \left(1 \pm \sqrt{R_S/R_L - (BR_S)^2}\right)/B - X_S \tag{11.42}$$

$$X_2 = \left(1 \pm \sqrt{R_L/R_S - (BR_L)^2}\right)/B - X_L \tag{11.43}$$

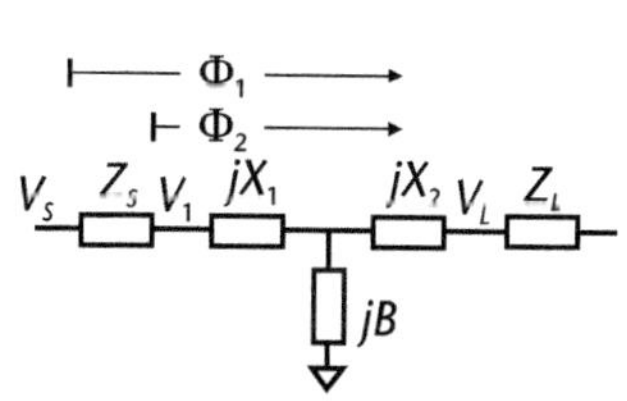

Figure 11.9 *T*-network reactive element definitions.

Case 11.6: Independent Variable = X_1

In terms of the parametric values in Figure 11.9, and those in (11.1) to (11.4), if the independent variable is taken to be the left shunt susceptance B_1, in order to match the complex source to the complex load impedance, we require

$$B = \frac{X_1 + X_S \pm \sqrt{\left(R_S^2 + (X_1 + X_S)^2\right) R_S/R_L - R_S^2}}{R_S^2 + (X_1 + X_S)^2} \quad (11.44)$$

$$X_2 = \left(1 \pm \sqrt{R_L/R_S - (BR_L)^2}\right)/B - X_L \quad (11.45)$$

Case 11.7: Independent Variable = X_2

In terms of the parametric values in Figure 11.9, and those in (11.1) to (11.4), if the independent variable is taken to be the right shunt susceptance B_2, in order to match the complex source to the complex load impedance, we require

$$B = \frac{X_2 + X_L \pm \sqrt{\left(R_L^2 + (X_2 + X_L)^2\right) R_L/R_S - R_L^2}}{R_L^2 + (X_2 + X_L)^2} \quad (11.46)$$

$$X_1 = \left(1 \pm \sqrt{R_S/R_L - (BR_S)^2}\right)/B - X_S \quad (11.47)$$

Case 11.8: Independent Variable = Network Phase Shift

Either the phase shift from the source to the load can be referenced, that is,

$$\Phi_1 = -\angle(V_L/V_S) = -\tan^{-1}(V_L/V_S) \quad (11.48)$$

in which case

$$B = \pm\frac{X_L \cos\Phi_1 + R_L \sin\Phi_1}{\sqrt{R_S R_L \left(R_L^2 + X_L^2\right)}} \quad (11.49)$$

or the phase shift through the network can be referenced, that is,

$$\Phi_2 = -\angle(V_L/V_1) = -\tan^{-1}(V_L/V_1) \quad (11.50)$$

in which case

$$B = \pm\frac{(G_S G_L - B_S B_L)\sin\Phi_2 - (G_S B_L + G_L B_S)\cos\Phi_2}{\sqrt{G_S G_L}} \quad (11.51)$$

For both phase references, as above,

$$X_1 = \left(1 \pm \sqrt{R_S/R_L - (BR_S)^2}\right)/B - X_S \quad (11.52)$$

$$X_2 = \left(1 \pm \sqrt{R_L/R_S - (BR_L)^2}\right)/B - X_L \quad (11.53)$$

For lowpass networks, the shunt susceptance B must be capacitive (i.e., $B > 0$). For highpass networks, the shunt susceptance B must be inductive (i.e., $B < 0$).

11.6 π and T-Network Characteristics

As an aid to understanding of the key distinguishing features between π and T-network impedance matches, some of their key characteristics are presented here with the design parameters listed in Table 11.3. To allow a direct comparison of between the two types of networks, all responses are determined using the following common design specifications.

Also, for each of the networks, the base design values listed correspond to a through-network phase shift $|\Phi| = 90°$.

11.6.1 Lowpass π-Network

An example of a lowpass π-network is shown in Figure 11.10. Note, in general, that the shunt elements may be either inductive or capacitive depending on the impedance terminations and through phase shift.

With element values:

$L = 1.348$ nH

$C_1 = 5.390$ pF

$C_2 = 5.390$ pF

and minimum passband insertion gain, $P_{del}/P_{avail} \approx$ 0.73 dB.

Table 11.3 Common Parameters Used for Evaluating Two-Port *LC* Networks

Design Parameters	
f_{lo} 1,710 MHz	f_{hi} 2,025 MHz
Phase shift	$\|\Delta\Phi\| = 90°$
$Z_S = 5\Omega$	$Z_L = 50\Omega$
Inductors, $Q_l = 60$	Capacitors, $Q_C \approx 100$

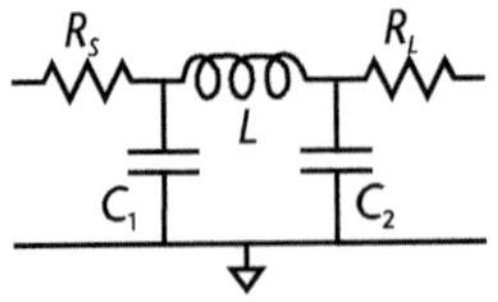

Figure 11.10 Lowpass π-network implementation.

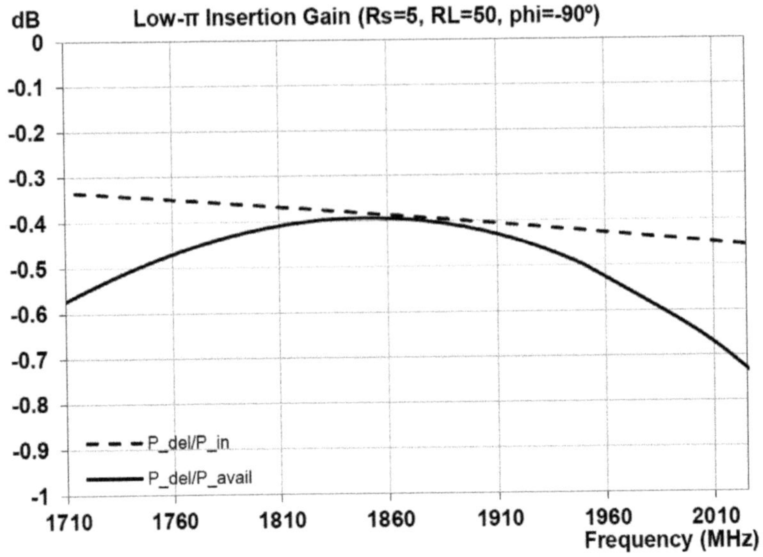

(a)

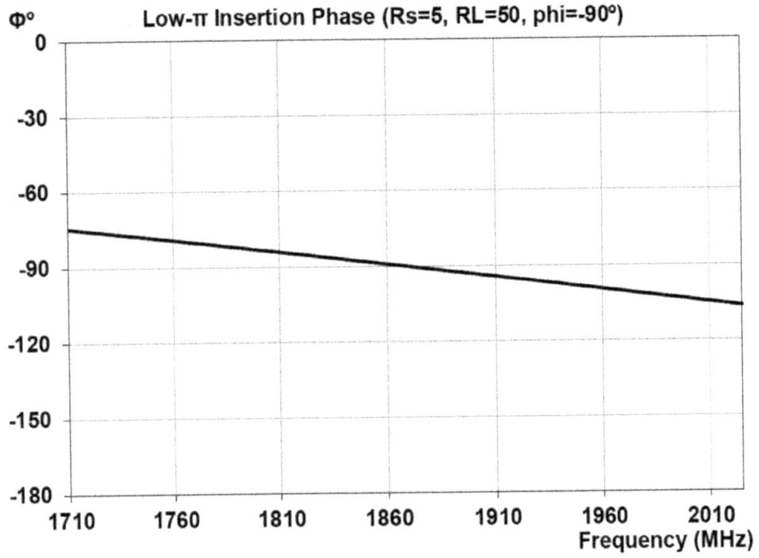

(b)

Figure 11.11 Lowpass π-network key characteristics: (a) insertion gain versus frequency, (b) insertion phase versus frequency, (c) insertion gain versus phase shift, and (d) element values versus phase shift.

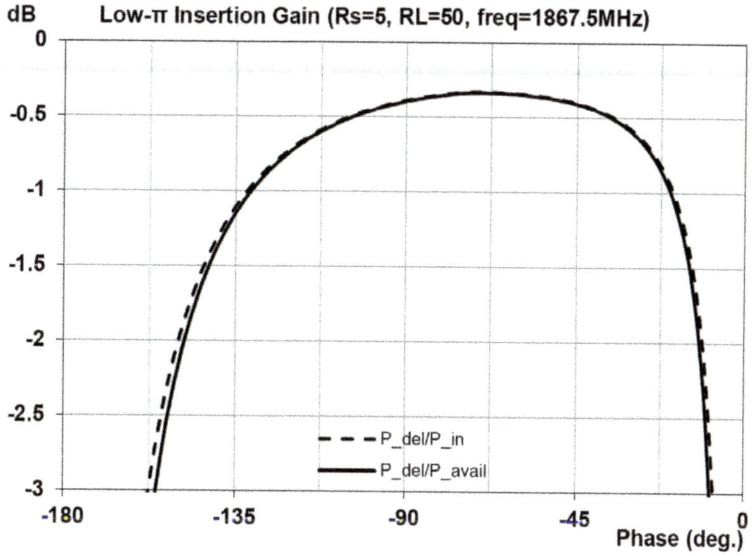

(c)

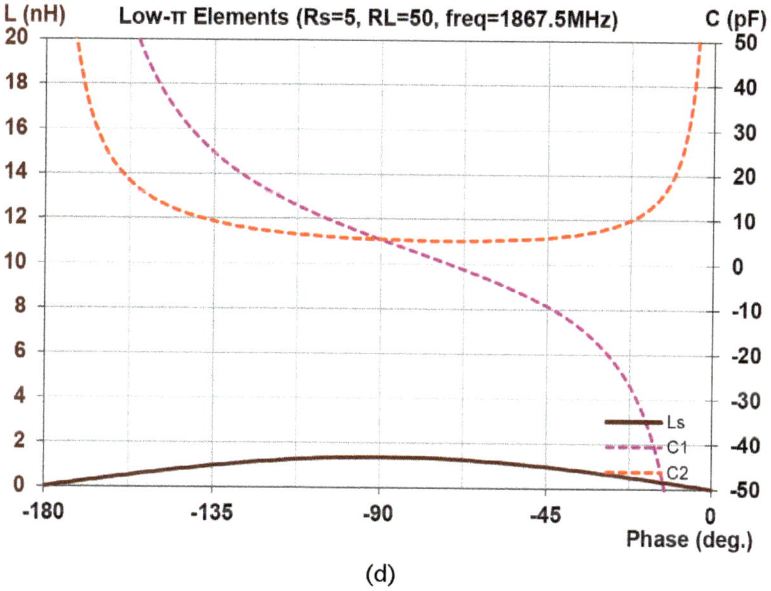

(d)

Figure 11.11 *(Continued)*

Note that, for insertion phase shifts in the range of 0° to ~ −71°, the corresponding value of C_1 is negative. For physical realization, to implement phase shifts in this range, C_1 must be replaced by a shunt inductor with an equivalent susceptance.

11.6.2 Highpass π-Network

An example of a highpass π-network is shown in Figure 11.12. Note, in general, that the shunt elements may be either inductive or capacitive depending on the impedance terminations and through phase shift.

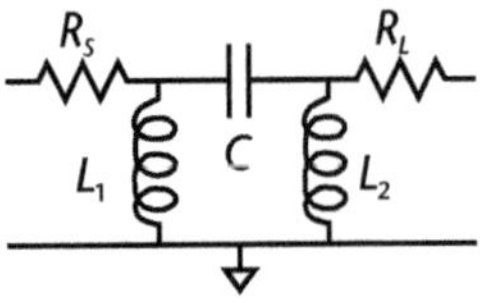

Figure 11.12 Highpass π-network implementation.

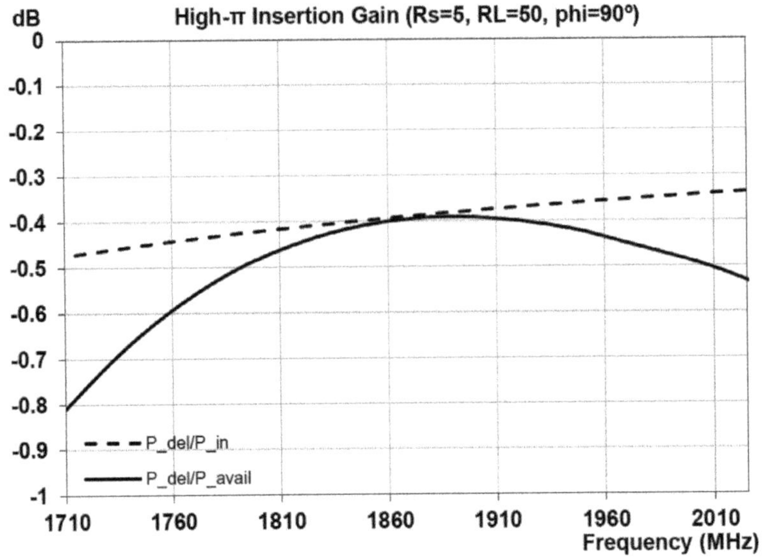

(a)

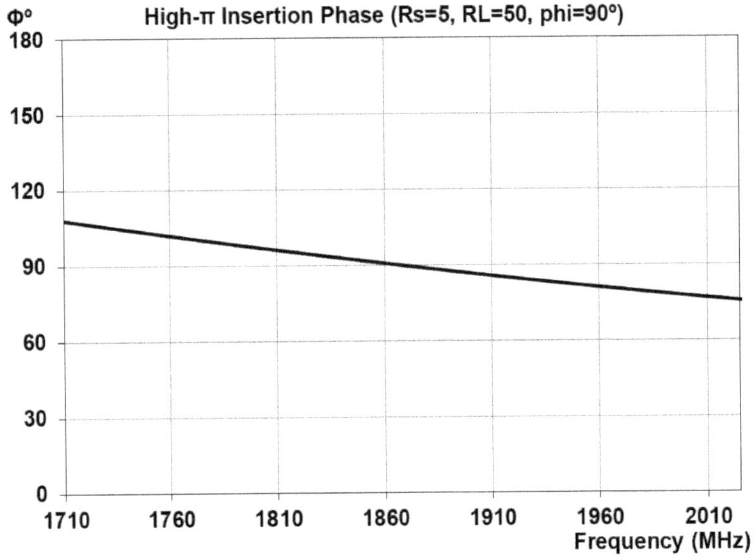

(b)

Figure 11.13 Highpass π-network key characteristics: (a) insertion gain versus frequency, (b) insertion phase versus frequency, (c) insertion gain versus phase shift, and (d) element values versus phase shift.

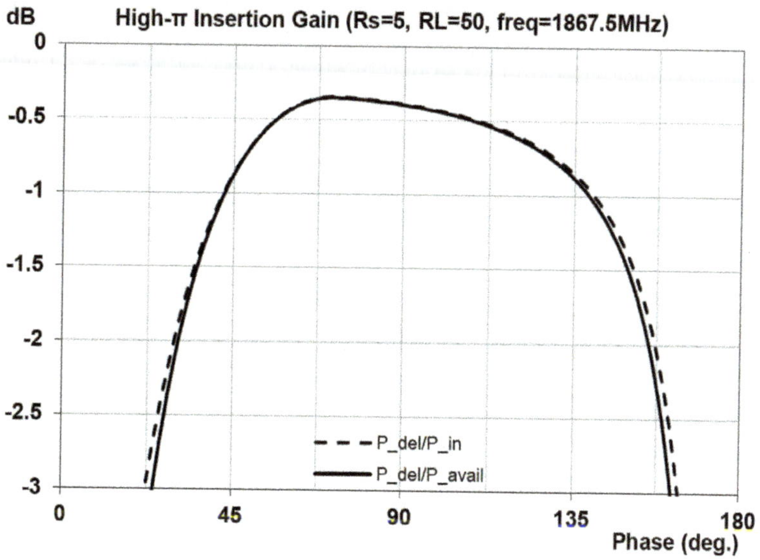

(c)

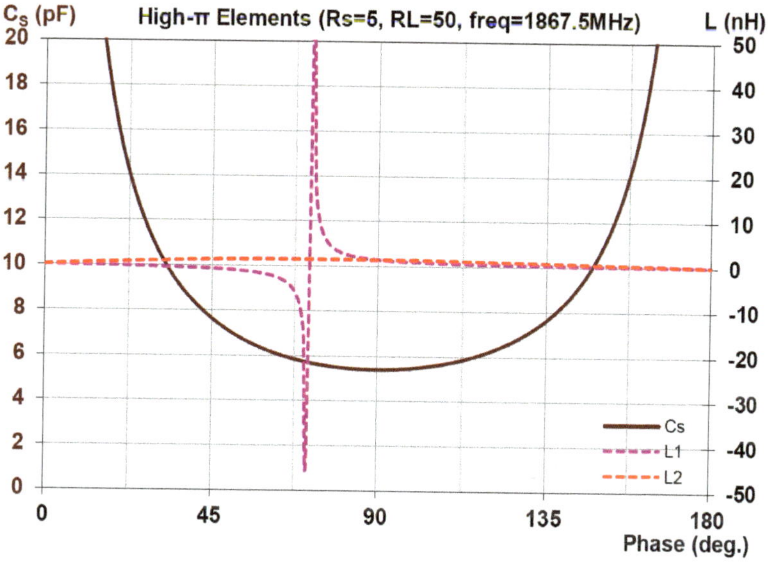

(d)

Figure 11.13 *(Continued)*

With element values:

$C = 5.39$ pF

$L_1 = 1.348$ nH

$L_2 = 1.348$ nH

and minimum passband insertion gain, $P_{del}/P_{avail} \approx$ 0.81 dB.

In this case, the value of L_1 is negative for insertion phase shifts in the range of 0° to ~71°. For physical realization, to implement phase shifts in this range, L_1 must be replaced by a shunt capacitor with an equivalent susceptance.

11.6.3 Lowpass *T*-Network

An example of a lowpass *T*-network is shown in Figure 11.14. Note, in general, that the series elements may be either inductive or capacitive depending on the impedance terminations and through phase shift.

With element values:

$C = 5.390$ pF

$L_1 = 1.348$ nH

$L_2 = 1.348$ nH

and minimum passband insertion gain, $P_{del}/P_{avail} \approx$ 0.73 dB.

In this case, the value of L_2 is negative for insertion phase shifts in the range of 0° to ~ −71°. For physical realization, to implement phase shifts in this range, L_2 must be replaced by a shunt capacitor with an equivalent susceptance.

11.6.4 Highpass *T*-Network

An example of a highpass *T*-network is shown in Figure 11.16. Note, in general, that the series elements may be either inductive or capacitive depending on the impedance terminations and through phase shift.

With element values:

$L = 1.348$ nH

$C_1 = 5.390$ pF

$C_2 = 5.390$ pF

and minimum passband insertion gain, $P_{del}/P_{avail} \approx$ 0.81 dB.

In this case, the value of C_2 is negative for insertion phase shifts in the range of 0° to ~71°. For physical realization, to implement phase shifts in this range, C_2 must be replaced by a shunt inductor with an equivalent susceptance.

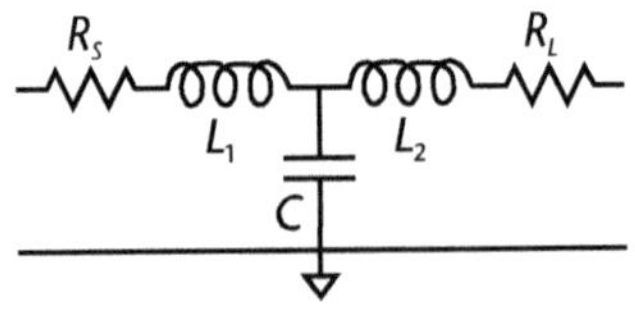

Figure 11.14 Lowpass *T*-network implementation.

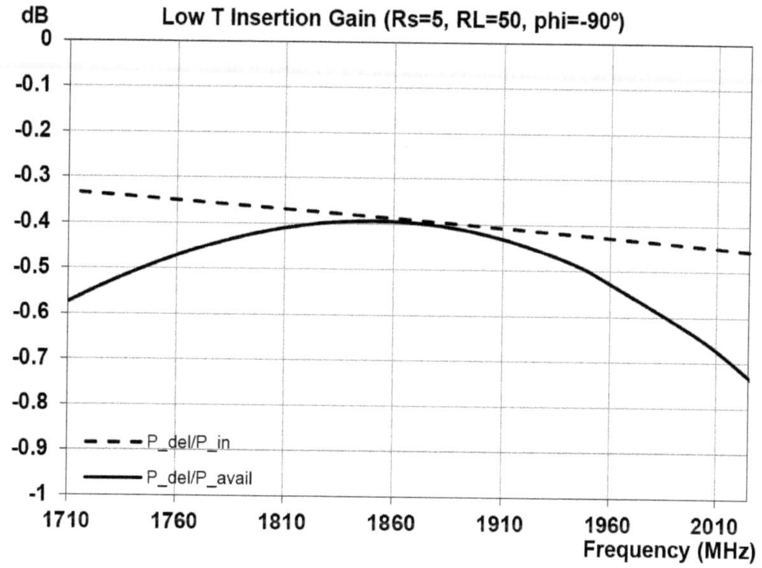

(a)

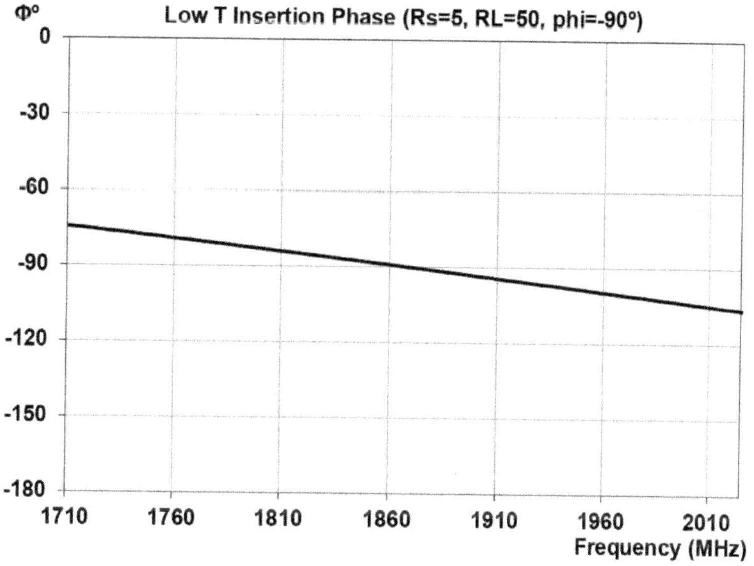

(b)

Figure 11.15 Lowpass *T*-network implementation: (a) insertion gain versus frequency, (b) insertion phase versus frequency, (c) insertion gain versus phase shift, and (d) element values versus phase shift.

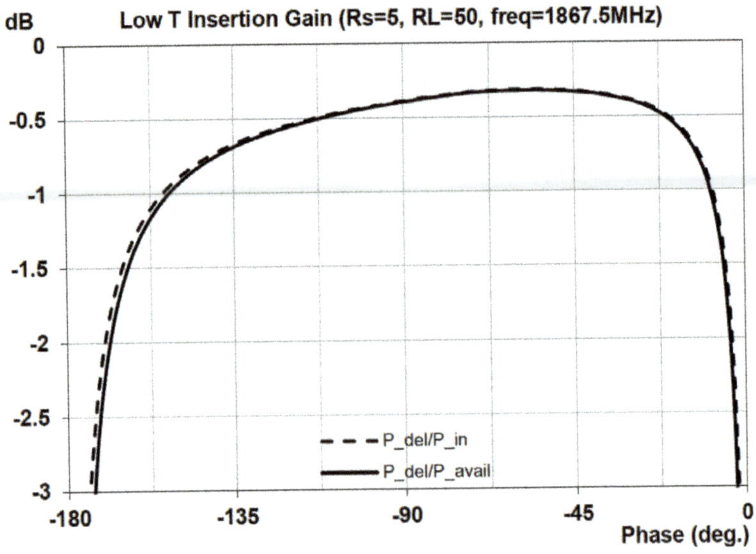

(c)

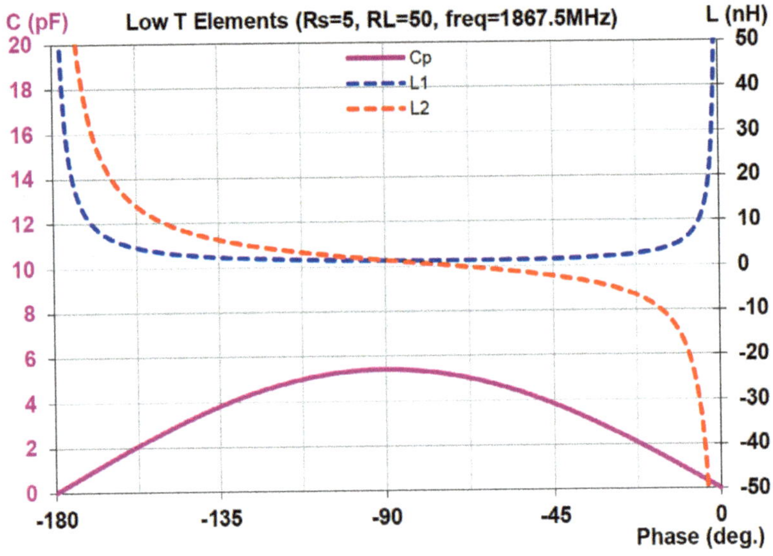

(d)

Figure 11.15 *(Continued)*

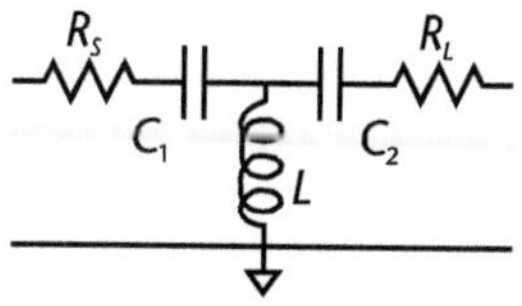

Figure 11.16 Highpass *T*-network implementation.

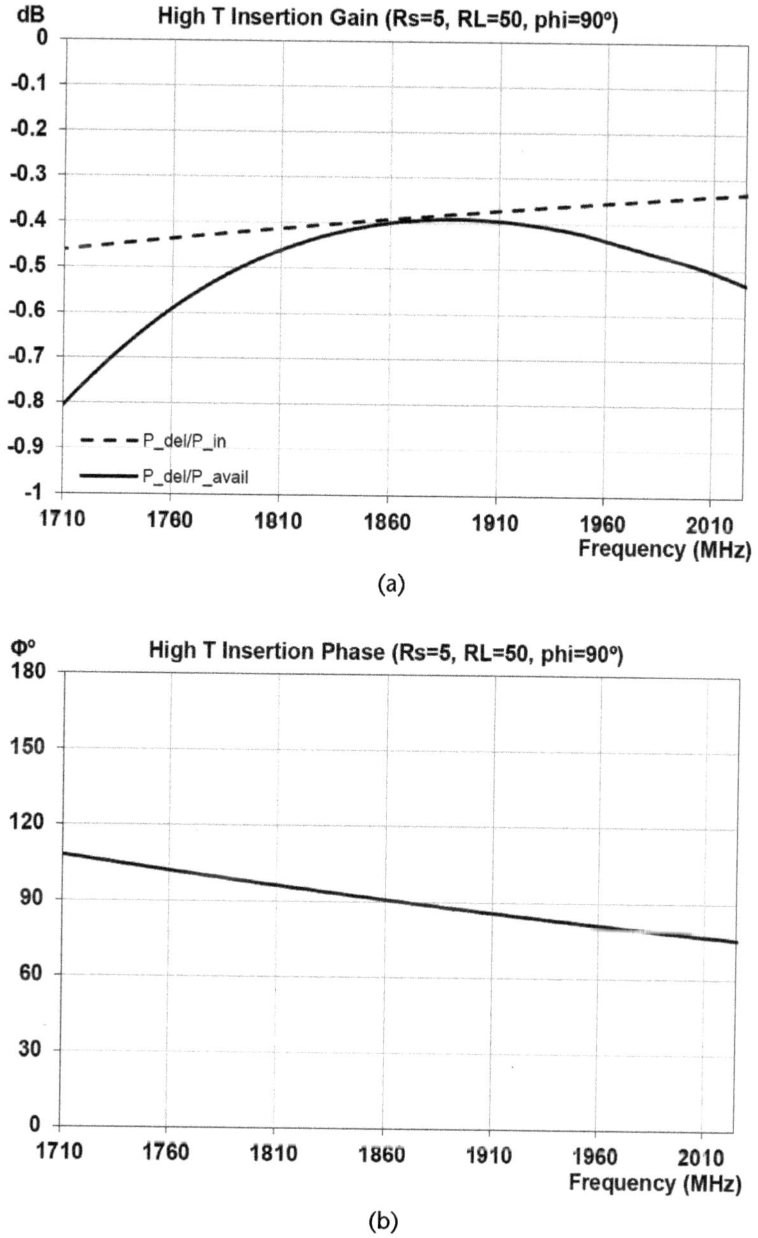

Figure 11.17 Highpass *T*-network key characteristics: (a) insertion gain versus frequency, (b) insertion phase versus frequency, (c) insertion gain versus phase shift, and (d) element values versus phase shift.

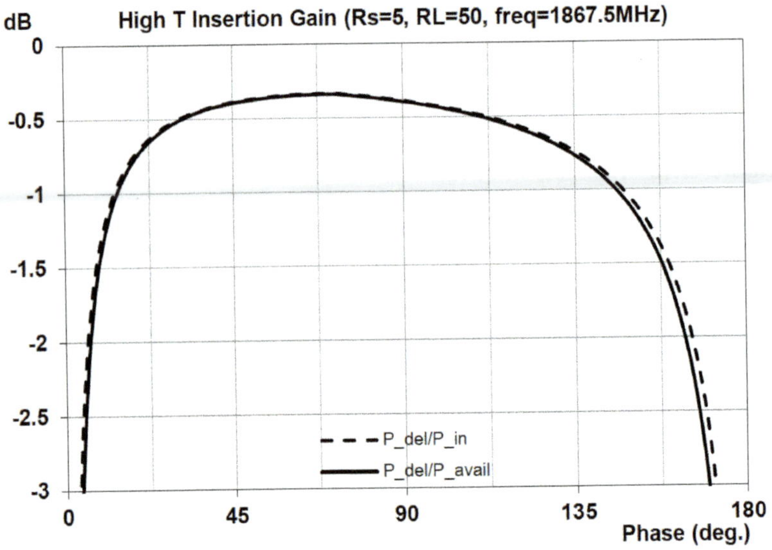

(c)

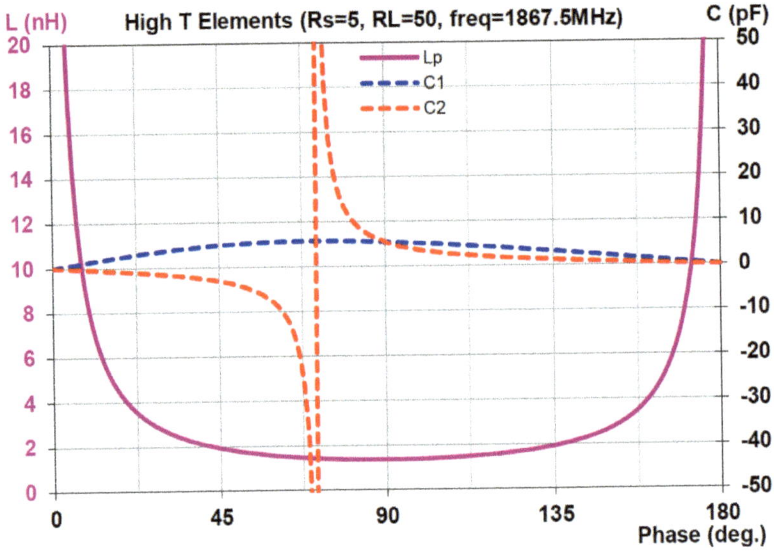

(d)

Figure 11.17 *(Continued)*

11.6.5 π and *T*-Network Equivalencies

Any π-network can be transformed into an equivalent *T*-network or vice versa. Figure 11.18 shows generic schematics for a π-network and a *T*-network.

The π-network will be equivalent to the *T*-network, given the following elemental relationships

$$Z_A = \frac{Z_1Z_2 + Z_1Z_3 + Z_2Z_3}{Z_2} \tag{11.54}$$

$$Z_B = \frac{Z_1Z_2 + Z_1Z_3 + Z_2Z_3}{Z_1} \tag{11.55}$$

$$Z_C = \frac{Z_1Z_2 + Z_1Z_3 + Z_2Z_3}{Z_3} \tag{11.56}$$

or, alternately,

$$Y_A = \frac{Z_2}{Z_1Z_2 + (Z_1 + Z_2)/Y_3} \tag{11.57}$$

$$Y_B = \frac{Z_1}{Z_1Z_2 + (Z_1 + Z_2)/Y_3} \tag{11.58}$$

$$Z_C = Z_1 + Z_2 + Y_3Z_1Z_2 \tag{11.59}$$

Reciprocally, the *T*-network will be equivalent to the π-network, given the following elemental relationships

$$Z_1 = \frac{Z_AZ_C}{Z_A + Z_B + Z_C} \tag{11.60}$$

$$Z_2 = \frac{Z_BZ_C}{Z_A + Z_B + Z_C} \tag{11.61}$$

$$Z_3 = \frac{Z_AZ_B}{Z_A + Z_B + Z_C} \tag{11.62}$$

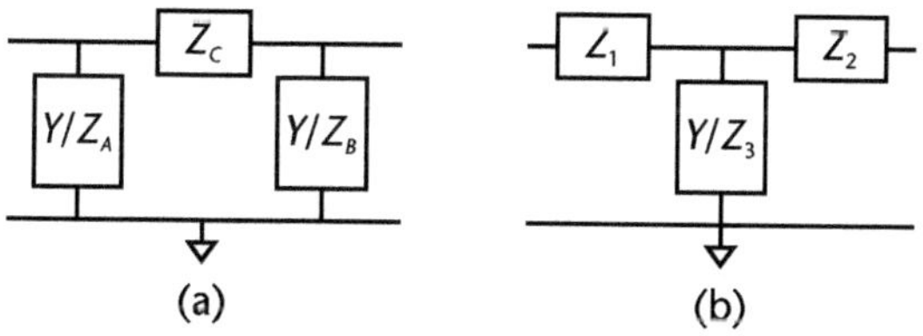

Figure 11.18 Equivalent π and *T*-networks: (a) generic π-network, and (b) generic *T* network.

or, alternately,

$$Z_1 = \frac{Y_B Z_C}{Y_A + Y_B + Y_A Y_B Z_C} \tag{11.63}$$

$$Z_2 = \frac{Y_A Z_C}{Y_A + Y_B + Y_A Y_B Z_C} \tag{11.64}$$

$$Y_3 = Y_A + Y_B + Y_A Y_B Z_C \tag{11.65}$$

11.7 Two-Element, Single-Ended Matching Networks

The π and T-matching networks each comprise three elements. This gives them extreme flexibility with an infinite set of possible element values available for providing an impedance match between two arbitrary complex impedances. Having an infinite set of solutions from which to choose gives the RF designer the freedom to choose the design set that most closely optimizes the parameter of most importance, typically insertion loss.

In some cases, however, this additional degree of freedom is not required and the designer would prefer to use a matching network with fewer elements (i.e., two). This has advantages for both footprint and cost. Unfortunately, it will inevitably result in incurring some degradation in performance. In addition, depending on the terminating impedances, a two-element solution might not even be possible.

The two-element match has two potential forms, either a series-shunt connection, or a shunt-series connection. These are shown in Figure 11.19, together with the element parameters.

Values for the network elements can be derived directly from those for either the π or T-networks. Simply equating one of the susceptances on either side of the π-network to zero yields solutions for the two-element networks. Conversely, equating one of the reactances on either side of the T-network to zero will also yield solutions for the two-element networks.

The formulae for determining the appropriate element values for each of the four two-element LC matches networks are given below (see Figure 11.20):

$$L = \left(-X_S + \sqrt{R_S\left(1/G_L - R_S\right)}\right)/\omega \tag{11.66}$$

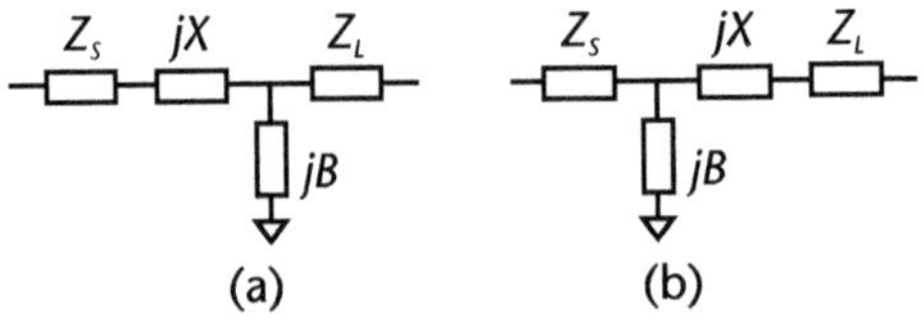

Figure 11.19 Two-element *LC* impedance matching networks: (a) series-shunt match, and (b) shunt-series match.

$$C = \left(-B_L + \sqrt{G_L(1/R_S - G_L)}\right)/\omega \tag{11.67}$$

$$C = \frac{1}{\omega\left(X_S + \sqrt{R_S(1/G_L - R_S)}\right)} \tag{11.68}$$

$$L = \frac{1}{\omega\left(B_L + \sqrt{G_L(1/R_S - G_L)}\right)} \tag{11.69}$$

$$C = \left(-B_S + \sqrt{G_S(1/R_L - G_S)}\right)/\omega \tag{11.70}$$

$$L = \left(X_L + \sqrt{R_L(1/G_S - R_L)}\right)/\omega \tag{11.71}$$

$$L = \frac{1}{\omega\left(B_S + \sqrt{G_S(1/R_L - G_S)}\right)} \tag{11.72}$$

$$C = \frac{1}{\omega\left(X_L + \sqrt{R_L(1/G_S - R_L)}\right)} \tag{11.73}$$

Details of the impedance contours for these networks are shown in Figure 11.21.

11.8 Dual-π Single-Ended PA Matching

The most common exploited matching network configuration for a single-ended mobile PA comprises a serial cascade of two π-networks. In general, such a network is easy to design, not challenging to lay out, and has good performance. A simplified schematic of a typical two-section π PA output match is shown in Figure 11.22.

In the schematic, the bias and harmonic trap taken together constitute the leftmost susceptance of the two-section π match. Including the latter, there are a

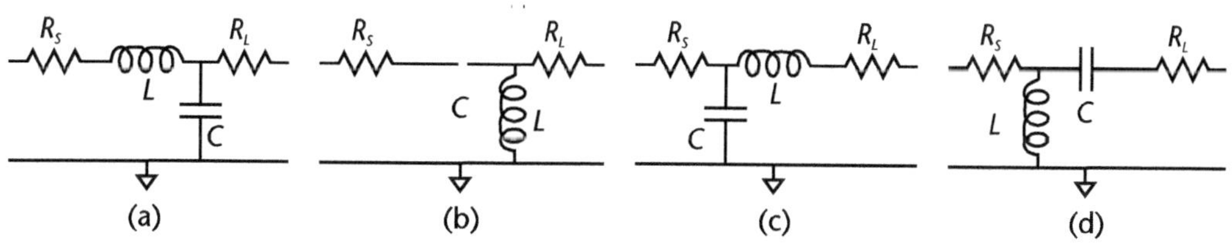

Figure 11.20 Two-element *LC* impedance matching networks: (a) series-shunt–lowpass (11.66) and (11.67), (b) series-shunt–highpass (11.68) and (11.69), (c) shunt-series–lowpass (11.70) and (11.71), and (d) shunt-series–highpass (11.72) and (11.73).

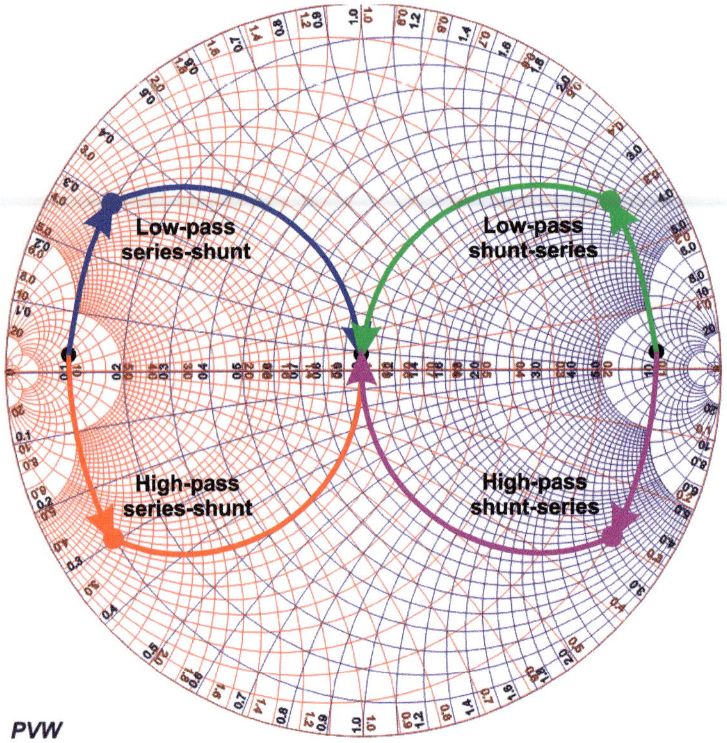

Figure 11.21 Two-element *LC* impedance matches.

total of five elements in the two cascaded sections. Capacitor C_L is included in the circuit to isolate the load from the DC bias, and it is accounted for in the analysis as an additional series impedance with the load. As only two equations need to be satisfied for the real and imaginary components of impedance at a single frequency, the element values are underconstrained and thus an infinite set of solutions is possible. To constrain the element values, additional boundary conditions must be applied. Examples of these are limits on the element values or passband variations in insertion loss, impedance match, and insertion phase, or out-of-band, rejection.

In addition to requiring an impedance match at the design frequency, three independent variables must be defined to determine a unique set of element values for the network. Sweeping these variables then permits contour maps, such as those in Section 7.1.3, to be generated of the performance parameters of most importance. A network configuration that offers the best performance compromise can then be selected.

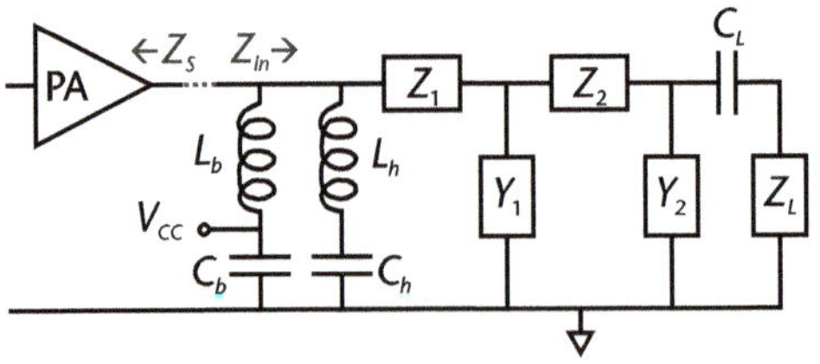

Figure 11.22 Cascaded two-section π PA match.

In Section 11.4.1, a convenient design approach was described based on choosing two of the swept variables as the phase shifts in the two π-sections. The phase shifts are advantageous choices as they are independent of frequency. An optimal choice for the third swept variable, again frequency-independent, is the conductance G_{in}, or resistance R_{in}, of the complex conjugate match at the common node between the two sections. Full design equations for this approach were given in Section 11.4.1.

Given that there are two cascaded π sections, each of which could be a high-pass or lowpass network (Section 11.6), there are four possible configurations to be considered. Some of the basic characteristics of each are analyzed in the following sections, with the common target specifications listed in Table 11.4.

The residual susceptance value, B_S of the harmonic termination elements at f_0, was chosen to achieve an acceptable second-harmonic response. The impact of this parameter on the matching network performance was described fully in Section 9.2. Succinctly, increasing B_S decreases the value of L_b, which reduces dispersion in the second-harmonic band, but increases insertion loss in the passband.

11.8.1 Lowpass + Lowpass π-Networks Cascade

Figure 11.23 shows a simplified schematic of the elements in a typical PA output match comprising a cascade of a lowpass π-network followed by a second lowpass π-network.

Because the impedance matching requirement for the network can be satisfied by an unlimited set of solutions, the designer must select, from among them, a compromise solution that comes closest to meeting the design goals. In order to

Table 11.4 Common Single-Ended PA Match Target Specifications

Design Specifications	
f_{lo} 1,710 MHz	f_{hi} 2,025 MHz
Phase shift	$\Delta\Phi = 180°$
Input impedance	4Ω
Output impedance	50Ω
Residual susceptance	$B_S = 0.11$
$C_b = 50$ pF	$C_L = 50$ pF
Inductors, $Q_L = 60$	Capacitors, $Q_C \approx 100$

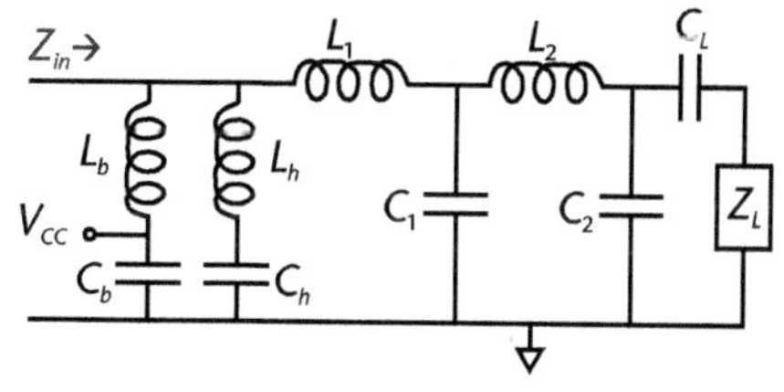

Figure 11.23 Lowpass π + lowpass π PA match.

make this choice, the designer must have a clear understanding of the performance trade-offs available for the network. An effective way to do this is to present the designer with visual presentations that clearly convey the sensitivities of the key parameters of concern to the network solutions available.

For the network "optimal" design approach proposed, therefore, a first step is to define the key parameters that are of most importance to the performance of the network. An example for such a list is:

1. Insertion gain.
2. Input impedance match $|Z_{in} - Z_S^*|$, where $Z_S = R_S + jX_S$ = source impedance.
3. Real input impedance match $|R_{in} - R_S|$.
4. Imaginary input impedance match $|X_{in} + X_S|$.
5. Second-harmonic rejection.

Two-dimensional contour plots for each of these parameters are then generated by sweeping the phase shifts in the first and second π-networks, $\delta\Phi_{\pi 1}$ and $\delta\Phi_{\pi 2}$, respectively. For each pair of phase shifts, the value of the real part of the complex impedance at the common node of the π-networks, R_{in}, is chosen to satisfy the parameter listed that is most critical. Because the optimum value for R_{in} is only typically weakly dependent on the choice of critical parameter, the choice has little effect on the contour plots generated. Furthermore, an optimal value of R_{in}, for any of the listed parameters, can be chosen once the designer has selected values for $\delta\Phi_{\pi 1}$ and $\delta\Phi_{\pi 2}$.

Figure 11.24 shows typical sensitivity plots of four key parameters on R_{in}. Data points corresponding to a selected value of $R_{in} \approx 17.62\Omega$ are denoted by the markers.

Any software suite capable of rendering visual presentations of data could be used for generating the parametric contour plots. As described in Chapter 7, Microsoft Excel provides one such option that is readily available to many. The contour plots below were generated in Excel, using the techniques described in Chapter 7.

In each of the contour plots shown in Figure 11.25(a–d), the green cells have the most desirable values for the corresponding dependent variable, while the dark red cells have the least desirable values. The cells in each chart that are dark green are the most favorable value in each chart. The cells in each chart that are yellow represent the current design selection, which can be simply changed by selecting a different cell in one of the charts. Each time the active cell is changed, the color of the corresponding cells in each of the charts is changed to yellow to highlight where the current design choice lies on each of the charts. The schematic and all network performance data and plots on the spreadsheet are determined based on the active cell selection. Thus, by simply moving the mouse around on the contour plot, the schematic and performance parameters displayed are instantly updated to reflect the design changes.

By way of example, Figure 11.25(e) provides one additional contour plot to aid in the selection of an optimal configuration for the network. In each cell, corresponding to a particular combination of phase shifts, the value of inductor L_1 (Figure 11.23) is displayed. Darker shading is used to denote an increase in value. Such data can be of benefit to the designer if there may be some limits on the practical size of the inductor in a compact module.

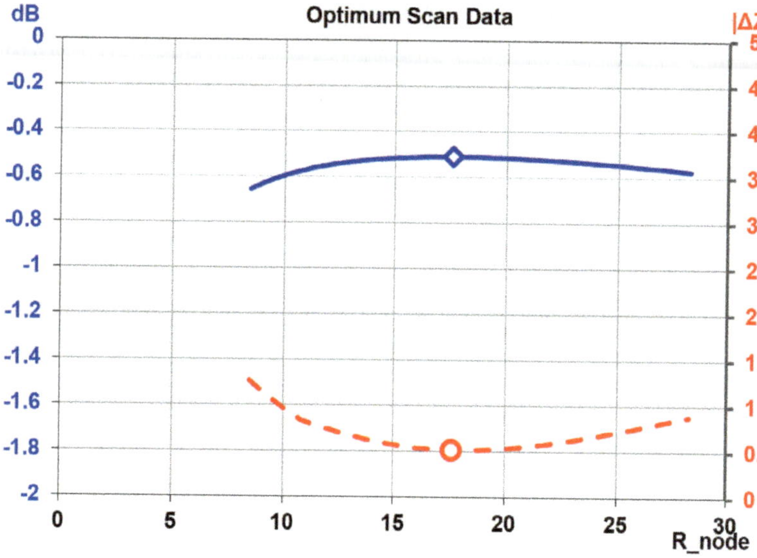

(a)

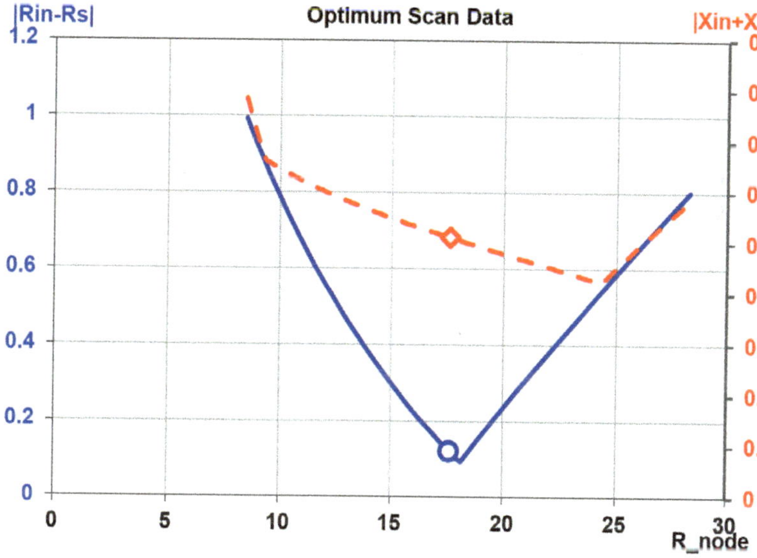

(b)

Figure 11.24 Lowpass π + lowpass π PA match R_{in} dependence: (a) insertion gain and $|Z_{in} - Z_S^*|$, and (b) $|R_{in} - R_S|$ and $|X_{in} + X_S|$.

Taken together, contour plots such as those shown in Figure 11.25 provide the designer with invaluable insight into the performance compromises that are available with the network.

Based on those plots, the optimal values were chosen as $R_{in} \approx 17.62\Omega$, $\delta\Phi_{\pi 1} = 61°$, and $\delta\Phi_{\pi 2} = 63°$. The corresponding set of optimal element values, to meet the design specifications of Table 11.4, are given in Table 11.5.

Using these element values, key plots that characterize the performance of the impedance matching network are shown in Figure 11.26.

The dual lowpass π impedance matching network exhibits good characteristics. Its input impedance across the passband is close to ideal, as is the output impedance. Across the second-harmonic bandwidth, the input impedance is close to the desired goal of an ideal short circuit. Angular dispersion of the impedance is low and could be further reduced at the expense of insertion loss, if desired.

The maximum insertion loss of the network over the passband is ~0.53 dB, and the rejection across the second-harmonic band is >25 dB.

	Selecting a cell in table below will automatically load phases in δΦπ1 & δΦπ2																		11	8
$\delta\Phi_{\pi 2}$	Insertion Gain (n_type=low-low,opt.=ins. loss)																			
80	-0.63	-0.63	-0.63	-0.62	-0.62	-0.62	-0.62	-0.62	-0.62	-0.62	-0.62	-0.62	-0.62	-0.62	-0.63	-0.63	-0.63	-0.64	-0.64	-0.65
78	-0.61	-0.61	-0.61	-0.6	-0.6	-0.6	-0.6	-0.6	-0.6	-0.6	-0.6	-0.6	-0.6	-0.61	-0.61	-0.61	-0.62	-0.62	-0.62	-0.63
77	-0.6	-0.6	-0.6	-0.6	-0.59	-0.59	-0.59	-0.59	-0.59	-0.59	-0.59	-0.59	-0.6	-0.6	-0.6	-0.6	-0.61	-0.61	-0.62	-0.62
75	-0.59	-0.58	-0.58	-0.58	-0.58	-0.58	-0.58	-0.57	-0.58	-0.58	-0.58	-0.58	-0.58	-0.58	-0.59	-0.59	-0.59	-0.6	-0.6	-0.61
74	-0.58	-0.57	-0.57	-0.57	-0.57	-0.57	-0.57	-0.57	-0.57	-0.57	-0.57	-0.57	-0.57	-0.58	-0.58	-0.58	-0.59	-0.59	-0.59	-0.6
72	-0.56	-0.56	-0.56	-0.56	-0.55	-0.55	-0.55	-0.55	-0.55	-0.56	-0.56	-0.56	-0.56	-0.56	-0.57	-0.57	-0.57	-0.58	-0.58	-0.59
71	-0.55	-0.55	-0.55	-0.55	-0.55	-0.55	-0.55	-0.55	-0.55	-0.55	-0.55	-0.55	-0.55	-0.56	-0.56	-0.57	-0.57	-0.57	-0.58	-0.58
69	-0.54	-0.54	-0.54	-0.54	-0.54	-0.53	-0.53	-0.54	-0.54	-0.54	-0.54	-0.54	-0.54	-0.55	-0.55	-0.55	-0.56	-0.56	-0.57	-0.57
67	-0.53	-0.53	-0.53	-0.52	-0.52	-0.52	-0.52	-0.52	-0.53	-0.53	-0.53	-0.53	-0.53	-0.54	-0.54	-0.55	-0.55	-0.55	-0.56	-0.56
66	-0.52	-0.52	-0.52	-0.52	-0.52	-0.52	-0.52	-0.52	-0.52	-0.52	-0.52	-0.53	-0.53	-0.53	-0.54	-0.54	-0.54	-0.55	-0.55	-0.56
64	-0.51	-0.51	-0.51	-0.51	-0.51	-0.51	-0.51	-0.51	-0.51	-0.51	-0.52	-0.52	-0.52	-0.53	-0.53	-0.53	-0.54	-0.54	-0.55	-0.55
63	-0.51	-0.51	-0.5	-0.5	-0.5	-0.5	-0.5	-0.51	-0.51	-0.51	-0.51	-0.51	-0.52	-0.52	-0.52	-0.53	-0.53	-0.54	-0.54	-0.55
61	-0.5	-0.5	-0.5	-0.5	-0.5	-0.5	-0.5	-0.5	-0.5	-0.5	-0.5	-0.51	-0.51	-0.52	-0.52	-0.52	-0.53	-0.53	-0.54	-0.54
59	-0.49	-0.49	-0.49	-0.49	-0.49	-0.49	-0.49	-0.49	-0.49	-0.5	-0.5	-0.5	-0.5	-0.51	-0.51	-0.52	-0.52	-0.53	-0.53	-0.54
58	-0.49	-0.48	-0.48	-0.48	-0.48	-0.49	-0.49	-0.49	-0.49	-0.49	-0.5	-0.5	-0.5	-0.51	-0.51	-0.52	-0.52	-0.53	-0.53	-0.54
56	-0.48	-0.48	-0.48	-0.48	-0.48	-0.48	-0.48	-0.48	-0.49	-0.49	-0.49	-0.49	-0.5	-0.5	-0.51	-0.51	-0.51	-0.52	-0.53	-0.53
55	-0.48	-0.48	-0.48	-0.48	-0.48	-0.48	-0.48	-0.48	-0.48	-0.48	-0.49	-0.49	-0.49	-0.5	-0.5	-0.51	-0.51	-0.52	-0.52	-0.53
53	-0.47	-0.47	-0.47	-0.47	-0.47	-0.47	-0.47	-0.48	-0.48	-0.48	-0.48	-0.49	-0.49	-0.5	-0.5	-0.51	-0.51	-0.52	-0.52	-0.53
52	-0.47	-0.47	-0.47	-0.47	-0.47	-0.47	-0.47	-0.47	-0.48	-0.48	-0.48	-0.48	-0.49	-0.5	-0.5	-0.5	-0.51	-0.52	-0.52	-0.53
50	-0.46	-0.46	-0.46	-0.46	-0.47	-0.47	-0.47	-0.47	-0.47	-0.48	-0.48	-0.48	-0.49	-0.49	-0.5	-0.5	-0.51	-0.51	-0.52	-0.53
$\delta\Phi_{\pi 1}$	50	52	53	55	56	58	59	61	63	64	66	67	69	71	72	74	75	77	78	80

(a)

$\delta\Phi_{\pi 2}$	|Z-Zs| (n_type=low-low,opt.=ins. loss)																			
80	0.5	0.5	0.5	0.4	0.4	0.4	0.4	0.3	0.3	0.3	0.3	0.3	0.3	0.3	0.3	0.4	0.4	0.4	0.4	0.5
78	0.5	0.4	0.4	0.4	0.4	0.3	0.3	0.3	0.2	0.2	0.3	0.3	0.3	0.4	0.4	0.4	0.4	0.5	0.5	0.5
77	0.4	0.4	0.4	0.3	0.3	0.3	0.3	0.2	0.3	0.3	0.3	0.3	0.3	0.4	0.4	0.4	0.5	0.5	0.5	0.5
75	0.4	0.3	0.3	0.3	0.3	0.2	0.2	0.3	0.3	0.3	0.3	0.4	0.4	0.4	0.4	0.5	0.5	0.5	0.6	0.6
74	0.3	0.3	0.3	0.3	0.2	0.2	0.3	0.3	0.3	0.3	0.4	0.4	0.4	0.4	0.5	0.5	0.5	0.6	0.6	0.6
72	0.3	0.2	0.2	0.3	0.3	0.3	0.3	0.3	0.4	0.4	0.4	0.4	0.5	0.5	0.5	0.5	0.6	0.6	0.6	0.6
71	0.2	0.3	0.3	0.3	0.3	0.3	0.3	0.4	0.4	0.4	0.4	0.4	0.5	0.5	0.5	0.6	0.6	0.6	0.6	0.7
69	0.3	0.3	0.3	0.3	0.3	0.4	0.4	0.4	0.4	0.4	0.5	0.5	0.5	0.6	0.6	0.6	0.6	0.7	0.7	0.7
67	0.4	0.4	0.4	0.4	0.4	0.4	0.4	0.4	0.5	0.5	0.5	0.5	0.6	0.6	0.6	0.6	0.7	0.7	0.7	0.7
66	0.4	0.4	0.4	0.4	0.4	0.4	0.4	0.5	0.5	0.5	0.5	0.6	0.6	0.6	0.6	0.7	0.7	0.7	0.7	0.8
64	0.4	0.4	0.4	0.5	0.5	0.5	0.5	0.5	0.5	0.6	0.6	0.6	0.6	0.7	0.7	0.7	0.7	0.8	0.8	0.8
63	0.5	0.5	0.5	0.5	0.5	0.5	0.5	0.5	0.6	0.6	0.6	0.6	0.6	0.7	0.7	0.7	0.7	0.8	0.8	0.8
61	0.5	0.5	0.5	0.5	0.5	0.6	0.6	0.6	0.6	0.6	0.6	0.7	0.7	0.7	0.7	0.8	0.8	0.8	0.8	0.9
59	0.6	0.6	0.6	0.6	0.6	0.6	0.6	0.6	0.6	0.7	0.7	0.7	0.7	0.8	0.8	0.8	0.8	0.8	0.9	0.9
58	0.6	0.6	0.6	0.6	0.6	0.6	0.6	0.6	0.7	0.7	0.7	0.7	0.7	0.8	0.8	0.8	0.8	0.9	0.9	0.9
56	0.6	0.6	0.6	0.7	0.7	0.7	0.7	0.7	0.7	0.7	0.7	0.8	0.8	0.8	0.8	0.8	0.9	0.9	0.9	0.9
55	0.7	0.7	0.7	0.7	0.7	0.7	0.7	0.7	0.7	0.7	0.8	0.8	0.8	0.8	0.8	0.9	0.9	0.9	0.9	1
53	0.7	0.7	0.7	0.7	0.7	0.7	0.7	0.8	0.8	0.8	0.8	0.8	0.8	0.9	0.9	0.9	0.9	0.9	1	1
52	0.8	0.7	0.7	0.7	0.7	0.8	0.8	0.8	0.8	0.8	0.8	0.8	0.9	0.9	0.9	0.9	0.9	1	1	1
50	0.8	0.8	0.8	0.8	0.8	0.8	0.8	0.8	0.8	0.8	0.9	0.9	0.9	0.9	0.9	1	1	1	1	1
$\delta\Phi_{\pi 1}$	50	52	53	55	56	58	59	61	63	64	66	67	69	71	72	74	75	77	78	80

(b)

Figure 11.25 Lowpass π + lowpass π PA match parametric contour plots: (a) insertion gain, (b) $|Z_{in} - Z_S^*|$, (c) $|R_{in} - R_S|$, (d) $|X_{in} + X_S|$, (e) second-harmonic rejection, and (f) inductor L_1 value.

|Rin-Rs| (n_type=low-low,opt.=ins. loss)

$\delta\Phi_{\pi 2}$																				
80	0.34	0.35	0.34	0.34	0.34	0.33	0.32	0.31	0.29	0.27	0.25	0.23	0.21	0.23	0.24	0.28	0.3	0.32	0.35	0.39
78	0.27	0.29	0.20	0.20	0.27	0.27	0.26	0.24	0.22	0.21	0.10	0.17	0.19	0.21	0.26	0.29	0.32	0.35	0.36	0.41
77	0.24	0.26	0.26	0.25	0.24	0.23	0.22	0.21	0.19	0.18	0.15	0.16	0.2	0.23	0.26	0.3	0.31	0.36	0.38	0.42
75	0.2	0.2	0.2	0.19	0.2	0.17	0.17	0.15	0.13	0.12	0.15	0.17	0.2	0.24	0.26	0.31	0.33	0.37	0.4	0.45
74	0.18	0.17	0.17	0.16	0.17	0.15	0.14	0.13	0.1	0.11	0.15	0.17	0.21	0.25	0.27	0.31	0.34	0.38	0.41	0.45
72	0.12	0.13	0.12	0.12	0.11	0.1	0.08	0.07	0.08	0.09	0.13	0.15	0.2	0.24	0.28	0.32	0.34	0.4	0.42	0.47
71	0.12	0.11	0.11	0.1	0.08	0.07	0.07	0.04	0.06	0.08	0.13	0.15	0.2	0.24	0.27	0.33	0.35	0.39	0.43	0.47
69	0.17	0.14	0.13	0.1	0.09	0.06	0.03	**0.02**	0.05	0.08	0.13	0.15	0.2	0.25	0.27	0.32	0.35	0.41	0.43	0.49
67	0.21	0.19	0.16	0.14	0.11	0.07	0.05	0.06	0.1	0.1	0.14	0.16	0.2	0.24	0.26	0.31	0.36	0.41	0.43	0.49
66	0.23	0.2	0.19	0.15	0.12	0.08	0.07	0.09	0.11	0.13	0.17	0.19	0.23	0.27	0.29	0.33	0.35	0.41	0.44	0.5
64	0.28	0.23	0.22	0.18	0.17	0.12	0.1	**0.12**	0.16	0.18	0.21	0.23	0.27	0.32	0.34	0.39	0.41	0.46	0.48	0.53
63	0.3	0.27	0.24	0.21	0.18	0.14	0.12	0.14	0.17	0.19	0.24	0.26	0.3	0.35	0.37	0.41	0.43	0.49	0.51	0.56
61	0.35	0.3	0.29	0.24	0.23	0.18	0.14	0.17	0.22	0.24	0.27	0.3	0.34	0.39	0.41	0.46	0.49	0.54	0.57	0.62
59	0.39	0.36	0.34	0.29	0.25	0.2	0.19	0.21	0.24	0.27	0.32	0.34	0.38	0.43	0.46	0.52	0.54	0.59	0.62	0.68
58	0.43	0.38	0.36	0.3	0.29	0.21	0.2	0.21	0.26	0.29	0.33	0.35	0.4	0.46	0.48	0.54	0.56	0.62	0.65	0.71
56	0.47	0.43	0.41	0.35	0.32	0.26	0.24	0.25	0.28	0.31	0.37	0.38	0.44	0.49	0.52	0.58	0.62	0.67	0.7	0.76
55	0.51	0.45	0.43	0.37	0.35	0.27	0.25	0.24	0.3	0.33	0.37	0.4	0.46	0.52	0.55	0.6	0.63	0.69	0.73	0.78
53	0.57	0.51	0.49	0.42	0.38	0.32	0.3	0.27	0.32	0.35	0.39	0.42	0.48	0.55	0.58	0.64	0.67	0.74	0.77	0.83
52	0.59	0.53	0.5	0.44	0.42	0.35	0.31	0.27	0.33	0.34	0.41	0.44	0.5	0.57	0.6	0.66	0.69	0.76	0.79	0.86
50	0.65	0.59	0.56	0.49	0.47	0.38	0.36	0.29	0.34	0.37	0.42	0.46	0.52	0.59	0.63	0.69	0.73	0.8	0.83	0.9
$\delta\Phi_{\pi 1}$	50	52	53	55	56	58	59	61	63	64	66	67	69	71	72	74	75	77	78	80

(c)

|Xin-Xs| (n_type=low-low,opt.=ins. loss)

$\delta\Phi_{\pi 2}$																				
80	0.4	0.3	0.3	0.3	0.3	0.2	0.2	0.1	**0.1**	0.2	0.2	0.2	0.2	0.2	0.2	0.2	0.2	0.3	0.3	0.3
78	0.4	0.3	0.3	0.2	0.2	0.2	0.2	0.2	0.2	0.2	0.2	0.2	0.3	0.3	0.3	0.3	0.3	0.3	0.3	0.3
77	0.3	0.3	0.3	0.2	0.2	0.2	0.2	0.2	0.2	0.2	0.3	0.3	0.3	0.3	0.3	0.3	0.3	0.3	0.3	0.3
75	0.3	0.3	0.2	0.2	0.2	0.2	0.2	0.3	0.3	0.3	0.3	0.3	0.3	0.4	0.4	0.4	0.4	0.4	0.4	0.4
74	0.3	0.2	0.2	0.2	0.2	0.2	0.3	0.3	0.3	0.3	0.3	0.3	0.4	0.4	0.4	0.4	0.4	0.4	0.4	0.4
72	0.2	0.2	0.2	0.2	0.3	0.3	0.3	0.3	0.4	0.4	0.4	0.4	0.4	0.4	0.4	0.4	0.4	0.4	0.4	0.4
71	0.2	0.2	0.2	0.3	0.3	0.3	0.3	0.4	0.4	0.4	0.4	0.4	0.4	0.5	0.5	0.5	0.5	0.5	0.5	0.5
69	0.2	0.3	0.3	0.3	0.3	0.4	0.4	0.4	0.4	0.4	0.5	0.5	0.5	0.5	0.5	0.5	0.5	0.5	0.5	0.5
67	0.3	0.3	0.3	0.4	0.4	0.4	0.4	0.4	0.5	0.5	0.5	0.5	0.5	0.5	0.6	0.6	0.6	0.6	0.6	0.6
66	0.3	0.3	0.3	0.4	0.4	0.4	0.4	0.5	0.5	0.5	0.5	0.5	0.6	0.6	0.6	0.6	0.6	0.6	0.6	0.6
64	0.3	0.4	0.4	0.4	0.4	0.5	0.5	**0.5**	0.5	0.6	0.6	0.6	0.6	0.6	0.6	0.6	0.6	0.6	0.6	0.6
63	0.3	0.4	0.4	0.4	0.5	0.5	0.5	0.5	0.6	0.6	0.6	0.6	0.6	0.6	0.6	0.7	0.7	0.7	0.7	0.6
61	0.4	0.4	0.4	0.5	0.5	0.5	0.5	0.6	0.6	0.6	0.6	0.6	0.7	0.7	0.7	0.7	0.7	0.7	0.7	0.7
59	0.4	0.4	0.5	0.5	0.5	0.6	0.6	0.6	0.6	0.7	0.7	0.7	0.7	0.7	0.7	0.7	0.7	0.7	0.7	0.7
58	0.4	0.5	0.5	0.5	0.5	0.6	0.6	0.6	0.7	0.7	0.7	0.7	0.7	0.7	0.8	0.8	0.8	0.8	0.8	0.8
56	0.4	0.5	0.5	0.5	0.6	0.6	0.6	0.7	0.7	0.7	0.7	0.8	0.8	0.8	0.8	0.8	0.8	0.8	0.8	0.8
55	0.4	0.5	0.5	0.6	0.6	0.6	0.6	0.7	0.7	0.7	0.8	0.8	0.8	0.8	0.8	0.8	0.8	0.8	0.8	0.8
53	0.5	0.5	0.5	0.6	0.6	0.7	0.7	0.7	0.8	0.8	0.8	0.8	0.8	0.8	0.9	0.9	0.9	0.9	0.9	0.9
52	0.5	0.5	0.5	0.6	0.6	0.7	0.7	0.7	0.8	0.8	0.8	0.8	0.9	0.9	0.9	0.9	0.9	0.9	0.9	0.9
50	0.5	0.5	0.6	0.6	0.6	0.7	0.7	0.8	0.8	0.8	0.9	0.9	0.9	0.9	0.9	0.9	0.9	0.9	0.9	0.9
$\delta\Phi_{\pi 1}$	50	52	53	55	56	58	59	61	63	64	66	67	69	71	72	74	75	77	78	80

(d)

Figure 11.25 *(Continued)*

11.8.2 Lowpass + Highpass π-Networks Cascade

Figure 11.27 shows a simplified schematic of the elements in a typical PA output match comprising a cascade of a lowpass π-network followed by a second highpass π-network.

Figure 11.28 shows typical sensitivity plots of four key parameters on the value of R_{in}. Data points corresponding to the optimal selected value of $R_{in} \approx 19.28\Omega$ are denoted by the markers.

Parametric contour plots of the most critical performance parameters were again generated, as in Section 11.8.1. Based on those plots, the optimal values were chosen

$\delta\Phi_{\pi 2}$	Rejection (n_type=low-low,opt.=ins. loss)																			
80	-28	-28	-28	-29	-29	-29	-29	-30	-30	-30	-31	-31	-31	-31	-31	-32	-32	-32	-32	**-32**
78	-27	-28	-28	-28	-28	-29	-29	-29	-30	-30	-30	-30	-31	-31	-31	-31	-31	-32	-32	-32
77	-27	-27	-28	-28	-28	-29	-29	-29	-29	-30	-30	-30	-30	-31	-31	-31	-31	-31	-32	-32
75	-26	-27	-27	-27	-28	-28	-28	-29	-29	-29	-29	-30	-30	-30	-30	-31	-31	-31	-31	-31
74	-26	-26	-27	-27	-27	-28	-28	-28	-29	-29	-29	-29	-30	-30	-30	-30	-30	-31	-31	-31
72	-25	-26	-26	-27	-27	-27	-27	-28	-28	-28	-29	-29	-29	-29	-30	-30	-30	-30	-30	-31
71	-25	-26	-26	-26	-27	-27	-27	-28	-28	-28	-28	-29	-29	-29	-29	-30	-30	-30	-30	-30
69	-25	-25	-25	-26	-26	-26	-27	-27	-27	-27	-28	-28	-28	-29	-29	-29	-29	-29	-30	-30
67	-24	-25	-25	-25	-25	-26	-26	-26	-27	-27	-27	-27	-28	-28	-28	-29	-29	-29	-29	-29
66	-24	-24	-24	-25	-25	-26	-26	-26	-27	-27	-27	-27	-27	-28	-28	-28	-28	-29	-29	-29
64	-23	-24	-24	-24	-25	-25	-25	**-26**	-26	-26	-26	-27	-27	-27	-27	-28	-28	-28	-28	-28
63	-23	-23	-23	-24	-24	-25	-25	-25	-26	-26	-26	-26	-27	-27	-27	-27	-28	-28	-28	-28
61	-22	-23	-23	-23	-24	-24	-24	-25	-25	-25	-26	-26	-26	-26	-26	-27	-27	-27	-27	-28
59	-21	-22	-22	-23	-23	-23	-24	-24	-24	-25	-25	-25	-25	-26	-26	-26	-26	-27	-27	-27
58	-21	-22	-22	-22	-23	-23	-23	-24	-24	-24	-25	-25	-25	-25	-26	-26	-26	-26	-26	-27
56	-20	-21	-21	-22	-22	-22	-23	-23	-23	-24	-24	-24	-24	-25	-25	-25	-25	-26	-26	-26
55	-20	-20	-21	-21	-22	-22	-22	-23	-23	-23	-24	-24	-24	-24	-25	-25	-25	-25	-25	-26
53	-19	-20	-20	-21	-21	-21	-21	-22	-22	-22	-23	-23	-23	-24	-24	-24	-24	-25	-25	-25
52	-19	-19	-20	-20	-20	-21	-21	-21	-22	-22	-22	-23	-23	-23	-23	-24	-24	-24	-24	-25
50	-18	-18	-19	-19	-20	-20	-20	-21	-21	-21	-22	-22	-22	-23	-23	-23	-23	-24	-24	-24
$\delta\Phi_{\pi 1}$	**50**	**52**	**53**	**55**	**56**	**58**	**59**	**61**	**63**	**64**	**66**	**67**	**69**	**71**	**72**	**74**	**75**	**77**	**78**	**80**

(e)

$\delta\Phi_{\pi 2}$	L1 (-ve=cap.) (n_type=low-low,opt.=ins. loss)																			
80	0.5	0.6	0.6	0.6	0.7	0.7	0.7	0.8	0.8	0.9	1	1	1.1	1.3	1.3	1.6	1.7	2	2.3	3.1
78	0.5	0.6	0.6	0.6	0.7	0.7	0.7	0.8	0.8	0.9	1	1	1.1	1.3	1.3	1.5	1.7	2	2.3	3
77	0.5	0.6	0.6	0.6	0.7	0.7	0.7	0.8	0.8	0.9	1	1	1.1	1.2	1.3	1.5	1.7	2	2.3	3
75	0.5	0.6	0.6	0.6	0.6	0.7	0.7	0.8	0.8	0.9	1	1	1.1	1.2	1.3	1.5	1.7	2	2.3	3
74	0.5	0.6	0.6	0.6	0.6	0.7	0.7	0.8	0.8	0.9	1	1	1.1	1.2	1.3	1.5	1.7	2	2.3	2.9
72	0.5	0.6	0.6	0.6	0.6	0.7	0.7	0.8	0.8	0.9	0.9	1	1.1	1.2	1.3	1.5	1.6	2	2.2	2.9
71	0.5	0.6	0.6	0.6	0.6	0.7	0.7	0.8	0.8	0.9	0.9	1	1.1	1.2	1.3	1.5	1.6	2	2.2	2.8
69	0.5	0.6	0.6	0.6	0.6	0.7	0.7	0.8	0.8	0.9	0.9	1	1.1	1.2	1.3	1.5	1.6	2	2.2	2.8
67	0.5	0.6	0.6	0.6	0.6	0.7	0.7	0.8	0.8	0.9	0.9	1	1.1	1.2	1.3	1.5	1.6	1.9	2.1	2.7
66	0.5	0.6	0.6	0.6	0.6	0.7	0.7	0.8	0.8	0.9	0.9	1	1.1	1.2	1.3	1.5	1.6	1.9	2.1	2.7
64	0.5	0.6	0.6	0.6	0.6	0.7	0.7	**0.8**	0.8	0.9	0.9	1	1.1	1.2	1.3	1.5	1.6	1.9	2.1	2.7
63	0.5	0.6	0.6	0.6	0.6	0.7	0.7	0.8	0.8	0.8	0.9	1	1.1	1.2	1.3	1.5	1.6	1.9	2.1	2.7
61	0.5	0.6	0.6	0.6	0.6	0.7	0.7	0.7	0.8	0.8	0.9	1	1.1	1.2	1.3	1.4	1.6	1.8	2	2.6
59	0.5	0.6	0.6	0.6	0.6	0.7	0.7	0.7	0.8	0.8	0.9	1	1	1.2	1.2	1.4	1.5	1.8	2	2.5
58	0.5	0.6	0.6	0.6	0.6	0.7	0.7	0.7	0.8	0.8	0.9	0.9	1	1.2	1.2	1.4	1.5	1.8	2	2.5
56	0.5	0.6	0.6	0.6	0.6	0.7	0.7	0.7	0.8	0.8	0.9	0.9	1	1.2	1.2	1.4	1.5	1.8	1.9	2.5
55	0.5	0.6	0.6	0.6	0.6	0.7	0.7	0.7	0.8	0.8	0.9	0.9	1	1.2	1.2	1.4	1.5	1.7	1.9	2.4
53	0.5	0.6	0.6	0.6	0.6	0.7	0.7	0.7	0.8	0.8	0.9	0.9	1	1.1	1.2	1.4	1.5	1.7	1.9	2.3
52	0.5	0.6	0.6	0.6	0.6	0.7	0.7	0.7	0.8	0.8	0.9	0.9	1	1.1	1.2	1.4	1.5	1.7	1.9	2.3
50	0.5	0.5	0.6	0.6	0.6	0.7	0.7	0.7	0.8	0.8	0.9	0.9	1	1.1	1.2	1.3	1.4	1.7	1.8	2.3
$\delta\Phi_{\pi 1}$	**50**	**52**	**53**	**55**	**56**	**58**	**59**	**61**	**63**	**64**	**66**	**67**	**69**	**71**	**72**	**74**	**75**	**77**	**78**	**80**

(f)

Figure 11.25 *(Continued)*

Table 11.5 Lowpass π + Lowpass π Optimal PA Match Elements

Element Values	
$L_b = 0.755$ nH	
$L_b = 0.273$ nH	$C_b = 6.934$ pF
$L_1 = 0.623$ nH	$C_1 = 9.731$ pF
$L_2 = 2.256$ nH	$C_2 = 2.295$ pF

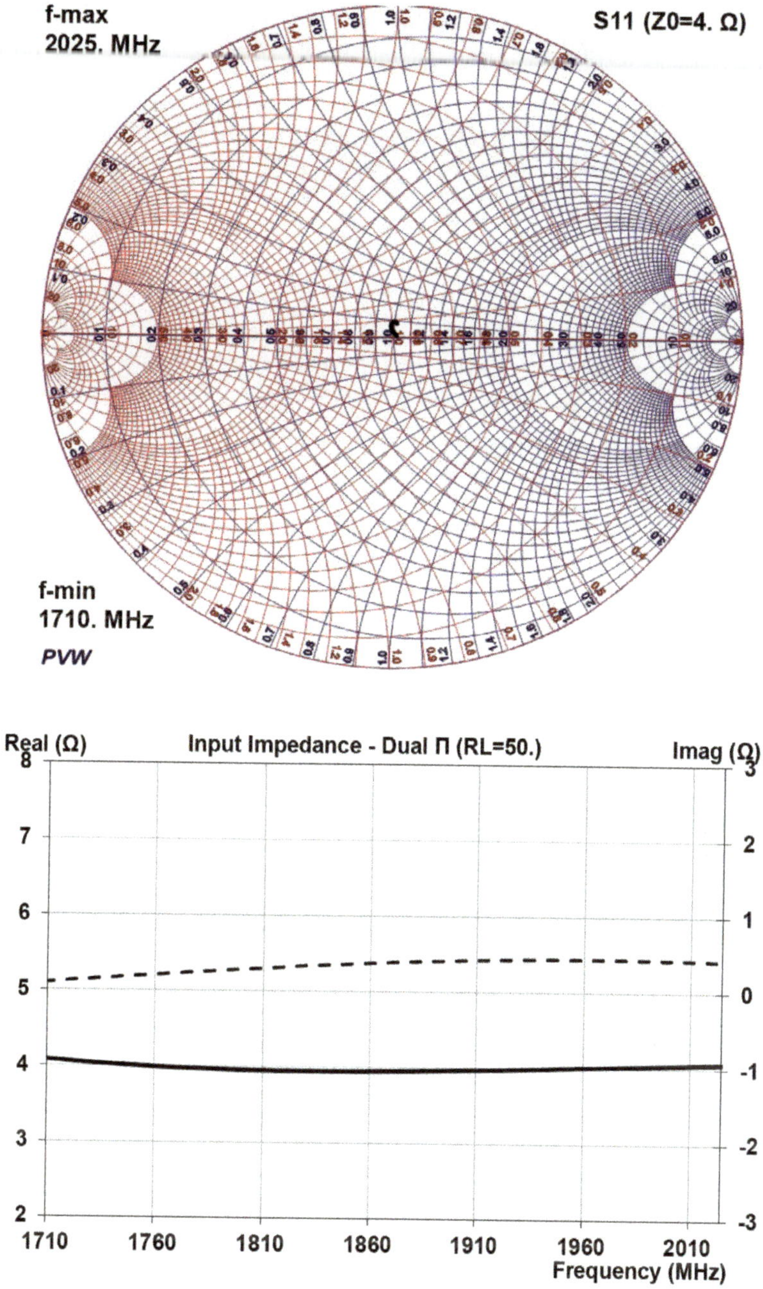

(a)

Figure 11.26 Lowpass π + lowpass π characterization plots: (a) input impedance, (b) passband output impedance, (c) passband insertion gain, (d) second-harmonic input impedance, and (e) second-harmonic insertion gain.

as $R_{in} \approx 19.28\Omega$, $\delta\Phi_{\pi 1} = 61°$, and $\delta\Phi_{\pi 2} = 64°$. The corresponding set of optimal element values, to meet the design specifications of Table 11.4, are given in Table 11.6.

Using these element values, key plots that characterize the performance of the impedance matching network are shown below.

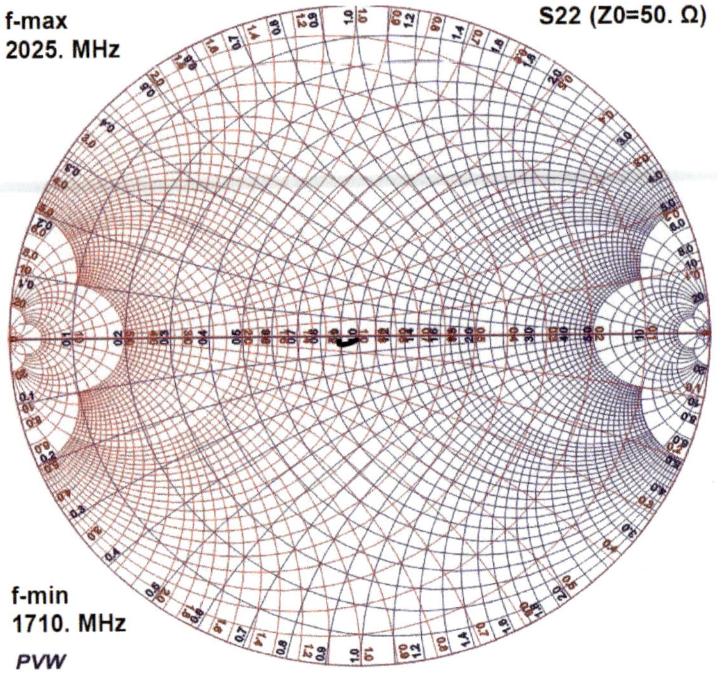

(b)

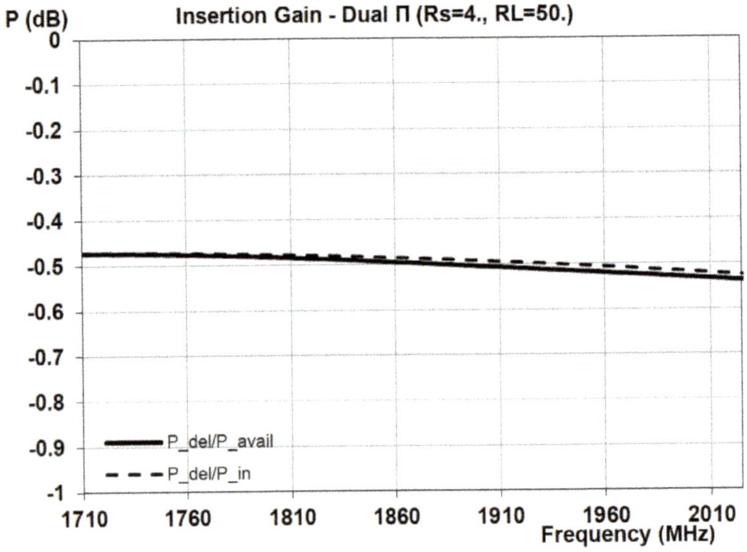

(c)

Figure 11.26 *(Continued)*

The dual-π lowpass + highpass impedance matching network exhibits good characteristics. Its input impedance across the passband is close to ideal, as is the output impedance. Across the second-harmonic bandwidth, the input impedance is close to the desired goal of an ideal short circuit. Angular dispersion of the latter is low and could be further reduced at the expense of insertion loss, if desired.

The maximum insertion loss of the network over the passband is ~0.52 dB, and rejection across the second-harmonic band is >18 dB.

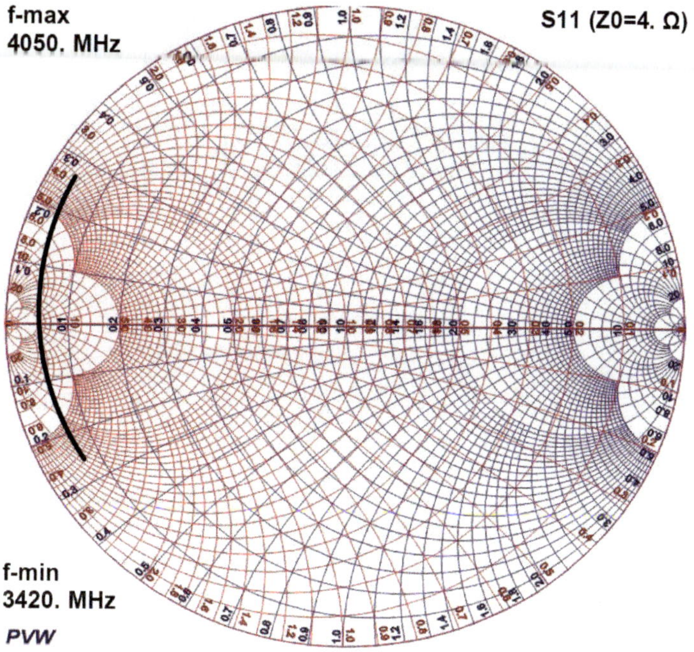

(d)

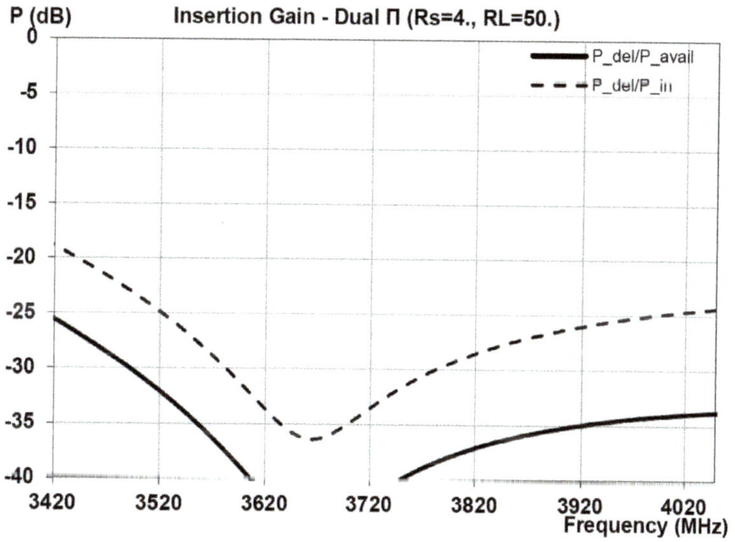

(e)

Figure 11.26 *(Continued)*

Table 11.6 Lowpass π + Highpass π Optimal PA Match Elements

Element Values	
$L_b = 0.716$ nH	
$L_h = 0.273$ nH	$C_h = 6.934$ pF
$L_1 = 0.652$ nH	$C_1 = 7.796$ pF
$C_2 = 3.051$ pF	$L_2 = 3.187$ pF

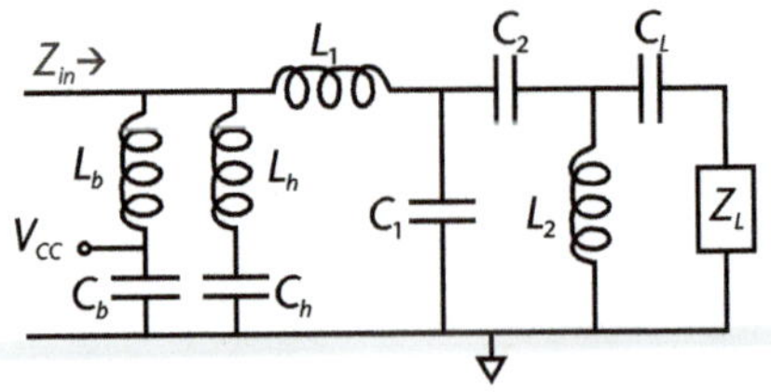

Figure 11.27 Lowpass π + highpass π PA match.

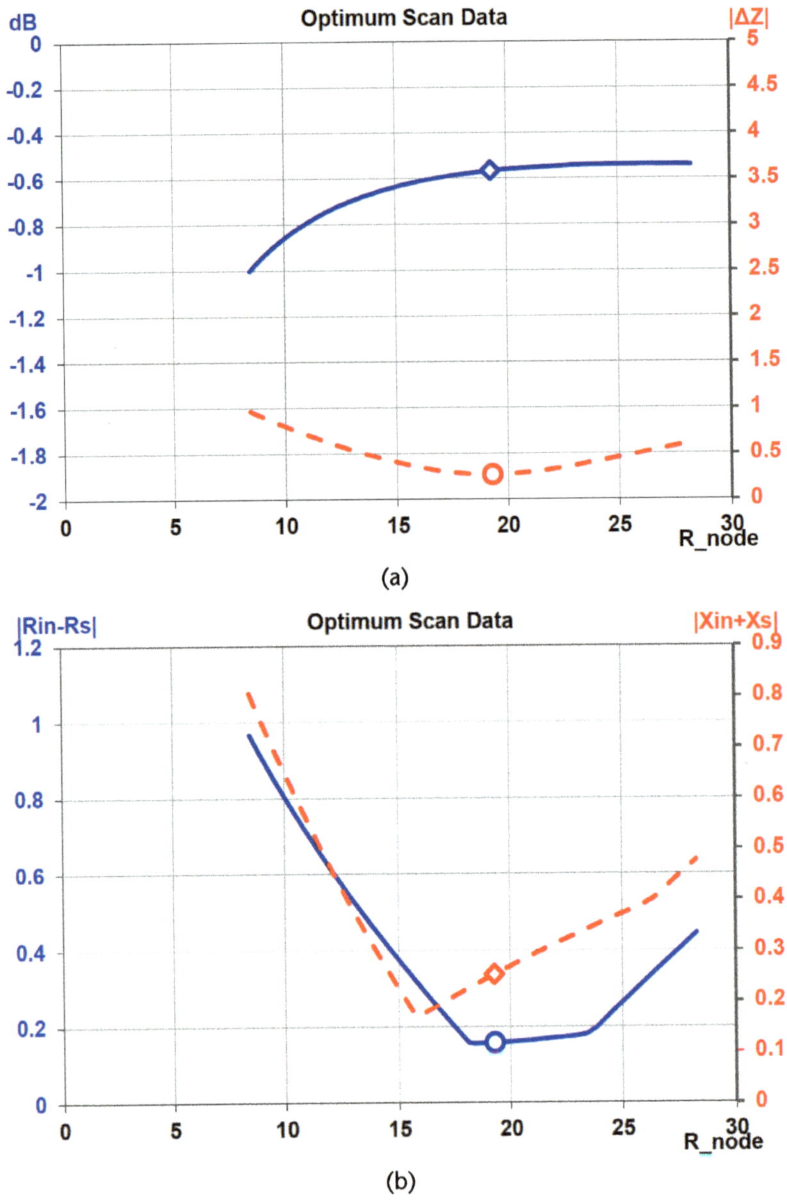

Figure 11.28 Lowpass π + highpass π PA match R_{in} dependence: (a) insertion gain and $|Z_{in} - Z_S^*|$, and (b) $|R_{in} - R_S|$ and $|X_{in} + X_S|$.

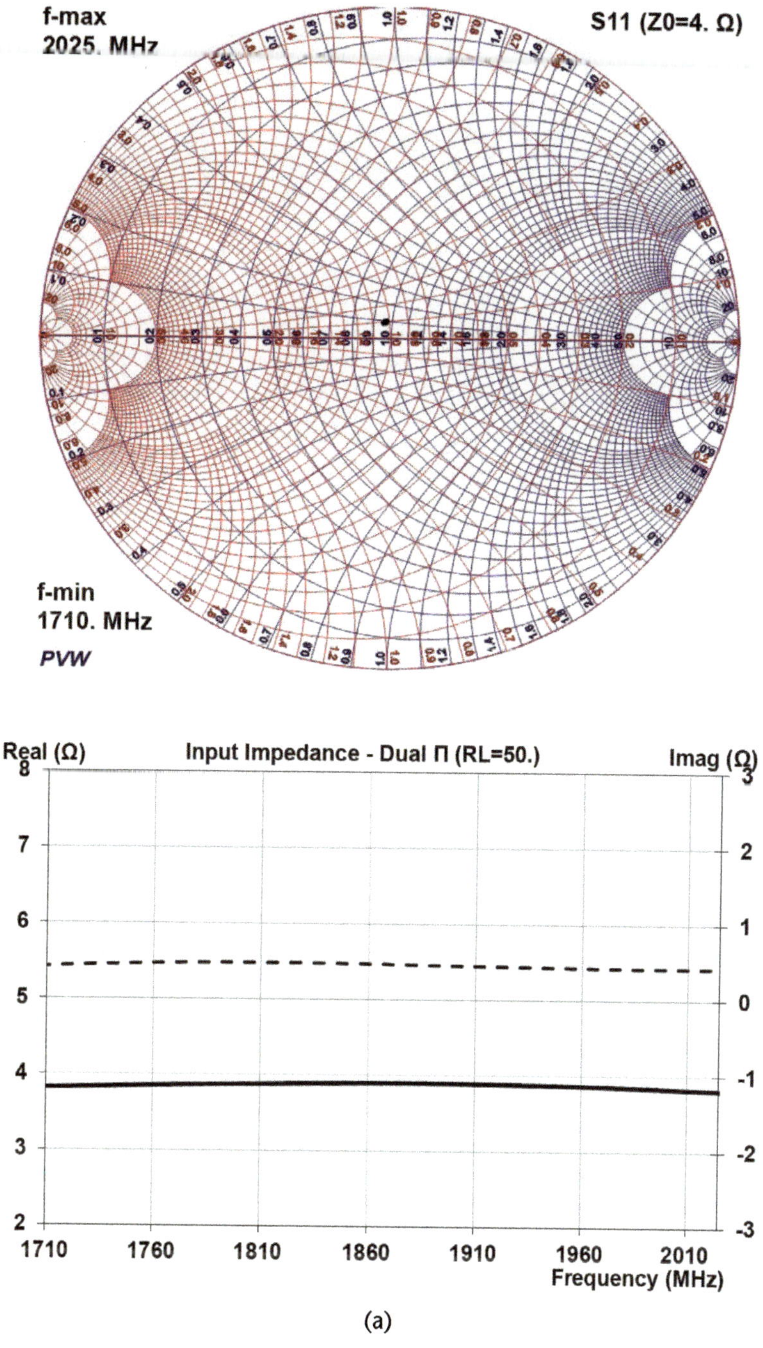

(a)

Figure 11.29 Lowpass π + highpass π characterization plots: (a) input impedance, (b) passband output impedance, (c) passband insertion gain, (d) second-harmonic input impedance, and (e) second-harmonic insertion gain.

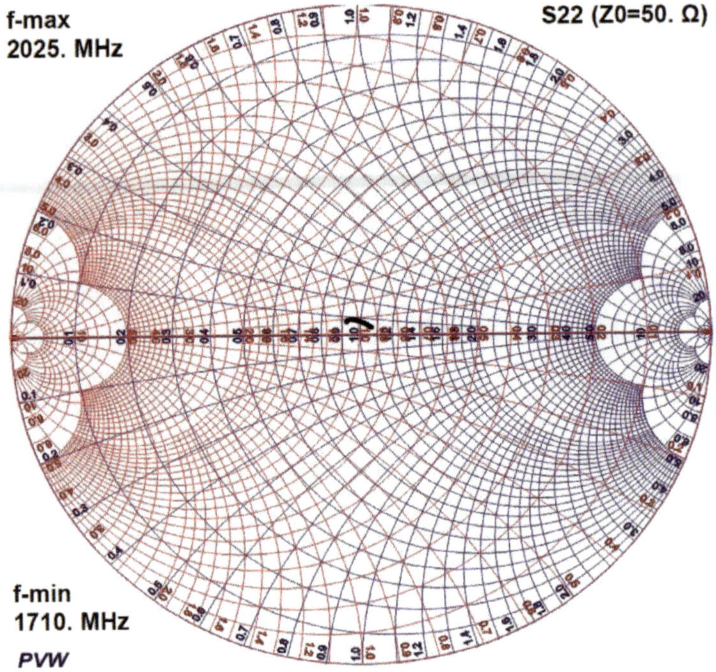

(b)

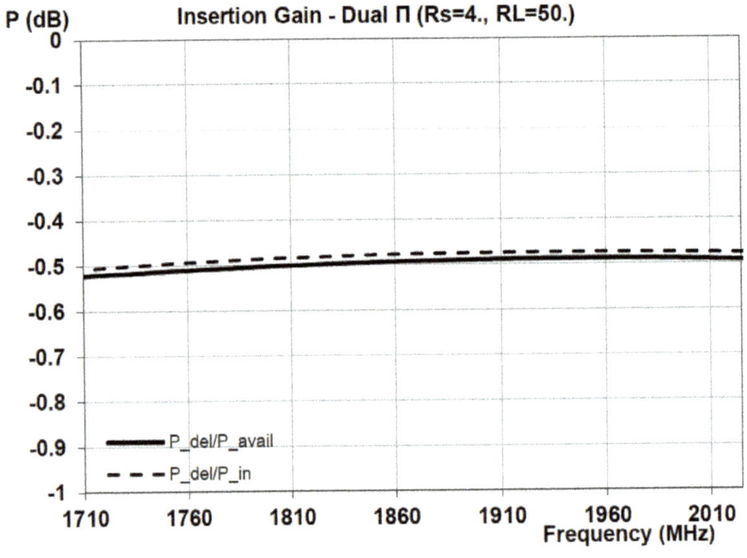

(c)

Figure 11.29 *(Continued)*

11.8.3 Highpass + Lowpass π-Networks Cascade

Figure 11.30 shows a simplified schematic of the elements in a typical PA output match comprising a cascade of a highpass π-network followed by a second lowpass π-network.

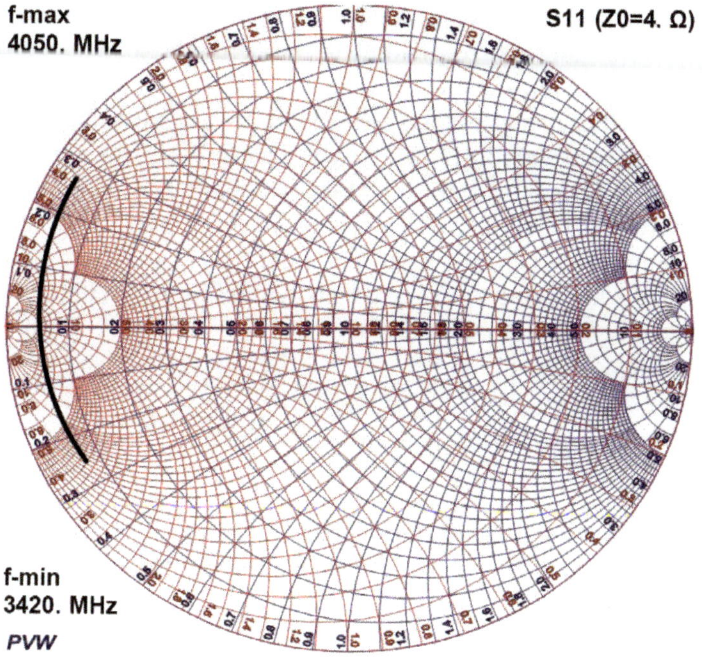

(d)

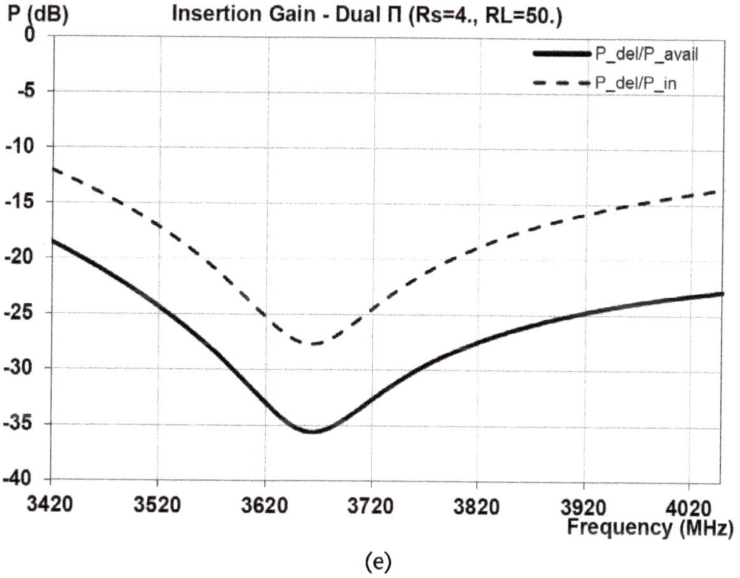

(e)

Figure 11.29 *(Continued)*

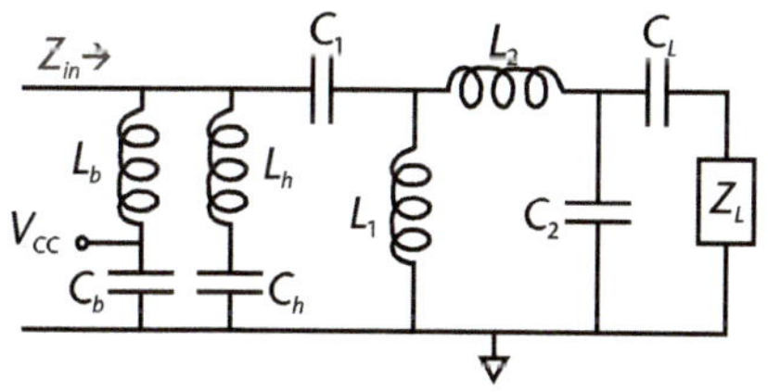

Figure 11.30 Highpass π + lowpass π PA match.

Figure 11.31 shows typical sensitivity plots of four key parameters on the value of R_{in}. Data points corresponding to the optimal selected value of $R_{in} \approx 18.41\Omega$ are denoted by the markers.

Parametric contour plots of the most critical performance parameters were again generated, as in Section 11.8.1. Based on those plots, the optimal values were chosen as $R_{in} \approx 18.41\Omega$, $\delta\Phi_{\pi 1} = 61°$, and $\delta\Phi_{\pi 2} = 64°$. The corresponding set of optimal element values, to meet the design specifications of Table 11.4, are given in Table 11.7.

Using these element values, key plots that characterize the performance of the impedance matching network are shown in Figure 11.32.

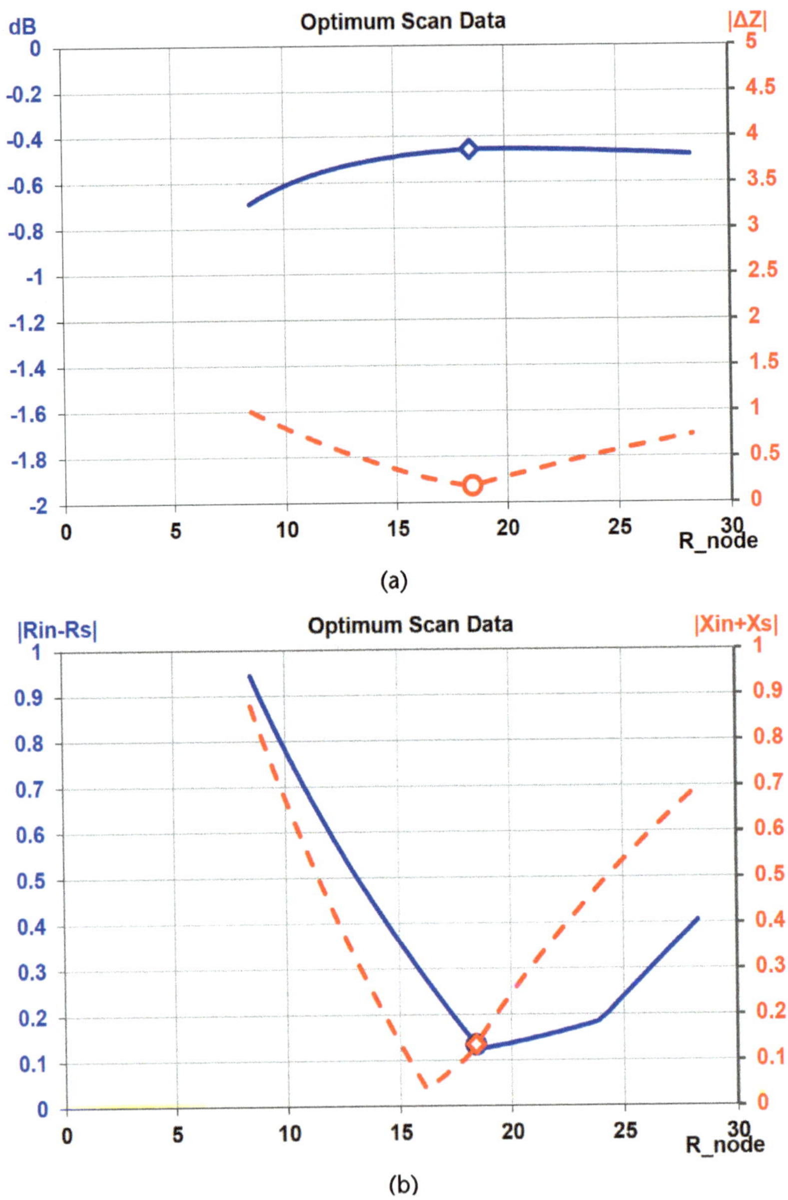

Figure 11.31 Highpass π + lowpass π PA match R_{in} dependence: (a) insertion gain and $|Z_{in} - Z_S^*|$, and (b) $|R_{in} - R_S|$ and $|X_{in} + X_S|$.

Table 11.7 Highpass π + Lowpass π "Optimal" PA Match Elements

Element Values	
$L_b = 0.819$ nH	
$L_b = 0.273$ nH	$C_b = 6.934$ pF
$L_1 = 0.911$ nH	$C_1 = 11.404$ pF
$C_2 = 2.235$ pF	$L_2 = 2.326$ pF

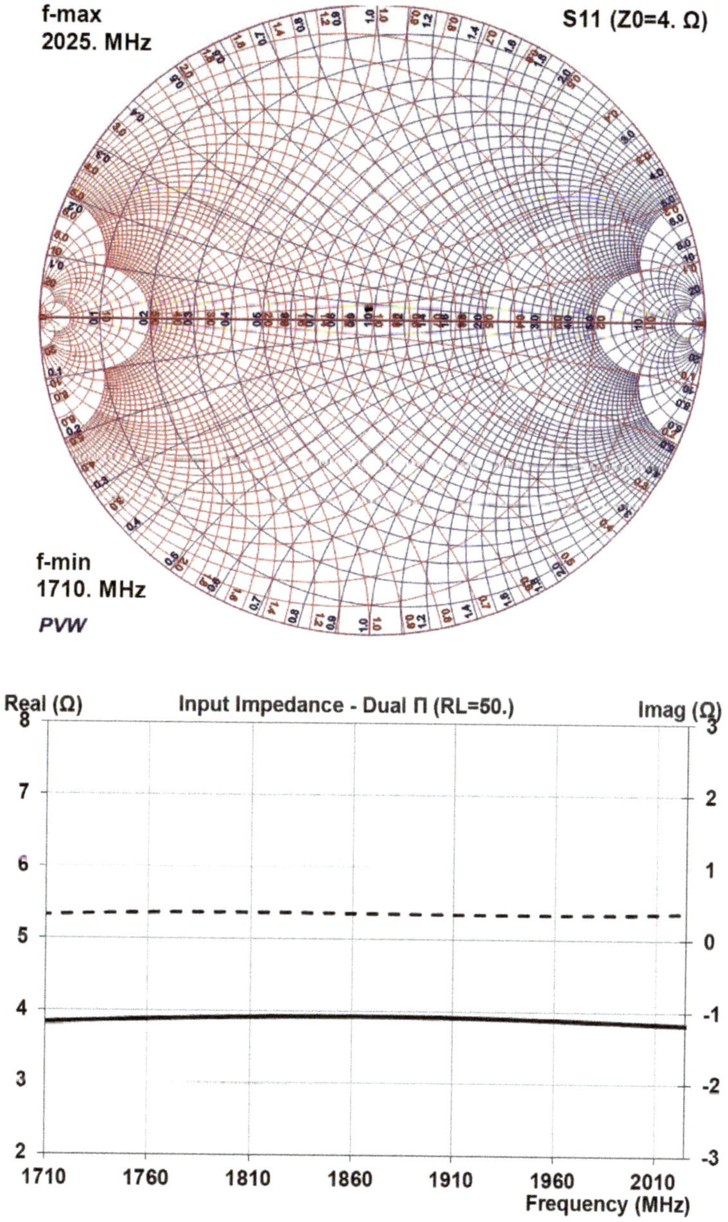

(a)

Figure 11.32 Highpass π + lowpass π characterization plots: (a) input impedance, (b) passband output impedance, (c) passband insertion gain, (d) second-harmonic input impedance, and (e) second-harmonic insertion gain.

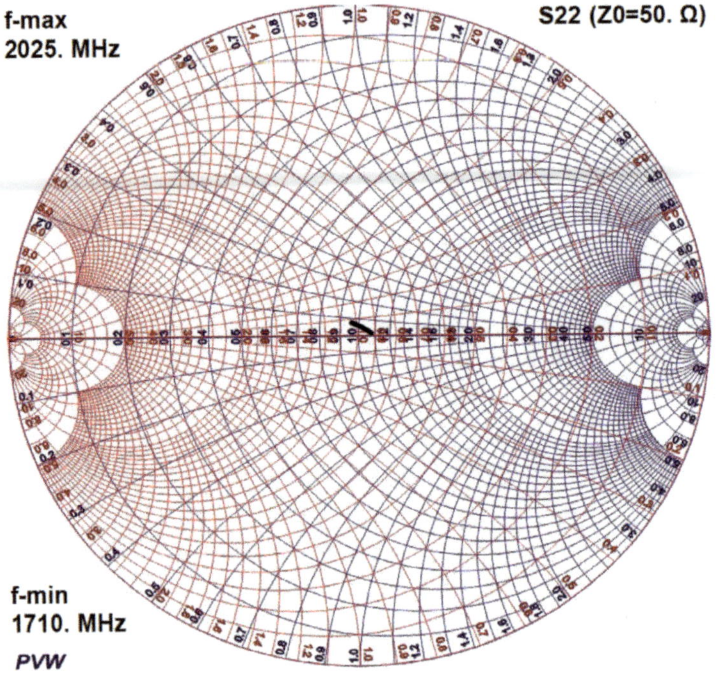

(b)

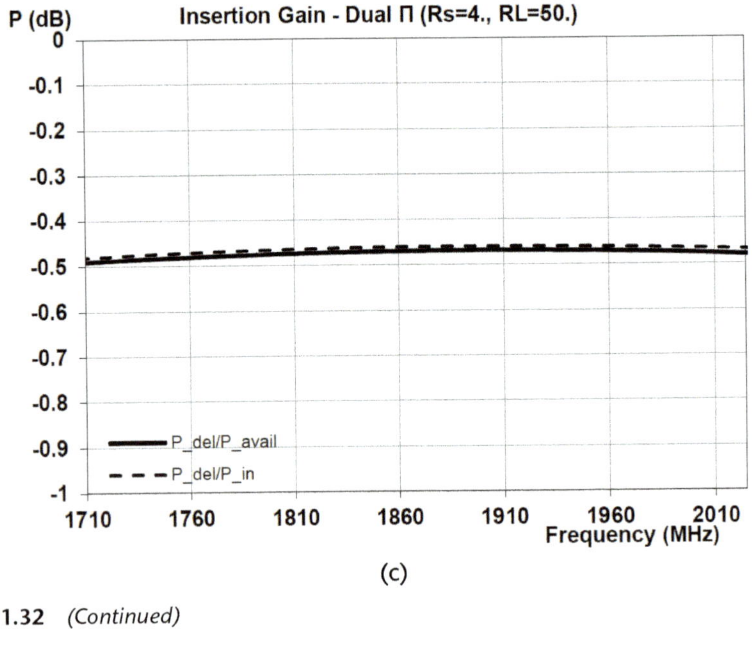

(c)

Figure 11.32 *(Continued)*

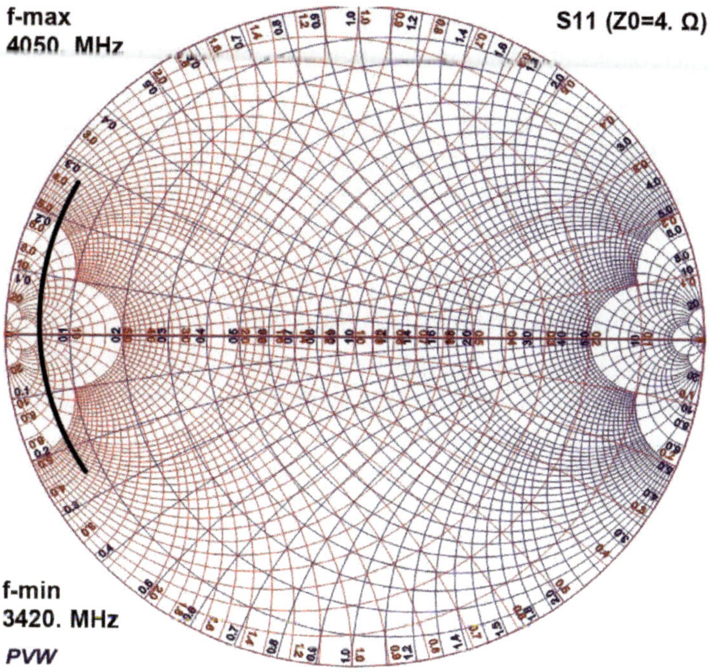

(d)

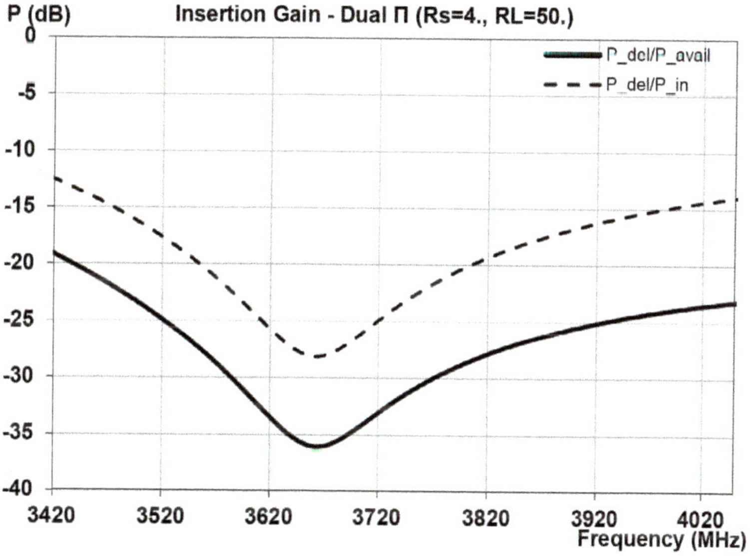

(e)

Figure 11.32 *(Continued)*

The dual-π highpass + lowpass impedance matching network exhibits good characteristics. Its input impedance across the passband is close to ideal, as is the output impedance. Across the second-harmonic bandwidth, the input impedance is close to the desired goal of an ideal short circuit. Angular dispersion across the band is low and could be further reduced at the expense of insertion loss, if desired.

Maximum insertion loss of the network over the passband is ~0.49 dB, and rejection across the second-harmonic band is >19 dB.

11.8.4 Highpass + Highpass π-Networks Cascade

Figure 11.33 shows a simplified schematic of the elements in a typical PA output match comprising a cascade of a highpass π-network followed by a second highpass π-network.

Figure 11.34 shows typical sensitivity plots of four key parameters on the value of R_{in}. Data points corresponding to the optimal selected value of $R_{in} \approx 21.7\Omega$ are denoted by the markers.

Parametric contour plots of the most critical performance parameters were again generated, as in Section 11.8.1. Based on those plots, the optimal values were chosen as $R_{in} \approx 21.7\Omega$, $\delta\Phi_{\pi 1} = 61°$, and $\delta\Phi_{\pi 2} = 64°$. The corresponding set of optimal element values, to meet the design specifications of Table 11.4, are given in Table 11.8.

Using these element values, key plots that characterize the performance of the impedance matching network are shown below.

The dual-π highpass + highpass impedance matching network exhibits good characteristics. Its input impedance across the passband is close to ideal, as is the output impedance. Across the second-harmonic bandwidth, the input impedance is close to the desired goal of an ideal short circuit. Angular dispersion across the band is low and could be further reduced at the expense of insertion loss, if desired.

The maximum insertion loss of the network over the passband is ~0.56 dB, and the rejection across the second-harmonic band is >14 dB.

The above data, related to the design of dual-π PA output impedance matching networks, are intended to convey the performance trade-offs required to decide upon a preferred configuration. First, there is the basic decision as to the nature of each of the two π-sections (i.e., either lowpass or highpass). The data presented above shows that any of the four possible network combinations can be designed to achieve relatively similar performance characteristics. Second, for each of the π-network pairs, no one configuration exists that will simultaneously optimize all the key parameters of the network. Thus, the designer must seek a compromise solution.

Table 11.8 Highpass π + Highpass π Optimal PA Match Elements

Element Values	
$L_b = 0.912$ nH	
$L_b = 0.273$ nH	$C_b = 6.934$ pF
$L_1 = 0.782$ nH	$C_1 = 10.503$ pF
$C_2 = 2.876$ pF	$L_2 = 3.452$ pF

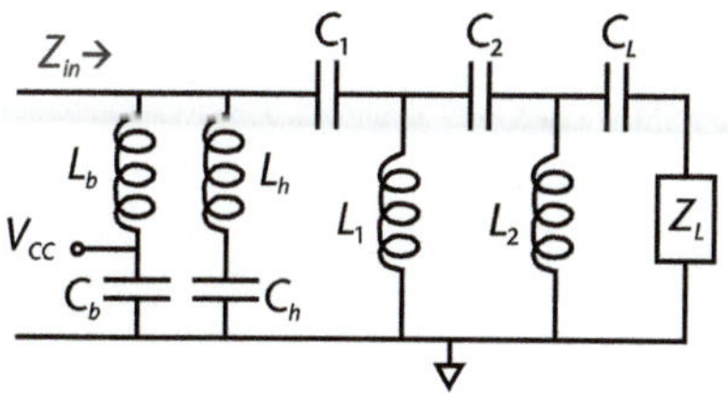

Figure 11.33 Highpass π + highpass π PA match.

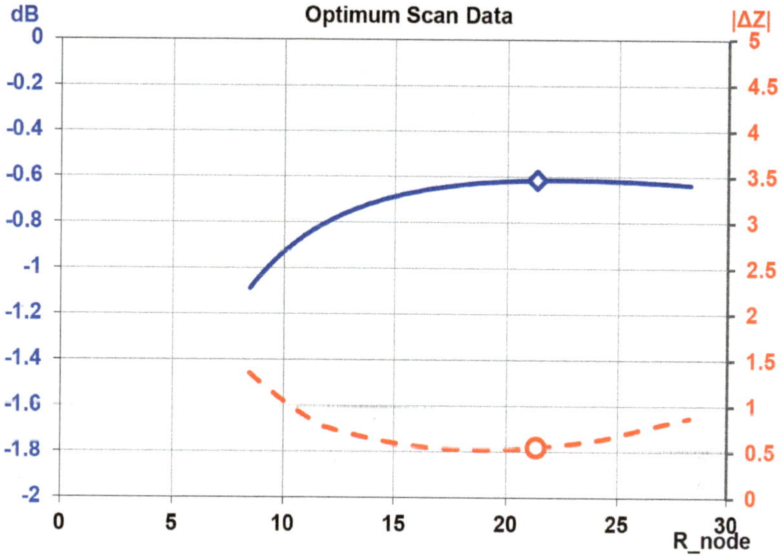

(a)

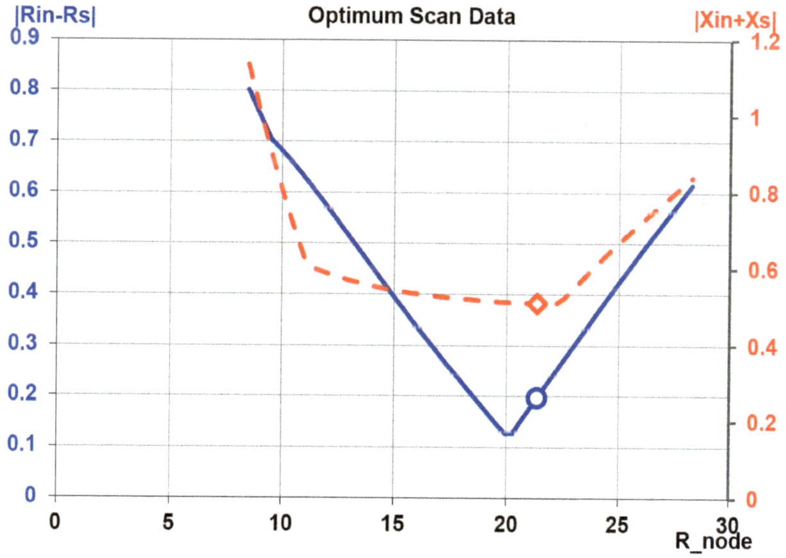

(b)

Figure 11.34 Highpass π + highpass π PA match R_N dependence: (a) insertion gain and $|Z_{in} - Z_S|$, and (b) $|R_{in} - R_S|$ and $|X_{in} - X_S|$.

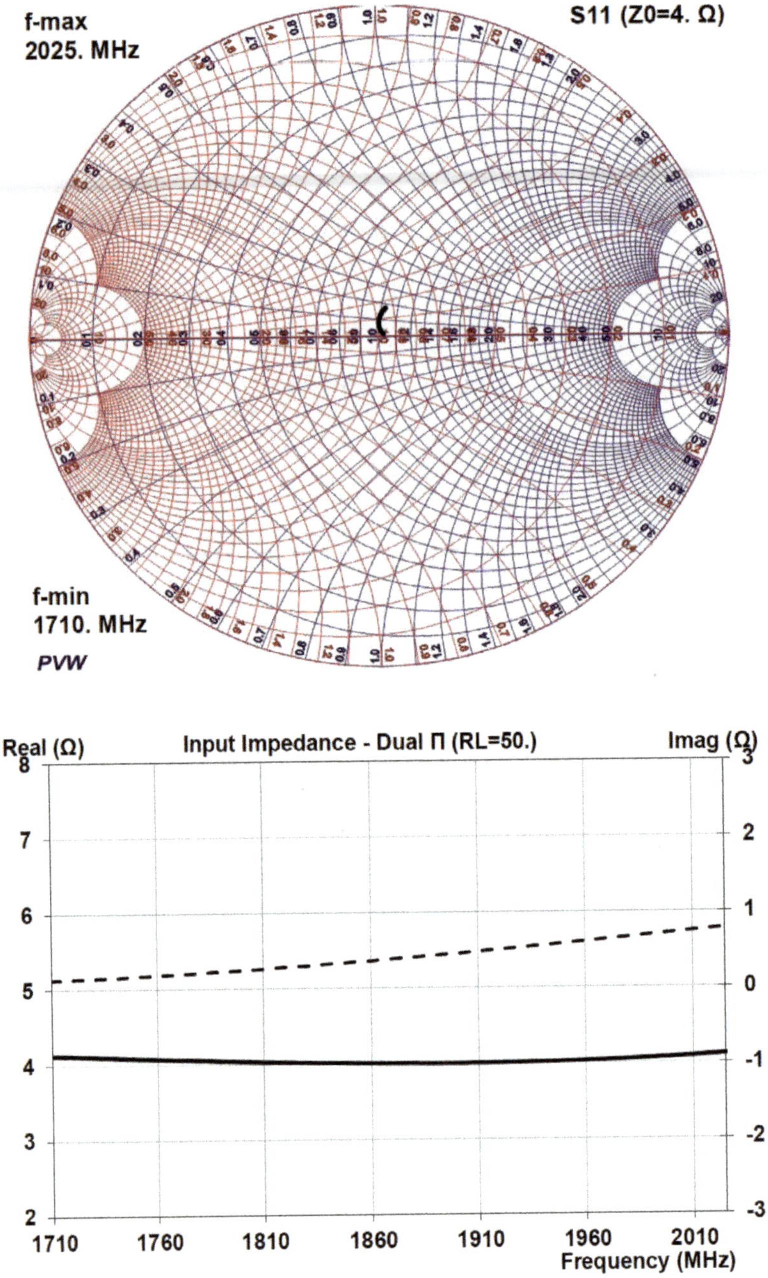

(a)

Figure 11.35 Highpass π + highpass π characterization plots: (a) input impedance, (b) passband output impedance, (c) passband insertion gain, (d) second-harmonic input impedance, and (e) second-harmonic insertion gain.

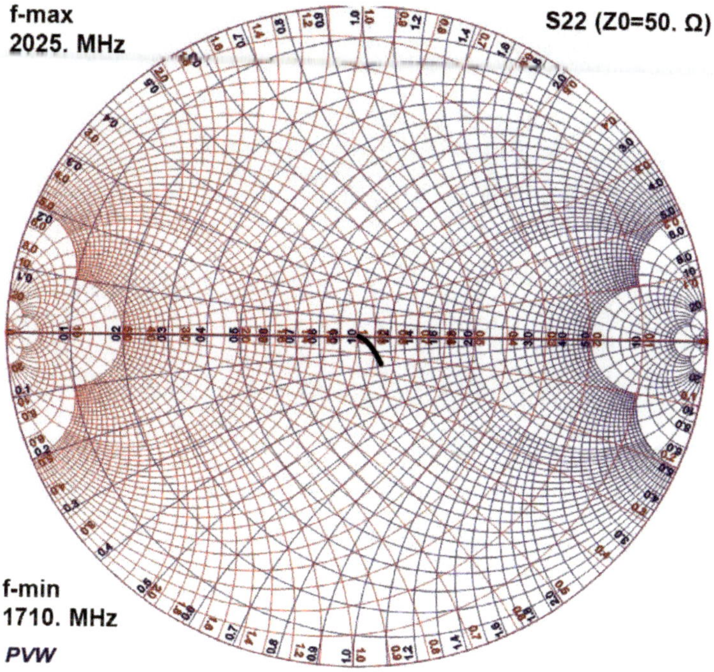

(b)

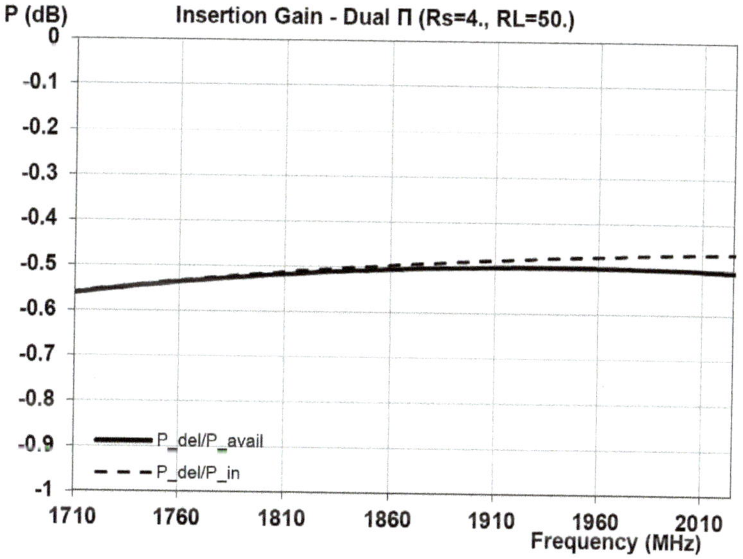

(c)

Figure 11.35 *(Continued)*

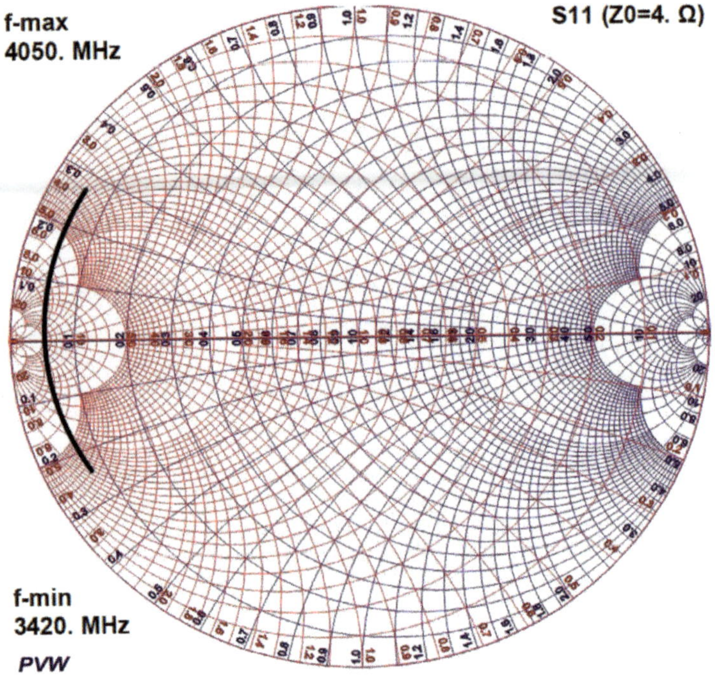

(d)

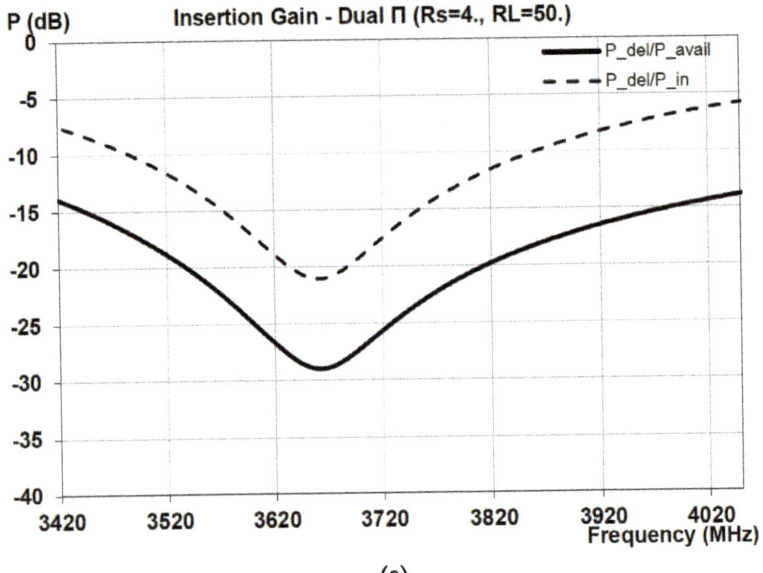

(e)

Figure 11.35 *(Continued)*

CHAPTER 12

Coupled-Inductor Single-Ended PA Matching

Chapter 11 explored the use of three-element and two-element *LC* networks for impedance matching. Such networks are widely used for the output matching of mobile PAs, where a high impedance transformation ratio, typically >10:1, is usually required. Such a high transformation ratio is challenging to achieve with low loss over wide frequency bandwidths. In the majority of cases, a cascade of two or more matching networks is required. Because, in addition to the two aforementioned parameters, size is equally a major constraint in any mobile device, designers are always looking for alternate solutions. One such alternative is the use of networks based on coupled inductors.

In the *LC* networks of Chapter 11, all the RF energy is coupled through the network via direct electrical connection. Current and voltage flows between the nodes of the network provide the sole path for energy transfer through the network from input to output. However, there is another potential physical means for transferring energy across a network and that is magnetic. Transformers, which are commonplace for converting voltages and currents at low frequencies, ideally convert all the energy through the network via magnetic coupling. There is no direct electrical path across the transformer.

Magnetic energy transfer can also be exploited in RF circuits. Two inductors in close proximity, but electrically isolated, can be constructed to efficiently transfer RF energy between them and change the voltage level of the signal. This action is somewhat analogous to the low-frequency transformer and thus a pair of coupled inductors used in an RF network is frequently referred to as a "transformer." However, there are some fundamental differences and use of the transformer can lead to incorrect assumptions being made regarding key performance characteristics of the network.

Coupled inductors are not widely used for the output impedance matching of single-ended PAs, for reasons that will become evident in this chapter. However, coupled inductors are widely used for the output matching and combining of differential PAs.

In this chapter, we examine the use of coupled inductors for single-ended impedance matching. Their use in differential PA architectures is examined in subsequent chapters.

12.1 Terminology: Coupled Inductors Versus Transformers

Transformers are one of the most basic and widely used electronic components in innumerable low-frequency (i.e., <1 MHz) applications. The most common application with which most people are familiar is their use in AC power conversion, where, they transform voltage levels, either up or down, and typically accomplish this function with extremely high efficiencies. Such low-frequency (<100 Hz) devices are physically implemented with two windings of copper (Cu) wire wrapped mutually and tightly together around a magnetic former, typically iron (Fe).

The two sets of turns, one for input and one for output, are electrically isolated. All energy transfer from input to output is via the magnetic coupling between the turns, which is greatly enhanced by the high magnetic permeability of the former. Typically, each set of windings employs over 100 turns. The ratio between the number of primary turns n_1 to the number of secondary turns n_2 sets the voltage transfer ratio and impedance ratio as

$$V_2/V_1 = n_2/n_1 \qquad \text{and} \qquad R_2/R_1 = (n_2/n_1)^2 \tag{12.1}$$

Although the classic low-frequency transformer from a physical viewpoint truly functions as a pair of coupled inductors, it is not usually thought of in this context. The reason for this is that, for an ideal transformer, the parametric values for the coupled inductors are inherently assumed to be limited value extrema that permit a simplified model.

Figure 12.1 shows the contrasting representations for an ideal transformer and an ideal pair of coupled inductors. Note that, for the ideal transformer, there is only one defining parameter, the turns ratio n_2/n_1. In contrast, for the coupled inductors, there are three defining parameters required, L_1, L_2, and the mutual coupling coefficient k.

For the ideal transformer, it is assumed that if the secondary winding is open-circuited, the input impedance on the primary winding will likewise be an open-circuit (i.e., $Z_1 = \infty$).

From the matrix representations for these two networks from Chapter 1, we find that this implies

(1) $L_1 = L_2 \approx \infty$

(2) $k = 1$

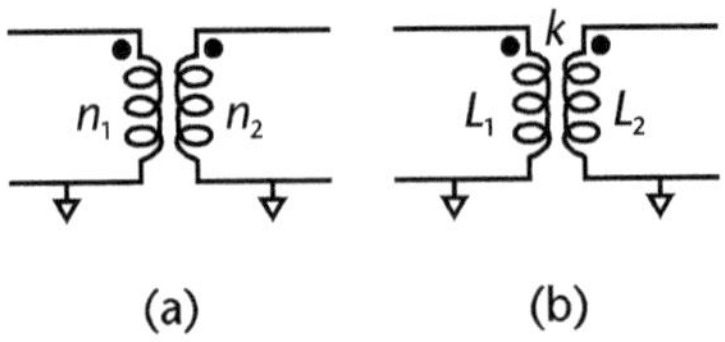

Figure 12.1 Circuit models for (a) ideal transformer and (b) pair of coupled inductors.

While these two limiting approximations may be close to valid for low-frequency transformers, they are far from reality for any realizable pair of coupled inductors for RF applications in the gigahertz region. One major challenge in trying to realize tight magnetic coupling between inductors at high frequencies is that most materials with high permeabilities are extremely lossy at high frequencies. Thus, in practice, they are rarely used in RF circuits, limiting the mutual coupling coefficient achievable. In addition, achieving very high inductance values at RF is extremely challenging due to the resistive skin effect and also interwinding capacitance.

In practice, coupled RF inductors are generally realized by printed planar traces either on a die or on a module substrate. Broadly speaking, such classes of coupled inductors can be divided into predominantly edge-coupled or overlap coupled topologies, although frequently a combination of the two may be used. In predominantly coupled inductor layouts, seeking to achieve strong magnetic coupling via edge coupling, the two inductor traces are tightly interwound on a single metal layer, as illustrated in Figure 12.2.

For enhanced quality factor, the line widths of the traces may not be constant along the length of the spirals as illustrated in Figure 12.2(b).

For coupled-inductor layouts, seeking to achieve strong magnetic coupling via overlap or broadside coupling, the two inductor traces are typically superposed on top of one another in two or more isolated metal layers. A representative layout for an overlap coupled-inductor pair is shown in Figure 12.3.

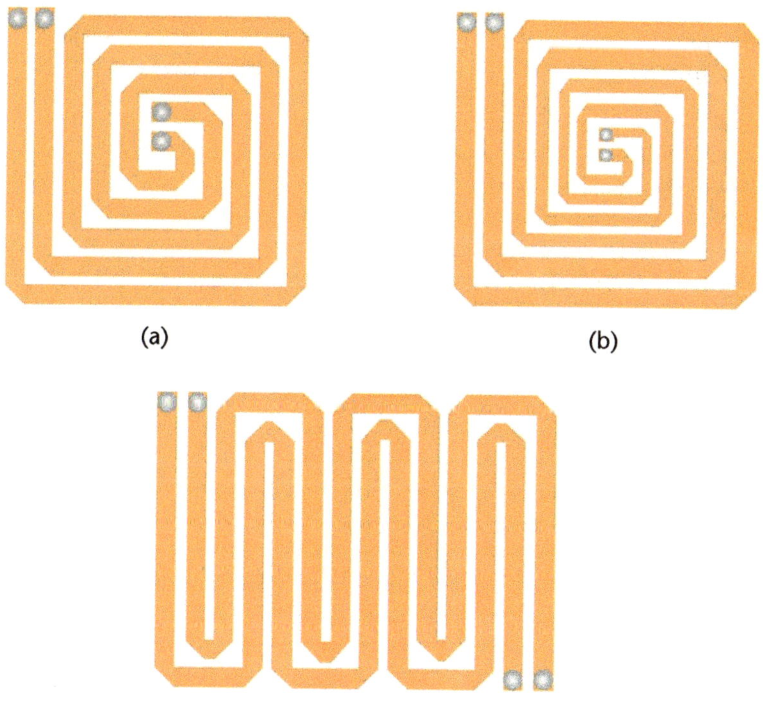

(a) (b) (c)

Figure 12.2 Edge-coupled printed inductors: (a) spiral, (b) high-*Q* spiral, and (c) meander lines.

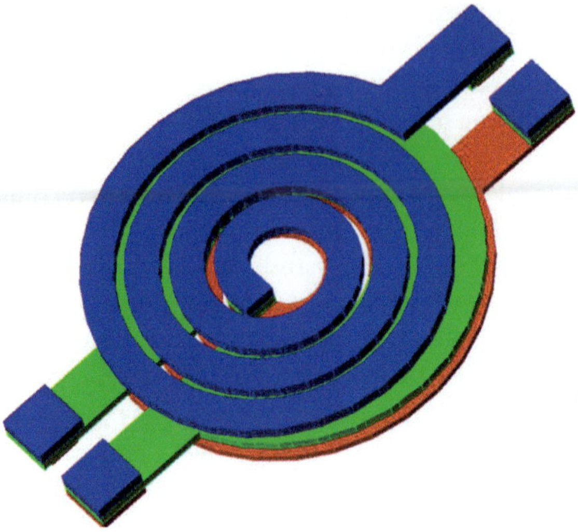

Figure 12.3 Overlay printed inductors.

In this three-layer implementation, inductor L_1 is realized in intermediate layer two. Coils on layers one and three are connected in series to realize output inductor L_2. Coupling between the two inductive traces results from the magnetic fields generated by the current flowing through the coils. The magnetic fields flow partially through air above the plane and partially through the supporting material below. Given that the magnetic fields of the inductors extend some distance away from the traces, care must be taken to exclude other components from the immediate neighborhood. Extraneous coupling to other circuit elements can degrade the performance of the coupler or otherwise degrade critical characteristics of the module.

Faced with the challenge of designing a set of coupled inductors to function as splitters and/or combiners in a differential PA architecture, it is not helpful for the designer to approach the problem as having to design a transformer. The single parameter that characterizes a classic transformer, that is, the turns ratio n_1/n_2 is an irrelevant parameter to the design. The challenge for the designer is to determine the optimum physical layout for the device, which requires determining the optimum value for the two inductors L_1 and L_2, given the coupling coefficient that can be achieved. Knowing these values also facilitates scaling the device in frequency if required in another application.

Lastly, another conceptual error in associating a pair of RF coupled inductors with a transformer is the expectation that a matching network based on coupled inductors will have wide bandwidth characteristics. This is because low-frequency transformers are widely regarded as nonresonant devices. However, as will be seen in this chapter, at RF inductors are predominantly reactive. Thus, to achieve real input and output impedances for the network, the inductors must be tuned with capacitive elements. Tuning with capacitors implies a resonant network with frequency-dispersive characteristics.

For these reasons, from hereon in this book, the term coupled inductors will be preferred to the term “transformer” when describing circuit elements as those shown in Figures 12.2 and 12.3.

12.2 Basic Single-Ended Coupled-Inductor Design

Coupled inductors can be on the output stage of a single-ended PA design to increase the impedance level towards that of the load. A single pair of coupled inductors is generally not sufficient to provide the full impedance transformation required, typically ~10:1. If used alone, the insertion loss and bandwidth would generally be inferior to those achievable with a two-stage matching network. Therefore, for optimum PA design, the coupled inductors would be followed by an additional matching stage employing a network of discrete elements.

Figure 12.4 shows three basic alternative circuit configurations for output matching a single-ended PA using coupled inductors. The coupled inductors serve to match the relatively low output impedance Z_S of the PA to a higher load impedance Z_L, which is typically an intermediate impedance between Z_S and the external impedance that the PA must drive, which is usually 50Ω. In general, both Z_S and Z_L may be complex. For an impedance match on the input to the network, we require

$$Z_{\text{in}} = Z_S^* \tag{12.2}$$

Because inductors, by their very nature, exhibit a positive reactance, to achieve an input match, with a real impedance component identical to that of the source (i.e., PA output), a capacitive tuning element on the input, output, or both must be included in the network as shown in Figure 12.4. Taken together, the inductive and capacitive elements comprise a resonant network, which will inevitably have band-limiting characteristics. This is in contrast with the popular equivalence of a coupled-inductor impedance match being equivalent to a low-frequency transformer.

In order to satisfy (12.2), two independent variables are required for the network because both the real and imaginary parts of the equality must be satisfied. In the networks of Figures 12.4(a, b), there are three independent variables, namely L_1, L_2, and C_S/C_L. If one of the inductor values is chosen as the independent variable, there are associated unique solutions for the remaining two elements of the circuit required to satisfy (12.2). For the circuit in Figure 12.4(c), there is an additional degree of flexibility allowing for the choice of two independent variables. In all cases, the matching network performance may be optimized by choosing appropriate values of the independent variables based on numerical analysis.

First, defining the input and output terminating impedances and admittances on the network as

$$Z_S = R_S + jX_S \quad \text{and} \quad Z_L = R_L + jX_L \tag{12.3}$$

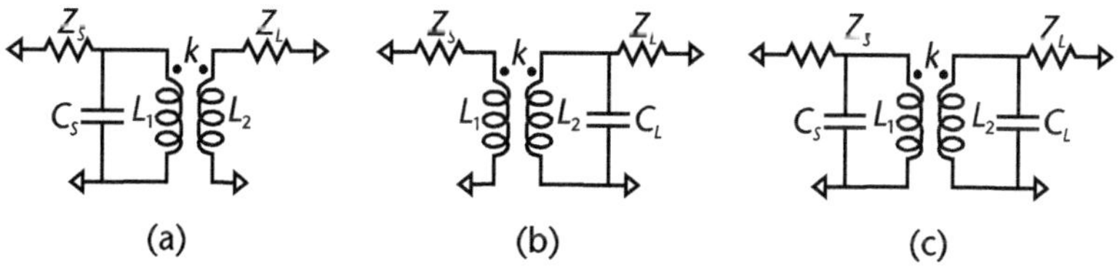

Figure 12.4 Basic single-ended, coupled-inductor matching alternatives: (a) input, (b) output, and (c) input and output.

$$Y_S = 1/Z_S = G_S + jB_S \qquad \text{and} \qquad Y_L = 1/Z_L = G_L + jB_L \tag{12.4}$$

with k = mutual coupling factor between the inductors and ω = angular design frequency.

Solutions for the dependent variables in the networks are as given below.

Case 12.1: Shunt Input Match Only (Figure 12.4(a)), with Independent Variable L_1

$$L_2 = \frac{k^2R_L/G_S - 2(1-k^2)\omega L_1 X_L - \sqrt{\left((k^2/G_S)^2 - \left(2(1-k^2)\omega L_1\right)^2\right)R_L^2 - 4(1-k^2)k^2\omega L_1 R_L X_L/G_S}}{2\omega^2(1-k^2)^2 L_1} \tag{12.5}$$

$$C_S = \frac{\left((1/\omega L_1 - (1-k^2)B_S)R_L - (1-k^2)G_S X_L\right)\omega L_2 - (R_L^2 + X_L^2)G_S}{\omega^2(1-k^2)R_L L_2} \tag{12.6}$$

For a solution to exist to (12.5), the argument of the square root must be ≥0. This leads to a constraint on the maximum value permissible for the independent variable L_1. We find

$$\omega L_1 \leq \frac{-X_L/R_L + \sqrt{1+(X_L/R_L)^2}}{2(1/k^2 - 1)G_S} \tag{12.7}$$

Thus, when seeking to explore the range of possible solutions for the coupled-inductor network of Figure 12.4, the value of the primary inductance L_1 cannot exceed the upper limit given by (12.7).

Case 12.2: Shunt Output Match Only (Figure 12.4(b)), with Independent Variable L_2

$$L_1 = \frac{k^2R_S/G_L - 2(1-k^2)\omega L_2 X_S - \sqrt{\left((k^2/G_L)^2 - \left(2(1-k^2)\omega L_2\right)^2\right)R_S^2 - 4(1-k^2)k^2\omega L_2 R_S X_S/G_L}}{2\omega^2(1-k^2)^2 L_2} \tag{12.8}$$

$$C_L = \frac{\left((1/\omega L_2 - (1-k^2)B_L)R_S - (1-k^2)G_L X_S\right)\omega L_1 - (R_S^2 + X_S^2)G_L}{\omega^2(1-k^2)R_S L_1} \tag{12.9}$$

For a solution to exist to (12.12), the argument of the square root must be ≥0. This leads to a constraint on the maximum value permissible for the independent variable L_2. We find

$$\omega L_2 = \frac{-X_S/R_S + \sqrt{1+(X_S/R_S)^2}}{2\left(1/k^2 - 1\right)G_L} \tag{12.10}$$

Thus, when seeking to explore the range of possible solutions for the coupled-inductor network of Figure 12.4, the value of the secondary inductance L_2 cannot exceed the upper limit given by (12.10).

As discussed in Chapter 9, a low impedance is frequently required on the output of the active devices across the second-harmonic frequency band. This is readily accomplished for the single-ended, coupled-inductor matches in Figure 12.4(a, c) by the addition of an inductor L_S in series with the input shunt capacitor C_S, as shown in Figure 12.5.

To preserve the impedance match in the passband, this trap must provide an identical susceptance to C_S, while it is series resonant at the second-harmonic frequency.

Denoting the frequency at the center of the passband by ω_0 and that at the center of the second-harmonic band by ω_2, we require

$$C_S' = \left(1 - (\omega_0/\omega_2)^2\right)C_S \tag{12.11}$$

$$L_S = \frac{1}{\left(\omega_2^2 - \omega_0^2\right)C_S} \tag{12.12}$$

Analysis of the coupled-inductor, single-ended match is most easily carried out using two-port $[ABCD]$ matrix parameters as described in Chapter 1.

Defining the finite Q impedances of the inductor elements as

$$Z_{L1} = R_{L1} + j\omega L_1 \quad \text{and} \quad Z_{L2} = R_{L2} + j\omega L_2 \tag{12.13}$$

and the finite Q admittances of the shunt input and output elements as

$$Y_{CS} = G_{CS} + jB_{CS} \quad \text{and} \quad Y_{CL} = G_{CL} + jB_{CL} \tag{12.14}$$

The network $[ABCD]$ parameters for the general form of the coupled-inductor match in Figure 12.5(b) are

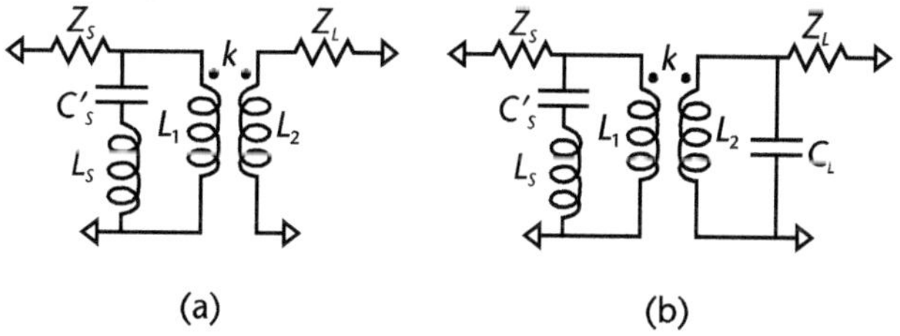

Figure 12.5 Basic single-ended, coupled-inductor matching networks with harmonic short: (a) input match, and (b) input and output matching.

$$A = \frac{(1 + Y_{CL}Z_{L2})Z_{L1} + (\omega k)^2 L_1L_2Y_{CL}}{j\omega k\sqrt{L_1L_2}} \tag{12.15}$$

$$B = \frac{Z_{L1}Z_{L2} + (\omega k)^2 L_1L_2}{j\omega k\sqrt{L_1L_2}} \tag{12.16}$$

$$C = \frac{(1 + Y_{CS}Z_{L1})(1 + Y_{CL}Z_{L2}) + (\omega k)^2 L_1L_2Y_{CS}Y_{CL}}{j\omega k\sqrt{L_1L_2}} \tag{12.17}$$

$$D = \frac{(1 + Y_{CS}Z_{L1})Z_{L2} + (\omega k)^2 L_1L_2Y_{CS}}{j\omega k\sqrt{L_1L_2}} \tag{12.18}$$

12.2.1 Basic Single-Ended, Coupled-Inductor Matching Configurations

In this section, we apply the design and analysis formulae of the previous section to a practical design, in order to illustrate the basic performance capabilities of this type of PA matching architecture.

For simplicity, we shall assume both source and load impedance are pure real, together with the additional network parameters given in Table 12.1.

Case 12.3: Shunt Input Match Only (Figure 12.4(a)), with Dependent Variable L_1

With the values from Table 12.1, from (12.7), we determine

$$L_1 \max \approx 0.1638 \text{ nH} \tag{12.19}$$

First, we investigate how the minimum passband insertion gain (P_{del}/P_{avail}) varies as a function of the independent variable L_1. This dependence is shown in Figure 12.6(a).

It can be seen that the lowest insertion loss ~0.73 dB occurs for a value of L_1 = ~0.14 nH, which is approximately 14.5% below the maximum possible value given

Table 12.1 Coupled-Inductor Impedance Match Parameters

Design Parameters	
Mutual coupling, $k = 0.7$	
f_{lo} 1,710 MHz	f_{hi} 2,025 MHz
$f2_{lo}$ 3,420 MHz	$f2_{hi}$ 4,050 MHz
$Z_S = 4\Omega$	$Z_L = 50\Omega$
Inductors, $Q_L = 60$	Capacitors, $Q_C \approx 100$

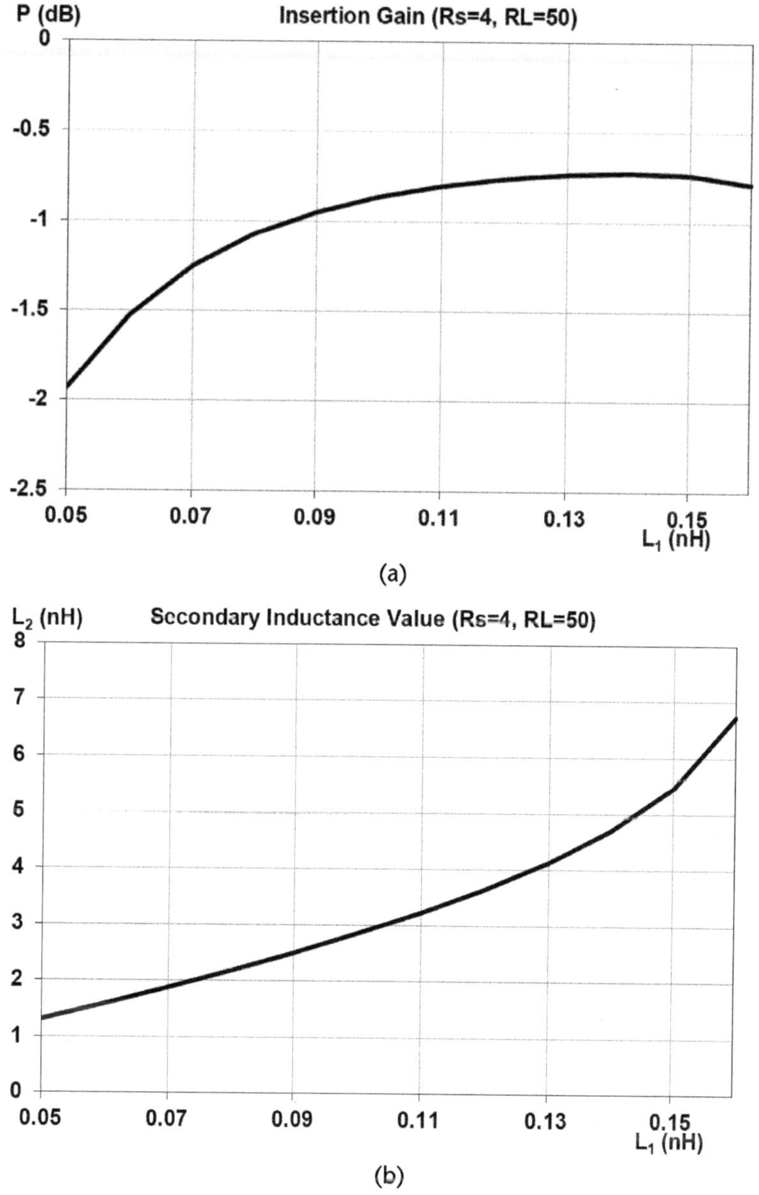

Figure 12.6 Insertion gain and secondary inductance dependence on primary inductance: (a) gain, and (b) secondary inductance.

by (12.19). Figure 12.6(b) shows the required value for the secondary inductance L_2, as a function of L_1.

The optimized circuit element values for the input shunt-tuned, coupled-inductor network are given in Table 12.2.

The corresponding input and output impedances of the network, in Cartesian form, are shown in Figure 12.7. The corresponding passband insertion gain dependence is shown in Figure 12.8.

Table 12.2 Optimum Input Shunt-Matched Coupled-Inductor Element Values

Element Values	
$L_1 = 0.142$ nH	$L_2 = 4.841$ nH
$C_S = 63.467$ pF	$L_s = 0$ nH

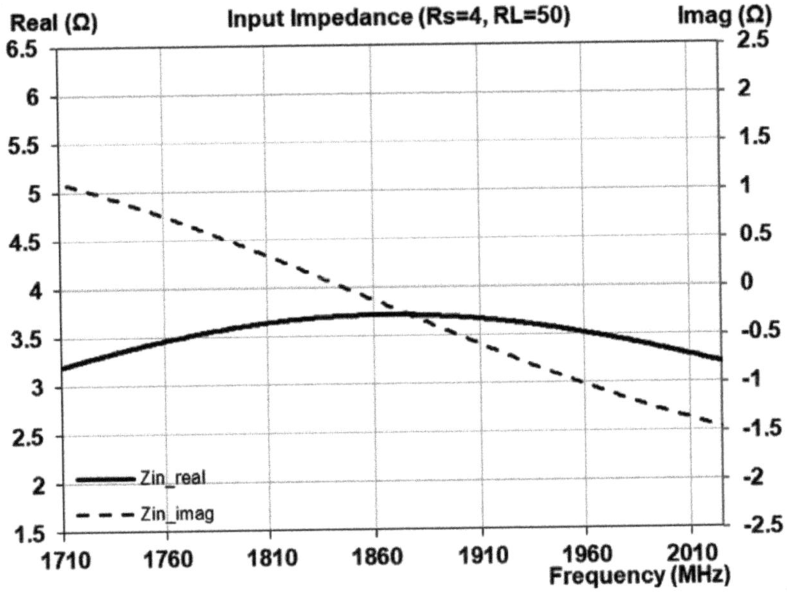

(a)

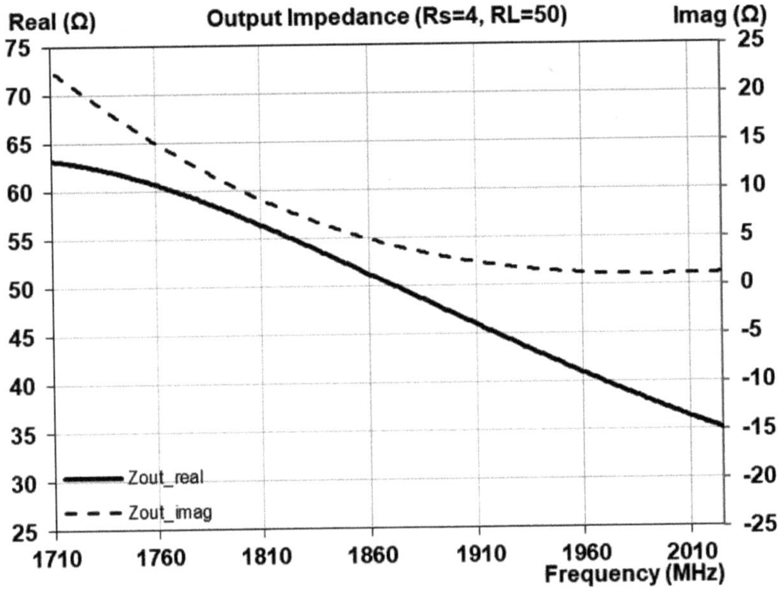

(b)

Figure 12.7 Optimized input shunt-matched coupled-inductor network impedances: (a) input impedance, and (b) output impedance.

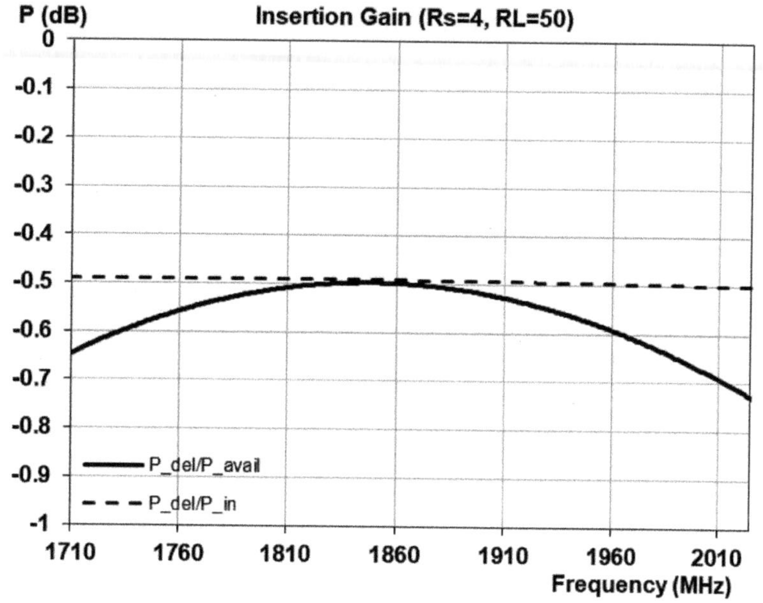

Figure 12.8 Optimized input shunt-matched coupled-inductor insertion gain.

For the ideal transformer, the ratio of the matched input and output impedances are related to the square of the turn's ratio, as expressed in (12.1). How is this ratio related to the inductor ratio for the coupled-inductor matching network of Figure 12.4(a)? Figure 12.9 shows the dependence of the inductor ratio on the impedance ratio for this network with input shunt tuning. It is a straight-line relationship. This is analogous to the transformer impedance relationship as the value of an inductor is proportional to square of the number of turns.

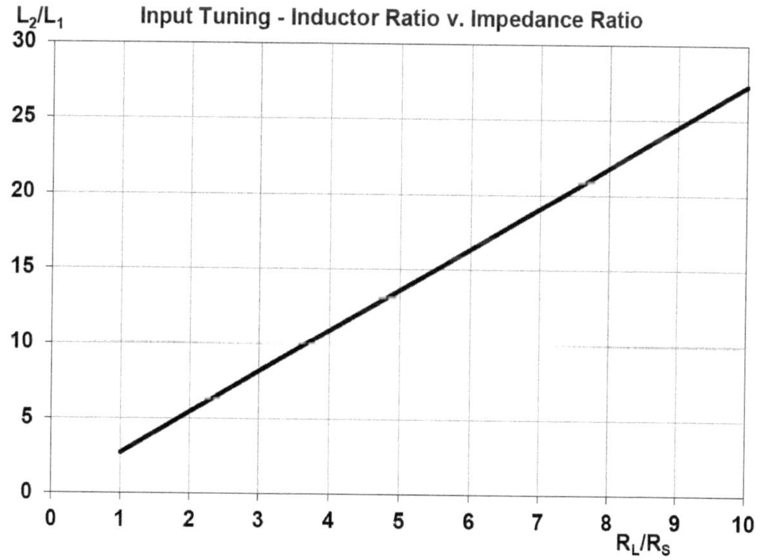

Figure 12.9 Optimized input shunt-matched coupled-inductor ratio dependence on impedance ratio.

Case 12.4: Shunt Output Match Only (Figure 12.4(b)), with Dependent Variable L_2

With the values from Table 12.1, from (12.10), we determine

$$L_2 \max \approx 2.047 \text{ nH} \tag{12.20}$$

First, we investigate how the minimum passband insertion gain (P_{del}/P_{avail}) varies as a function of the independent variable L_2. This dependence is shown in Figure 12.10(a).

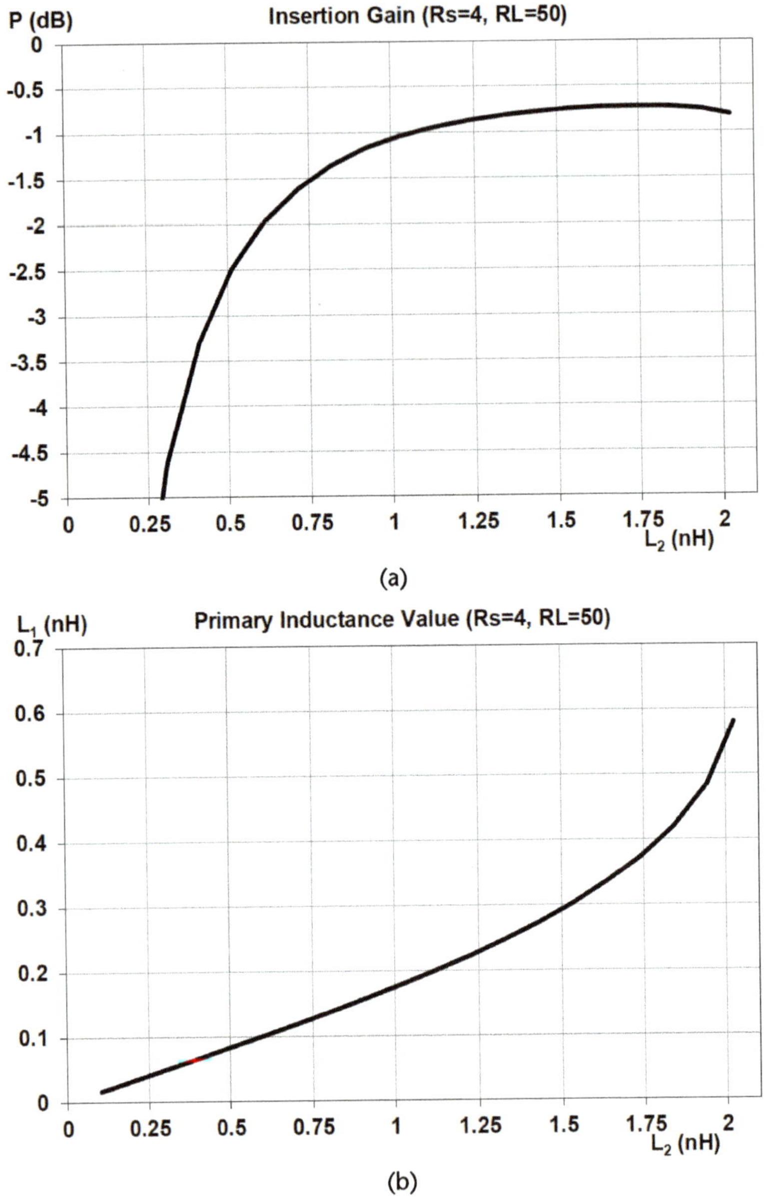

Figure 12.10 Insertion gain and primary inductance dependence on secondary inductance: (a) gain, and (b) secondary inductance.

It can be seen that the lowest insertion loss ~0.73 dB occurs for a value of L_2 = 1.78 nH, which is approximately 13.2% below the maximum possible value given by (12.19). Thus, the minimum passband insertion loss achievable with shunt-matching the output of the coupled-inductor pair is near identical to that achievable with input shunt-matching. Figure 12.10(b) shows the required value for the primary inductance L_1 as a function of L_2.

Table 12.3 lists the optimum circuit element values for maximum passband gain with output shunt matching.

The corresponding input and output impedances of the network, in Cartesian form, are shown in Figure 12.11.

The corresponding passband insertion gain dependence is shown in Figure 12.12. Note that the passband insertion dependence is almost identical to that for input shunt tuning in Figure 12.8.

Figure 12.13 shows the dependence of the inductor ratio on the impedance ratio for this network with output shunt tuning. As with shunt input tuning, it is a straight-line relationship. However, comparing Figure 12.13 with Figure 12.9, it can be seen that the optimized inductor ratio for output shunt tuning is significantly less than that required for input shunt tuning. The reason for this is that a capacitor C, in parallel with a load R_L, reduces the real part of the impedance, as is evident in the following relationship

$$Z = \frac{R_L - j\omega C R_L^2}{1 + (\omega C R_L)^2} \tag{12.21}$$

Thus, output shunt tuning reduces the impedance ratio that must be matched by the coupled inductors. Conversely, shunt tuning on the input increases the needed transformation ratio.

Case 12.5: Shunt Input and Output Match (Figure 12.4(c))

For this case, both L_1 and L_2 may be taken as separate independent variables and the optimum combination for achieving maximum passband insertion gain determined by a numerical approach. For the parameters of Table 12.1, the optimum network element values are as listed in Table 12.4.

The corresponding input and output impedances of the network, in Cartesian form, are shown in Figure 12.14. The corresponding passband insertion gain dependence is shown in Figure 12.15. The maximum passband insertion loss is ~0.48 dB. This is substantially lower than ~0.73 dB for the solely input or output shunt tuning configurations considered above.

Table 12.3 Optimum Output Shunt-Matched, Coupled-Inductor Element Values

Element Values	
$L_1 = 0.387$ nH	$L_2 = 1.776$ nH
$C_L = 5.077$ pF	$L_s = 0$ nH

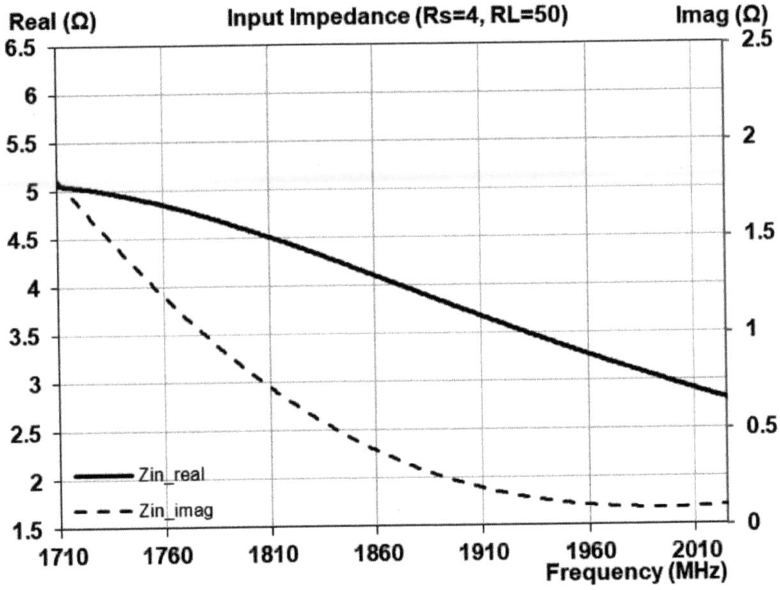

(a)

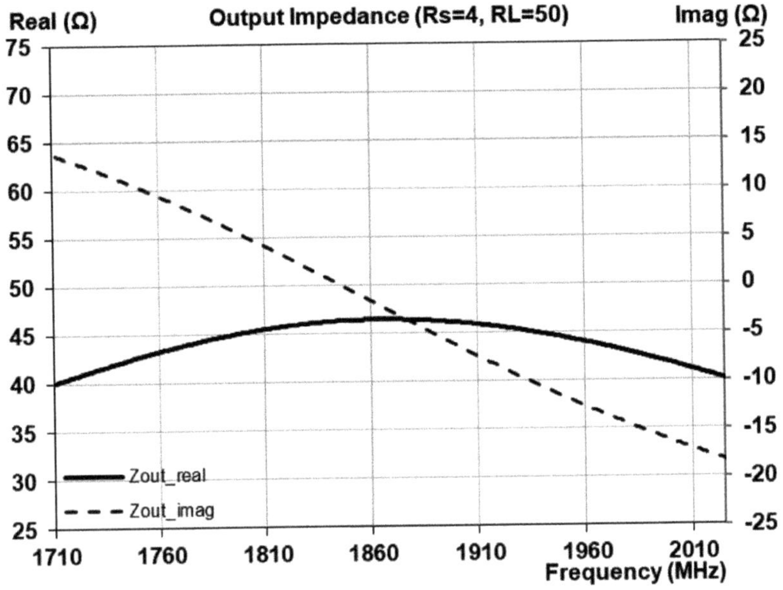

(b)

Figure 12.11 Optimized output shunt-matched, coupled-inductor network impedances: (a) input impedance, and (b) output impedance.

Figure 12.16 shows the dependence of the inductor ratio on the impedance ratio for this network with output shunt tuning. Again, it is essentially linear. Comparing Figure 12.16 with Figures 12.9 and 12.13, the optimized inductor ratio is intermediary between the two.

The impedance plots in Figure 12.7, with tuning only on the input port of a pair of coupled inductors, show a good input match can be obtained, but the output

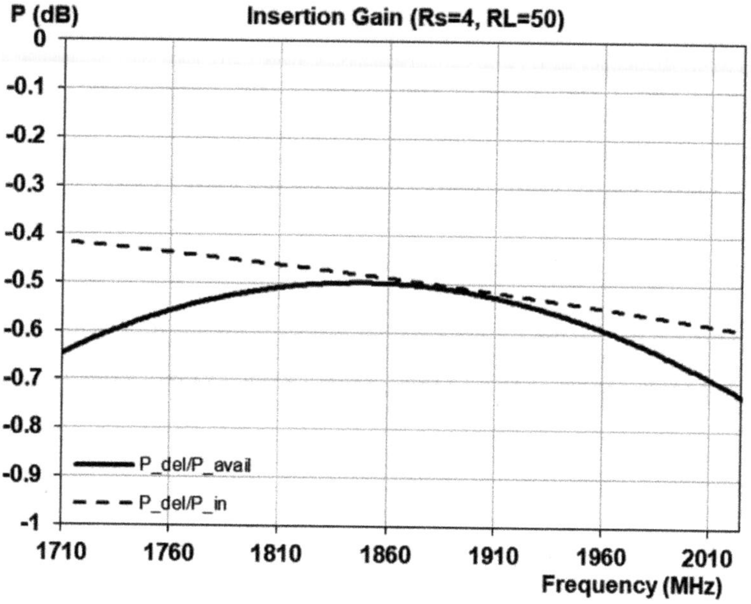

Figure 12.12 Optimized output shunt-matched, coupled-inductor insertion gain.

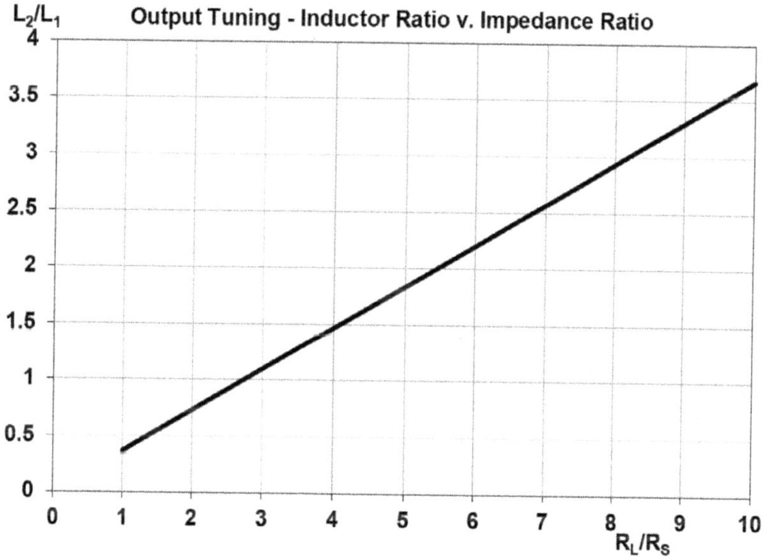

Figure 12.13 Optimized output shunt-matched, coupled-inductor ratio dependence on impedance ratio.

Table 12.4 Optimum Input and Output Shunt-Matched, Coupled-Inductor Element Values

Element Values	
L_1 = 0.162 nH	L_2 = 2.727 nH
C_S = 39.526 pF	C_L = 1.328 pF

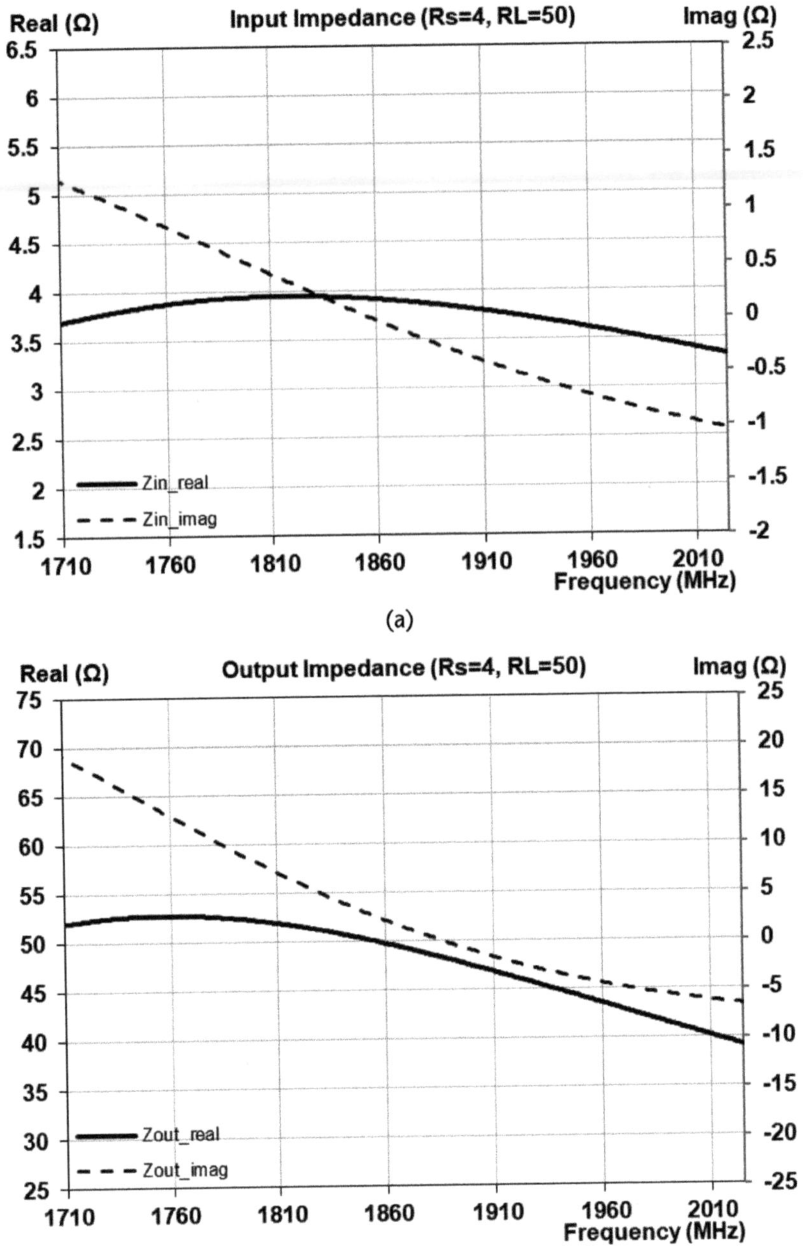

(a) (b)

Figure 12.14 Optimized input and output shunt-matched, coupled-inductor network impedances: (a) input impedance, and (b) output impedance.

match is poor. Conversely, the impedance plots in Figure 12.11, with tuning only on the output port of a pair of coupled inductors, show that a good output match can be obtained, but the input match is poor. In contrast, the impedance plots in Figure 12.14, with both input and output shunt-tuning, show a well-balanced compromise between the two. However, the significantly increased inductor ratio required in the latter case may not be achievable in a practical device layout.

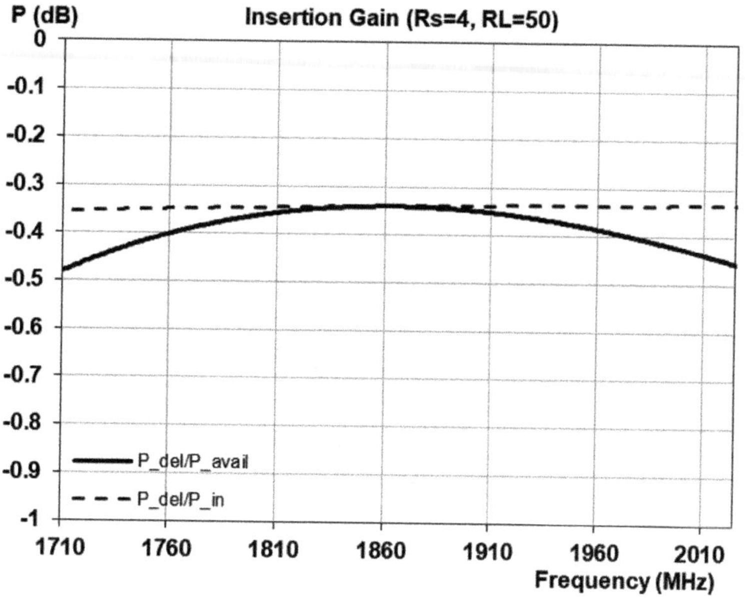

Figure 12.15 Optimized input and output shunt-matched, coupled-inductor insertion gain.

12.2.2 Basic Single-Ended, Coupled-Inductor Matching Complexities

There are two significant disadvantages requiring that compromises be made associated with the coupled-inductor matching architectures examined in Section 12.2.1. The first one is related to the relative values of the inductors in the networks. The second is related to the second-harmonic termination impedances presented to the PAs.

Figures 12.6(b) and 12.10(b) show the required value of the secondary inductance L_2, as a function of L_1, for the input and output shunt-tuned scenarios, respectively.

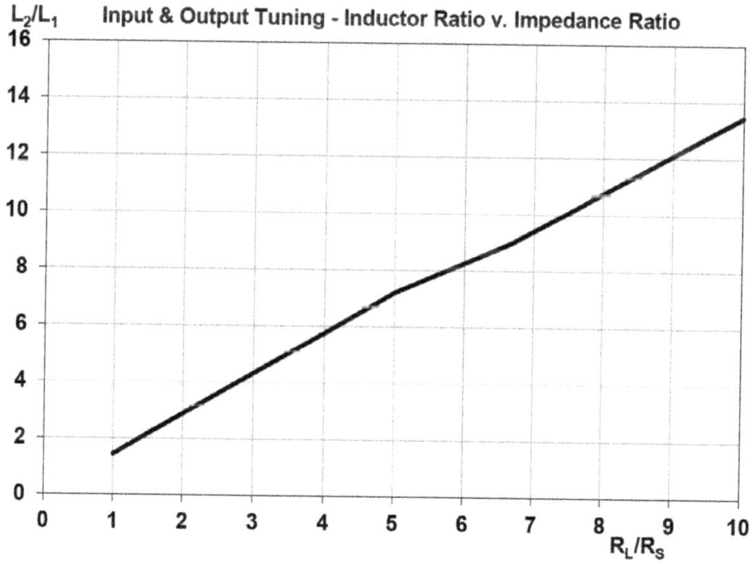

Figure 12.16 Optimized output shunt-matched, coupled-inductor ratio dependence on impedance ratio.

It is instructive to replot this data in a form showing how the ratio L_2/L_1 varies as a function of the independent variable, either L_1 or L_2.

Figure 12.17(a) plots the required ratio L_2/L_1 as a function of the independent variable L_1, for the shunt input tuned network of Figure 12.4(a). For all values of L_1, it is >25:1 everywhere. Such a high ratio would make the layout of the two coupled inductors extremely difficult to realize in practice, making the input shunt tuning only option physically challenging.

Another important takeaway from Figure 12.17 is that the appropriate inductor ratio L_2/L_1 depends dramatically on how the coupled-inductor pair is tuned.

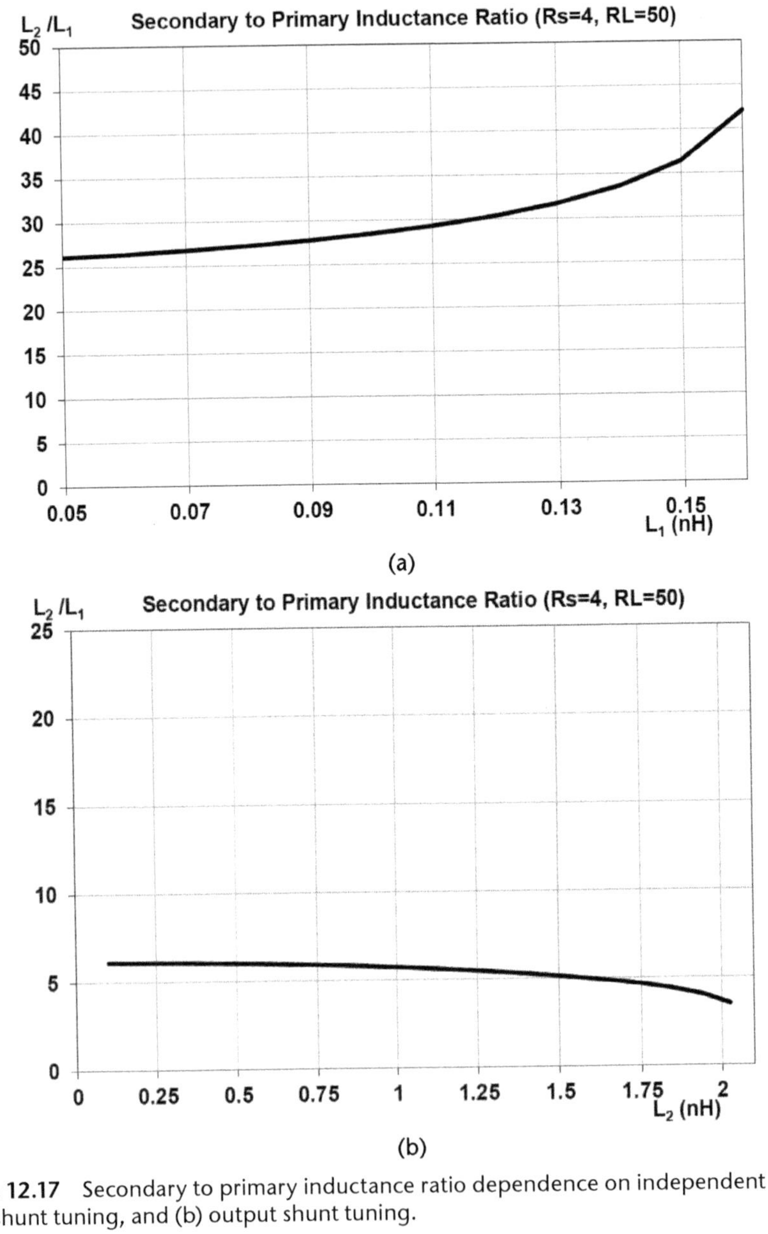

Figure 12.17 Secondary to primary inductance ratio dependence on independent variable: (a) input shunt tuning, and (b) output shunt tuning.

The impedance transformation ratio for both scenarios is 12.5:1 from Table 12.1. If the coupled inductors were naïvely regarded as the equivalent to a transformer, as is frequently done in practice, the secondary to primary inductance ratio would also be expected to be about this value. However, for the input shunt-tuned case, the ratio is approximately double that. For the output shunt-tuned case, it is about one-half that.

Figure 12.17(b) plots the required ratio L_2/L_1 as a function of the independent variable L_2 for the shunt output tuned network of Figure 12.4(b). For all values of L_2, it is ~5:1. This ratio is much more attractive than that for the input shunt-tuned option and should not be problematic for layout. However, there is another aspect of any PA output matching network that must be considered. Namely, its characteristics across the second-harmonic band.

By virtue of the mixing of nonlinear products generated in the amplification process in a PA, there are inevitably frequency components within the second-harmonic frequency band at the output of the PA. Any output matching network is required to accomplish two functions: (1) correctly terminate these components to maximize the efficiency of the PA, and (2) prevent them from reaching the output.

Figure 12.18(a) shows the input reflection coefficient of the output shunt-tuned, coupled-inductor match across the second-harmonic band. It is entirely inductive and not easy to modify. This is not ideal. For maximum PA efficiency, typically, the second-harmonic PA terminating impedances are desired to be close to a short-circuit. Figure 12.18(b) shows the corresponding network gain across the second-harmonic band. It is ~ −10 dB. This is far from ideal. A much higher attenuation is generally required.

Higher second-harmonic rejection can be readily achieved by adding an inductor in series with the output shunt-tuning capacitor C_L, in a manner like that shown for the input tuning configuration in Figure 12.5(a). Figure 12.19 shows the network characteristics equivalent to those in Figure 12.18 with such an output shunt trap.

It can be seen in Figure 12.19(a) that the network input admittance remains inductive and, in fact, is further away from presenting a low impedance termination on the output of the PA, as is desirable. As expected, the network does provide higher rejection on the output of undesired second-harmonic components.

Turn now to Case 12.5, considered above, with both input and output shunt tuning, as in Figure 12.4(c). This architecture exhibited overall superior characteristics to isolated input or output shunt tuning. For the optimum case, from Table 12.4, the ratio L_2/L_1 is ~17:1, which would be difficult to implement in a practical layout.

Figure 12.20 shows the input reflection coefficient of the input and output shunt-tuned, coupled-inductor match, with parameters in Table 12.4, across the second-harmonic band.

With this configuration, the network input impedance in Figure 12.20(a), which is presented to the output of the PA, is capacitive with a significant real component. This is consistent with the insertion gain plot in Figure 12.20(b), which shows little isolation of the load from second-harmonic output from the PA.

To address these problems, the input shunt capacitor C_S may be resonated a series inductor L_S, as shown in Figure 12.5(b). The network elements for this case are listed in Table 12.5.

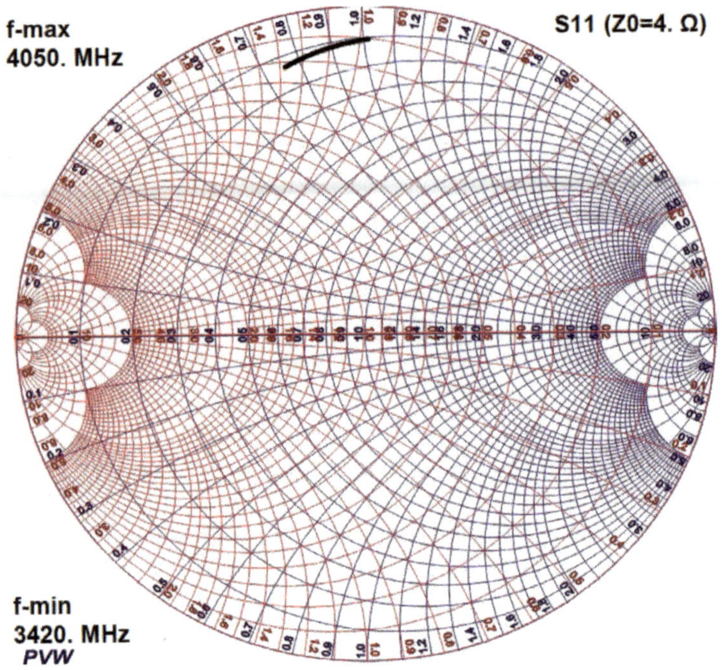

(a)

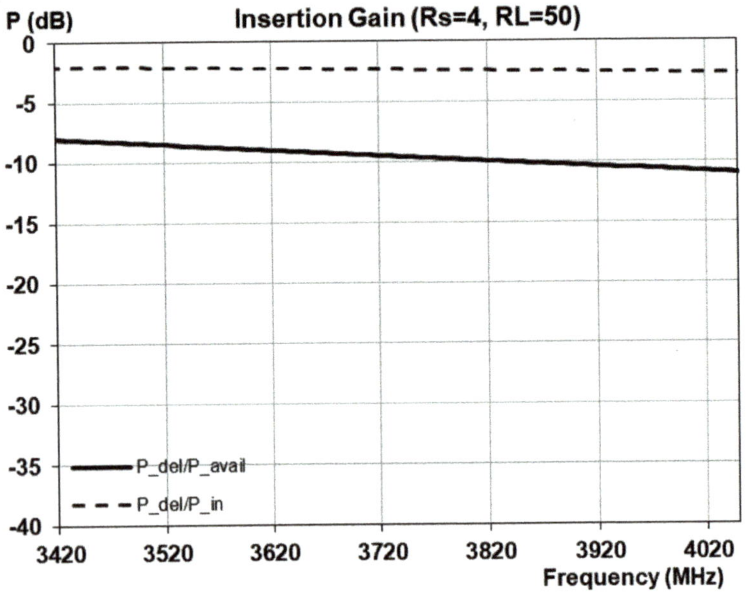

(b)

Figure 12.18 Output shunt-tuned, coupled-inductor match second-harmonic characteristics: (a) input reflection coefficient, and (b) insertion gain.

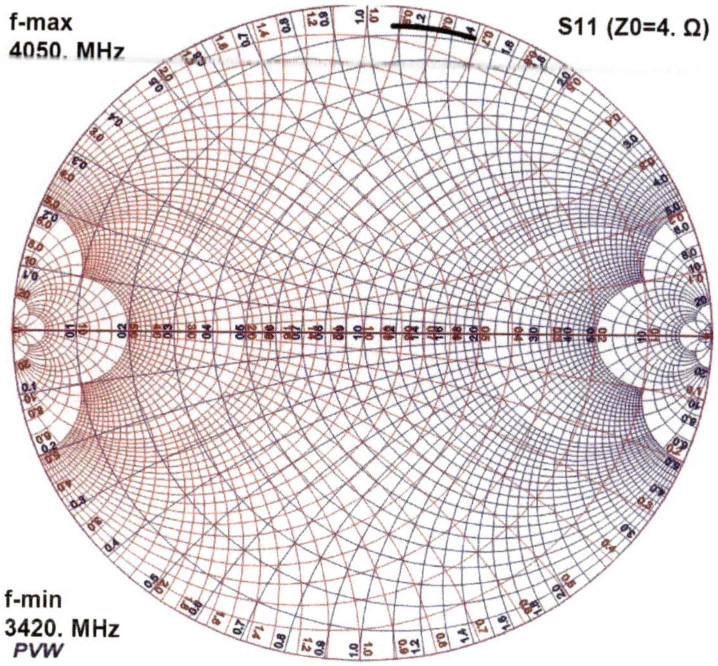

(a)

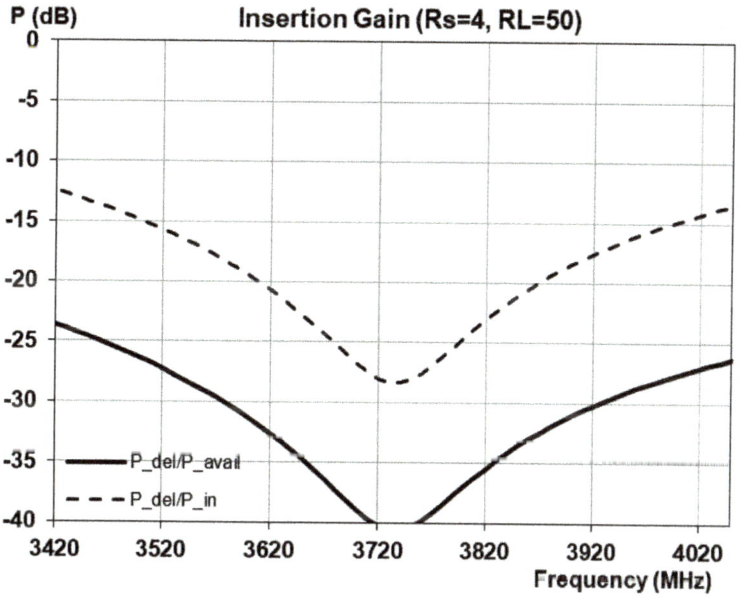

(b)

Figure 12.19 Output shunt-tuned, coupled-inductor match second-harmonic characteristics, with trap: (a) input reflection coefficient, and (b) insertion gain.

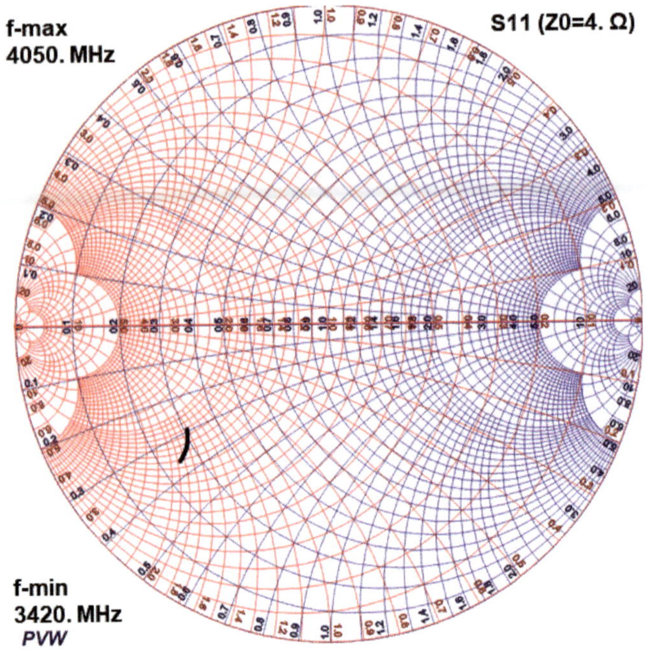

(a)

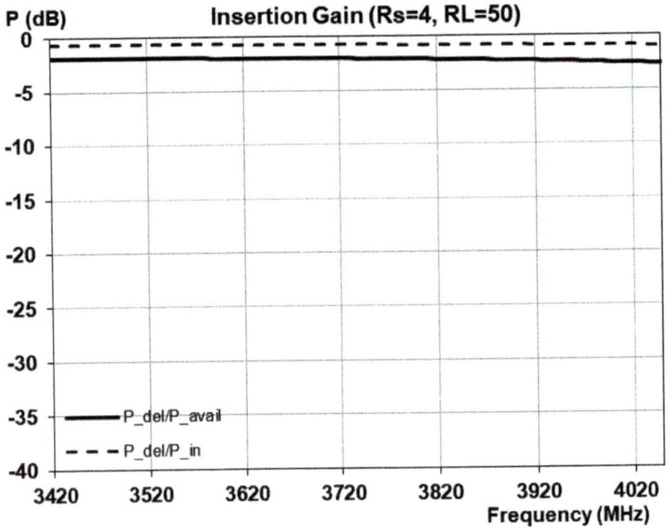

(b)

Figure 12.20 Input and output shunt-tuned, coupled-inductor match second-harmonic characteristics: (a) input reflection coefficient, and (b) insertion gain.

Table 12.5 Optimum Input and Output Shunt-Matched, Coupled-Inductor Element Values with Trap

Element Values	
$L_1 = 0.162$ nH	$L_2 = 2.179$ nH
$C_S = 24.618$ pF	$L_S = 0.074$ nH
$C_L = 2.086$ pF	

Figure 12.21 shows the resulting input and gain characteristics for the matching network.

The network input impedance is close to a pure reactive short across the second-harmonic passband, as generally desired. The actual value of the reactance can be skewed slightly towards inductive or capacitive by recentering the trap. In Figure 12.21(b), the second-harmonic isolation across the network is seen to exceed ~15 dB.

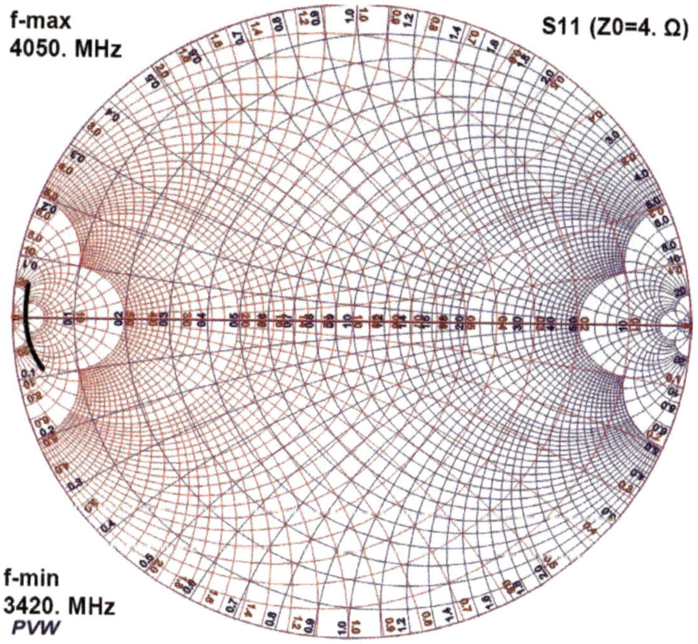

(a)

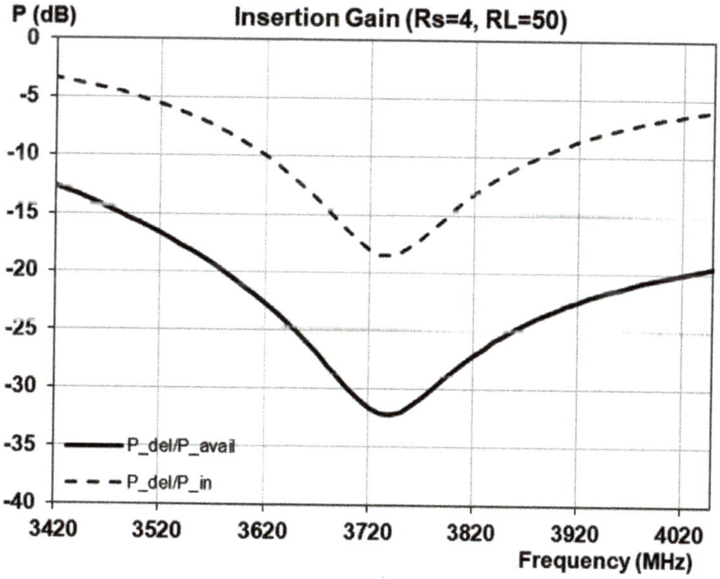

(b)

Figure 12.21 Input and output shunt-tuned, coupled-inductor match second-harmonic characteristics with trap: (a) input reflection coefficient, and (b) insertion gain.

The corresponding insertion gain for this network, shown in Figure 12.22(a), is ~ −0.6 dB.

From Table 12.5, the inductor ratio L_2/L_1 is ~13.4:1. This is higher than would be desired and would typically prove very challenging for layout. To avoid this, it is common to reduce the input impedance ratio required of the network by following the coupled-inductor match by a single-section LC network.

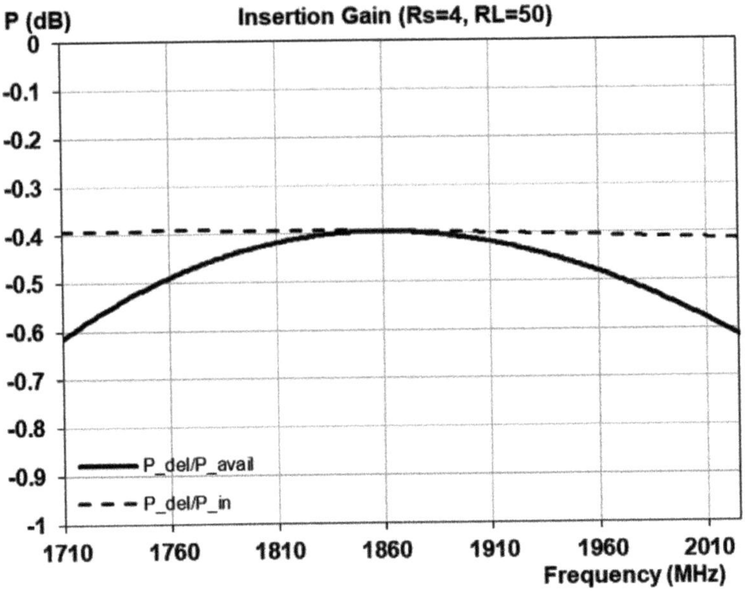

(a)

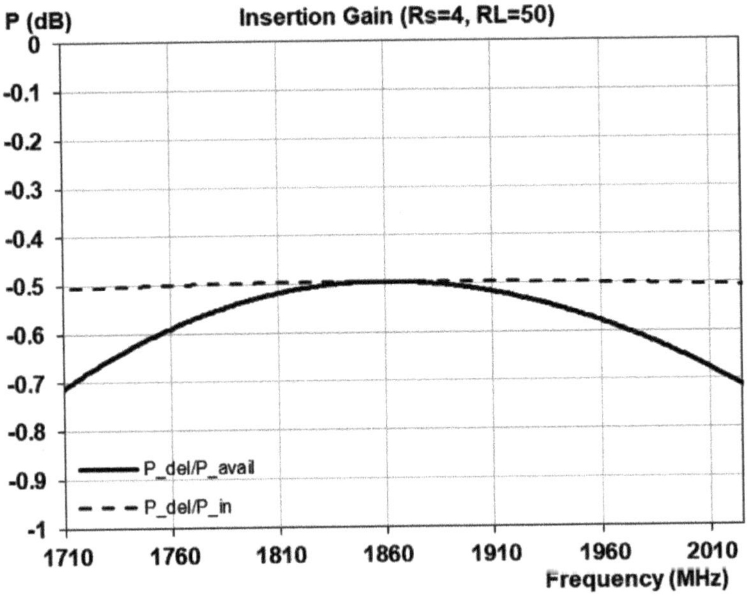

(b)

Figure 12.22 Optimized input and output shunt-matched coupled inductors with trap, insertion gains: (a) $Q_L = 60$, and (b) $Q_L = 40$.

In general, it is increasingly difficult to achieve a high inductor Q factor for lower values of inductance. Thus, achieving a Q-factor of 60, for an inductance value of 0.162 nH, would be extremely problematic. Figure 12.22(b) shows the modified insertion gain characteristic for the matching network with the quality factor of inductor L_1 reduced to 40. The minimum insertion gain in the passband is reduced to ~ −0.66 dB.

Lastly, the appropriate inductor values for a coupled-inductor matching network are highly dependent on the mutual coupling factor k, which can be achieved. This greatly complicates the challenges for both design and layout. Figure 12.23

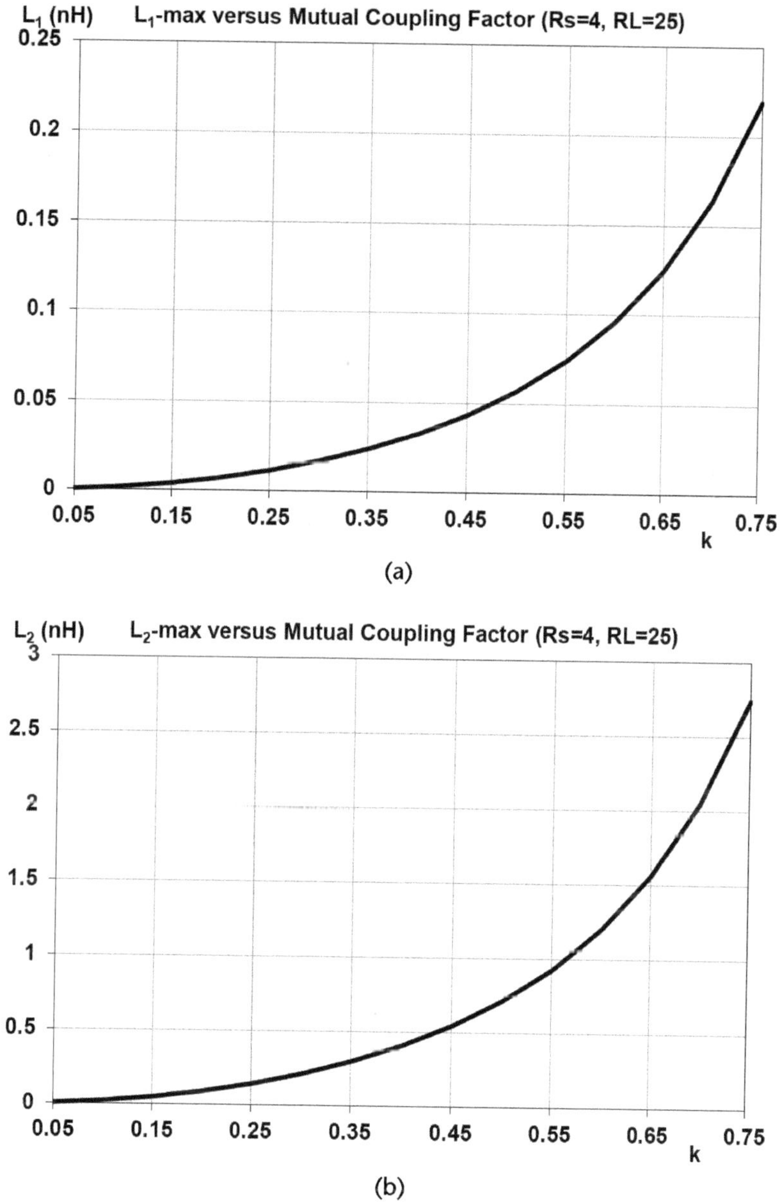

Figure 12.23 Dependence of coupled-inductor independent inductor value on mutual coupling: (a) L_1 for input shunt tuning, and (b) L_2 for output shunt tuning.

shows the dependence of the independent inductor value on the mutual coupling coefficient.

The deleterious effect on insertion gain of a coupled-inductor impedance matching network with reduced coupling coefficient is equally dramatic. Consider the simple input and output shunt-tuned coupled-inductor matching network of Figure 12.4(c). If the coupling coefficient is reduced from 0.7 to 0.5, inductor quality factors are all maintained as $Q_L = 60$, and no required input trap is included, the optimized required element values are as shown in Table 12.6.

Two problems are immediately apparent in Table 12.6. The primary inductance value is too small to be practical and the inductance ratio $L_2/L_1 = \sim 49:1$ would also be similarly very difficult to achieve in a physical layout. More disappointing, if both these challenges could be overcome, the insertion gain dependence for the network would be as shown in Figure 12.24(a).

The minimum passband gain = ~ −1.7 dB. This compares to a minimum passband gain = ~ −0.48 dB for $k = 0.7$, as shown in Figure 12.15.

The extreme sensitivity of the simple input and output shunt-tuned, coupled-inductor matching network, to the mutual coupling coefficient, is shown in Figure 12.24(b). It can be seen that without a strong mutual coupling coefficient between the two inductors such a network would not be a viable candidate for PA output impedance matching for $k < \sim 0.55$.

12.2.3 Single-Ended, Coupled-Inductor Matching Compared with Conventional *LC*-Output Matching

To understand why coupled inductors are rarely used for PA-output matching, it is instructive to compare their key performance characteristics to those of an equivalent conventional dual lowpass *LC* network. Such a network was presented in Section 11.8.1. Key performance parameters from that data, placed side by side with those for the optimized coupled-inductor match of the previous section, are provided below.

In all cases, it can be seen that the conventional dual lowpass match has superior performance parameters to those of the coupled-inductor matching network, which, in general, exhibits increased insertion loss and more dispersion across the passband. Second-harmonic characteristics are also inferior. This demonstrates that any expectations that a coupled-inductor network will have superior characteristics, as they are akin to a transformer, are not supported by the data.

The inferior performance characteristics of a single-ended, coupled-inductor impedance match, compared to a conventional *LC* match, might surprise those who consider an RF coupled-inductor pair as being the equivalent of a transformer. As discussed in Section 12.1, however, this is a misguided conception that is likely to misdirect. Rather than achieving the low loss and very wide bandwidth, which would

Table 12.6 Optimum Input and Output Shunt-Matched, Coupled-Inductor Element Values for $k = 0.5$

Element Values	
$L_1 = 0.056$ nH	$L_2 = 2.727$ nH
$C_S = 134.040$ pF	$C_L = 0.497$ pF

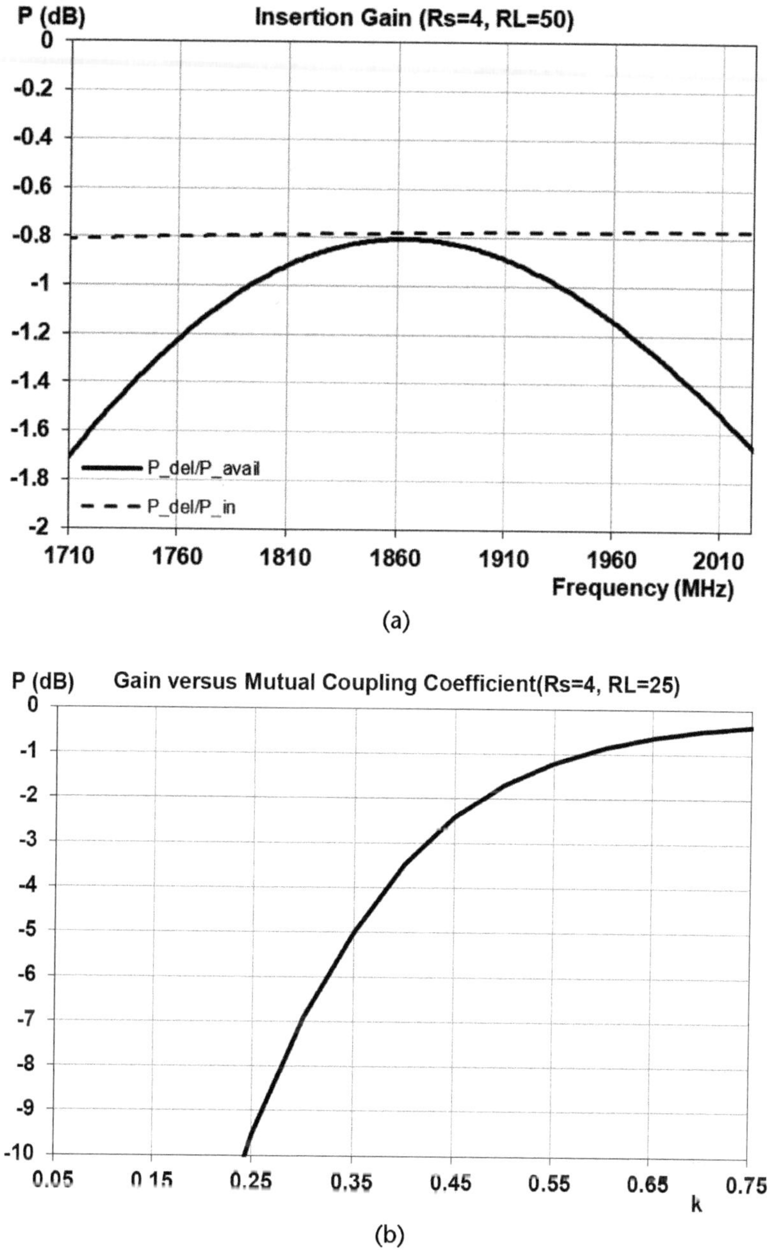

Figure 12.24 Mutual coupling factor impact on gain of coupled-inductor matching networks: (a) $k = 0.5$, and (b) input and output shunt tuning gain.

be expected of a true transformer, a coupled-inductor network alone generally has significantly inferior characteristics to the *LC* networks presented in Section 11.8. For this reason, and others given earlier in the section, it is invariably a poor choice for impedance matching of single-ended networks and therefore almost never used.

In general, impedance matching networks employing coupled inductors, which depend solely on magnetic transduction for RF energy transport through the network, have inferior performance characteristics compared to those that also have

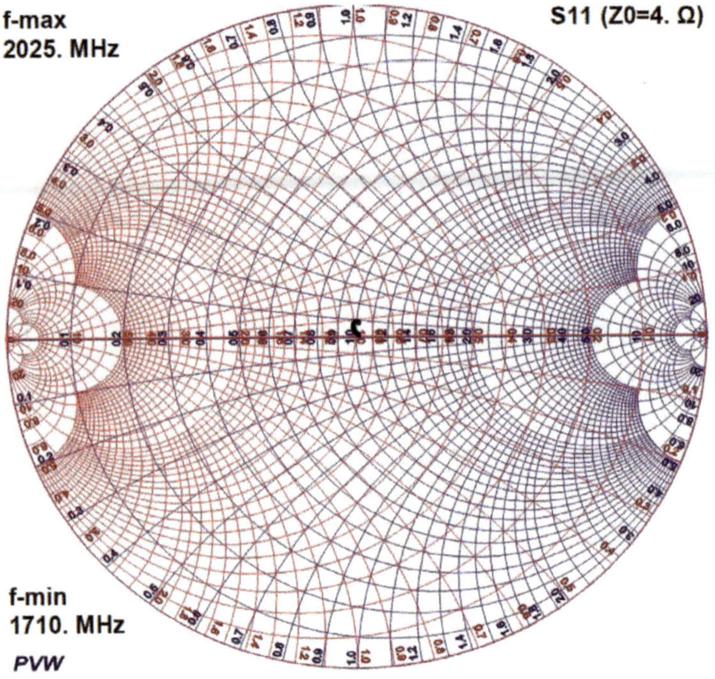

(a)

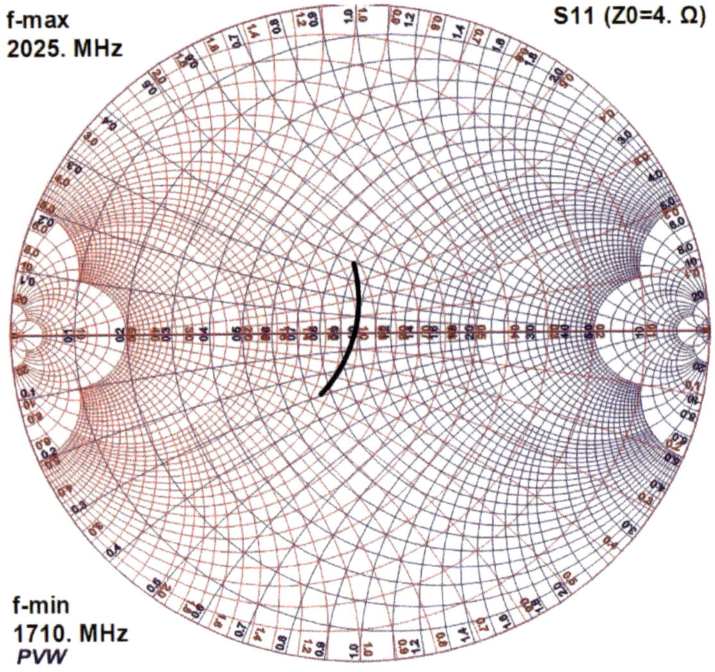

(b)

Figure 12.25 Network input reflection coefficients: (a) dual lowpass, and (b) optimized coupled inductors.

direct electrical conduction through the network. Efficient RF magnetic energy transduction between a pair of inductors invariably relies upon a resonance-type condition between the inductors. This resonance increases resistive losses in the inductors and results in a relatively narrow bandwidth for efficient energy transfer.

A detailed examination of the use of coupled inductors for impedance matching of both single-ended and multiphase amplifier networks is given in the following two chapters.

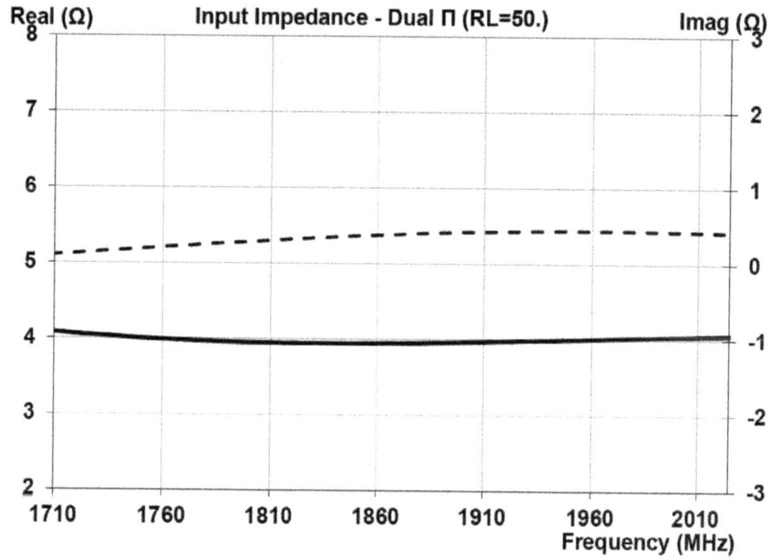

(a)

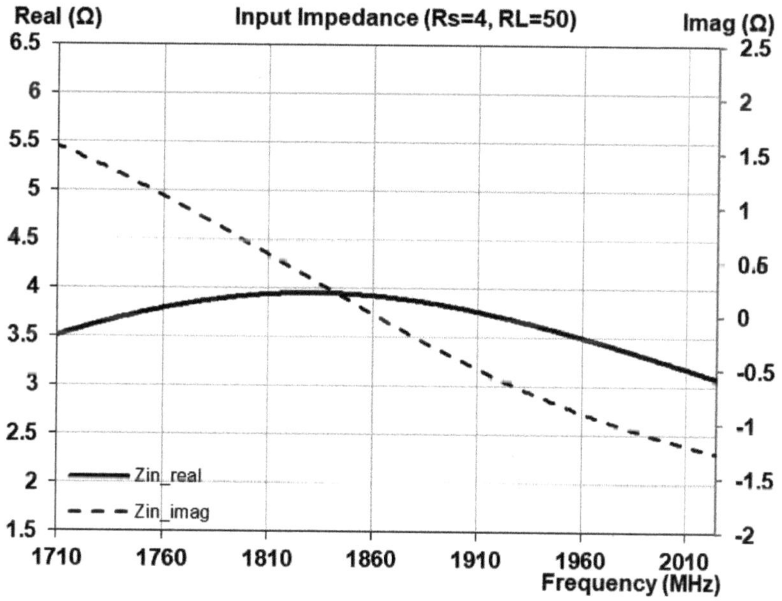

(b)

Figure 12.26 Network input impedances: (a) dual lowpass, and (b) optimized coupled inductors.

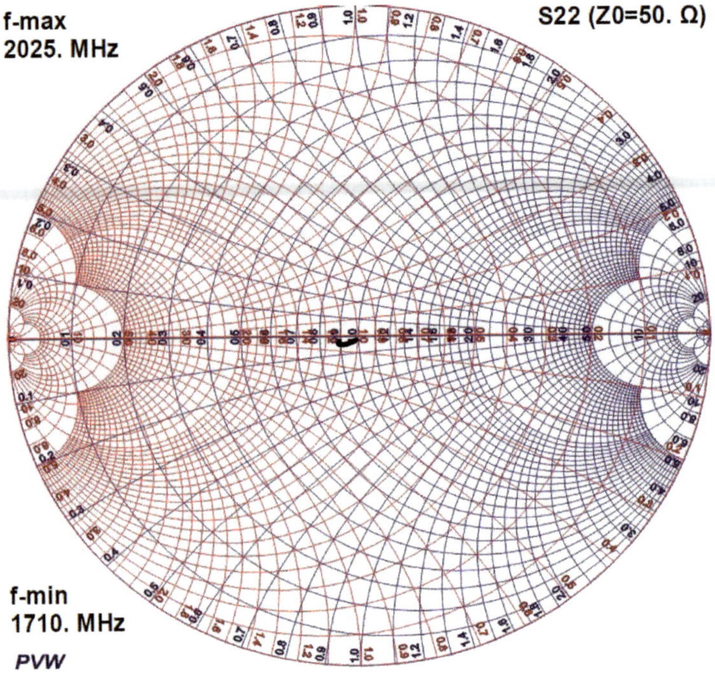

(a)

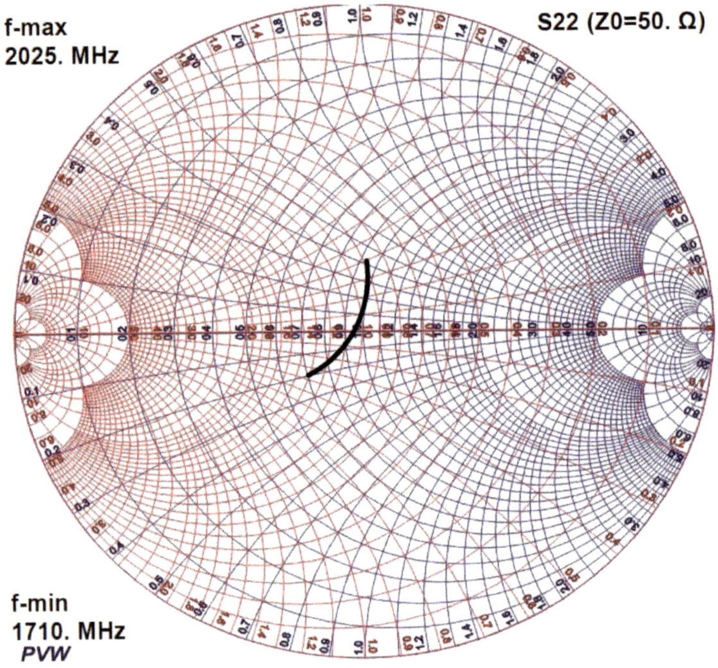

(b)

Figure 12.27 Network output reflection coefficients: (a) dual lowpass, and (b) optimized coupled inductors.

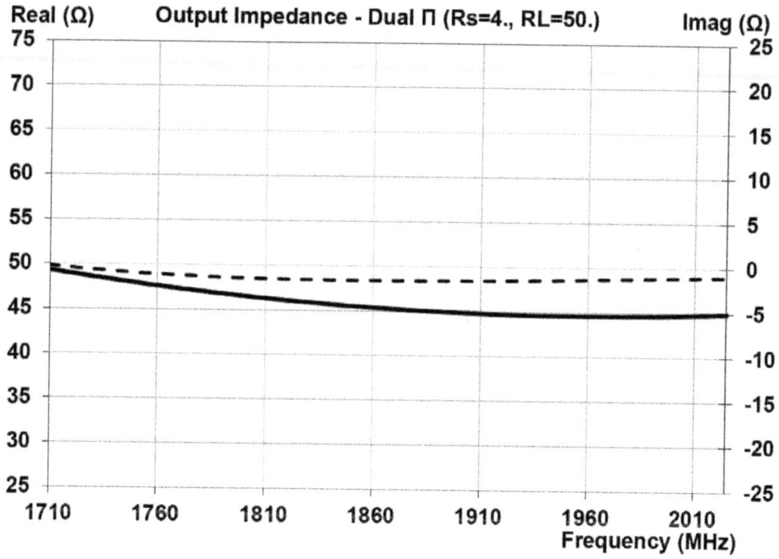

(a)

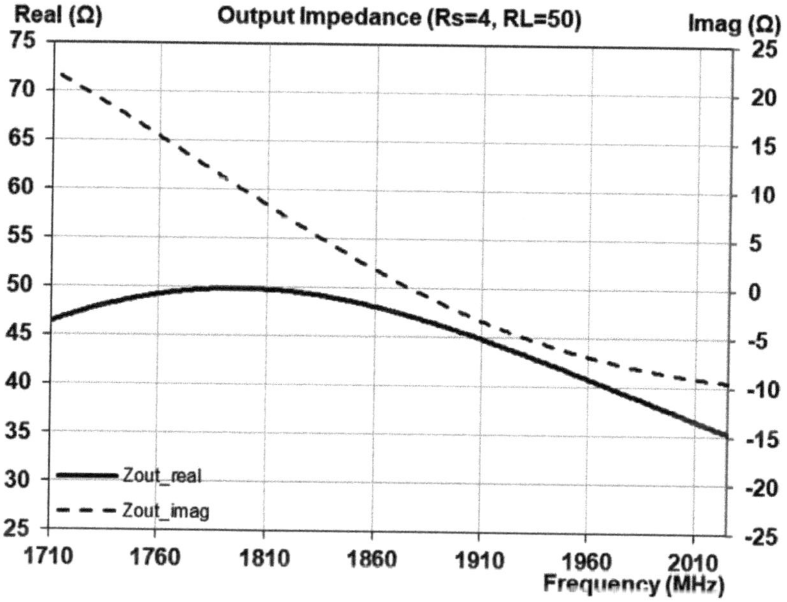

(b)

Figure 12.28 Network output impedances: (a) dual lowpass, and (b) optimized coupled inductors.

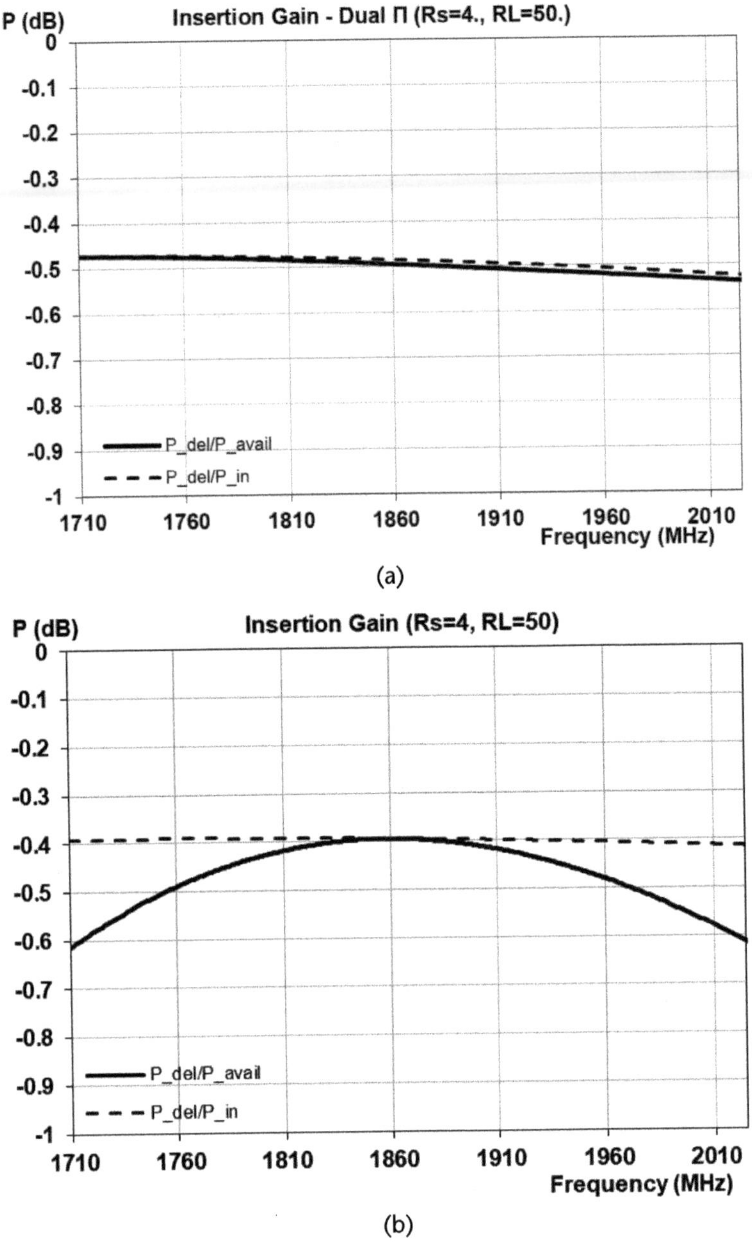

Figure 12.29 Network insertion gains: (a) dual lowpass, and (b) optimized coupled inductors.

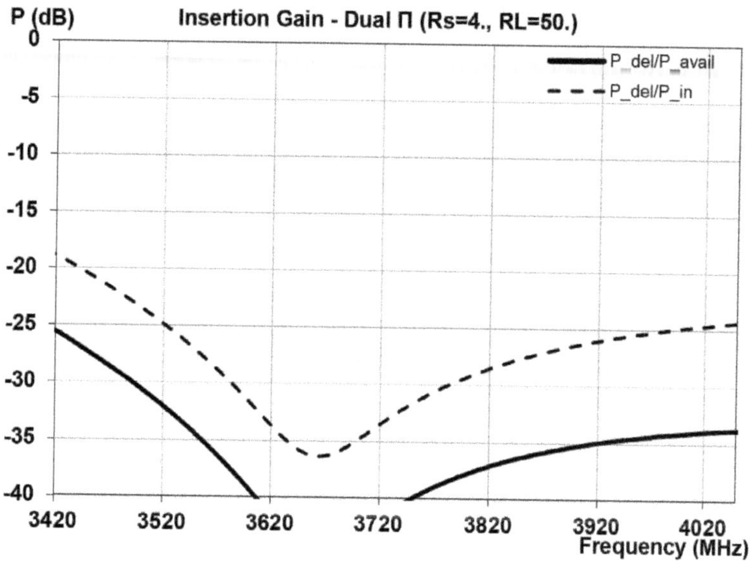

(a)

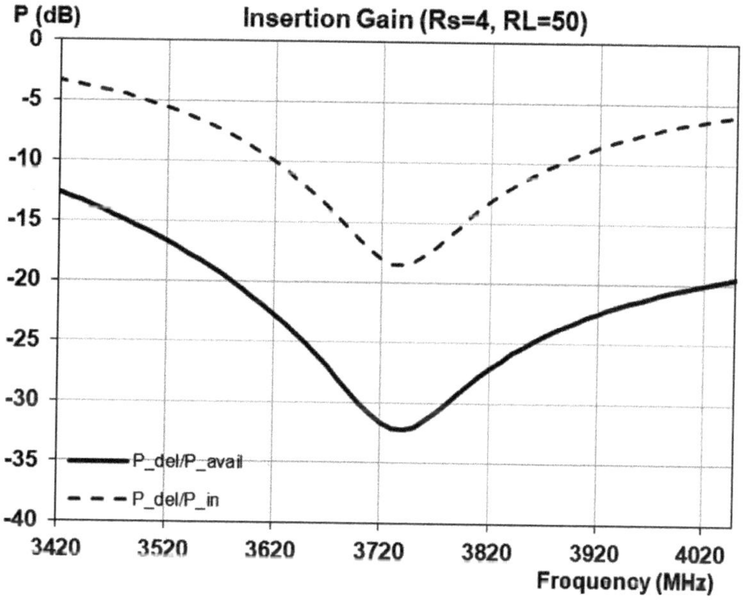

(b)

Figure 12.30 Networks second-harmonic insertion gains: (a) dual lowpass, and (b) optimized coupled inductors.

12.3 Single-Ended Coupled Inductors with an Interwinding Capacitance Analysis Approach

As discussed in Section 12.2, a high magnetic coupling coefficient k is essential in any type of coupled inductors impedance match; $k = 1$ would be ideal. Unfortunately, this is not achievable in practice, with printed inductor layouts typically being limited to $|k| \leq \sim 0.7$.

To achieve higher magnetic coupling factors, the inductor traces must be very close together in order for the magnetic fields associated with the currents flowing in each inductor to overlap. This has the unfortunate side effect of also resulting in significant interwinding capacitance between the traces, which can have important consequences for the performance of the impedance match, particularly at frequencies above the passband.

If the two coupled inductors are of equal value, closed-form expressions for the four-port S-parameters can be obtained, as presented in Section 2.5. However, in general, this is not the case. For matching applications, it is generally required that the inductors be of different values. Accounting for the cross-coupling capacitance with a discrete lumped-element model is challenging. Two such models that attempt to account for the effects of cross-coupling are shown in Figure 12.31. However, in order to most accurately account for the effects of cross-coupling on the performance characteristics of a pair of coupled inductors, an incremental model is required.

Figure 12.32 shows the schematic representation to be modeled. The admittance elements across both inductors are possible matching elements. Typically, when present, they are predominantly capacitive in the passband.

In Appendix F, a detailed analytic analysis is provided for a pair of coupled inductors with uniformly distributed interwinding capacitance. Closed-form analytical expressions are provided for all key network performance parameters. The performance parameters for the single-ended, coupled-inductor schematic shown in Figure 12.32 can be derived from these solutions.

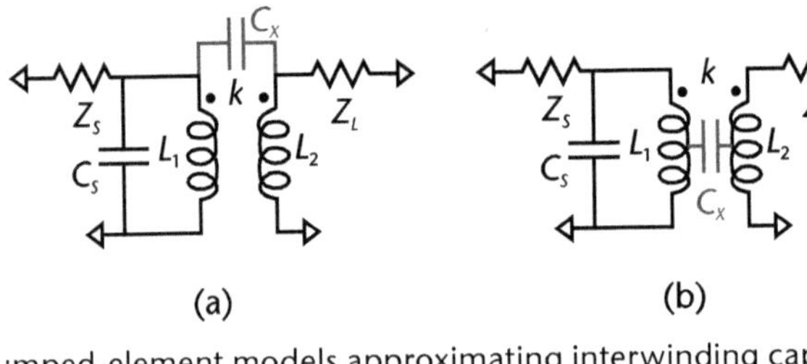

Figure 12.31 Lumped-element models approximating interwinding capacitance: (a) direct capacitive coupling, and (b) intermediate capacitive coupling.

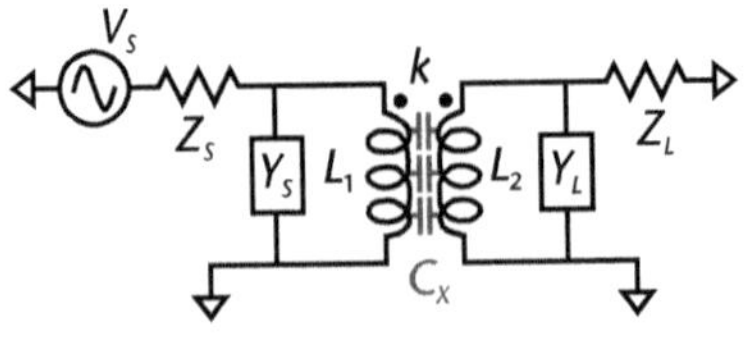

Figure 12.32 Single-ended, coupled-inductor schematic including interwinding capacitance and external matching.

12.3.1 Impact of Interwinding Capacitance on Single-Ended, Coupled-Inductor Characteristics

The focus of this section is to illustrate how the interwinding capacitance between two coupled inductors impacts key performance characteristics. The network performance characteristics are derived applying the generalized analytic solutions provided in Appendix F. With reference to Figure A.3, in this case, $V_{S2} = V_{S3} = V_{S4} = 0$, and $Z_2 = Z_4 = 0$. Unfortunately, the analytic solutions of Appendix F are not readily invertible for synthesis. Thus, the design equations of Section 12.2 will again be used for network synthesis.

To clearly distinguish the effects of interwinding capacitance on coupled-inductor performance, design parameters identical to those in Table 12.1 are used, except for the addition of an interwinding capacitance value of 3 pF, as listed in Table 12.7.

Element values are identical to those in Table 12.2, as no allowance is made in the design equations for the interwinding capacitance. The network insertion gain and input impedance are shown in Figure 12.33.

The input impedance observed in Figure 12.33(a) is ~2.5Ω across the passband. This is substantially below the target input impedance of 4Ω. It is inferior to that without the interwinding capacitance shown in Figure 12.7(a). Likewise, comparing the insertion gain response with finite interwinding capacitance in Figure 12.33(b) to that without interwinding capacitance in Figure 12.8, the insertion gain is seen to be significantly reduced across the passband. There are two potential causes of the performance degradation: (1) it is directly attributable to the presence of capacitive coupling, or (2) it results from not accounting for the capacitance in synthesizing the network elements. To investigate the latter cause, the network synthesis equations of Section 12.2 were recalculated with a higher target input impedance than the desired 4Ω for the network.

A design input target impedance of 6Ω was used in applying the synthesis equations of Section 12.2 for the network element values. The modified element values are listed in Table 12.8.

The equivalent performance parameters to Figure 12.33 are shown in Figure 12.34.

The input impedance in Figure 12.34(a) is now seen to very close to the desired value of 4Ω across the full passband. Perhaps, even more importantly, the insertion

Table 12.7 Coupled Inductors with Capacitive Coupling Impedance Match Parameters

Design Parameters	
Mutual coupling, $k = 0.7$	
Cross-coupling capacitance, $C_X = 3$ pF	
f_{lo} 1,710 MHz	f_{hi} 2,025 MHz
$f2_{lo}$ 3.420 MHz	$f2_{hi}$ 4,050 MHz
$Z_S = 4\Omega$	$Z_L = 50\Omega$
Inductors, $Q_L = 60$	Capacitors, $Q_C \approx 100$

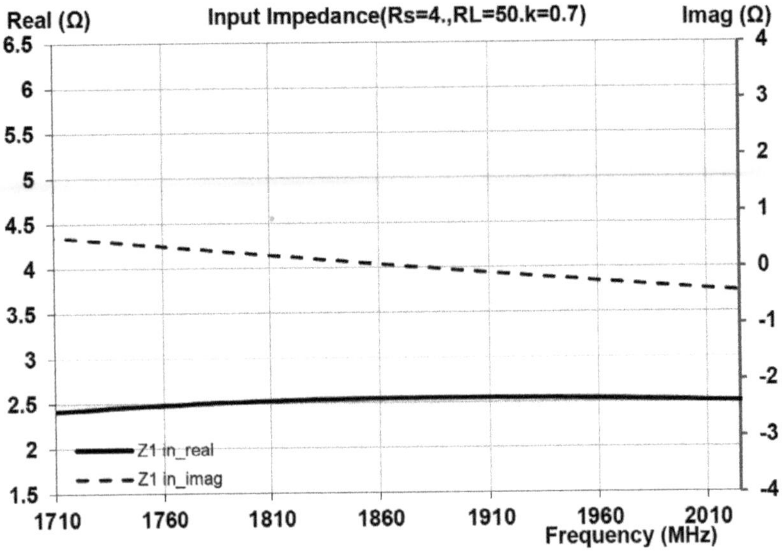

(a)

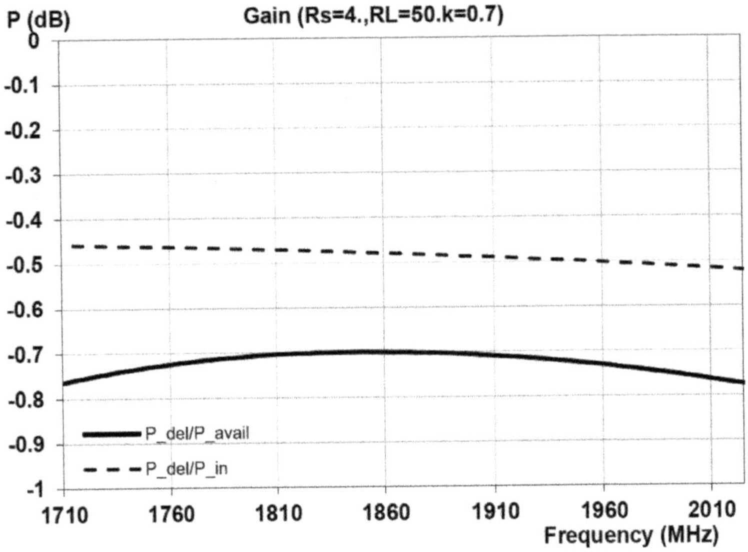

(b)

Figure 12.33 Optimized input shunt-matched, coupled-inductor network characteristics with 3-pF interwinding capacitance: (a) input impedance, and (b) insertion gain.

Table 12.8 Single-Ended, Coupled-Inductor Match Elements with 3-pF Interwinding Capacitance and Target $Z_{in} = 6\Omega$

Element Values	
$L_1 = 0.205$ nH	$L_2 = 4.532$ nH
$C_S = 43.195$ pF	$L_S = 0$ nH

gain in Figure 12.34(b) is also seen to be relatively flat across the passband and significantly higher than that in Figure 12.33(b). Moreover, the insertion gain is higher than that in Figure 12.8 with no interwinding capacitance. Thus, for this case, the presence of the interwinding capacitance actually leads to improved coupled-inductor performance.

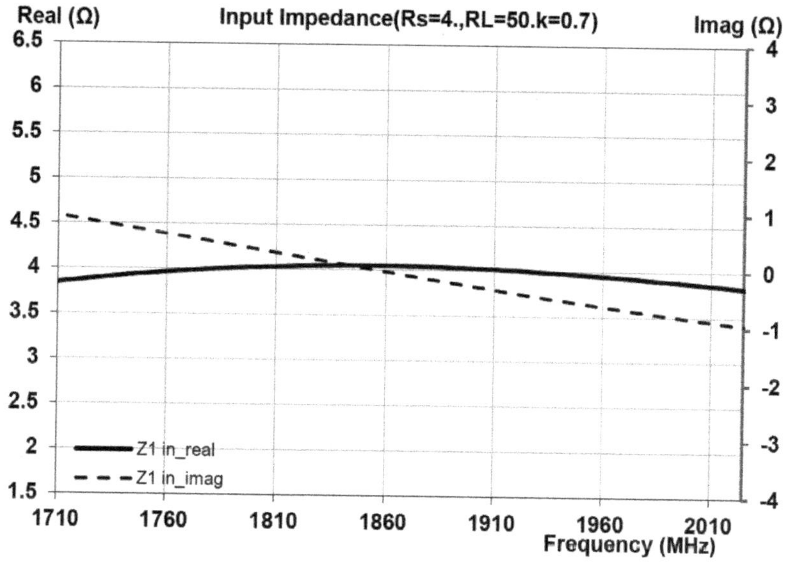

(a)

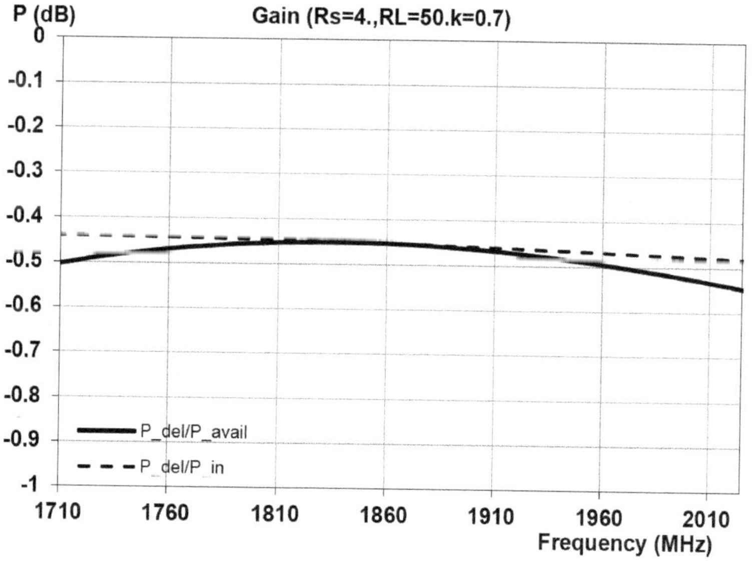

(b)

Figure 12.34 Optimized input shunt-matched, coupled-inductor network characteristics with 3-pF interwinding capacitance and target Z_{in} = 6Ω: (a) input impedance, and (b) insertion gain.

As a further example, consider a configuration designed similarly to meet the specifications of Table 12.7, but with optimized input and output matching. For optimum performance, in this case, a design input target impedance of 5.6Ω was used. The optimized element values were determined as listed in Table 12.9.

The corresponding input and insertion gain characteristics are shown in Figure 12.35. The insertion gain in Figure 12.35(b) is flatter, and overall higher across the passband, than that in Figure 12.15 for the equivalent network with no interwinding capacitance. Impedance plots in Figure 12.35(a) and Figure 12.14(a) are comparable. Thus, for this matching scenario also, the interwinding capacitance is not deleterious.

One important aspect of the coupled-resonator performance that must be explored to gain a full appreciation of the impact of interwinding capacitance is an examination of how network characteristics are affected above the passband. Due to the capacitive nature of the coupling, it might be expected that the coupling will lead to increased transmission above the passband, thereby decreasing high-side rejection. To evaluate this, the insertion gains of the coupled-resonator network, considered above, were determined without and with the 3-pF interwinding capacitance. The wideband high-frequency insertion gain for both scenarios is presented for comparison in Figure 12.36.

Possibly somewhat counterintuitively, introducing the interwinding capacitance does not lead to significantly reduced rejection above the passband. In fact, for the majority of frequencies, there is higher rejection. Moreover, at multiple frequencies, there are deep transmission nulls. In the data shown, one of these nulls occurs only slightly above the center of the second-harmonic band. If value of the interwinding capacitance C_X is increased to ~4 pF, the null could be located exactly at the center of the second-harmonic band. The complex transmission characteristics above the passband result from the interplay of the distributed inductance and capacitance of the network.

To summarize, the presence of some interwinding capacitance in a coupled-inductor matching network can result in improved performance characteristics, in particular, improved input impedance characteristics and reduced insertion loss. An additional, perhaps unanticipated, benefit is the potential for increased rejection above the passband and specifically within the second-harmonic band. However, too much interwinding capacitance will inevitably harm the network characteristics. In the example given, if C_X is increased above 4 pF, the null at the second-harmonic frequency moves down towards the passband. As it does so, the insertion gain and input impedance characteristics are increasingly degraded from the high end of the passband.

Table 12.9 Optimum Input and Output Shunt-Matched, Coupled-Inductor Element Values with 3-pF Interwinding Capacitance and Target Z_{in} = 5.6Ω

Element Values	
L_1 = 0.232 nH	L_2 = 3.159 nH
C_S = 30.514 pF	C_L = 0.948 pF

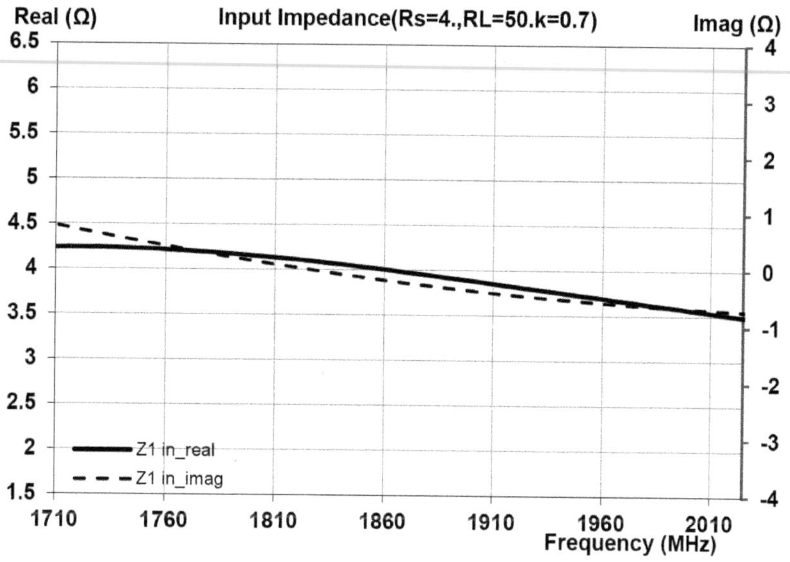

(a)

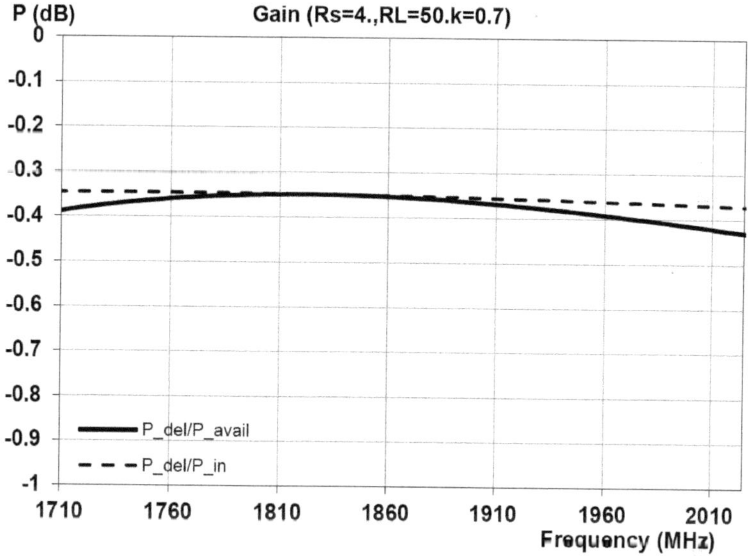

(b)

Figure 12.35 Optimized input and output shunt-matched, coupled-inductor network characteristics with 3-pF interwinding capacitance and target $Z_{in} = 5.6\Omega$: (a) input impedance, and (b) insertion gain.

12.4 Lowpass π-Network with Auto-Transformer Action

In Section 12.2.1, it was demonstrated that networks relying solely on magnetic coupling for energy transfer generally have performance characteristics inferior to those that rely entirely on direct electrical transfer. In Section 12.3.1, it was demonstrated that the added presence of interwinding capacitance in a coupled-inductor pair significantly improved the overall characteristics of the matching network.

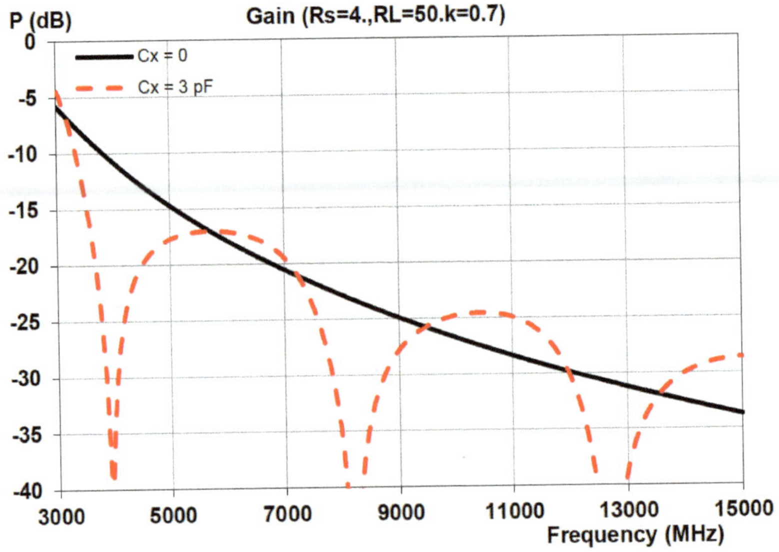

Figure 12.36 Optimized input shunt-matched, coupled-resonator network insertion gain without, and with 3 pF, interwinding capacitance.

This suggests the possibility of creating a matching network with optimal features that combines both magnetic and direct electrical coupling. A way to implement such an approach is suggested by a device technology commonly employed in low-frequency applications and, in particular, in the field of power voltage conversion.

Section 12.1 discussed the basics of classic transformers, which comprise two ideally infinite lossless inductors whose magnetic fields are wholly coupled together. The two inductors, the primary and secondary, are electrically isolated, and the voltage ratio on the two inductors is given by the turns' ratio of the inductors (Equation (12.1)). In addition to the classic transformer for voltage conversion, however, there is another class of transformers that are referred to as "auto-transformers."

The differentiating feature of an auto-transformer is that it comprises only one inductance coil, again ideally infinite, in place of two. One terminal of the inductor is generally ground, while the other is one of the two network terminals. The remaining network terminal is connected to the inductor at some intermediate point along its length. Figure 12.37(a) shows a basic schematic of an auto-transformer.

Unlike a classic transformer, the auto-transformer does not provide electrical isolation between its two terminals. This can be a safety issue in power distribution networks. However, it has the advantage that energy transfer thorough the network results from both magnetic and direct electrical couplings. In consequence, an auto-transformer is typically more efficient than a classic transformer.

One further advantage of the auto-transformer is that the intermediate contact point along the inductor for the one terminal can often be moved. This makes it relatively simple to adjust the voltage transformation ration of the device. In contrast, the turns ratio in a classic transformer is fixed and cannot be adjusted after manufacture.

Similar to the classic transformer and referring to Figure 12.37(a), the voltage transformation ratio for the auto-transformer is

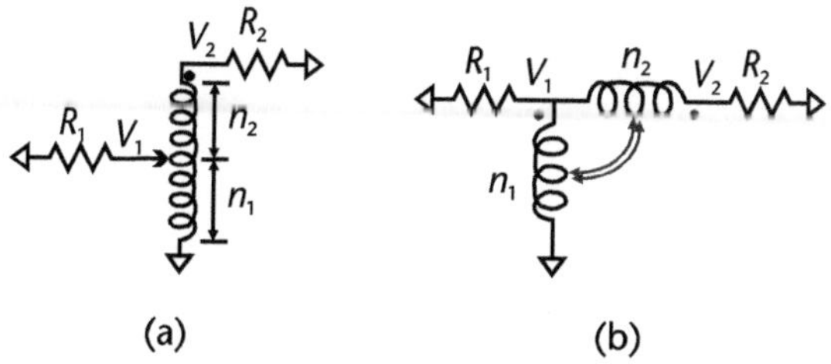

Figure 12.37 Auto-transformer schematics: (a) auto-transformer, and (b) folded auto-transformer.

$$V_2 = ((n_1 + n_2)/n_1)V_1 \tag{12.22}$$

and the resistances on the two ports are related by

$$R_2 = ((n_1 + n_2)/n_1)^2 R_1 \tag{12.23}$$

The voltage and impedance levels can thus be stepped up across an auto-transformer in a manner similar to that of a classic transformer. However, it has the potential advantage of improved characteristics due to its direct electrical coupling.

To see how an auto-transformer type action might be used to advantage in an impedance matching network, consider redrawing the auto-transformer in a folded topology as shown in Figure 12.37(b). No change to the elements or magnetic coupling of the inductor are implied; thus, the transfer properties will be unchanged. However, in the folded form, it can now be identified as similar in form to the left-most part of a lowpass π-network.

Figure 12.38(a) shows the lowest-loss form of a lowpass π-network, for implementing an impedance transformation from 4Ω to 10Ω. For the latter network, the through insertion phase $\Delta\Phi \approx -69°$. Figure 12.38(b) shows a generalization of this network in which magnetic coupling is introduced between the inductive elements. The mutual coupling coefficient between the inductors is denoted by k. In the limit $k = 0$, the network reverts to the classic form of Figure 12.38(a). The objective is to determine whether having $k > 0$ can augment the performance of the lowpass network via the auto-transformer action, resulting from the magnetic coupling.

Just as with the coupled-inductor correspondence to a classic transformer, the coupled inductors in Figure 12.38(b) do not correspond exactly to a true

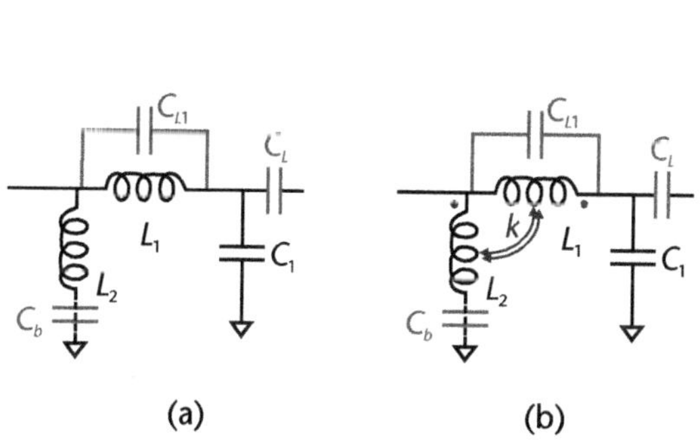

Figure 12.38 Lowpass π-networks with and without magnetic coupling: (a) classic lowpass π-network, and (b) lowpass π-network with coupling.

auto-transformer. The magnetic coupling between the inductor is less than ideal and neither of the inductances are infinite. Nonetheless, the magnetic coupling does introduce an auto-transformer like magnetic coupling aspect into the network.

In Figure 12.38, three additional capacitors C_b, C_{L1}, and C_L have been included as optional components in the schematics. As such, they are all accounted for in the circuit analysis matrix element formulae given at the end of this section.

Capacitor C_b is only required when, for reasons to be explained in Section 12.4.2, the designer desires to replace the hard ground, at the bottom of inductor L_2, by a virtual one. In this case, C_b should ideally be large enough to present only a negligible reactance across the passband. If this is not the case, it will generally be deleterious to the operation of the matching network. The elemental synthesis equations in this section thus do not account for the presence of C_b. However, if present, the analysis formulae permit any consequences of C_b in the circuit to be determined.

For reasons linked to those related to the inclusion of capacitor C_b in the circuit, a second capacitor C_L is frequently required in the circuit on the output side of series inductor L_1. In this position, capacitor C_L serves to block any DC flow across the network. Again, the utility of including these two capacitors in the circuit is fully elucidated in Section 12.4.2.

Lastly, capacitor C_{L1}, included in the schematics across series inductor L_1, is also not an essential part of the lowpass networks. However, for a lowpass π-network such as shown in Figure 12.38(a), it is common practice to add a capacitor in shunt across L_1 to provide attenuation for frequency bands above the passband. Determining the appropriate value for this capacitor and L_1 is relatively simple in the uncoupled case. Design equations for the latter are given in Section 8.6. However, designing a resonant trap for inductor L_1 in the presence of magnetic coupling, as in Figure 12.38(b), is more complex. The precise values for L_1 and C_{L1} are dependent upon the coupling. Thus, it is important to include this dependency in the synthesis equations for them to be useful.

Defining the basic network parameters (with $C_b = \infty$) as

$$Z_S = R_S + jX_S \tag{12.24}$$

$$Z_L = R_L + jX_L \tag{12.25}$$

$$Z_{L1} = R_{L1} + j\omega L_1 \tag{12.26}$$

$$Z_{L2} = R_{L2} + j\omega L_2 \tag{12.27}$$

$$Z_{CL1} = R_{CL1} - j/\omega C_{L1} \tag{12.28}$$

$$Z_{CL} = R_{CL} - j/\omega C_L \tag{12.29}$$

also,

$$\omega_r = \text{angular frequency at the center of } L_1 \text{ stopband} \tag{12.30}$$

Synthesis equations for the elements of the auto-transformer-like form of the lowpass π-network shown in Figure 12.38(b) are

$$C_{L1} = \frac{1 + k\sqrt{L_1/L_2}}{\omega_r^2 (1-k)^2 L_1} \tag{12.31}$$

$$\omega C_1 = \frac{\begin{pmatrix} R_S R_L + X_S X_L + \omega^2\left(1-k^2\right)L_1 L_2 + \omega L_2 X_L + \omega\left(L_1 + L_2 + 2k\sqrt{L_1 L_2}\right)X_S \\ -(\omega/\omega_r)^2\left(L_2 + k\sqrt{L_1 L_2}\right)\left(\omega(X_S + X_L) + \dfrac{R_S R_L + X_S X_L}{(1-k)^2 L_2}\right) \end{pmatrix}}{\omega\begin{pmatrix} \omega\left(1-k^2\right)L_1 L_2 X_L \\ +\left(L_1 + L_2 + 2k\sqrt{L_1 L_2} - (\omega/\omega_r)^2\left(L_2 + k\sqrt{L_1 L_2}\right)\right)(R_S R_L + X_S X_L) \end{pmatrix}} \tag{12.32}$$

which are dependent upon solving the governing equation

$$\begin{aligned} &L_1\left(\left(1-k^2\right)L_2 - (\omega/\omega_r)^2\left(L_2 + k\sqrt{L_1 L_2}\right)\right)\left(R_L^2 + X_L^2\right)R_S + \left(\omega\left(1-k^2\right)L_1 L_2\right)^2 R_L \\ &\quad + \left(L_1 + L_2 + 2k\sqrt{L_1 L_2} - (\omega/\omega_r)^2\left(L_2 + k\sqrt{L_1 L_2}\right)\right) \\ &\quad \bullet \begin{pmatrix} 2\omega\left(1-k^2\right)L_1 L_2 R_L X_S + \left(L_1 + 2k\sqrt{L_1 L_2}\right)\left(R_S^2 + X_S^2\right)R_L \\ +\left(L_2 - (\omega/\omega_r)^2\left(L_2 + k\sqrt{L_1 L_2}\right)\right)\left(\left(R_S^2 + X_S^2\right)R_L - \left(R_L^2 + X_L^2\right)R_S\right) \end{pmatrix} = 0 \end{aligned} \tag{12.33}$$

Given a value for the independent variable L_1, (12.33) must be solved to determine the corresponding value for L_2. Unfortunately, a closed-form solution for L_2 is not available; however, a solution for L_2 that satisfies (12.33) is readily derivable by numeric techniques.

Having determined the element values for the network, all electrical parameters are readily determined from an $[ABCD]$ matrix analysis, using the equations given in Chapter 1. For the network, the individual $[ABCD]$ parameters are given by

$$A = 1 + \frac{\left((Z_{L2} + Z_{Cb})Z_{L1} + (\omega k)^2 L_1 L_2\right)/Z_1 - j\omega k\sqrt{L_1 L_2}}{(1 + Z_{L1}/Z_{CL1})(Z_{L2} + Z_{Cb}) + (\omega k)^2 L_1 L_2/Z_{CL1} + j\omega k\sqrt{L_1 L_2}} \tag{12.34}$$

$$B = \frac{(Z_{L2} + Z_{Cb})Z_{L1} + (\omega k)^2 L_1 L_2}{(1 + Z_{L1}/Z_{CL1})(Z_{L2} + Z_{Cb}) + (\omega k)^2 L_1 L/Z_{CL1} + j\omega k\sqrt{L_1 L_2}} \tag{12.35}$$

$$C = \frac{1}{Z_1} + \frac{1 + \left(Z_{L1} + j\omega k\sqrt{L_1 L_2}\right)/Z_1 + Z_{L1}/Z_{CL1}}{(1 + Z_{L1}/Z_{CL1})(Z_{L2} + Z_{Cb}) + (\omega k)^2 L_1 L/Z_{CL1} + j\omega k\sqrt{L_1 L_2}} \tag{12.36}$$

$$D = 1 + \frac{Z_{L1} + j\omega k\sqrt{L_1 L_2}}{(1 + Z_{L1}/Z_{CL1})(Z_{L2} + Z_{Cb}) + (\omega k)^2 L_1 L/Z_{CL1} + j\omega k\sqrt{L_1 L_2}} \tag{12.37}$$

12.4.1 Lowpass π-Network with Auto-Transformer Action Example

In order to determine whether adding magnetic coupling between the two inductors in a lowpass π-network is beneficial, consider the design of an impedance matching network with identical parameters to those specified in Table 12.1.

The lowest-loss lowpass π-network gain with no inductor coupling, as in Figure 12.38(a), is shown in Figure 12.39(a). In contrast, allowing for mutual coupling between the two inductors, as in Figure 12.38(b), the lowest network loss achievable is shown in Figure 12.39(b).

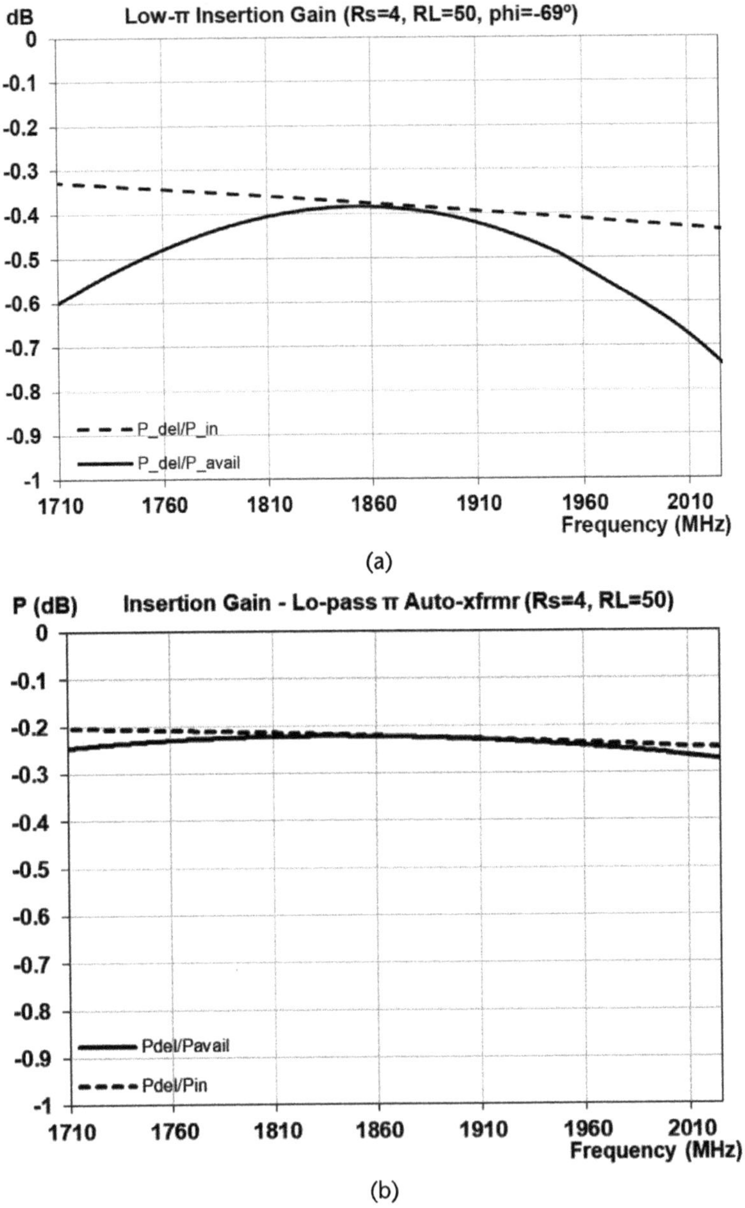

Figure 12.39 Lowpass π-matches without and with mutual inductive coupling: (a) uncoupled lowpass π-match, and (b) coupled-inductor lowpass π-match.

Element values for the two networks are provided in Table 12.10.

The maximum passband insertion loss for the uncoupled lowpass π match is ~0.74 dB, while the maximum passband insertion loss for the coupled-inductor lowpass π-match is only ~0.27 dB. This is a dramatic reduction in insertion loss, attributable entirely to the introduction of magnetic coupling between the inductors. In addition to improving overall insertion loss, the coupled network also shows considerably less dispersion across the passband. The increase in gain (i.e., lower insertion loss) for the network, as a function of the increased coupling factor, is shown in Figure 12.40.

Another important consequence of the added coupling between the inductors, which can be noted in Table 12.10, is a large reduction in the required value of the shunt inductor L_2. The dependence of L_2 on the coupling coefficient is shown in Figure 12.40. While this might be seen as an advantage, unfortunately, low values

Table 12.10 Lowpass π-Network Element Values

Uncoupled Lowpass π-Match
$k = 0.$
$L_1 = 1.125$ nH
$L_2 = 4.214$ nH
$C_1 = 5.801$ pF
Coupled-Inductor Lowpass π-Match
$k = 0.7$
$L_1 = 2.230$ nH
$L_2 = 0.557$ nH
$C_1 = 2.399$ pF

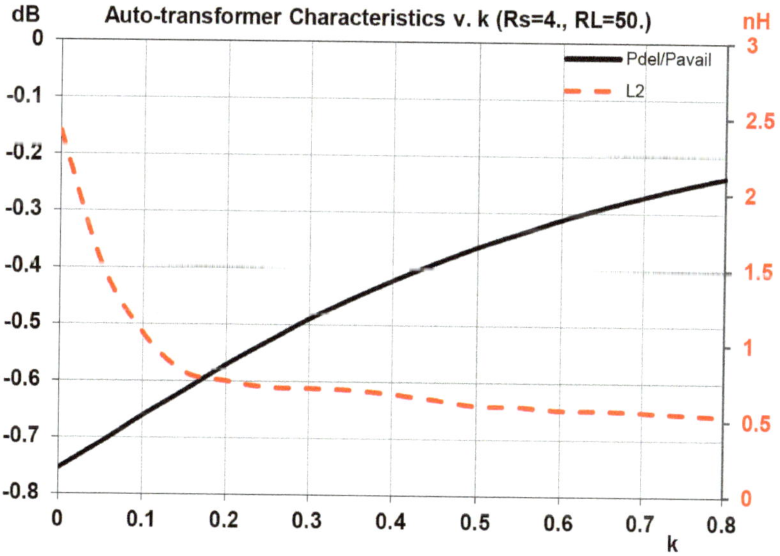

Figure 12.40 Auto-transformer characteristics' dependence on inductive coupling.

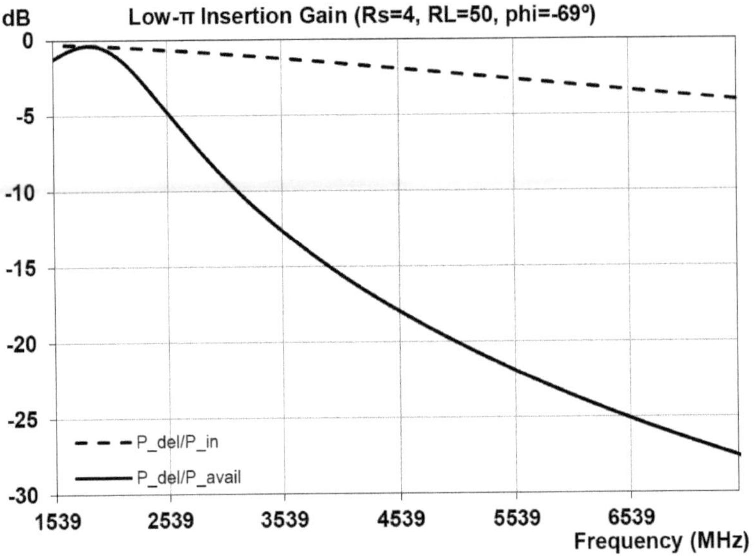

(a)

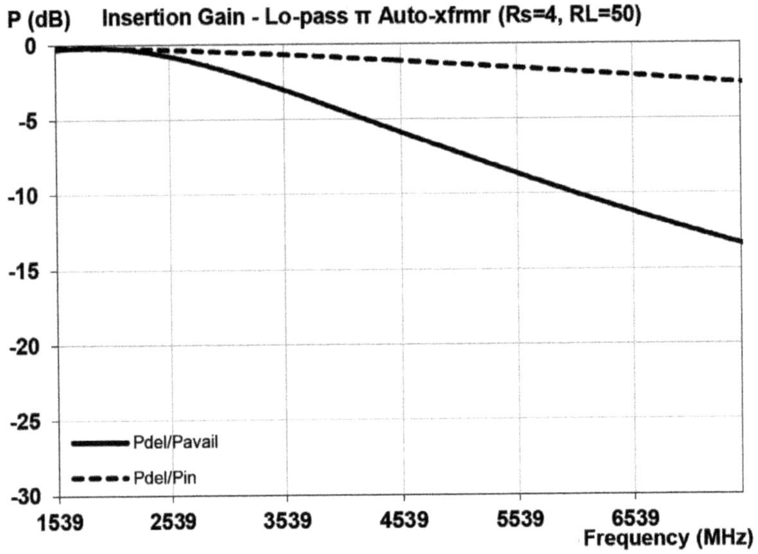

(b)

Figure 12.41 Lowpass π-matches without and with mutual inductive coupling: (a) uncoupled lowpass π-match, and (b) coupled-inductor lowpass π-match.

for the shunt inductor can be problematic if the network is to be used simultaneously for matching and supplying bias current to a PA output stage. The impact of a very low value shunt inductor in a PA output matching circuit is examined fully in Section 12.4.2.

While the reduction in passband insertion loss for the lowpass π-network, with inductor coupling, is highly desirable, unfortunately, this also continues above the

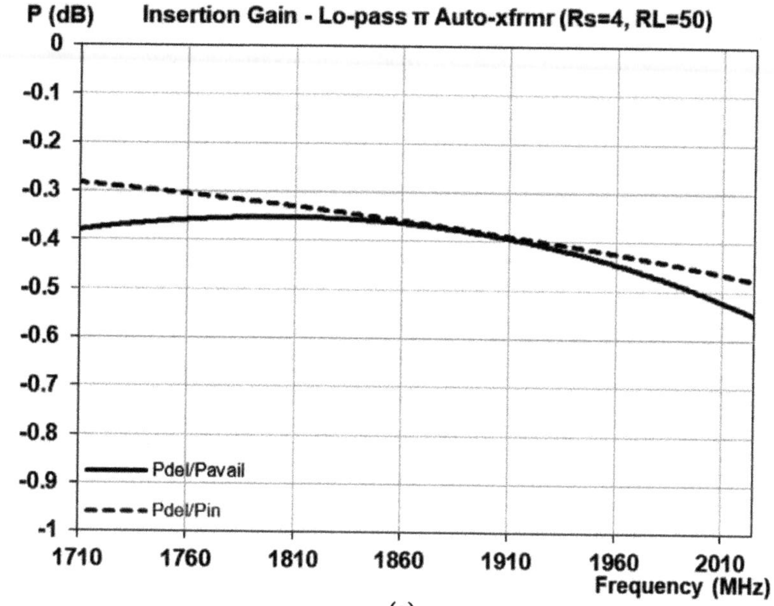

(a)

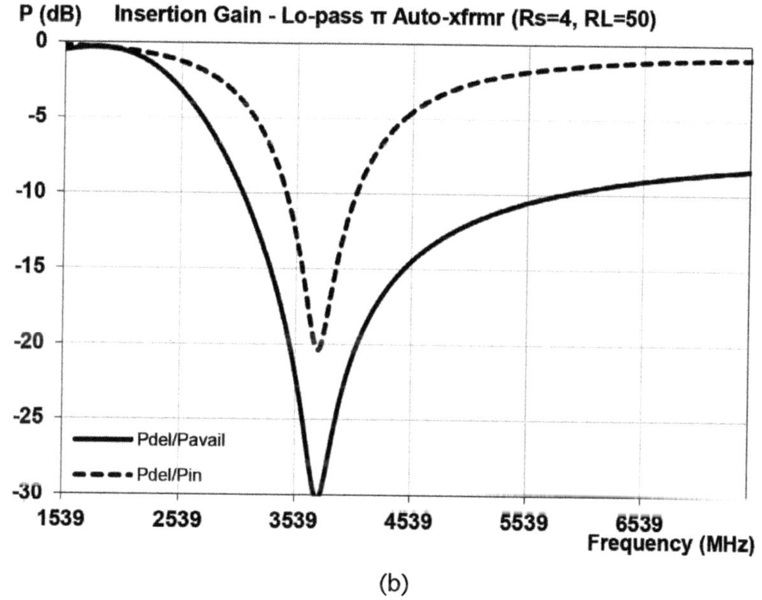

(b)

Figure 12.42 Coupled-inductor lowpass π match with a second-harmonic trap on l_1: (a) coupled lowpass π-match passband, and (b) coupled lowpass π-match wideband.

passband. Figure 12.41 shows wide-bandwidth plots for the insertion loss of the network, without and with inductor coupling.

Harmonic rejection above the passband can be seen to be significantly reduced as a result of the inductor coupling. In a PA matching application, this will commonly require some additional "trapping" to be added to the network to further reduce harmonic emissions arising from nonlinear mixing in the PA. This will incur some additional insertion loss, which will somewhat reduce the passband insertion loss advantage of coupling in the network.

Figure 12.42 shows the passband and wide-bandwidth response of the lowpass π-match with inductor coupling and a second-harmonic harmonic trap implemented on inductor L_1 by inclusion of C_{L1} in Figure 12.38(b).

The modified element values for the network are given in Table 12.11.

Comparing with the bottom part of Table 12.10, it can be seen that implementing the frequency trap on inductor L_1 requires significant modification of all the element values in the matching network. Also, comparing Figure 12.38(b) and Figure 12.42(a), the maximum passband insertion loss is seen to increase by ~0.28 dB. While the second-harmonic response is significantly attenuated, rejection of the higher harmonics remains very poor compared to the uncoupled lowpass π-network.

In addition to parallel resonating inductor L_1, a second possibility for trapping frequencies above the passband exists by series resonating shunt inductor C_1, as shown in Figure 12.43.

With C_1 again determined from (12.32), the modified capacitor value C_1' and series inductor are given by

$$C_1' = \left(1 - (\omega_0/\omega_r)^2\right)C_1 \tag{12.38}$$

$$L_{C1} = \frac{1}{\left(\omega_r^2 - \omega_0^2\right)C_1} \tag{12.39}$$

For a trap at the second-harmonic frequency, the modified element values for this network are given in Table 12.12.

Figure 12.44 shows the passband and wide-bandwidth response of the coupled lowpass π-network with a second-harmonic harmonic trap implemented on capacitor C_1.

Table 12.11 Coupled Lowpass π-Match Element Values with a Second-Harmonic Trap on L_1

$k = 0.7$
$L_1 = 1.533$ nH
$C_{L1} = 4.826$ pF
$L_2 = 0.646$ nH
$C_1 = 0.738$ pF

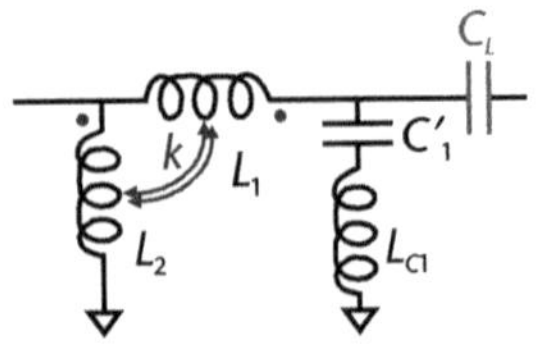

Figure 12.43 Coupled lowpass π-network with a second-harmonic trap on C_1.

Table 12.12 Coupled Lowpass π Match Element Values with a Second-Harmonic Trap on C_1

$k = 0.7$
$L_1 = 2.317$ nH
$L_2 = 0.591$ nH
$C'_1 = 1.774$ pF
$L_{C1} = 1.365$ pF

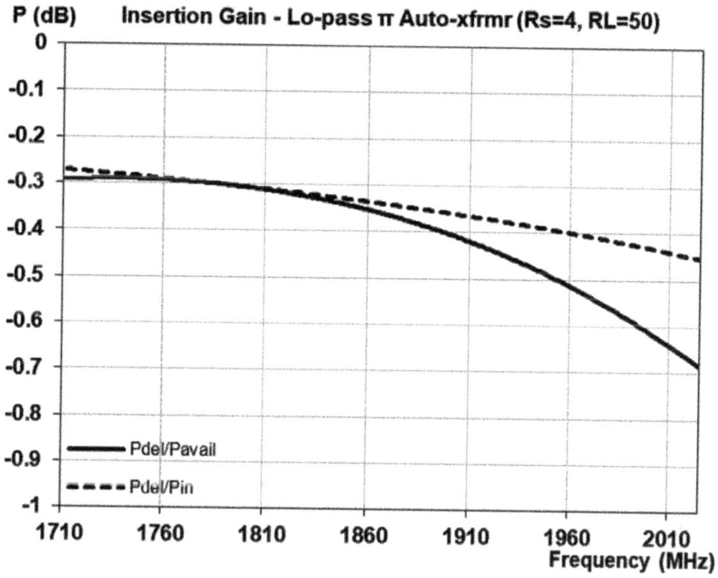

(a)

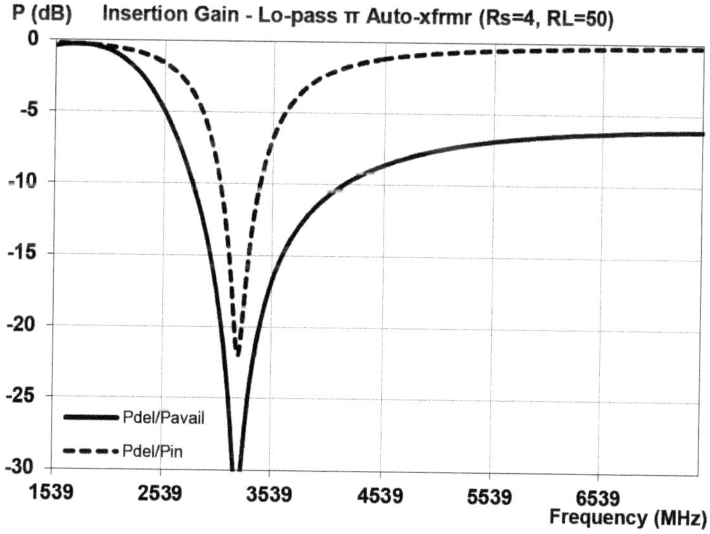

(b)

Figure 12.44 Coupled-inductor lowpass π-match with second-harmonic trap on C_1: (a) coupled lowpass π-match passband, and (b) coupled lowpass π-match wideband.

Comparing Figure 12.44 with Figure 12.42, it can be seen that with the trap on C_1, rather than on L_1, there is significantly more rolloff in insertion loss across the passband. Also, the harmonic rejection characteristic is narrower than that with the trap on L_1. Thus, any harmonic trapping in the network is best accomplished across L_1 rather than C_1.

12.4.2 Lowpass π-Network with Auto-Transformer Action and Virtual Inductor Ground

This section provides the synthesis equations and expressions necessary for determining the electrical characteristics of a lowpass π-network with auto-transformer action. The network was generalized to permit the inclusion of a capacitor C_b between inductor L_2 and ground. Because such a capacitor is not normally required in a π-network capacitor, what was the motivation for including it? That will be explained in this section. Some potential pitfalls in the auto-transformer approach will also be examined.

A detailed discussion on the critical aspects of matching a single-ended PA is provided in Chapter 9. Figure 9.1 shows a simplified schematic of the output match on a single-ended PA. The initial elements of the matching network are invariably required to provide harmonic termination and bias to the active output cells of the PA.

The purpose of harmonic termination is twofold. First, by optimally terminating the loading impedance on the PA across the harmonic frequency band, particularly the second harmonic, the PA efficiency may be optimized. Second, the harmonic termination reduces harmonic emissions from the PA that arise from nonlinear mixing. Most commonly, the harmonic termination presents a near short-circuit impedance at the second harmonic. As a consequence, the harmonic termination presents a capacitive loading in the passband.

As in Figure 9.1, the PA output stage is typically biased from the supply rail V_{CC} via an inductor L_b, which is, in turn, RF grounded through a bias capacitor C_b. In order to limit the RF leakage on to the supply line through L_b, the latter must present a sufficiently large reactance in the passband. Capacitor C_b serves to further reduce the leakage that does occur through the inductor. This RF leakage onto the supply rail is a critical parameter in a mobile device that typically includes lots of digital circuitry. If there is too much ripple on the supply rail, it will interfere with the operation of the digital components. Typically, any RF module in a mobile device is required to meet a Power-Supply Rejection (*PSR*) requirement that specifies a maximum value for the RF leakage on the supply rail.

Following the harmonic termination and bias networks, the matching network then comprises one or two sections of inductive and capacitive elements to increase the line impedance to that of the load. These matching sections are most commonly implemented with series inductor–shunt capacitor lowpass networks, as they provide increasing rejection above the passband. Replacing the generic "matching network" in Figure 9.1 by a single lowpass *LC* matching section the output matching elements close to the PA comprise those shown schematically in Figure 12.45.

Figure 12.45 incorporates the possibility of magnetic coupling between the bias inductor and the series inductor of the *LC* matching section. Such a coupling may

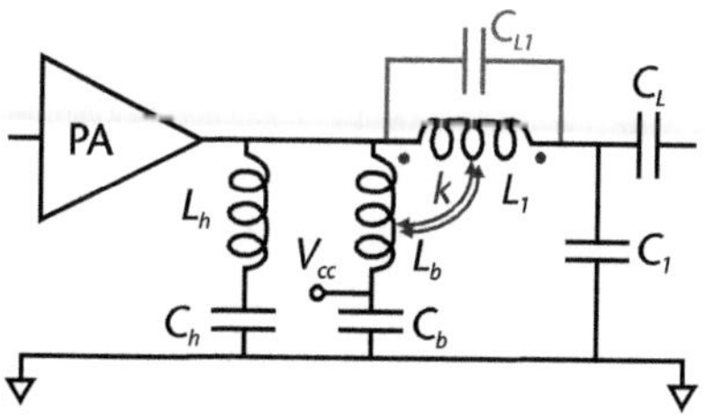

Figure 12.45 Coupled lowpass π-network with virtual ground on L_b.

be incidental or deliberate. As discussed in Section 12.4, any magnetic coupling will result in an auto-transformer impedance matching contribution to the network. Given the minimalistic footprints of modern PA modules, some coupling between the two inductors is almost inevitable, as the two inductors must be located in close proximity to each other. Thus, even unintentionally, most mobile PA module layouts with classic harmonic, bias, and *LC* matching elements likely have an auto-transformer element inherent in their matching networks.

The focus of this section is to examine how the performance of a lowpass π-network with intentional strong magnetic coupling between the inductors may be influenced by exploiting the shunt inductor for PA biasing. In order to implement this, the inductor ($L_b \equiv L_2$) must have virtual ground through a capacitor C_b, rather than a direct connection to ground.

Absorbing the bias inductor into the π-network has a significant size advantage for the module, but unfortunately has some potential disadvantages that may limit the full benefits of the technique. In a typical PA output matching network, the bias inductor must do double duty. As discussed in Chapter 9, in addition to biasing the PA output stage, the bias inductor must also provide some passband compensation for the reactive loading of the harmonic terminations. Any remaining reactive compensation must be provided by the matching network. In Section 9.2, it was shown that the trade-off between how much reactive compensation is provided by the bias inductor and how much is provided by the matching network can critically impact the network insertion loss. For the schematic of Figure 12.45, the residual reactance of the harmonic termination in the passband must be accommodated for entirely by the lowpass π-network, which can impact both the bandwidth and the loss of the network.

As mentioned above, RF leakage onto the supply rail is a critical parameter that must be limited. Increasing the value of either L_b or C_b will reduce the amount of RF leakage onto the supply rail, thus increasing the *PSR* for the network. To achieve a sufficiently high *PSR* value for the matching network, the following requirement must be met:

$$\omega^2 L_b C_b >> 1 \tag{12.40}$$

Lower values of the bias inductor will thus require a larger bypass capacitor to provide sufficient rejection. Unfortunately, in many of today's mobile platforms, there are limits on how large a bypass capacitor is practical. A technique known as Envelope Tracking (*ET*) is frequently employed to increase PA efficiency. In this technique, the supply rail is amplitude-modulated with the envelope bandwidth. Unfortunately, such modulators can only effectively modulate supply rail if the capacitance to ground on the rail is severely limited. In a modern mobile device, there are typically multiple RF PA modules, which permit the device to operate across multiple frequency bands. Although the device may only operate within one frequency band at a time, the bypass capacitors for each transmit module continuously load the supply line. Thus, for this reason, the maximum bypass capacitance of any individual PA module is typically limited to ≤~50 pF.

Figure 12.40 illustrated the very significant reduction required, in the lowpass π-network, for the value of the shunt inductance L_2 ($\equiv L_b$) as a function of the inductive coupling factor. For the practical example given in Table 12.10, the shunt inductor value is reduced by more than a factor of 5 in the presence of a coupling factor of $k = 0.7$. Given the inequality requirement of (12.40), this implies that, to achieve the same *PSR* for the module, the bypass capacitor C_b must be >5 times larger for the π-network with inductor coupling than without. In an *ET* PA module, if there is a limit on the capacitance that can be tolerated, this can potentially significantly limit the degree to which the auto-transformer action can be exploited. Effectively, this would mean limiting the mutual coupling factor k between the inductors.

In addition to limits on the bias capacitor C_b, there is another potential issue that must be accounted for when considering any potential benefits of exploiting the auto-transformer action in a lowpass π-network. Again, the required value for the shunt inductor in the π-network is reduced substantially when strong coupling is employed. In response to the reduction in inductor value, the ground currents through the inductor increase in an inverse manner. Thus, with strong mutual coupling and associated reduction in shunt inductor value, there will be increased power dissipation resulting from any resistance in the inductive path to ground. This implies that greater attention will be needed to be taken to minimize any resistance in the RF return path from the shunt inductor to the active PA module ground, in the case of strong mutual coupling in the network. Regrettably also in this respect, if the shunt inductor is virtually grounded through C_b, rather than being directly grounded, any series resistance in the capacitor can contribute significantly to network resistive losses. Some of these issues are quantized in Figure 12.46.

In Figure 12.46, relative network insertion loss dependences are plotted as a function of the mutual coupling coefficient, k. The dashed line shows the increased insertion loss dependency for the network resulting from a simple resistor of 200 mΩ in series with the shunt inductor to ground. Such a resistance could be associated with a capacitor C_b to ground or even with a via in the circuit board. Whatever the cause, the loss due to any series shunt resistance clearly increases with coupling, as expected.

The solid curve in Figure 12.46 shows the increased network insertion loss, as a function of inductor coupling, resulting from a bias capacitor C_b in the shunt arm having the following characteristics:

$$C_b = 40\ \text{pF} \qquad R_{Cb} = 200\ \text{m}\Omega \tag{12.41}$$

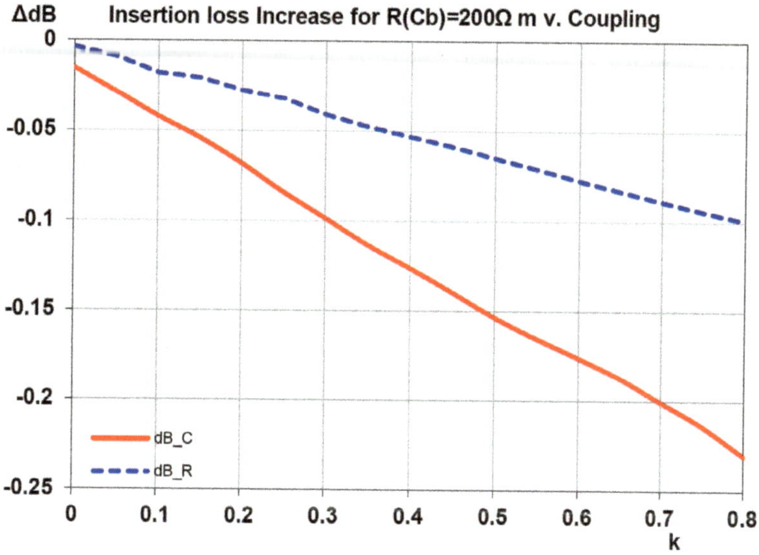

Figure 12.46 Relative π-network losses due to series elements as a function of coupling.

With a coupling coefficient k = 0.7, as in the example of Section 12.4.1, the introduction of the capacitor results in an additional insertion loss of ~0.2 dB for the network, relative to the loss that would occur in the network without coupling. There are two components to the additional insertion loss: (1) is mismatch loss because the reactance of the 40-pF capacitor is too large to sufficiently ground the reduced value shunt inductor; and (2) results from dissipative losses due to the series resistance of the capacitor.

Figure 12.47 shows the network gains of the coupled lowpass π-network of Section 12.4.1 with $k = 0$, without and with $C_b = 40$ pF ($R_{Cb} = 200$ mΩ) to permit the shunt inductor to be used for PA biasing.

There is no perceptible change in the network gains resulting from the virtual ground on the bias inductor. This is what is desirable as it indicates that introducing the virtual ground is not significantly affecting the characteristics of the matching/bias network. This, in turn, implies that the capacitor is effectively grounding the bias inductor.

For comparison, Figure 12.48 shows the corresponding network gains of the coupled lowpass π-network with k = 0.7. In this case, with strong magnetic coupling, there is a noticeable increase in insertion loss in the network. This is highly undesirable, as it indicates that the introduction of the capacitor is perturbing the performance of the network. This implies that the virtual ground node to which the supply rail V_{CC} is attached is not effectively grounded and thus has an appreciable RF component on the node. This is further evidenced by Table 12.13.

The top of Table 12.13 shows the optimized element values for the coupled network with inductor L_2 connected directly to ground (i.e., $C_b = 0$). The bottom of Table 12.13 shows the optimized element values for the coupled network with the inductor L_2 connected directly to ground via capacitor $C_b = 40$ pF. The fact

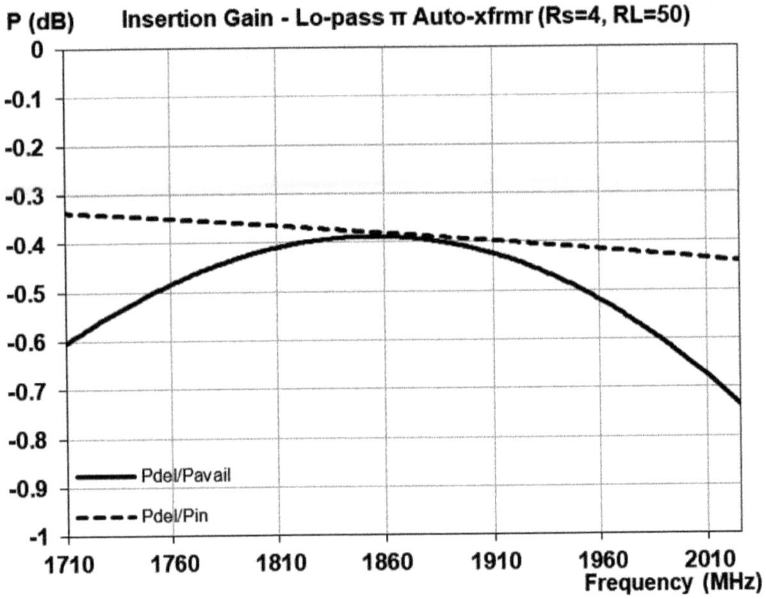

(a)

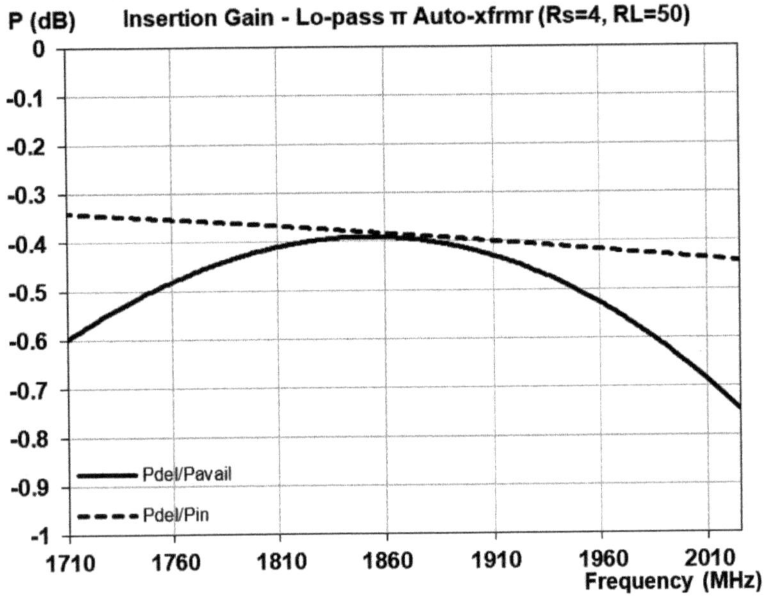

(b)

Figure 12.47 Coupled lowpass π-network gains for $k = 0$, without and with virtual ground: (a) network gain without C_b, and (b) network gain with $C_b = 40$ pF.

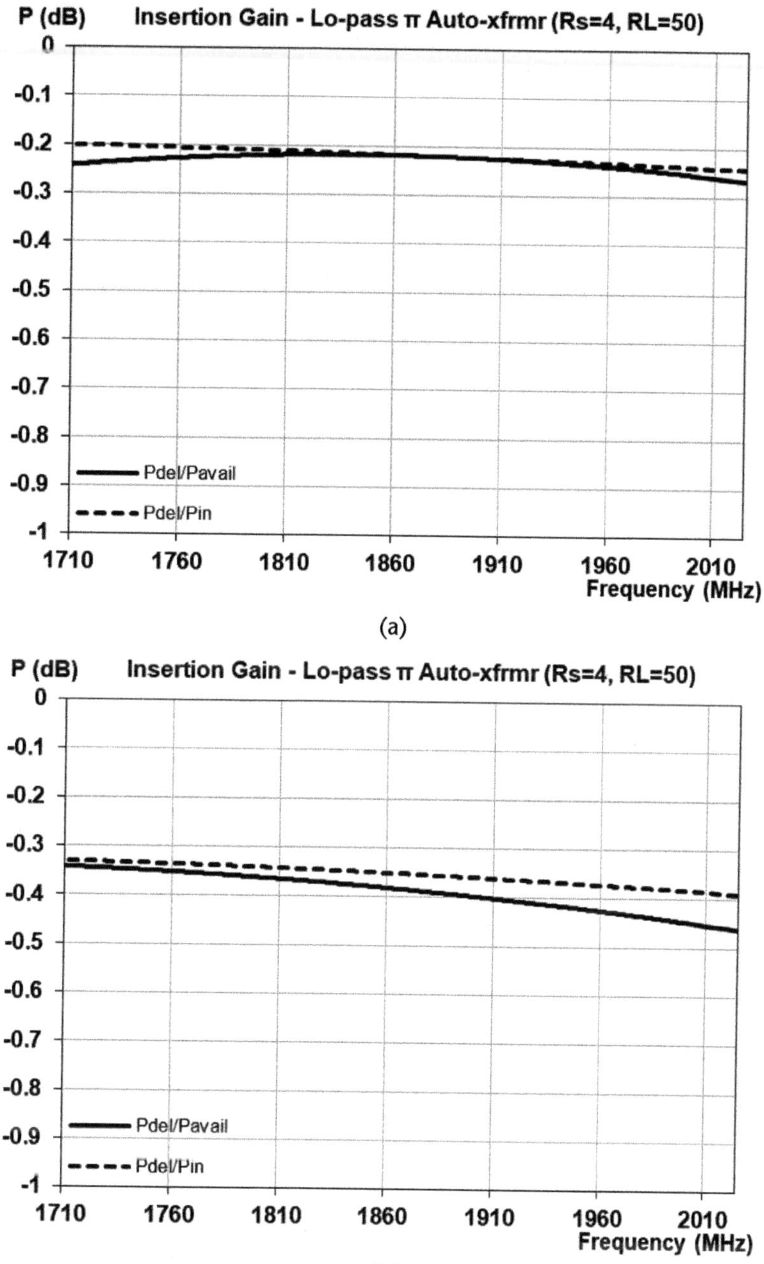

Figure 12.48 Coupled lowpass π-network gains for $k = 0.7$, without and with virtual ground: (a) network gain without C_b and (b) network gain with $C_b = 40$ pF.

Table 12.13 Lowpass π-Network Element Values

Uncoupled Lowpass π-Match
$k = 0.7$
$C_b = 0$ pF
$L_1 = 2.230$ nH
$L_2 = 0.557$ nH
$C_1 = 2.399$ pF
Coupled-Inductor Lowpass π-Match
$k = 0.7$
$C_b = 40$ pF
$L_1 = 2.752$ nH
$L_2 = 0.812$ nH
$C_1 = 2.291$ pF

that the two sets of element values are different is indicative that C_b is insufficiently large to effectively ground the inductor.

The voltage on capacitor C_b can be readily calculated in terms of the $[ABCD]$ matrix parameters given in Section 12.4. In terms of the supply voltage V_S, we find

$$\frac{V_{Cb}}{V_S} = \frac{CZ_L + D - 1 - Z_L/Z_1}{AZ_L + B + Z_S(CZ_L + D)} \tag{12.42}$$

Using (12.42), the voltages on the bias bypass capacitor C_b = 40 pF and R_{Cb} = 200 mΩ were determined across the passband and are shown in Figure 12.49.

The dashed dependency is the voltage on the capacitor (i.e., V_{CC} node) when the inductors in the π-network are uncoupled ($k = 0$), which is the classic case with no auto-transformer action. The solid dependency in Figure 12.49 is the voltage on the capacitor when the mutual coupling coefficient between the bias and series inductors $k = 0.7$. As expected from the ratio of the shunt inductor values in the two cases (Table 12.10) in the auto-transformer case with strong coupling, the RF voltage on the V_{CC} node is increased ~7 times. Such a large increase in RF leakage onto the supply line V_{CC} will generally not be acceptable. To reduce the leakage, either the coupling coefficient in the network must be reduced, thereby reducing the auto-transformer action, or the bypass capacitor V_{Cb} must be increased. According to (12.40), the capacitor will have to be increased by the same ratio that the inductor is decreased, in this example, by ~7 times. Thus, the capacitor C_b must be increased from 40 pF to ~280 pF in the presence of strong inductor coupling in order to achieve the same *PSR* for the uncoupled network.

So far in this section, no consideration has been given to any potential impact that the harmonic network might have on the perceived benefits of exploiting inductor coupling in the lowpass π-network. In the large majority of PA designs, a low or slightly inductive impedance is required at the second-harmonic frequency on the output of the PA.

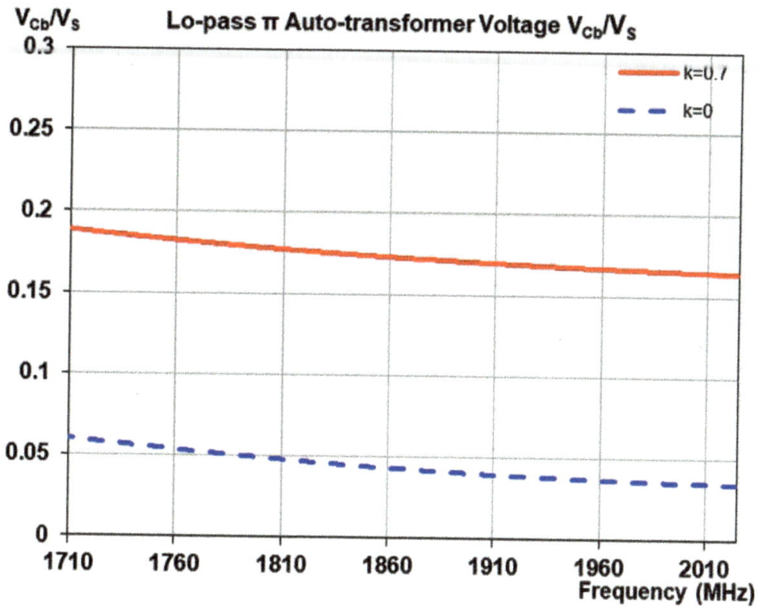

Figure 12.49 Bypass bias capacitor voltage.

Figure 12.50(a) shows the input network reflection coefficient across the second-harmonic band with no harmonic network present. In Figure 12.45, this would imply L_h and C_h not being present in the circuit.

With the harmonic network present and $L_h = 0.3$ nH, the input reflection coefficient becomes as shown in Figure 12.50(b). As desired, the impedance remains very low across the full second-harmonic band.

With no inductor coupling, the lowest passband insertion loss for the network of Figure 12.45 is obtained with the network parameters shown in Table 12.14.

The network gain for such a network is shown in Figure 12.51. Comparing Figure 12.47(b) and Figure 12.51(a), there is an increase of ~0.13 dB in insertion loss in the passband due to the presence of the harmonic trap.

With inductor coupling, the lowest passband insertion loss for the network of Figure 12.45 is obtained with the network parameters shown in Table 12.15.

The network gain for this network is shown in Figure 12.52(a), while the corresponding second-harmonic reflection coefficient is shown in Figure 12.52(b).

Comparing Figure 12.48(b) and Figure 12.52(a), there is an increase >0.2 dB in insertion loss in the passband. Comparing Figure 12.50(b) and Figure 12.52(b), the inductor coupling has resulted in significantly more variation in the desired low impedance across the second-harmonic band compared to the uncoupled case. To rectify this problem, the harmonic inductor L_h would have to be reduced in value, but this would lead to yet higher network insertion loss.

Plots of the RF voltage on the V_{CC} node are shown in Figure 12.53, which again indicate that, for the coupled-inductor case, the bypass capacitor must be increased to ~280 pF to effectively ground the bias inductor.

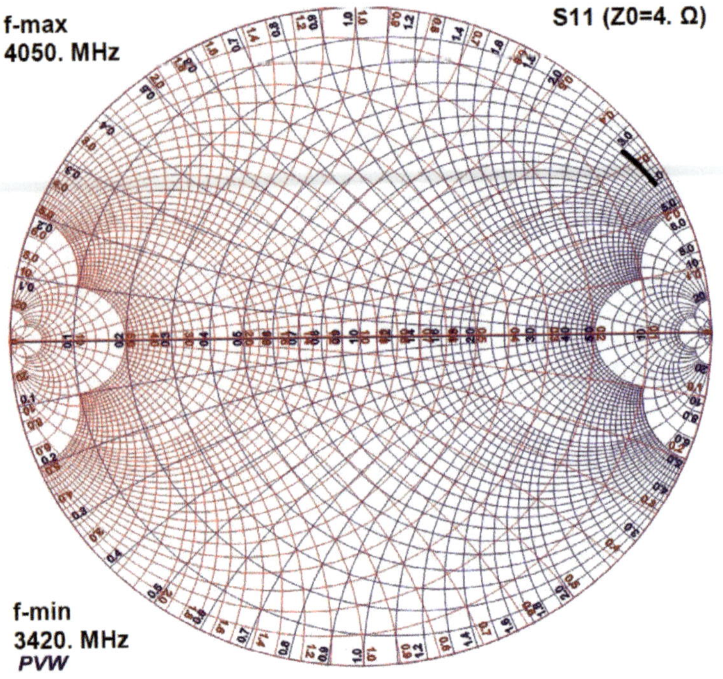

(a)

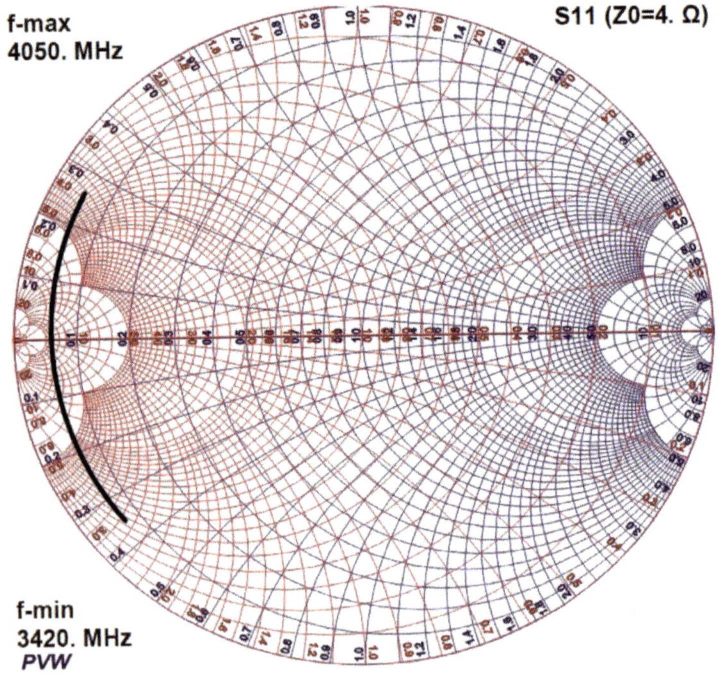

(b)

Figure 12.50 π-network second-harmonic reflection coefficient for $k = 0$, without and with trap: (a) no harmonic trap, (b) with harmonic trap.

Table 12.14 Lowpass π-Network Parameters for $k = 0$, with a Second-Harmonic Trap

$k = 0$
$C_b = 40$ pF
$L_h = 0.3$ nH
$C_h = 6.053$ pF
$L_1 = 1.187$ nH
$L_2 = 1.372$ nH
$C_1 = 5.855$ pF

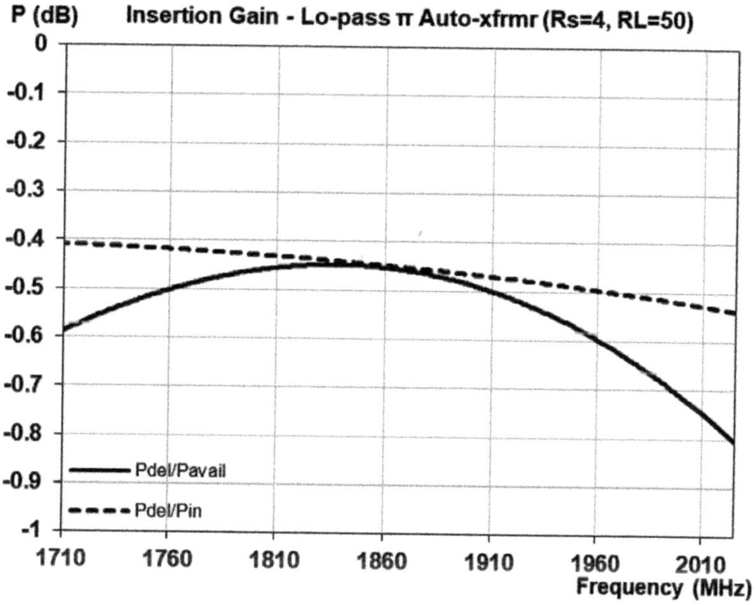

Figure 12.51 Lowpass π-network gain for $k = 0$, with second-harmonic trap.

Table 12.15 Coupled Lowpass π-Network Parameters for $k = 0.7$, with a Second-Harmonic Trap

$k = 0.7$
$C_b = 40$ pF
$L_h = 0.3$ nH
$C_h = 6.053$ pF
$L_1 = 2.810$ nH
$L_2 = 0.490$ nH
$C_1 = 1.645$ pF

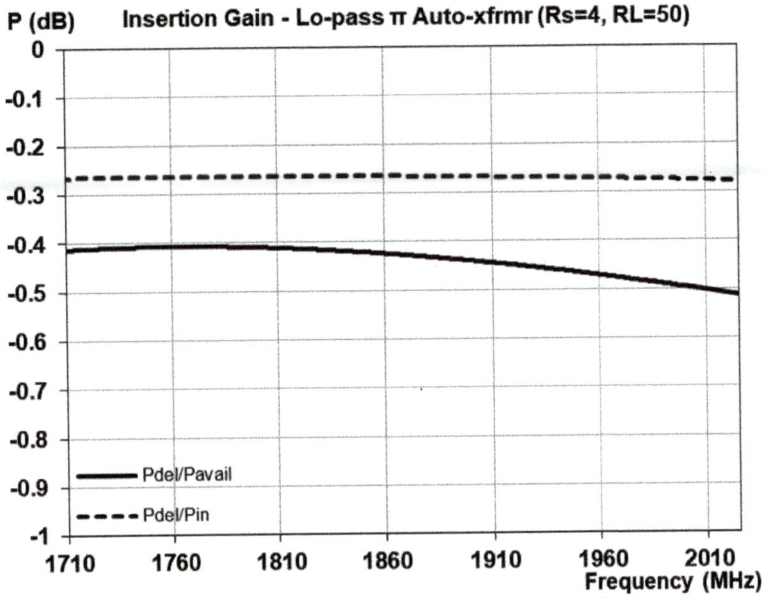

(a)

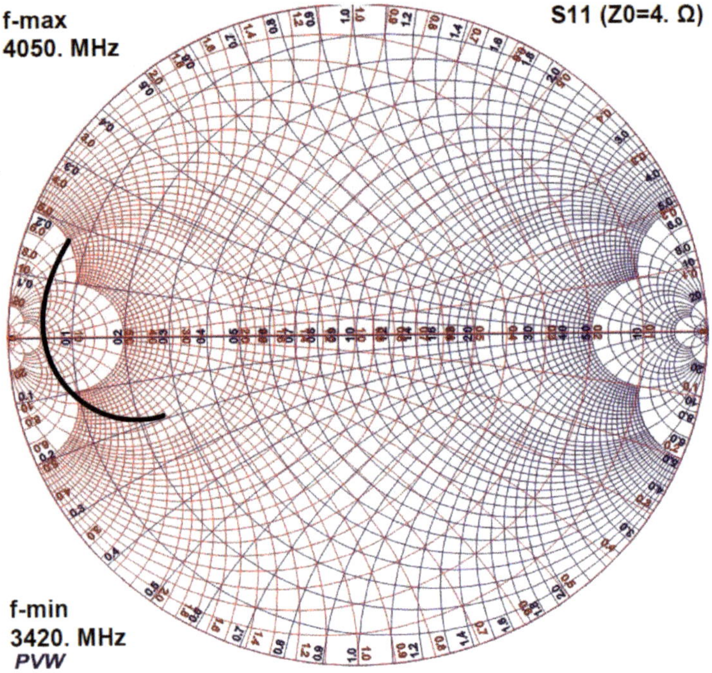

(b)

Figure 12.52 Coupled lowpass π-network characteristics for $k = 0.7$, with a second-harmonic trap: (a) network gain, and (b) second-harmonic reflection coefficient.

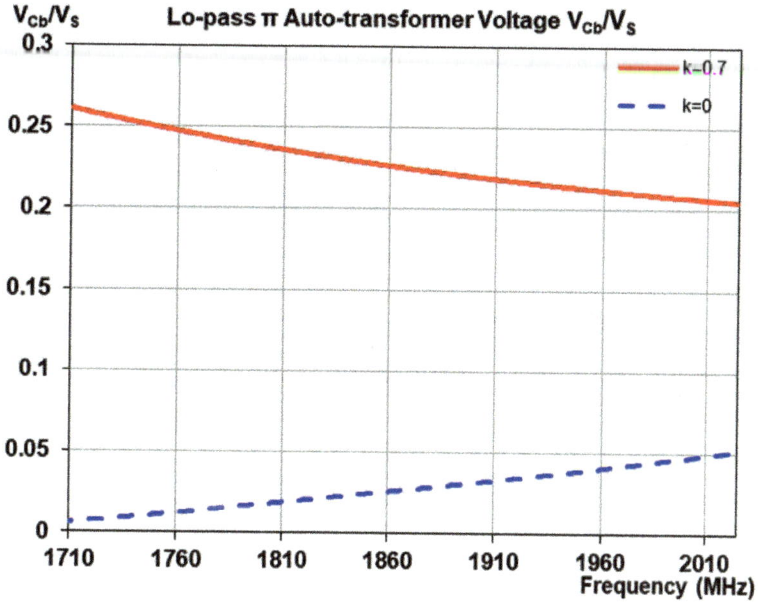

Figure 12.53 Bypass bias capacitor voltage with a second-harmonic trap.

If size permits, optimum advantage can be taken of the auto-transformer action in a lowpass π impedance matching network if the shunt inductor is not simultaneously used for PA biasing. This requires a separate bias inductor and capacitor as in a conventional match. The shunt inductor in the π-network can then be grounded directly, eliminating the issues associated with the capacitor required to achieve a virtual ground for the inductor.

12.5 Highpass *T*-Network with Auto-Transformer Action

A highpass *T*-network is another possible network architecture that can be used to transform a low impedance to a higher one. When transforming from a very low impedance to a much higher one, such as a transformation from 5Ω to 25Ω, as with the lowpass π-network, the highpass *T*-network comprises both a shunt and a series inductor. The schematic of such a highpass *T*-network is shown in Figure 12.54(a).

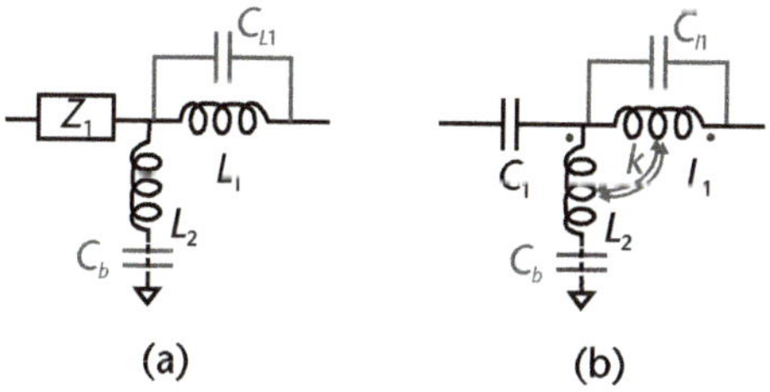

Figure 12.54 Highpass *T*-networks with and without magnetic coupling: (a) classic highpass *T*-network, and (b) highpass *T*-network with coupling.

Just as with the lowpass π-network, the possibility exists for magnetically coupling the two inductors to take advantage of an auto-transformer action, as shown in Figure 12.54(b). For completeness in analysis, and in a like manner to the lowpass π-network analysis, a capacitor in shunt with the series inductor and a capacitor in series with the shunt inductor are included in the schematics.

Lowpass networks are generally preferred for PA output matching because of their inherent ability to increasingly attenuate frequencies above the passband. Thus, highpass networks are seldom used. However, for completeness, and for those less common instances in which a highpass network may be preferred, full synthesis and analysis solutions are presented in this section. It should be noted from Figure 12.54 that in this architecture the shunt inductor cannot be used to bias the PA as the DC path to the PA is blocked by the series capacitor C_1. Despite this, a capacitor C_b is still included in the analysis to permit the evaluation of how ineffective grounding of the shunt inductor L_2 might affect the characteristics of the network.

Defining the basic network parameters as

$$Z_S = R_S + jX_S \tag{12.43}$$

$$Z_L = R_L + jX_L \tag{12.44}$$

$$Z_{L1} = R_{L1} + j\omega L_1 \tag{12.45}$$

$$Z_{L2} = R_{L2} + j\omega L_2 \tag{12.46}$$

$$Z_{CL1} = R_{CL1} - j/\omega C_{L1} \tag{12.47}$$

$$Z_1 = R_{C1} - j/\omega C_1 \tag{12.48}$$

also,

$$\omega_r = \text{angular frequency at the center of } L_1 \text{ stopband} \tag{12.49}$$

Synthesis equations for the elements of the auto-transformer-like form of the highpass T-network shown in Figure 12.54(b) are

$$C_{L1} = \frac{1 + k\sqrt{L_1/L_2}}{\omega_r^2 (1-k)^2 L_1} \tag{12.50}$$

$$\frac{-1}{\omega C_1} = \left(\frac{R_S}{R_L}\right) X_L - X_S - \omega L_2 + \frac{\omega\left(\left(L_1 + L_2 + 2k\sqrt{L_1 L_2}\right) R_S - \left(\left(1-k^2\right) R_S + k^2 R_L\right)\omega^2 L_1 L_2 C_{L1}\right)}{\left(1 - \omega^2 L_1 C_{L1}\right) R_L} \tag{12.51}$$

which are dependent upon solving the governing equation

$$\left(X_L - \omega^2 L_1 C_{L1}\left(X_L + \omega\left(1-k^2\right)L_2\right) + \omega\left(L_1 + L_2 + 2k\sqrt{L_1L_2}\right)\right)^2 R_S$$
$$+\left(\begin{array}{l}\left(1-\omega^2 L_1 C_{L1}\right)\left(\left(1-\omega^2 L_1 C_{L1}\right)\left(R_S R_L - (\omega L_2)^2\right) - 2\left(\omega^2 k L_2\right)^2 L_1 C_{L1}\right) \\ -\omega^2 L_2\left(k^2 L_1 + 2k\sqrt{L_1L_2} + \left(\omega^2 k^4 L_1 L_2 C_{L1} - 2\left(1-k^2\right)k\sqrt{L_1L_2}\right)\omega^2 L_1 C_{L1}\right)\end{array}\right) R_L = 0 \quad (12.52)$$

Given a value for the independent variable L_1, (12.52) must be solved to determine the corresponding value for L_2. Unfortunately, a closed-form solution for L_2 is not available; however, a solution for L_2 that satisfies (12.52) is readily derivable by numeric techniques.

Having determined the element values for the network, all electrical parameters are readily determined from an $[ABCD]$ matrix analysis, using the equations given in Chapter 1. For the network, the individual $[ABCD]$ parameters are given by

$$A = 1 + \frac{(1 + Z_{L1}/Z_{CL1})Z_1 - j\omega k\sqrt{L_1L_2}}{(1 + Z_{L1}/Z_{CL1})(Z_{L2} + Z_{Cb}) + (\omega k)^2 L_1L_2/Z_{CL1} + j\omega k\sqrt{L_1L_2}} \quad (12.53)$$

$$B = Z_1 + \frac{(Z_1 + Z_{L2} + Z_{Cb})Z_{L1} + (\omega k)^2 L_1L_2 + j\omega k\sqrt{L_1L_2}Z_1}{(1 + Z_{L1}/Z_{CL1})(Z_{L2} + Z_{Cb}) + (\omega k)^2 L_1L_2/Z_{CL1} + j\omega k\sqrt{L_1L_2}} \quad (12.54)$$

$$C = \frac{1 + Z_{L1}/Z_{CL1}}{(1 + Z_{L1}/Z_{CL1})(Z_{L2} + Z_{Cb}) + (\omega k)^2 L_1L_2/Z_{CL1} + j\omega k\sqrt{L_1L_2}} \quad (12.55)$$

$$D = 1 + \frac{Z_{L1} + j\omega k\sqrt{L_1L_2}}{(1 + Z_{L1}/Z_{CL1})(Z_{L2} + Z_{Cb}) + (\omega k)^2 L_1L_2/Z_{CL1} + j\omega k\sqrt{L_1L_2}} \quad (12.56)$$

12.5.1 Highpass *T*-Network with Auto-Transformer Action Example

In order to examine how adding magnetic coupling between the two inductors in a highpass *T*-network can impact its performance as an impedance matching network, consider a design with parameters identical to those specified in Table 12.1.

The lowest-loss highpass *T*-network gain with no inductor coupling, as in Figure 12.54(a), is shown in Figure 12.55(a). Allowing for mutual coupling between the two inductors, as in Figure 12.54(b), the lowest network loss achievable is shown in Figure 12.55(b).

Element values for the two networks are provided in Table 12.16.

Comparing the network characteristics in Figure 12.39 for the lowpass π and those in Figure 12.55 for the highpass *T*-network, they are seen to be very comparable. As with the π-network, the *T*-network also shows a significant improvement in performance with the introduction of mutual inductor coupling. As previously, the magnetic coupling component markedly reduces the network insertion loss and significantly widens the bandwidth of the impedance match.

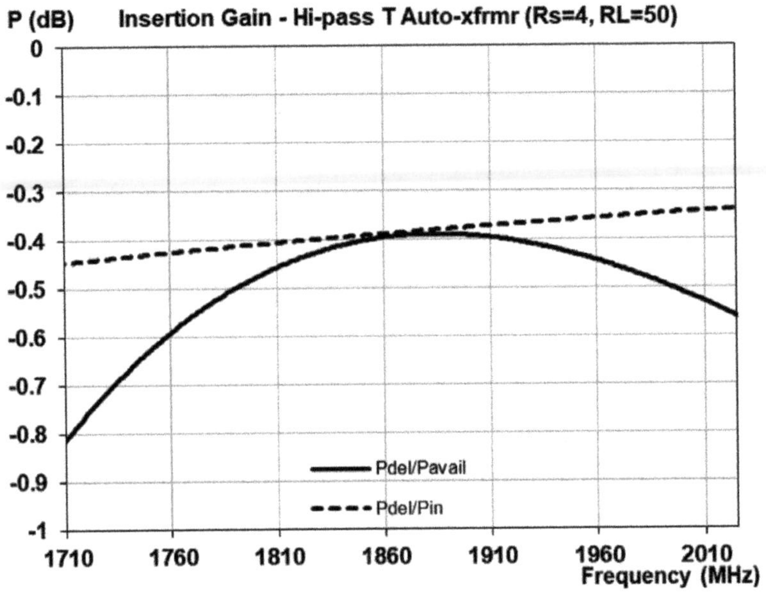

(a)

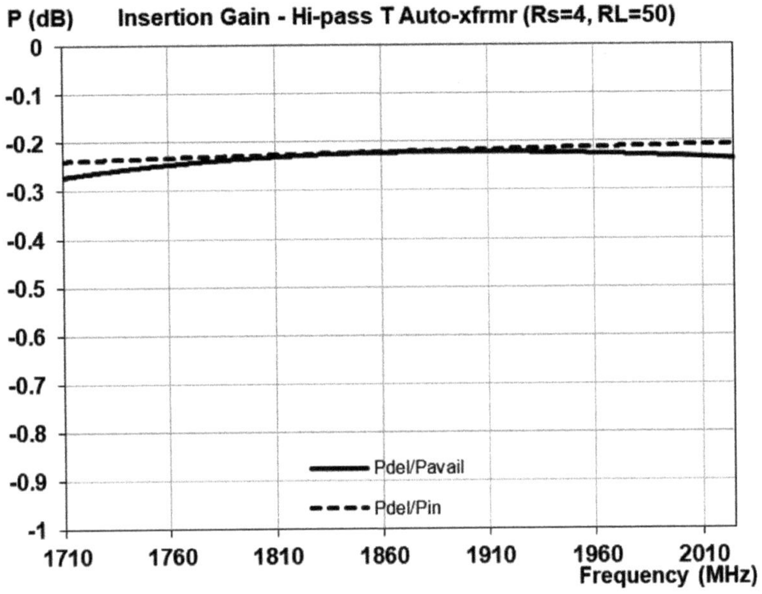

(b)

Figure 12.55 Highpass *T*-matches without and with mutual inductive coupling: (a) uncoupled highpass *T*-match, and (b) coupled-inductor highpass *T*-match.

Because this network has highpass characteristics, its attenuation characteristics are expectedly poor, as shown in Figure 12.56(a). Unfortunately, as with the π-network, the high-side attenuation is further reduced by the coupling, as shown in Figure 12.56(b).

In a PA matching application, this will commonly require some additional "trapping" to be added to the network to further reduce harmonic emissions arising from

Table 12.16 Highpass *T*-Network Element Values

Uncoupled Highpass T Match
$k = 0$
$C_1 = 6.166$ pF
$L_1 = 0.8$ nH
$L_2 = 1.35$ nH
Coupled-Inductor Highpass T Match
$k = 0.7$
$C_1 = 15.226$ pF
$L_1 = 2.4$ nH
$L_2 = 0.907$ nH

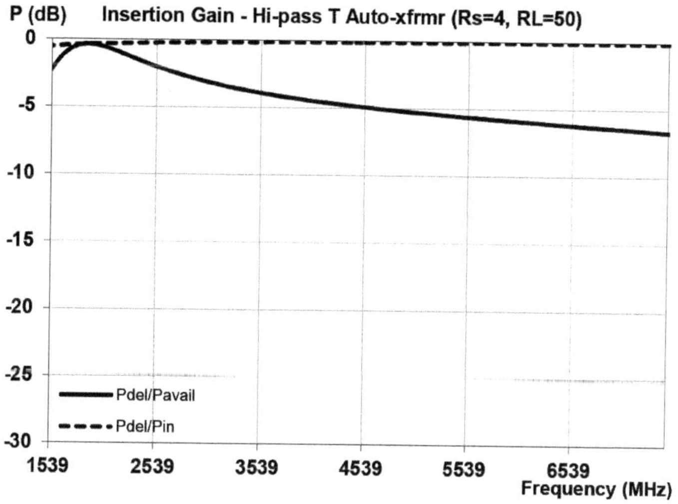

(a)

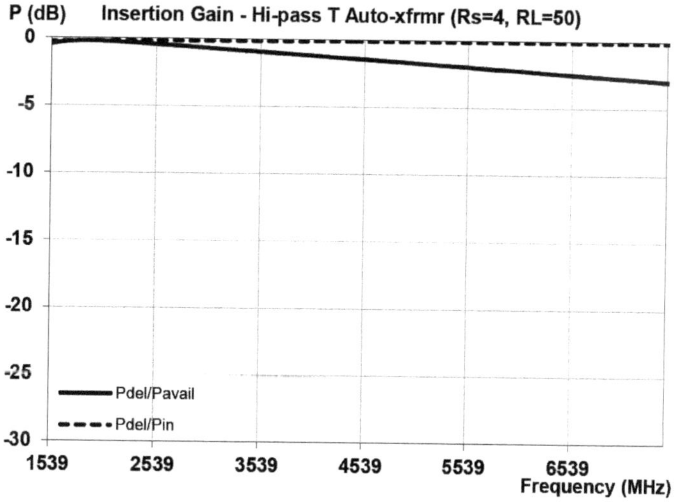

(b)

Figure 12.56 Highpass *T*-matches without and with mutual inductive coupling: (a) uncoupled highpass *T*-match, and (b) coupled-inductor highpass *T*-match.

nonlinear mixing in the PA. This will incur some additional insertion loss, which will somewhat reduce the passband insertion loss advantage of coupling in the network.

Figure 12.57 shows the passband and wide-bandwidth response of the highpass *T*-network with a second-harmonic trap implemented on inductor L_1.

The modified element values for the network are given in Table 12.17.

Comparing with the bottom of Table 12.16, it can be seen that implementing the frequency trap on inductor L_1 again requires significant modification in the element values in the matching network. Also, comparing Figure 12.55(b) and Figure

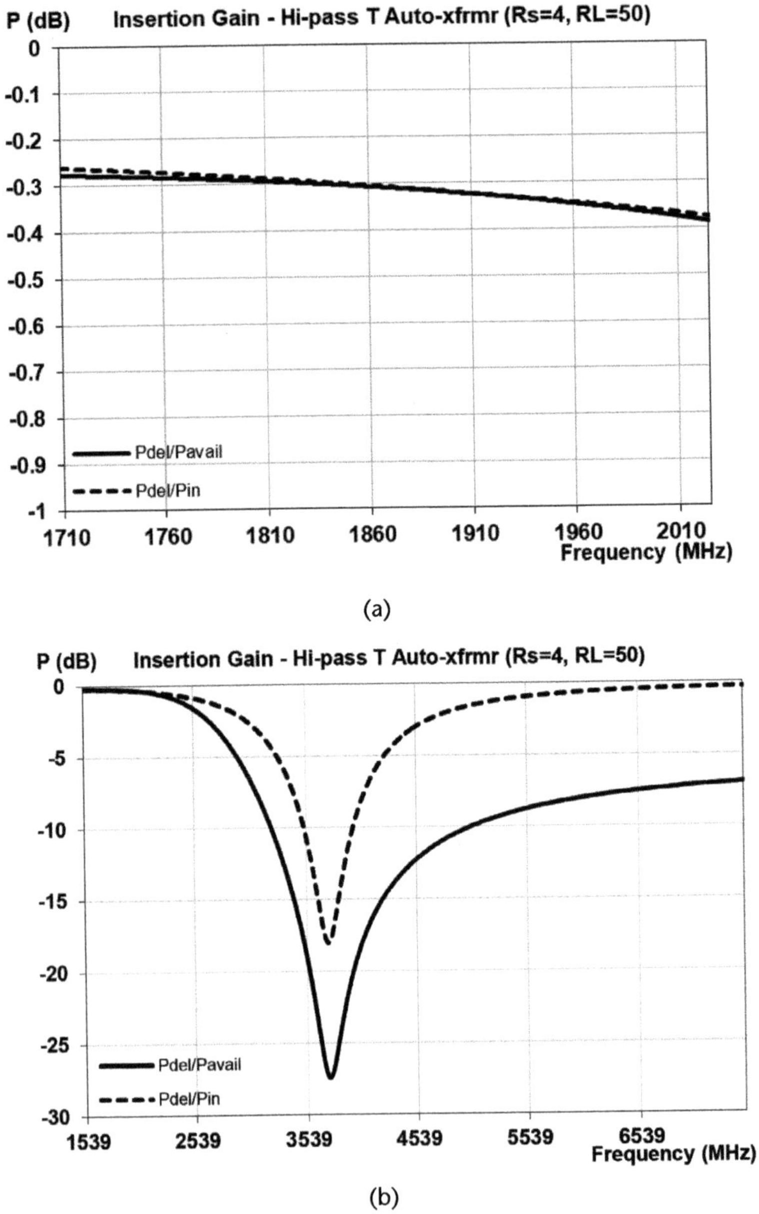

Figure 12.57 Coupled-inductor highpass *T*-match with a second-harmonic trap on L_1: (a) coupled highpass *T*-match passband, and (b) coupled highpass *T*-match wideband.

Table 12.17 Highpass *T*-Network Element Values with a Second-Harmonic Trap on L_1

$k = 0.7$
$C_1 = 39.38$ pF
$L_1 = 2.4$ nH
$C_{L1} = 3.274$ pF
$L_2 = 0.807$ nH

12.57(a), the maximum passband insertion loss is seen to increase by >0.1 dB. This is a much smaller increase in insertion loss than was observed with the comparable π-network case. Comparing Figure 12.42(a) with Figure 12.57(a), the insertion loss for the *T*-network is seen to be ~0.15 dB less than for the equivalent π-network.

Comparing the high-side responses for the two networks with L_1 traps, the rejection characteristics are virtually the same. Thus, if the shunt inductor is not required to be used for PA bias, with second-harmonic trapping on the series inductor, the highpass *T* has a performance advantage over the lowpass π when the inductors are mutually coupled.

These results taken together with the observations at the end of the previous section suggest that the highest-performance matching network architecture would be one with a conventional harmonic termination and bias inductor, followed by a highpass *T*-network with an L_1 trap.

CHAPTER 13

Considerations of Single-Phase Versus Multiphase Power Amplifiers

RF power amplifiers (PAs) for mobile applications may employ a single chain of amplifiers or multiple chains of amplifiers with varying phase relationships. As mentioned in Chapter 9, the single-chain architecture is commonly referred to as single-ended. An example of a simplified single-ended PA output architecture is shown in Figure 9.1.

For the multiphase configurations, the two most popular phase relationships employed between the varying amplifier chains are differential (180°) and quadrature (90°). An example of a simplified differential output matching architecture employing coupled inductors for the 180° phase splitting and combining is shown in Figure 9.2.

Quadrature PA architectures are commonly referred to as balanced PAs. A simplified block diagram of a balanced PA architecture is shown in Figure 13.1.

Common choices for the quadrature splitter and the quadrature combiner are the branch-line coupler or the Lange coupler. The balanced architecture has the advantage of extremely low input and output reflection coefficients across the operational bandwidth of the couplers, independent of the amplifier characteristics (assuming the two PAs are identical). Power incident on the inputs or outputs of the amplifiers is absorbed by the two loads Z_0. The balanced PA architecture is not widely employed in handset applications as the couplers typically require a relatively large area to implement. Nonetheless, it has found some application as it has another advantage, namely, reduced power output variation with load mismatch. This aspect is explored in some detail in Chapter 4 of Volume 2 of this series.

In addition to the three principal classes of PA architectures above, there are also limitless other variants with two or more PA chains and varying phase angles between each of the chains. Several of these are covered in later chapters in this book. Each has its own advantages and disadvantages.

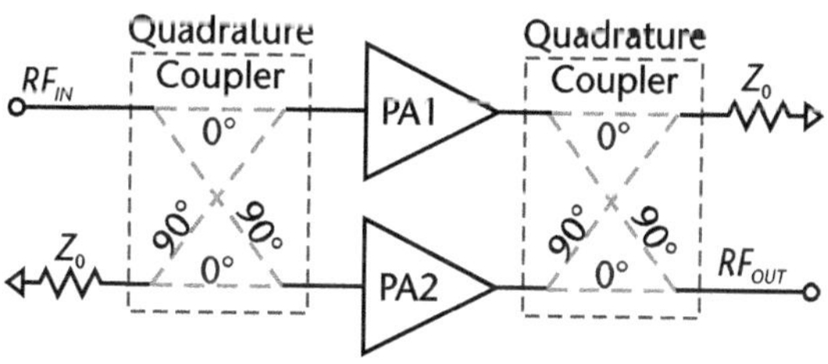

Figure 13.1 Simplified block diagram of balanced PA architecture.

Having selected a particular PA architecture for an application, the design engineer must then decide upon the most effective way to implement the design. Invariably, there is no single way to implement an RF circuit function and the engineer must make a trade-off between size and performance. For mobile applications, such as cellular handsets, size is critical, but because battery power is limited, PA efficiency is equally important. The engineer is therefore always faced with the challenge of trading off size versus performance.

Because single-ended and differential architectures are by far the most common choices for PA matching output networks, in today's mobile applications, we focus on the characteristics of each and how they compare in the next section.

13.1 Considerations of Single-Ended Versus Differential PA Architectures

Even among veteran RF PA designers, frequently the relative merits and drawbacks of single-ended versus differential output PA matching is poorly understood. More often than not, this is due to a prejudice or familiarity that one group of designers has for one of the architectures, which results in the other never being considered. This can be a missed opportunity for the design. In this section, we try to clarify the similarities and differences between the two architectures to enable the engineer to make a more informed choice.

The single-ended lowpass PA output matching network of Figure 9.3 is implemented with discrete reactive elements. In comparison, the differential PA output matching architecture of Figure 9.2 is implemented with coupled inductors, frequently referred to as "transformers." The use of "transformers" for implementing a differential design is often taken for granted by many designers. However, this need not be the case, as a transformer can also be used in a single-ended design. Equally, a differential phasing network comprising only discrete (i.e., uncoupled) reactive elements can also be used in place of a transformer in a differential design. One example of a lumped-element differential combiner/splitter is shown in Figure 13.2. Other versions, their design details, and their potential use in PA matching networks are detailed in Chapter 15.

Before delving into a detailed comparison of the basic characteristics of single-ended versus differential designs, it is instructive to first consider how a transformer is implemented in a typical high-frequency RF circuit.

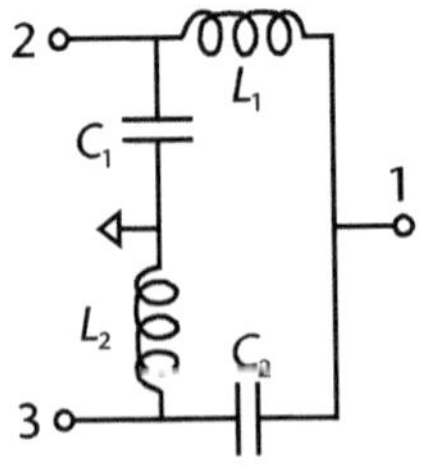

Figure 13.2 One version of a lattice differential combiner/splitter.

13.2 Comparative Aspects of Single-Ended Versus Differential PA Architectures

To be able to make the optimum design choice between a single-ended or a differential PA architecture, it is important to understand fully the relative merits of each. Unfortunately, this is frequently not the case and conclusions are based on incorrect assumptions.

Two of the most common misconceptions, frequently cited in claims, that a differential PA architecture has inherent advantages over a single-ended architecture are:

1. A differential solution in which the combining and splitting networks are implemented using coupled inductors is inherently wider bandwidth than a single-ended version.
2. A differential design has 4 times the input impedance of a single-ended design.

Neither of these statements is correct.

The first misconception is largely attributable to the poor but widely-repeated description of a coupled-resonator RF network as a transformer. This was covered in detail in the previous section. The term "transformer" is associated with a wide (ideally infinite) bandwidth device since their sole defining parameter (i.e., turns ratio) is frequency-independent. In actuality, the coupled-resonator network has frequency-limiting characteristics similar to those of any other reactive RF component.

The first misconception is additionally inappropriate for the simple reason that a coupled-resonator network can equally be used for implementing an impedance transformation in a single-ended architecture. A differential PA architecture is not required for a coupled-resonator network to be made use of for impedance matching purposes.

Figure 13.3 shows a single-ended PA output network with a coupled-resonator network for impedance matching on the output of the amplifier. The voltage at the output of the coupled inductors network is double that on the input at the output of the PA. From power conservation, this implies the impedance level at the output of the resonators is 4 times that on the input (i.e., the PA output impedance).

The second misconception concerns input impedance. The source of the confusion here is more subtle. The network input impedance is the output loading impedance on the PA. This is a critical parameter for all RF PAs used in mobile devices. As discussed in Chapter 9, due to the low battery voltages available in mobile devices, sufficient RF output power can only be generated at a low impedance level. Because the power must be fed to an antenna with a usual input impedance of 50Ω, this

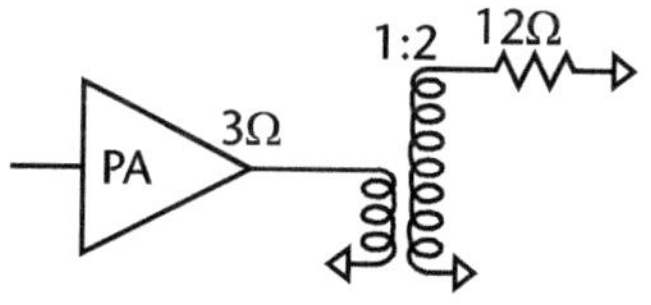

Figure 13.3 Single-ended PA with coupled inductors output impedance setup.

requires a matching network between the PA output and antenna to match these disparate impedances. Typically, a setup in impedance > 10:1 is required.

In general, the greater the impedance setup required for the network, the lower its operational bandwidth and the higher its insertion loss. It would therefore be a significant advantage for a differential network if it could truly provide a PA output impedance 4 times higher than a single-ended architecture. Unfortunately, this advantage is illusory and results from a misinterpretation of circuit theory.

By way of example, consider a PA with an output impedance of 3Ω. If a coupled-resonator network with a voltage setup ratio of n:1 is connected directly on its output, as in Figure 13.4(a), the transformed impedance at the output of the resonators will be $3\,n^2\,\Omega$. If we now take the same PA output periphery (i.e., active cell output area) and divide the output cells into two groups, I and II, each group will have an output impedance of 6Ω. If the signal input to one of the PA groupings is then phase-shifted by 180°, the two outputs will form a differential drive that can be applied directly to the same coupled-resonator pair as in the single-ended case. Such a network is shown in Figure 13.4(b). Note that, for this case, the output impedance is $12\,n^2\,\Omega$, an increase of 4 times over the single-ended configuration of Figure 13.4(a).

Figure 13.4(a, b) might seem to support the conjecture that a differential PA architecture has a 4 times increase in output impedance over a single-ended one. However, the increase here is not due to the differential nature of the network, but simply to dividing up the output cell groupings. To understand this, refer to Figure 13.5(a, b).

Figure 13.5(a) shows a single-ended PA and coupled inductor network essentially equivalent to the differential network of Figure 13.4(b) and having the same higher output impedance of $12\,n^2\,\Omega$. The PA periphery is the same, as are the two inductors and mutual coupling coefficients assumed for the two. The sole difference between

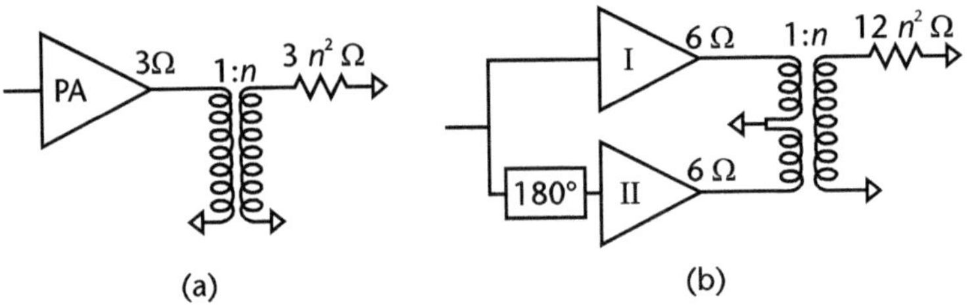

Figure 13.4 PA with single-ended and differential output configurations: (a) single-ended PA, and (b) equivalent differential implementation.

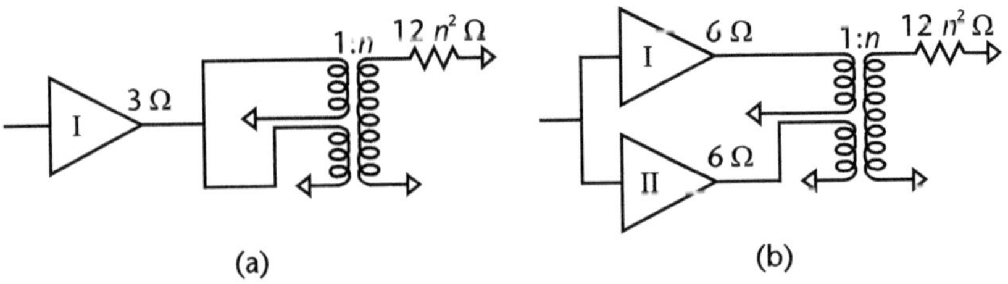

Figure 13.5 Single-ended PA with split outputs to coupled inductors: (a) single-ended PA, and (b) coupled inductors.

Figure 13.4(a) and Figure 13.5(a) is that the primary inductor has been split in two, with the two halves being driven in anti-phase by the single PA. The periphery of the single PA could even split into two groups, as in the differential case, with each output similarly driving the two halves of the split primary inductor in anti-phase. This configuration is shown in Figure 13.5(b). Comparing Figure 13.4(b) to Figure 13.5(b), we can see that there is little difference between the two. The active output cells of the PA experience identical voltage and current excursions in both cases and have exactly the same loading. The true impedance on the cells is unchanged between Figure 13.4(b) and Figure 13.5(a, b).

It is easy to draw the wrong conclusion from Figure 13.4(b) regarding the potential impedance advantages of a differential PA architecture over a single-ended one. The impedance across the primary resonator is 4 times higher than a single-ended PA with equal total periphery. However, this is not comparing like impedances. The higher impedance is a differential impedance; it should not be compared directly with a single-ended one. Because the load is generally single-ended, not balanced, any differential impedance must be transformed back into a single-ended impedance for valid comparison.

In the differential case, rather than viewing the drive to the primary resonator as a single balanced impedance with an impedance 4 times that of a single PA, it is more appropriate to view the drive as two independent single-ended PA chains having 2 times the impedance of a single PA, as shown in Figure 13.6. Note, however, that the coupled inductors in each track only provide an impedance transformation ratio of n^2, which is identical to the single-ended case. For this reason, we should not expect any bandwidth advantage of the differential PA architecture, with coupled-inductor matching, over a similar single-ended one.

In summary, the sole reason that the primary impedance in a differential PA architecture is higher results from dividing the output periphery of the PA. It is not inherent in the differential architecture itself. Regardless of the relative phasing, the PA output periphery can be and sometimes is actually divided into more than two groups and each of the outputs combined via multiple pairs of coupled inductors or discrete elements. If the output periphery is divided into N groups, the output impedance of each will be increased by a factor of N.

13.3 Additional Comments on Single-Ended Versus Differential PA Architectures

As discussed in the section above, differential PA architectures have no inherent bandwidth or impedance advantages over single-ended ones. Coupled inductors

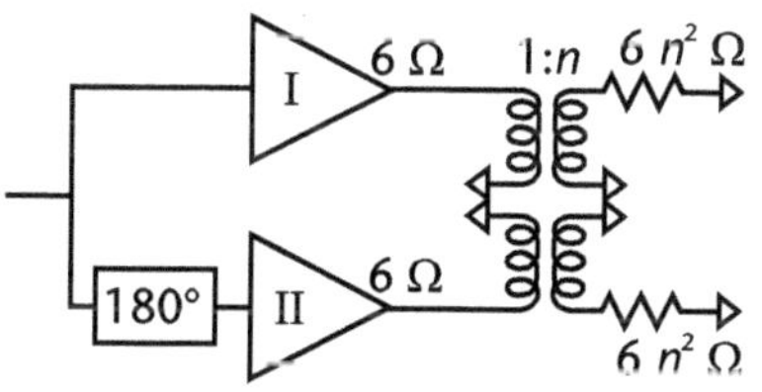

Figure 13.6 Differential amplifiers transformation ratios.

can equally be used for impedance transformation in both types of architecture. Likewise, differential PA architectures can be very effectively realized with *LC* networks, entirely devoid of any magnetic coupling.

If differential PA architectures do not effectively raise the PA loading impedances, are there alternatives that could? There are two, but, unfortunately, neither is practical. The two possibilities that exist are: (1) reduce the PA output power, or (2) raise the supply voltage. The PA output power is invariably a requirement of the application and thus cannot be arbitrarily reduced. The low battery voltage in a mobile device can be increased by an inverter, but these have challenges with noise, size, and efficiency.

What, if any, then are the potential advantages for a differential PA architecture over a single-ended one? Three possibilities are:

1. The bias inductor for supplying DC power to the output stages can be very small, without impacting insertion loss or degrading power-supply rejection on the supply rails.
2. The differential architecture permits the second-harmonic responses to be terminated without having to trade off passband insertion loss. This was demonstrated in Section 10.7 with the all-pass bridge-*T* lowpass network.
3. In Figure 13.4(b), it can be observed that ideally there is no returning ground current at the center of the primary inductor, as it is a voltage null. This eliminates any voltage drop attributable to finite resistivity of ground vias that could potentially improve PA gain and reduce insertion loss.

Unfortunately, there are also potential disadvantages to a differential architecture:

1. A differential PA architecture typically requires a larger active die area than single-ended in order to implement the two preliminary and output amplifier chains.
2. Coupled inductors for output combining may occupy larger substrate or die (active or passive) area than discrete elements in single-ended design. There are similar size concerns for the differential input splitter.
3. For optimum performance, care must be taken in layout to ensure that splitter and combiner coupled inductors are both truly differential.
4. Given size constraints, it is not practical to achieve tight magnetic coupling between printed inductor traces without also having significant interwinding capacitance. The latter can have important impacts on design and performance that are examined in detail in volume 2 of this book.

Because differential PA designs commonly employ coupled inductors for combining the two out-of-phase signal paths, one further disadvantage of the architecture might be added. Namely, the design, layout, and optimization of coupled-inductor elements is invariably more complex than a design based on discrete components. Lengthy EM simulation is invariably required. However, in volume 2, it is shown how differential designs can also be implemented very effectively with only discrete components, in which case, this disadvantage does not apply.

One final point to bear in mind is that a coupled-inductor combiner and impedance transformer generally do not provide the full impedance transformation required to match a mobile PA output to the output/antenna. As mentioned earlier, this ratio is typically ~10:1. A second single-ended impedance match, typically using discrete elements, must be added on the output of the secondary inductor to provide the additional output impedance match. For an optimized performance, the designer must decide the proportion of the impedance match to assign to the coupled inductors, and the portion to the single-ended match.

CHAPTER 14

Classic Coupled-Inductor Matching for Differential PAs

In this chapter, we consider the "classic" application of coupled inductors to impedance matching and combining of the outputs of differential PAs. The term "classic" here refers to the transformer-like manner in which the coupled inductors are connected transverse to the signal path as in Figure 14.1. The outputs of both PAs are applied to the primary inductor and all RF energy transfer to the output is via magnetic coupling between the inductors. This is similar to their use for the single-ended case matching as described in the previous chapter.

The following chapter details how coupled inductors may also be used in an in-line architecture for the output matching of differential PAs.

14.1 Basic Differential Coupled-Inductor Design

Figure 14.1 shows a basic schematic of a differential PA matching combiner architecture employing coupled inductors. As with the similar single-ended PA architectures in Figure 12.4, shunt input and/or output tuning elements are generally required to match the network to the external impedances.

In the schematic, the two input drive voltages, V_{S1} and V_{S2}, represent the output signal voltages from two PAs. In the neighborhood of the passband, if the input voltages have a differential relationship, then

(fundamental and odd) $$V_{S1} = -V_{S2} \tag{14.1}$$

This relationship is relatively true for all the odd harmonics, generated as a result of internal nonlinear mixing.

In contrast, the even-harmonic frequencies, also generated by nonlinear internal mixing in the amplifiers, will generally be in-phase in the two tracks. Thus, for even-harmonic frequencies,

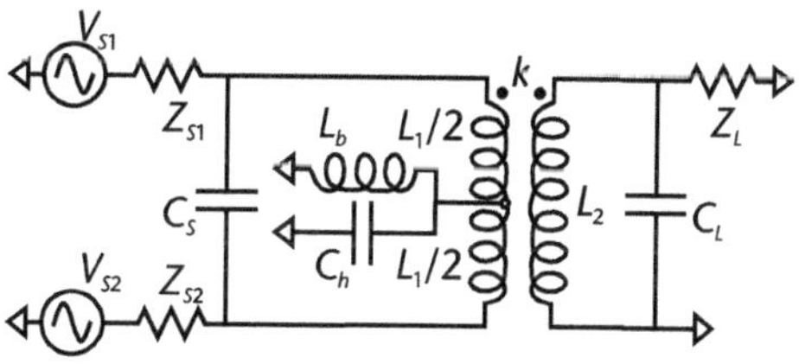

Figure 14.1 Differential PA matching architecture with coupled inductors.

$$V_{S1} = V_{S2} \tag{14.2}$$

For the general case, the two source impedances are assumed to be equal, that is,

$$Z_{S1} = Z_{S2} \tag{14.3}$$

In some less common examples, both the voltages and impedances of the two amplifiers might differ between the tracks. One motivation for this is if there is a need for improved backed-off efficiency of the PA, in which case the amplifiers might be sized differently and one turned off in low-power mode.

As with the single-ended matching analyses in Chapter 12, an inductor L_S may be added in series the input tuning shunt capacitor, C_S, to implement a desired harmonic termination. For the differential case, however, rather than being able to terminate the second-harmonic frequencies, the trap can only be used for terminating odd harmonics.

Neglecting any interwinding capacitance, as discussed in Chapters 10 and 12, the differential input voltages (Equation (14.1)) result in the center point of the input resonator L_1 being a virtual ground in the passband. This thus provides a convenient network node for applying an external DC voltage to bias the output stages of the two PAs, without significantly perturbing the RF characteristics of the network. It also has the advantage that the bias inductor does not need to be as large as required in the single-ended case to avoid causing passband insertion loss (see Chapter 9). As shown in Figure 14.1, this is typically via a bias inductor L_b, frequently in combination with a shunt capacitor C_b. The capacitance C_b is included to provide a low impedance on the output of the amplifiers for frequencies around the second harmonics of the passband.

For the differential PA case, the coupled inductors have two functions:

1. They combine the two out-of-phase inputs into a single output. As discussed in Section 5.4, a three-port network used for this purpose is generally referred to as a balun.
2. They increase the line impedance on the output from that on the two input lines.

In Section 2.9, expressions for all the key performance characteristics of a three-port network are given in terms of the network S-parameters. Moreover, in Appendix D, the full four-port S-parameters are provided for a pair of coupled inductors, including a shunt capacitor across the primary inductor. Unfortunately, the S-parameters are very unwieldly and not useful for network synthesis. Fortunately, such complexity is not required as, due to the top-to-bottom symmetry of the architecture, the design and analysis can be reduced to that of only a two-port, single-ended PA. Effectively, each of the two amplifier chains can be viewed as operating independently into a single-ended load.

The analysis and synthesis of the differential network of Figure 14.1 are reducible to a two-port, single-ended problem using an even/odd-mode analysis, as described in Section 10.4. For the odd-mode drive (Equation (14.1)), the center point of the primary inductor, L_1, is a virtual ground. Thus, the odd-mode equivalent circuit is as shown in Figure 14.2(a).

For the even-mode drive (Equation (14.2)), no current flows between the two amplifier chains. Equally, because the currents flowing into the primary at ports 1 and 2 are equal and opposed, there will be no net current flow in the secondary inductor L_2. Thus, the even-mode equivalent circuit for the network is as shown in Figure 14.2(b).

For element synthesis, first consider the odd-mode circuit. By analogy with Section 12.2, if k is the mutual coupling factor between the inductors, ignoring elemental resistive components, L_2 is determined as

$$L_2 = \frac{k^2 R_L / G_S - \left(1-k^2\right)\omega L_1 X_L - \sqrt{\left(\left(k^2/G_S\right)^2 - \left(\left(1-k^2\right)\omega L_1\right)^2\right) R_L^2 - 2\left(1-k^2\right)k^2 \omega L_1 X_L R_L / G_S}}{\omega^2 \left(1-k^2\right)^2 L_1} \tag{14.4}$$

and the shunt input tuning capacitor C_S is determined as

$$C_S = \frac{\left(\left(2/\omega L_1 - \left(1-k^2\right)B_S\right)R_L - \left(1-k^2\right)G_S X_L\right)\omega L_2 - \left(R_L^2 + X_L^2\right)G_S}{2\omega^2\left(1-k^2\right)R_L L_2} \tag{14.5}$$

where ω = angular design frequency.

For a solution to exist to (14.4), the argument of the square root must be ≥ 0. This leads to a constraint on the maximum value permissible for the independent variable L_1. We find

$$\omega L_1 \leq \frac{-X_L/R_L + \sqrt{1 + X_L^2/R_L^2}}{\left(1/k^2 - 1\right)G_S} \tag{14.6}$$

Analogous to Section 12.2, if a low input impedance is required at the third harmonic, an inductor L_S may be added in series with capacitor C_S. Denoting the frequency at the center of the passband by ω_0 and that at the center of the third-harmonic band by ω_3, the modified capacitor value, and that of the inductor, are given by value

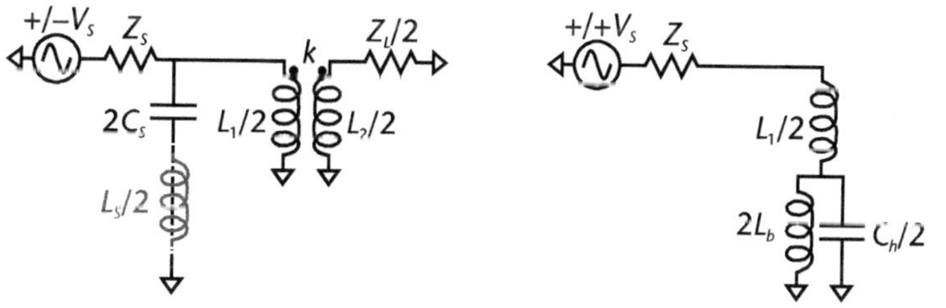

(a) (b)

Figure 14.2 Differential PA architecture odd-mode and even-mode schematics: (a) odd mode, and (b) even mode.

$$C_S' = \left(1 - (\omega_0/\omega_3)^2\right)C_S \tag{14.7}$$

$$L_S = \frac{1}{\left(\omega_3^2 - \omega_0^2\right)C_S} \tag{14.8}$$

Finally, considering the even-mode circuit, achieving a low impedance at the second-harmonic frequency requires

$$C_b = \frac{L_1 + 4L_b}{4\omega_2^2 L_1 L_b} \tag{14.9}$$

As with the single-ended, coupled-inductor match of Chapter 12, analysis of the network is most easily achieved by determining the $[ABCD]$ parameters for each of the two equivalent networks in Figure 14.2.

14.1.1 Basic Differential Coupled-Inductor Design Example

In this section, we apply the design and analysis formulae of the previous section to a practical differential PA design, in order to illustrate the basic performance capabilities of this type of matching architecture.

For simplicity, we take both source and load impedance as purely real, together with the additional network parameters given in Table 14.1. Note that the output impedances of the two differential amplifiers, Z_S, are both set to 8Ω, while the single-ended load impedance is assumed to be 50Ω. This is equivalent to the single-ended impedance of 4Ω for the coupled-inductor design analysis in Chapter 12. Both amplifiers are assumed to have the same output power (i.e., $|V_{S1}| = |V_{S2}|$).

With these values, from (14.6), we determine

$$L_1 \max \approx 0.655 \text{ nH} \tag{14.10}$$

Table 14.1 Differential Coupled-Inductor Impedance Match Parameters

Design Parameters	
Mutual coupling, $k = 0.7$	
f_{lo} 1,710 MHz	f_{hi} 2,025 MHz
$f2_{lo}$ 3,420 MHz	$f2_{hi}$ 4,050 MHz
$f3_{lo}$ 5,130 MHz	$f3_{hi}$ 6,075 MHz
$Z_{S1} = Z_{S2} = 8\Omega$	$Z_L = 50\Omega$
Inductors, $Q_L = 60$	Capacitors, $Q_C \approx 100$

which is ~4 times L_1 maximum for the comparable single-ended case, given in (12.19). This is to be expected as the effective source impedance across the primary inductor $2 \times 8 = 16\Omega$, is 4 times that of the single-ended case.

First, we investigate how the maximum passband insertion gain (P_{del}/P_{avail}) varies as a function of the independent variable L_1. This dependence is shown in Figure 14.3(a).

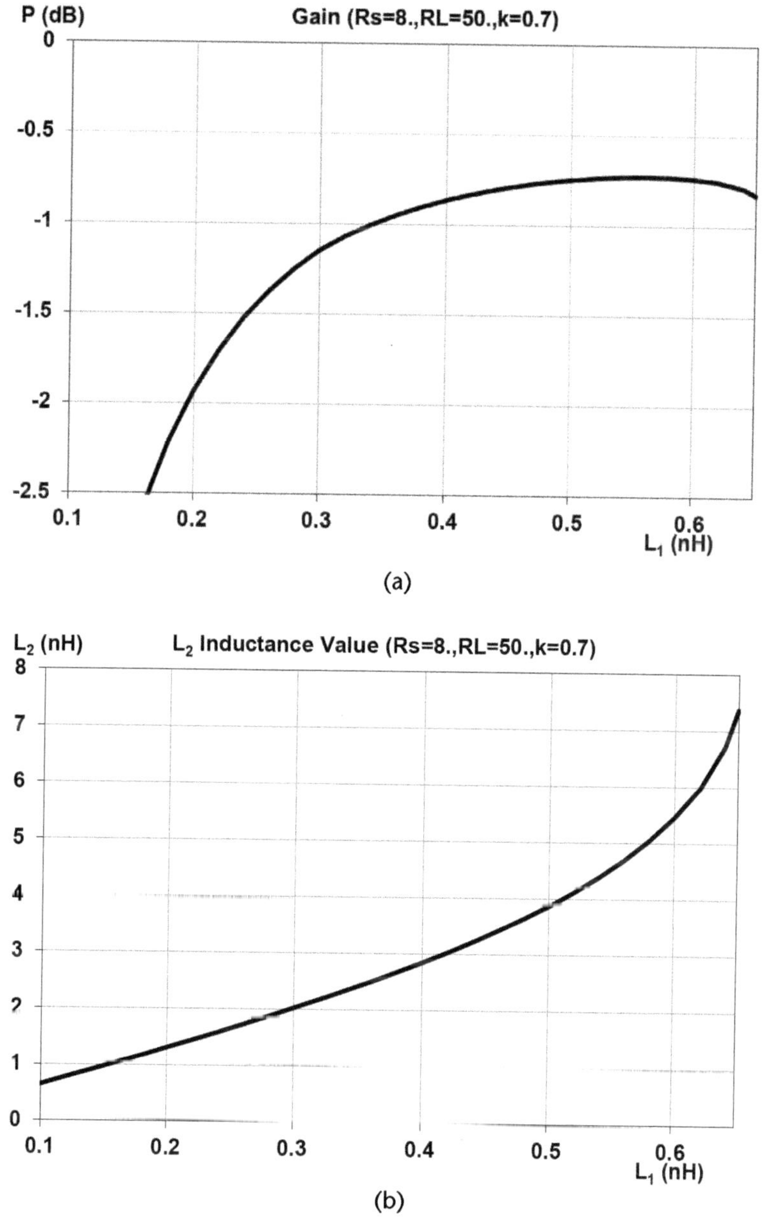

Figure 14.3 Differential insertion gain and secondary inductance dependence on primary inductance L_1: (a) gain, and (b) secondary inductance.

It can be seen that the lowest insertion loss ~0.73 dB occurs for a value of L_1 = ~0.56 nH, which is approximately 15% below the maximum possible value given by (14.6). The optimum gain is close to that of the single-ended case. The optimum value for L_1 is ~4 times that of the single-ended case in Section 12.2.1.

Figure 14.3(b) shows the required value for the secondary inductance L_2, as a function of L_1. The dependence is almost identical to that for the single-ended case, except for the primary inductor value being scaled by 4 times.

The ratio L_2/L_1, as a function of L_1, is presented in Figure 14.4 for the differential case. This can be compared to the comparable single-ended data in Figure 12.17(a). Due to the equivalent source impedance being increased by a factor of 4, L_1 is increased by the same factor. This results in the ratio L_2/L_1 being decreased by 4 times. The reduced L_2/L_1 ratio helps to make the differential layout more practical. However, it should be remembered, as noted elsewhere, that this advantage derives from dividing the output cells into two groupings, not from the differential phase relationship per se.

For the optimum value of L_1, the network element values are as shown in Table 14.2.

The corresponding passband insertion gain and input impedance Cartesian dependencies are shown in Figure 14.5. The input and output reflection coefficients for the network are shown in Figure 14.6.

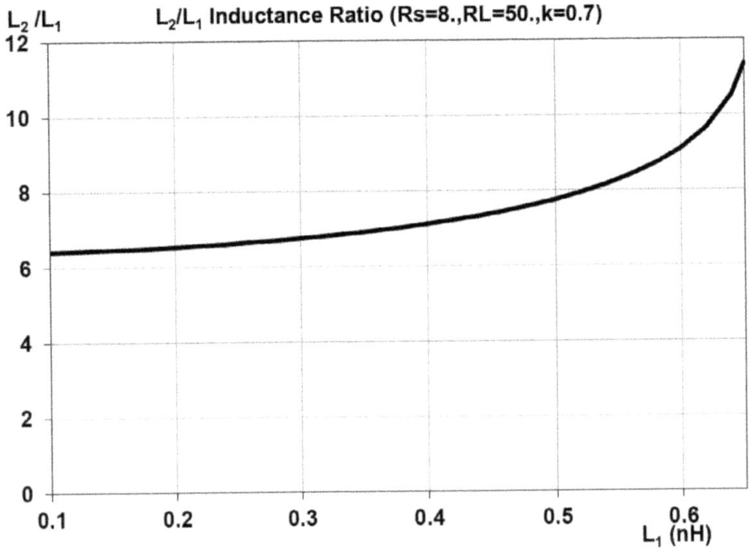

Figure 14.4 Differential secondary to primary inductance ratio.

Table 14.2 Optimum Differential Coupled-Inductor Element Values

Element Values	
L_1 = 0.56 nH	L_2 = 4.703 nH
C_S = 15.968 pF	

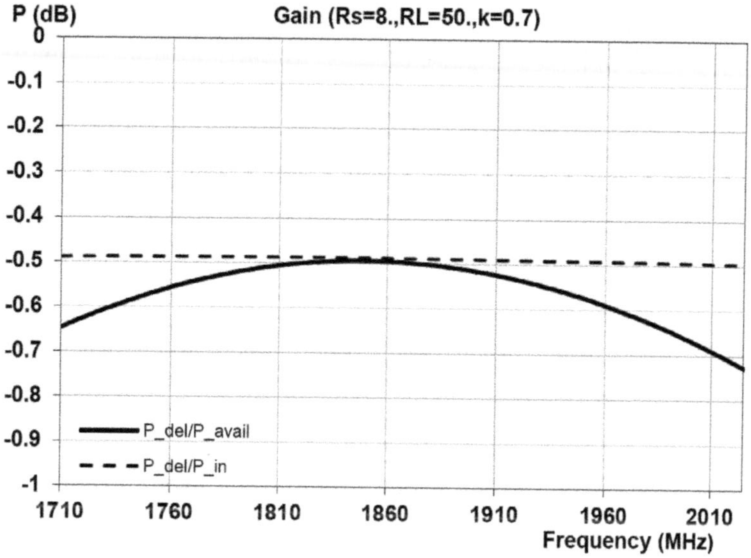

(a)

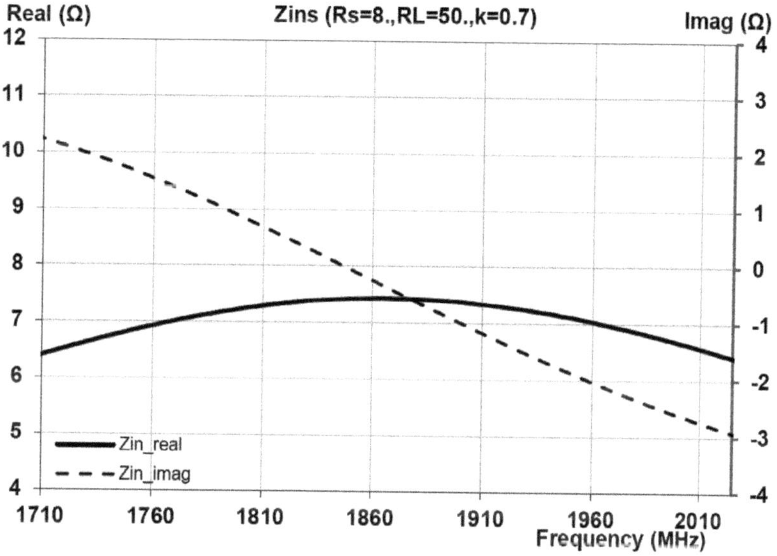

(b)

Figure 14.5 Differential coupled-inductor network gain and impedance: (a) gain, and (b) input impedance.

These responses demonstrate that a differential coupled-inductor network is capable of providing the desired impedance transformation across ~17% fractional bandwidth with good characteristics. The insertion loss is relatively flat and the input and output reflection coefficients are well behaved.

With the differential impedance network of Figure 14.1, the termination across the PA outputs across the second-harmonic band can be designed close to a short circuit. This is achieved by setting the value of C_b according to (14.9). Because the

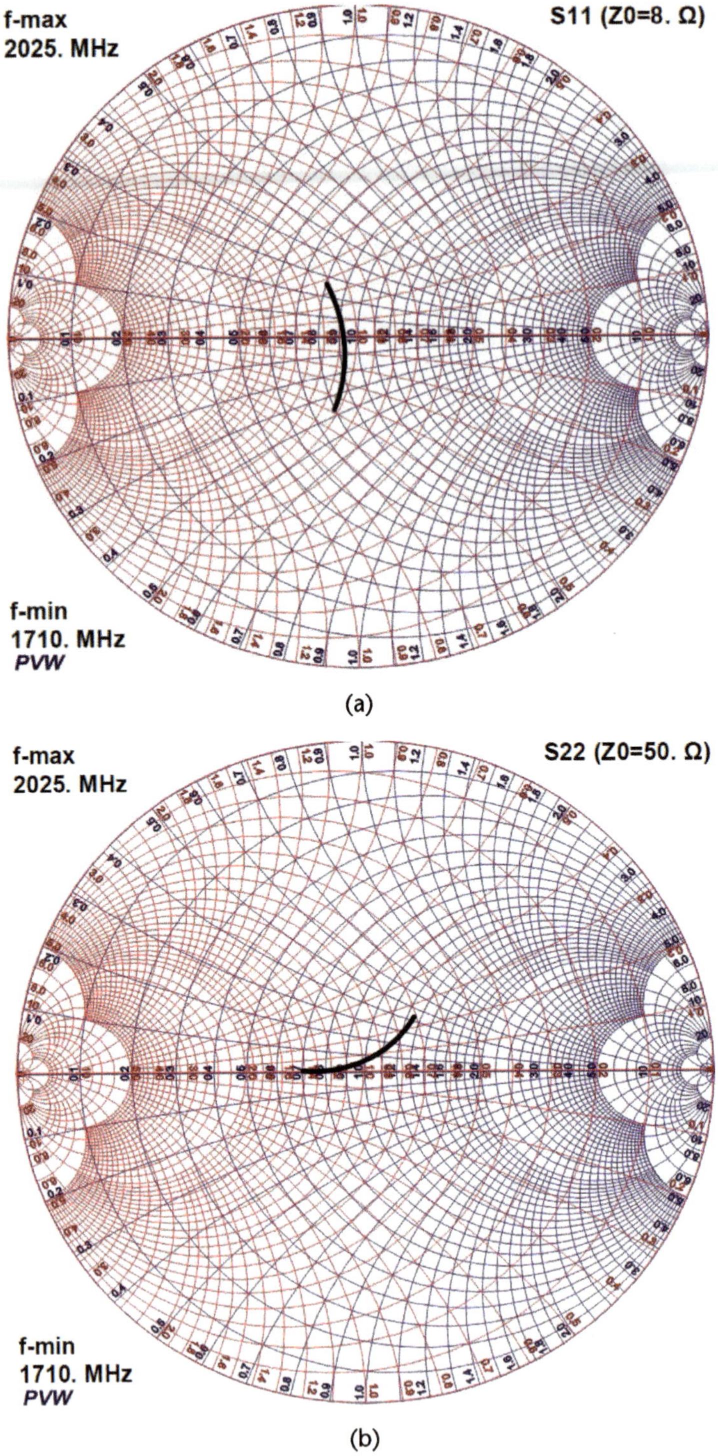

Figure 14.6 Differential coupled-inductor network input and output impedances: (a) input impedance, and (b) output impedance.

center point of primary inductor L_1 is effectively a ground for the odd-mode passband frequencies, this does not result in additional passband loss, as in the single-ended case. The input reflection coefficient of the network across the second-harmonic band is shown in Figure 14.7 for $L_b = 0.3$ nH and $C_b = 19.02$ pF.

The responses above are all without the additional trap inductor, L_S, in Figure 14.1, to provide an optional low impedance to the PA output stage across the third-harmonic band. The associated input third-harmonic reflection coefficient of the network is shown in Figure 14.8(a).

If the value of C_S is modified according to (14.7), and the inductor L_S is introduced in series as determined by (14.8), the reflection coefficient across the third-harmonic band becomes that shown in Figure 14.8(b). As desired, the impedance across the full third-harmonic band is now effectively a short circuit across the full band.

Figure 14.9(a, b) show the corresponding passband characteristics of the network with the third-harmonic trap. Introducing the trap increases the passband insertion loss and impedance variation across the passband, although only modestly. This can be seen by comparing these figures with those in Figure 14.5.

Because the trap resonance is at the third harmonic and not the second as in the single-ended case, the deleterious effects in the passband are much reduced compared to the latter.

In determining the above responses, a mutual coupling factor $k = 0.7$ was assumed throughout. The bandwidth of the impedance match achievable with the coupled-inductor architecture is critically dependent on this parameter. For the real 6.25:1 impedance transformation ratio specified by the parameters in Table 14.1,

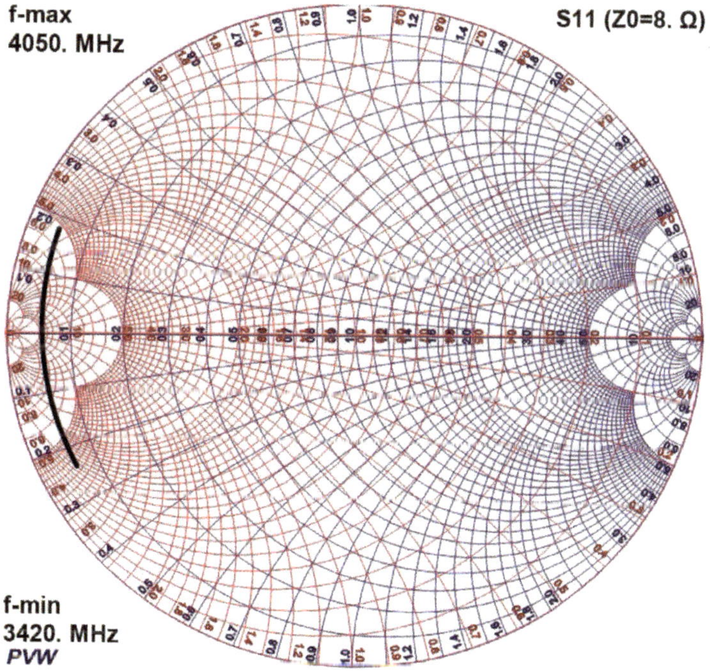

Figure 14.7 Differential second-harmonic band input reflection coefficient.

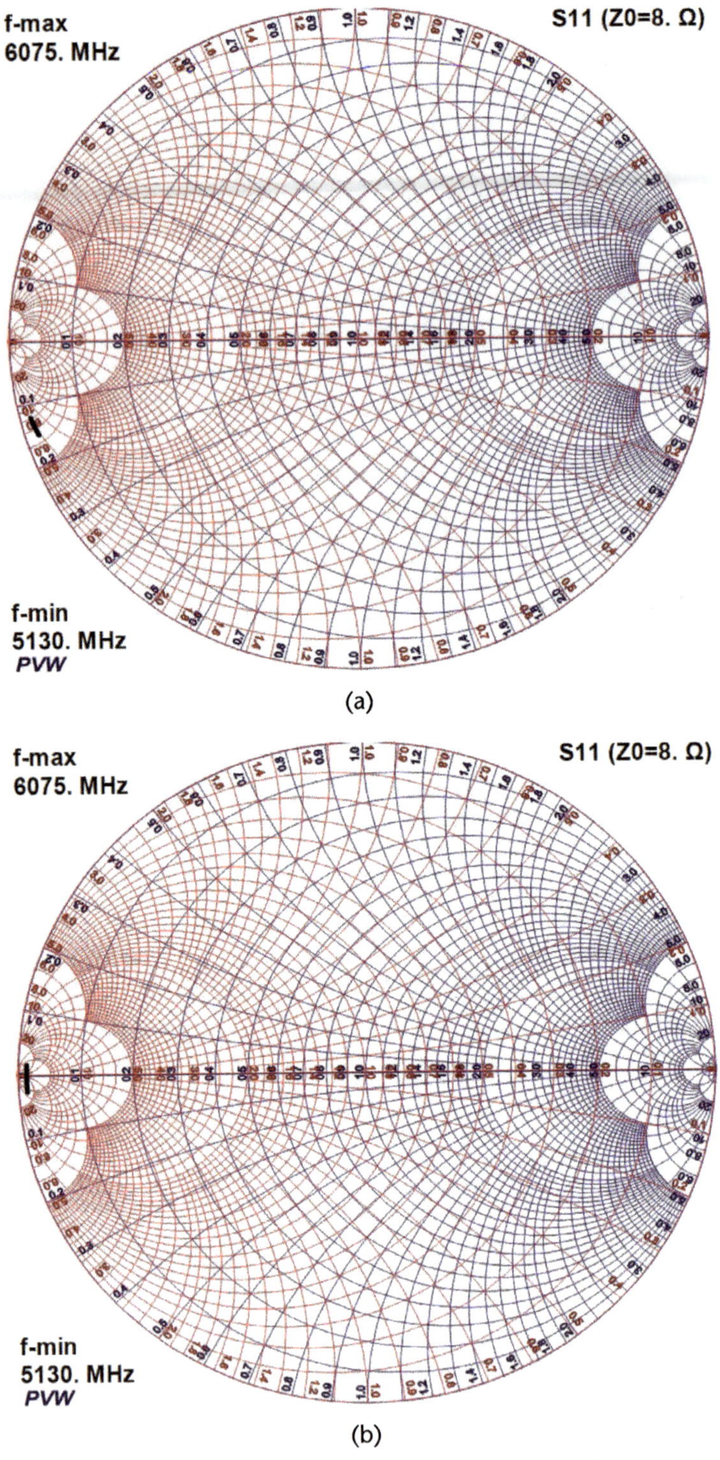

Figure 14.8 Differential network third-harmonic input reflection coefficients: (a) no trap, C_S only, and (b) with trap, $C_S + L_S$.

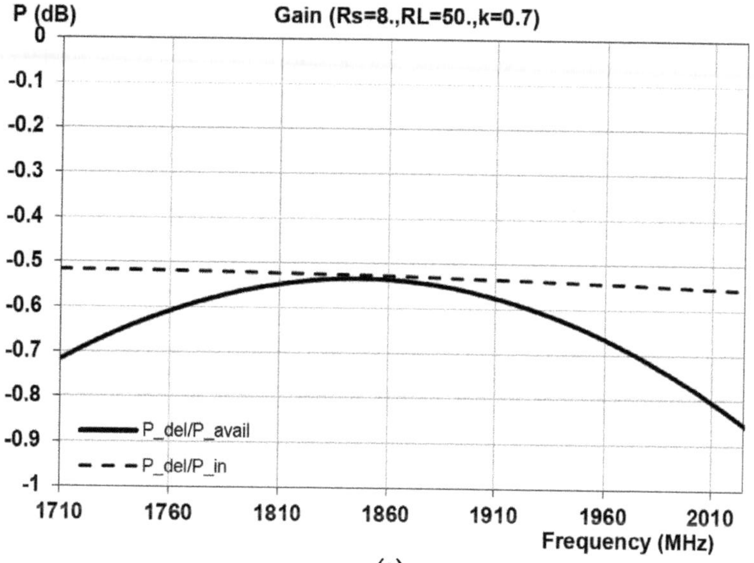

(a)

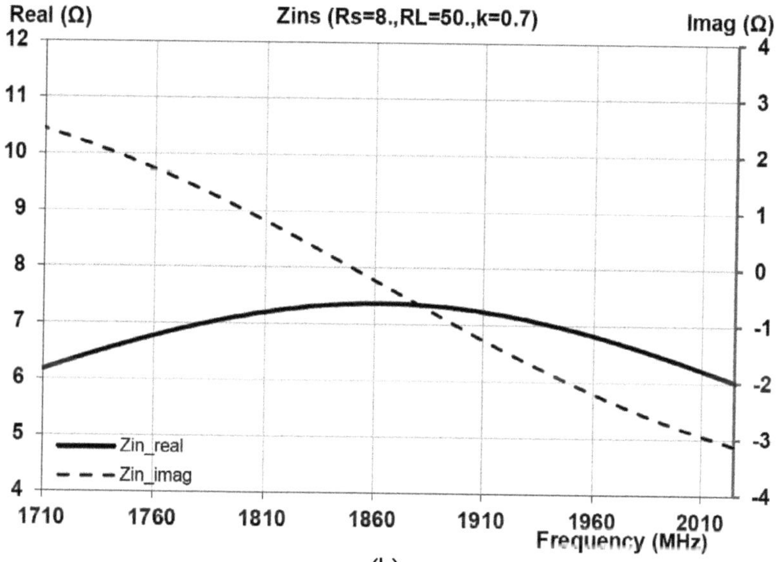

(b)

Figure 14.9 Differential network gain and impedance with third-harmonic trap: (a) gain, and (b) input impedance.

the maximum possible value for the primary inductance L_1 can be determined from (14.6) as a function of the coupling factor. This dependence is shown in Figure 14.10.

Note that the maximum possible value for L_1 drops precipitously with a reduced coupling factor. The consequence of a reduced primary inductance is a significant reduction in the bandwidth of the impedance matching characteristics. To illustrate this, consider the network response with identical parameters to those of Table 14.1, but with a mutual inductance value, $k = 0.55$. In this case, the maximum value for

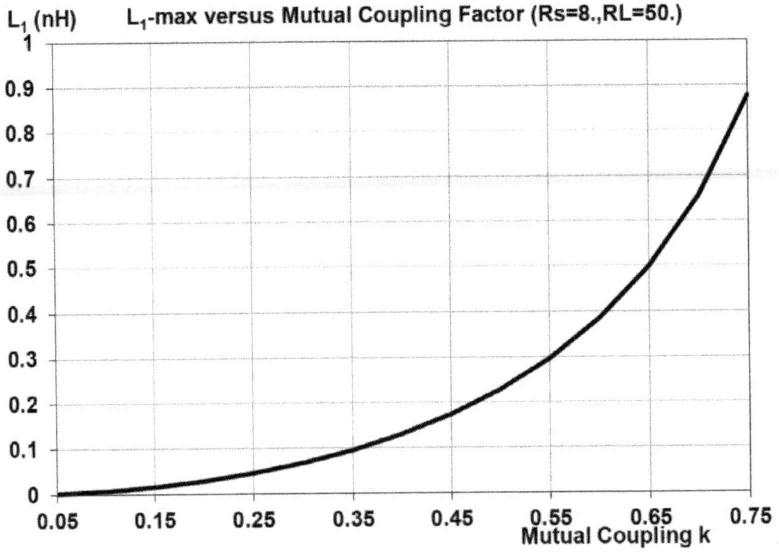

Figure 14.10 Maximum primary inductance versus coupling factor.

the primary inductance L_1 is limited to ~0.30 nH. A value $L_1 \approx 0.28$ nH is found to result in the best characteristics.

For this reduced coupling factor, insertion impedance and input impedance responses, corresponding to those of Figure 14.5, are shown in Figure 14.11.

Comparing these figures reveals a significant degradation of the characteristics of the differential network with the reduced coupling factor. This illustrates how critical it is to be able to achieve a high magnetic coupling factor for the coupled inductor impedance match for the network to be viable.

14.2 Differential Coupled Inductors with Interwinding Capacitance

As discussed previously, a high magnetic coupling coefficient, k, is essential for a coupled-inductor impedance match, whether for single-ended or differential. Ideally, $k = 1$ would be ideal, but, unfortunately, this is not achievable. In practice, with typical printed inductor layouts, $|k| \leq \sim 0.7$.

To achieve the higher coupling values, the inductor traces must be very close together in order for the magnetic fields associated with the currents flowing in each inductor to overlap. As mentioned in Section 13.1.2 and elsewhere, this has the unfortunate side effect of also resulting in significant interwinding capacitance between the traces. This can have important consequences for the performance of the impedance match, particularly at frequencies above the passband.

A differential coupled-inductor impedance network in addition to transforming impedance must also function as a balun. Looking at the network in reverse, this means that a signal entering the output (single-ended) port must be split by the network into two equal-amplitude but anti-phase outputs on the differential ports. With no interwinding capacitance, as in the previous section, the top-to-bottom

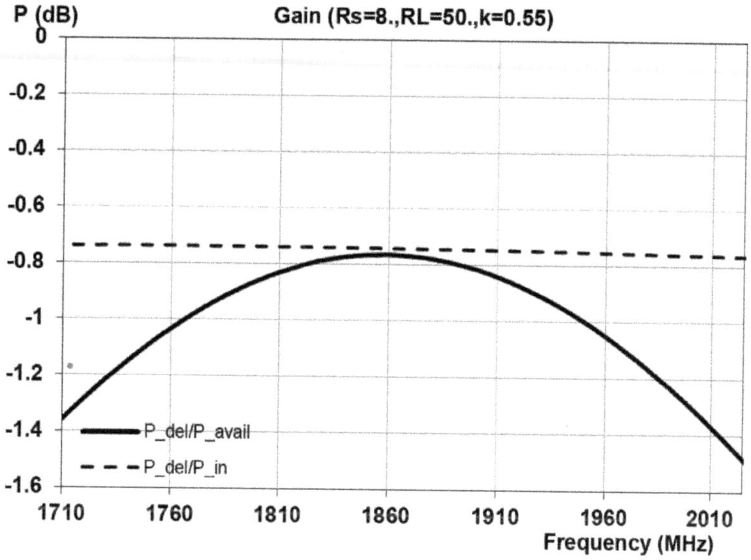

(a)

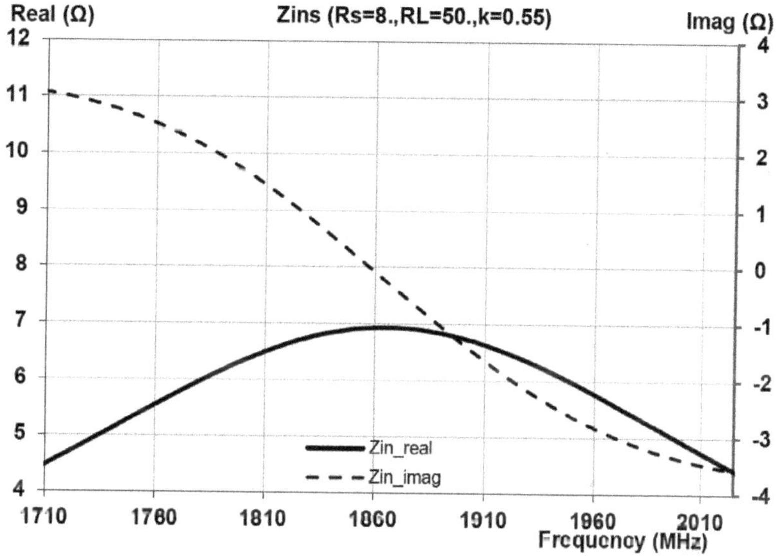

(b)

Figure 14.11 Differential coupled inductor network gain and impedance: (a) gain, and (b) input impedance.

symmetry of the network ensures symmetry of the reverse outputs. There is a voltage null at the center of the primary inductor L_1, for all odd-mode inputs, which includes the passband. For even-mode inputs, which includes the second harmonic, there is no net current flow down L_1 and thus no induced current flow in the secondary inductor L_2. These characteristics greatly simplify both the analysis and synthesis of the network. Unfortunately, they no longer hold in the presence of capacitive coupling across the inductors.

With interwinding capacitance between the two inductors, there is a direct signal transmission path between the two differential ports and the single-ended port independent on the magnetic coupling. On the primary inductor side, there is top-to-bottom symmetry. However, on the secondary inductor side, this is not the case as one terminal is grounded. There are several consequences to the lack of secondary symmetry in the presence of capacitive coupling:

1. The mid-point of the primary inductor L_1 is generally no longer a virtual ground.
2. The input impedances on the two terminals of L_1 will, in general, be unequal.
3. RF energy incident on the secondary single-ended terminal of L_2 will not, in general, be split with equal amplitude and 180° phase on the two terminals of L_1.
4. With a lack of top-to-bottom symmetry, both the network analysis and synthesis are significantly more complex.

In the presence of an interwinding capacitance C_X and the lack of an overall top-to-bottom symmetry, even with a pure differential drive, the virtual ground location on the primary inductor L_1 does not necessarily coincide with the center of the inductor. However, ideally, the bias network should be located at the virtual ground. A further consequence of the asymmetry is that the input impedances on ports 1 and 2 can no longer assumed to be equal. In addition, if the bias node is not appropriately located at the virtual ground location, the two input impedances will become dependent upon the bias network.

Referring to Figure 14.1, it is clear that, in the presence of any capacitive coupling between inductors L_1 and L_2, port 1 will be more tightly coupled to the ground than port 2, due to the single-ended nature of the output load. Hence, it should be expected that the minimum voltage point along L_1 in the presence of interwinding capacitance will be shifted down from its center.

Any useful circuit model to be the basis for analyzing the characteristics of a coupled-inductor pair network model, with interwinding capacitance, must allow for a noncentral location of the bias network on L_1. To this end, the architecture of Figure 14.1 is generalized to that shown in Figure 14.12.

In the generalized differential configuration of Figure 14.12, the bias network node is located a distance αL_1 from the bottom of the inductor, $(0 \le \alpha \le 1)$.

The top coupled-inductor sections have inductances $(1 - \alpha)L_1$ and $(1 - \alpha)L_2$, with an interwinding capacitance C_{X2}, where

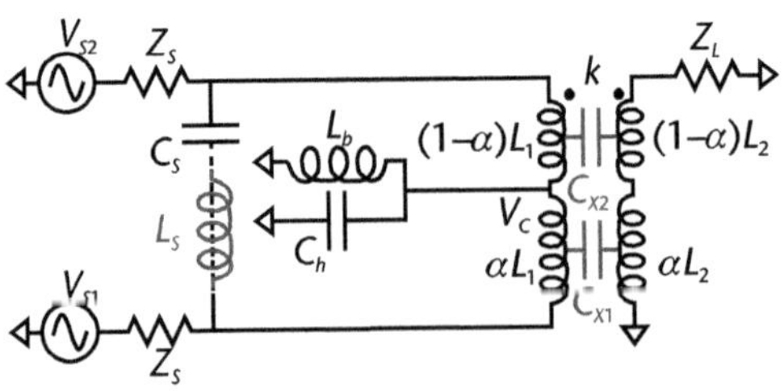

Figure 14.12 Differential PA matching architecture with coupled inductors including interwinding capacitance and input bias node.

$$C_{X2} = (1-\alpha)C_X \tag{14.11}$$

The bottom coupled inductor sections have inductances αL_1 and αL_2, with an interwinding capacitance C_{X1}, where

$$C_{X1} = \alpha C_X \tag{14.12}$$

and C_X = total interwinding capacitance.

These two discrete capacitors are assumed to be located at the center of each of their respective sections. The coupled-inductor network in Figure 14.12 is assumed to be identical to that in Figure 14.1 except for the inclusion of the interwinding capacitance C_X.

As stated above, synthesis and analysis of the differential coupled-inductor network with interwinding capacitance are greatly complicated by the lack of symmetry. An important consequence of this is the number of dependent variables required to synthesize the network element values. With no interwinding capacitance, by virtue of the symmetry and the assumption of a differential drive, the input impedances on the two differential ports are necessarily identical. Thus, to synthesize the network elements required for an input match, only two dependent variables are required. The two elements are sufficient to allow both the real and imaginary input matching equations to be met on ports 1 and 2, that is,

$$\Re\{Z_{\text{in}}\} = R_S \tag{14.13}$$

and

$$\Im\{Z_{\text{in}}\} = -X_S \tag{14.14}$$

where source impedance, $Z_S = R_S + jX_S$, and $\Re\{\ \}$ $\Im\{\ \}$ represent the real and imaginary parts of the argument, respectively.

With interwinding capacitance between the inductors, however, and the resulting loss of top-to-bottom symmetry, a minimum of four, rather than two, equations must now be satisfied. Equations (14.13) and (14.14) must now be independently satisfied for the unequal impedances on ports 1 and 2, at the top and bottom of L_1. Assuming that the primary inductance L_1 is again taken as an independent variable and L_2 and C_S remain dependent variables, a minimum of two more variables must be introduced in order to be able to satisfy the four equations.

With no interwinding capacitance, the center point of inductor L_1 is a virtual ground when the inductor is driven by a differential input, as in the passband. Thus, the impedances on ports 1 and 2 are completely independent of any loading at the center point of the inductor. This attribute is usually taken advantage of to apply bias to the amplifier output stages through a small inductor L_b, as in Figure 14.12.

With interwinding capacitance, however, the center point of inductor L_1 is no longer a virtual ground, and thus reactive loading at this point will generally affect the input impedances on ports 1 and 2. The effect on the two impedances will depend on both the value of the reactive loading and where it is located along the inductor. There should be no expectation that the optimum location for the reactive loading is at the mid-point of L_1. Referring to Figure 14.12, it is clear that, in the presence of any capacitive coupling between inductors L_1 and L_2, port 1 will be more tightly coupled to the ground than port 2, due to the single-ended nature of the output load. Hence, it should be expected that the minimum voltage point along L_1 in the presence of interwinding capacitance will be shifted down from its center.

Therefore, the two additional variables to try to achieve a simultaneous input match on ports 1 and 2 are α and X_b, where α quantifies the location along the inductor L_1 where a susceptive loading is located and X_b is its value. The susceptance comprises elements L_b and C_b.

It is all too often the case that a designer seeking to optimize a coupled-inductor match on the output of a PA limits the scope of the optimization by assuming that the bias feed inductor must be located at the center of the primary inductor. This predisposition is driven by the belief that the center of the inductor is a voltage null. Because this assumption is incorrect in any practical layout, in which there is finite interwinding capacitance, a truly optimized design cannot be achieved with even the best design software package. It is essential that the value of the bias inductor and its position on the primary inductor must be input as variables to arrive at a truly optimized design. Failing this, the design will have to incorporate some design asymmetry between ports 1 and 2 to achieve a simultaneous match on the two ports.

Equations for the full analysis and synthesis of the coupled-inductor network of Figure 14.12 are given in Appendix G. As discussed above, due to the lack of symmetry, their computation is very involved relying on the definition of many intermediate variables. While the equations might require too much effort for a typical RF design engineer to implement in practice, they can actually be of great use in understanding key characteristics of a coupled-inductor match with interwinding capacitance. The improved understanding of the network that results can significantly benefit any design approach to optimizing the performance of the structure.

The equations of Appendix G were implemented in an Excel workbook and applied to a coupled inductor matching network analogous to that in the previous section. The key aspects of that analysis are presented in the following section.

14.2.1 Differential Coupled Inductor with an Interwinding Capacitance Design Example

This section focuses on the ways in which interwinding capacitance between a pair of coupled inductors can have significant impacts on network performance. To illustrate these, consider the design of a matching network with identical parameters to those of Section 14.1.1, but with the addition of an interwinding capacitance C_X. These parameters are listed in Table 14.3.

With no interwinding capacitance and no susceptive loading on L_1, the impedances on the two input ports 1 and 2 are necessarily identical. The key characteristics of the network are those shown in Figure 14.6. However, with finite interwinding capacitance between the inductors, the network characteristics become much more complex. Maintaining $L_1 = 0.56$ nH, and with no susceptive loading on L_1, the synthesis equations can only be used to determine optimized values of L_2 and C_S for an input match on either port 1 or port 2. There is an insufficient number of variables to allow for a simultaneous match on both input ports.

Table 14.3 Differential Coupled-Inductor Impedance Match Parameters

Design Parameters	
Mutual coupling, $k = 0.7$	
Cross-coupling capacitance, $C_X = 4$ pF	
f_{lo} 1,710 MHz	f_{hi} 2,025 MHz
$f2_{lo}$ 3,420 MHz	$f2_{hi}$ 4,050 MHz
$f3_{lo}$ 5,130 MHz	$f3_{hi}$ 6,075 MHz
$Z_{S1} = Z_{S2} = 8\Omega$	$Z_L = 50\Omega$
Inductors, $Q_L = 60$	Capacitors, $Q_C \approx 100$

The top of Table 14.4 shows the network values determined to achieve an optimum impedance match on port 1. Conversely, the bottom of Table 14.4 shows the network values determined to achieve an optimum impedance match on port 2.

Note that these two sets of matching requirements are significantly different. The result is that, when one port is well matched in the passband, the other is significantly mismatched. This can be seen in Figure 14.13.

As a result of one of the ports being badly mismatched in either scenario, the network exhibits high insertion loss in each case. The corresponding insertion gains for the network are shown in Figure 14.14.

Rather than optimizing the match on either port 1 or port 2, the network may be optimized instead to simultaneously minimize the mismatch on the two ports. The element values for such a compromise match are shown in Table 14.5.

The corresponding key network performance characteristics are shown in Figure 14.15. It can be observed that this does result in significantly reduced insertion loss.

While illustrating some of the complications arising from the presence interwinding capacitance in a coupled-inductor matching network, the above elements

Table 14.4 Network Values for Optimum Match on Port 1 and Port 2, for $C_X = 4$ pF

Match on Port 1	
Element Values	
$k = 0.7$	$C_X = 4$ pF
$L_b = \infty$ nH	$C_S = 9.499$ pF
$L_1 = 0.56$ nH	$L_2 = 1.858$ nH
Match on Port 2	
Element Values	
$k = 0.7$	$C_X = 4$ pF
$L_b = \infty$ nH	$C_S = 15.14$ pF
$L_1 = 0.56$ nH	$L_2 = 3.398$ nH

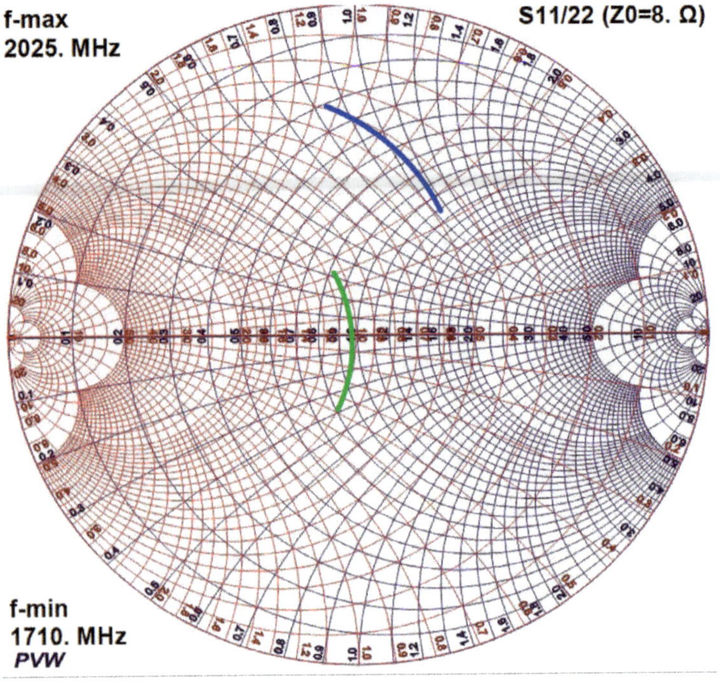

(a)

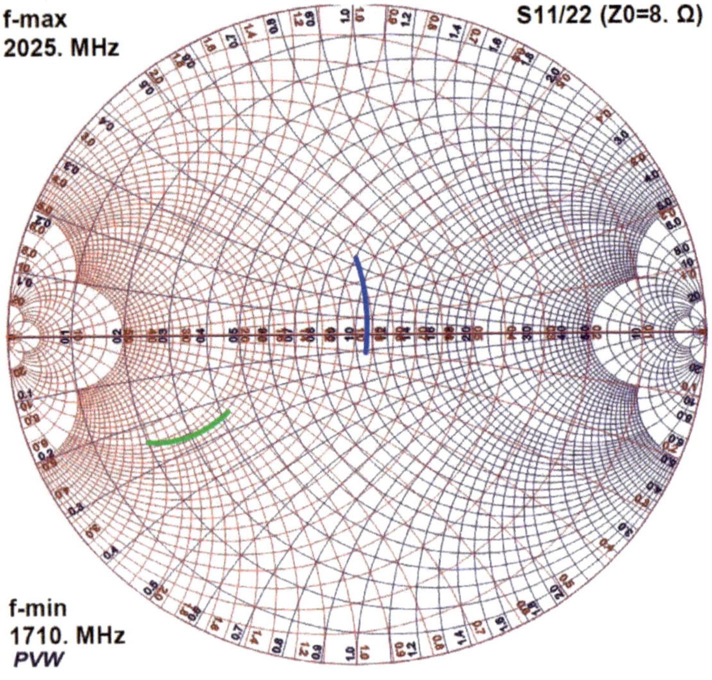

(b)

Figure 14.13 Differential input reflection coefficients for $C_X = 4$ pF: (a) match on port 1, and (b) match on port 2.

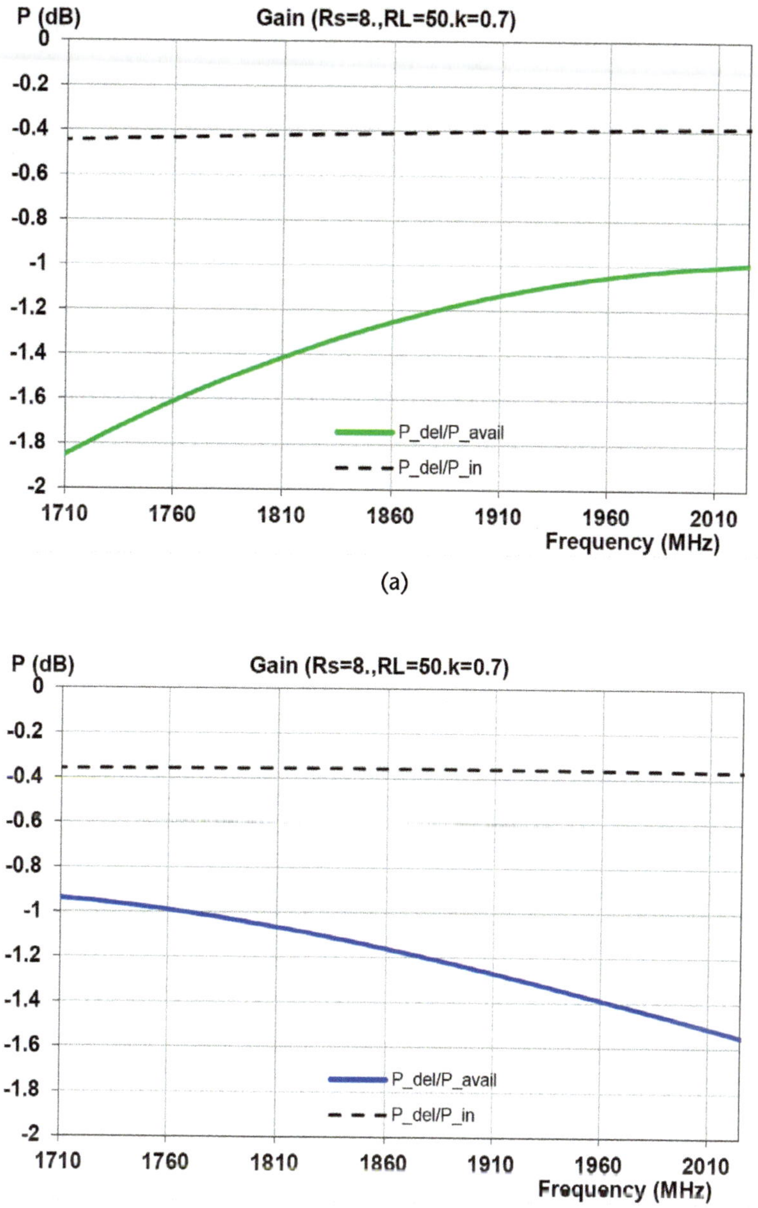

(a)

(b)

Figure 14.14 Differential insertion gains for $L_1 = 0.56$ pF and $C_X = 4$ pF: (a) match on port 1, and (b) match on port 2.

Table 14.5 Network Values for Compromise Match on Ports 1 and 2, for $C_X = 4$ pF

Element Values	
$k = 0.7$	$C_X = 4$ pF
$L_h = \infty$ nH	$C_S = 12.32$ pF
$L_1 = 0.56$ nH	$L_2 = 2.628$ nH

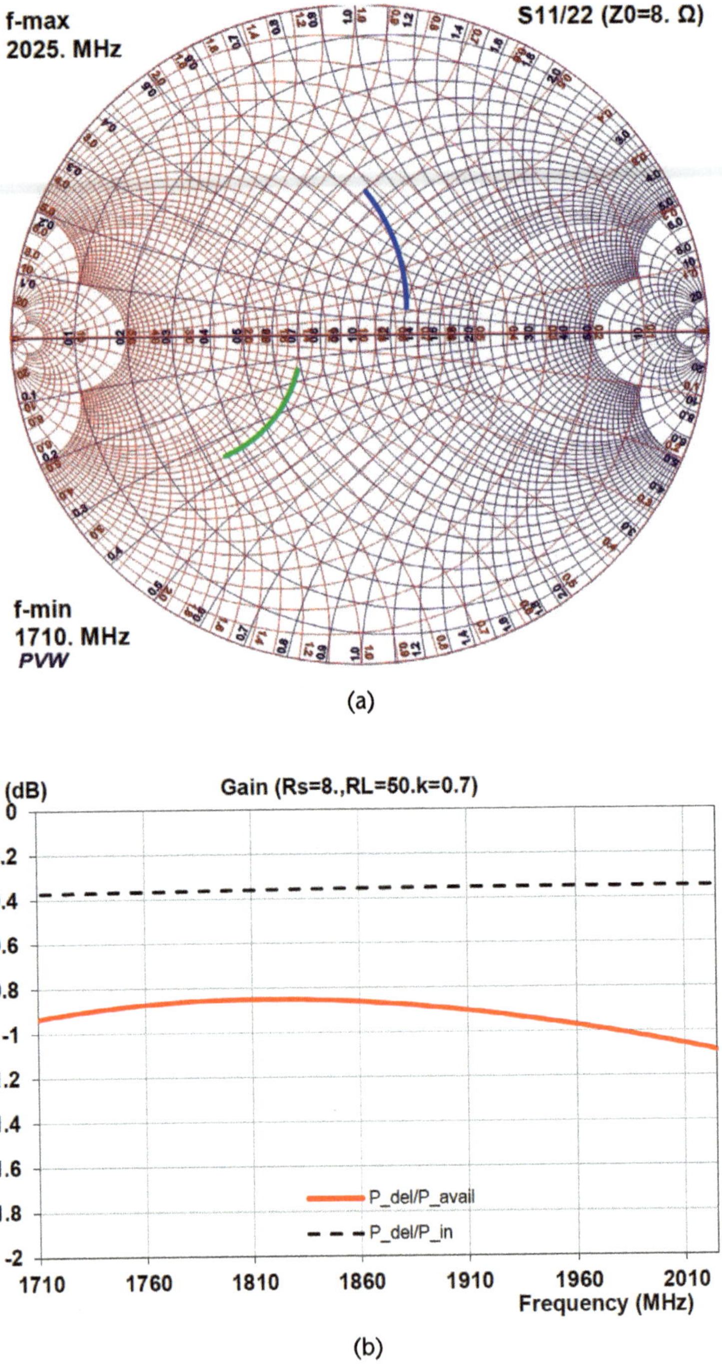

Figure 14.15 Compromise match performance parameters, for $C_X = 4$ pF: (a) input reflection coefficients, and (b) gain.

and performance parameters neglect any influence of the bias network. In practice, a finite bias network is invariably required on L_1. The bias is usually applied at the mid-point of the primary inductor L_1. However, if interwinding capacitance is important, such a choice will typically be nonoptimum. Instead, careful choice of the bias network, and its position along inductor L_1, can provide significant compensation for the deleterious effects of the capacitance, as is demonstrated below.

Before considering how the bias loading on L_1 might aid the performance of the coupled-inductor match, it is instructive to first examine how the capacitance affects the voltage distribution along primary inductor L_1. Equations are provided in Appendix G to allow the voltage V_C to be determined as a function of α (see Figure 14.12). The actual voltage distribution along L_1 with finite interwinding capacitance will vary somewhat with matching conditions. To understand qualitatively the influence that interwinding capacitance has on the voltage distribution, however, it is sufficient to maintain the same matching element values for $C_X = 0$, for the cases $C_X > 0$.

Figure 14.16 shows the voltage V_C/V_S along primary inductor $L_1 = 0.56$ nH, for $C_X = 0$, 2, and 4 pF. These dependencies were again determined with no bias loading on L_1 (i.e., $X_b = 0$). With no interwinding capacitance ($C_X = 0$), the voltage is zero at the mid-point, as expected. However, for increasing values of C_X, the voltage null on the inductor moves down closer to port 1. This is consistent with expectations, because the lower terminal of secondary inductor L_2 is grounded and is more tightly coupled to port 1.

With no interwinding capacitance, it is ideal to place the bias feed at the voltage null in the center of the primary inductor, where it has no effect on the passband characteristics. With finite interwinding capacitance, if the bias feed were located at the voltage null on the inductor, although now further down the inductor, similar advantages would accrue. However, without additional circuit elements to permit a simultaneous match on both ports 1 and 2, high insertion losses such as those shown in Figure 14.14 would be unavoidable.

With interwinding capacitance and no bias loading on L_1, ports 1 and 2 cannot be matched simultaneously as evidenced by the data in Figure 14.13. The two variables L_2 and C_S provide an insufficient degree of freedom to satisfy the four

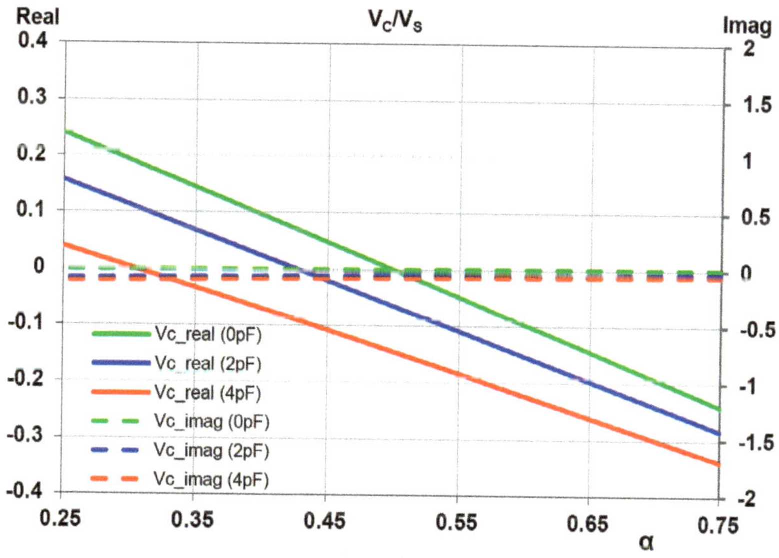

Figure 14.16 Voltage along the primary inductance for differing values of C_X.

equations required for a simultaneous complex match on the two ports. In an attempt to remedy this, two new design variables were introduced, that is, the bias reactance X_b, and its position along the inductor, α. Unfortunately, the equations of Appendix G reveal that an exact simultaneous match on the two ports cannot, in general, be obtained. This is due to the complex interplay of the parameters. Fortunately, a compromise match can usually be obtained with good characteristics that greatly exceed those that can be obtained without optimization of the bias feed.

A set of element values for the network that compromises the impedances on the two input ports with the placement of an inductive loading strategically located along L_1 is listed in Table 14.6.

With these values, the input reflection coefficients on ports 1 and 2 are shown in Figure 14.17(a), and the corresponding network insertion gain is shown in Figure 14.17(b).

Corresponding input impedances are shown in Figure 14.18.

These results verify that by optimally choosing the value of the reactive loading of the bias feed on the primary inductor, and adjusting its location, the deleterious effects of interwinding capacitance can be largely compensated for. In any optimization approach to the network design, therefore, it is important that these two parameters be incorporated as variables.

Table 14.6 Compromise Differential Coupled-Inductor Element Values for C_X = 4 pF

Element Values	
$k = 0.7$	$C_X = 4$ pF
$\alpha = 0.38$	$L_b = 0.27$ nH
$L_1 = 0.6$ nH	$L_2 = 2.59$ nH
$C_S = 11.617$ pF	

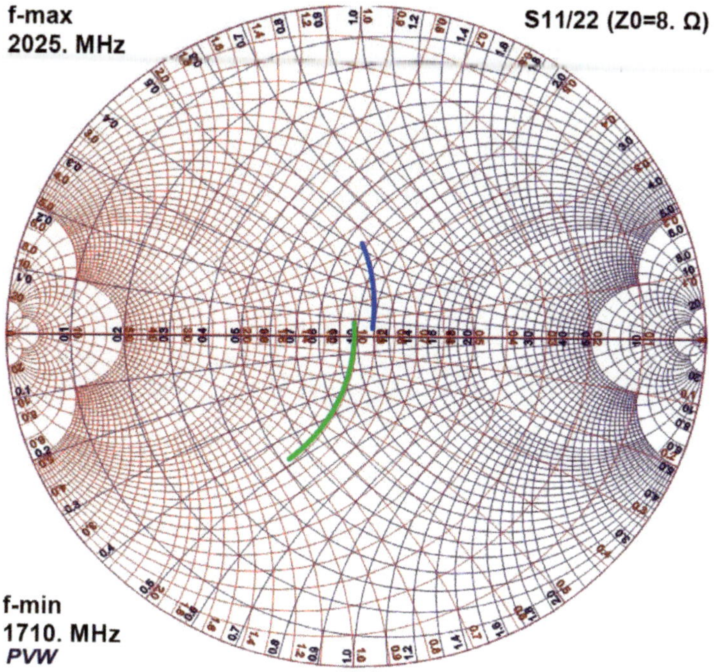

(a)

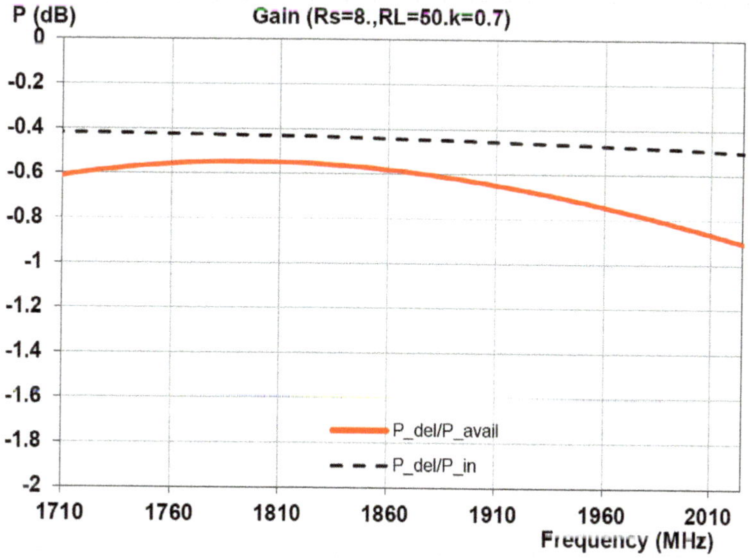

(b)

Figure 14.17 Compromise differential input reflection coefficients and gain for $C_X = 4$ pF: (a) input reflection coefficients, and (b) insertion gain.

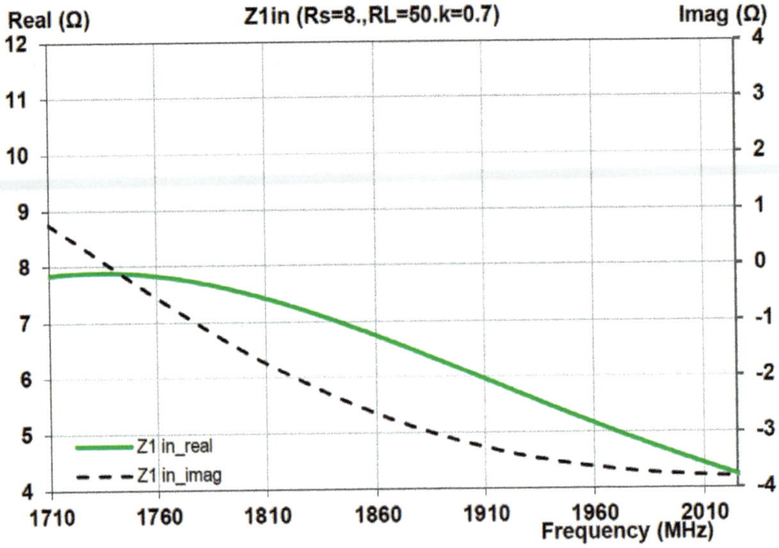

(a)

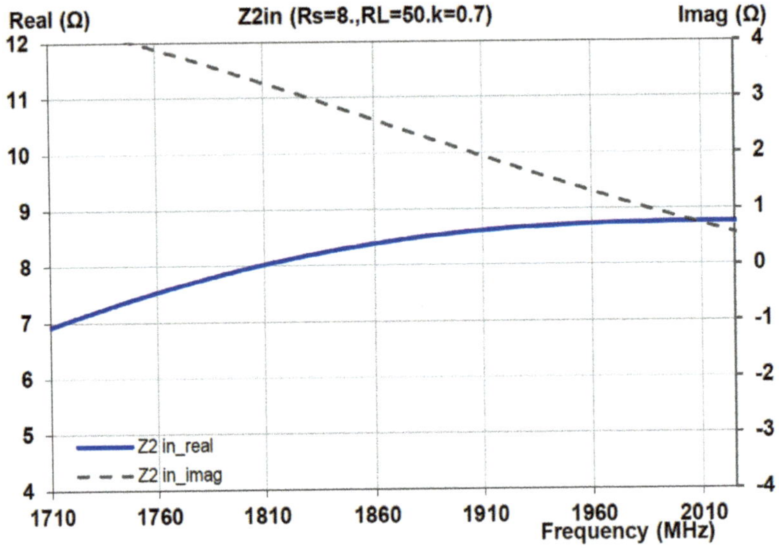

(b)

Figure 14.18 Compromise port input impedances for $C_X = 4$ pF: (a) match on port 1, and (b) match on port 2.

CHAPTER 15

Lattice Splitter/Combiner

Figure 15.1 shows the common form of a schematic for a lumped-element three-port network that has found an application for combining or splitting differential signals. Given the form of the schematic, the network is commonly referred to as a "lattice balun." The network comprises two equal capacitors and two equal inductors. The appropriate values for these elements are determined from

$$L = \sqrt{2R_1R_2}/\omega \qquad \text{and} \qquad C = 1/\omega\sqrt{2R_1R_2} \tag{15.1}$$

where R_1 = impedance at port 1, and $R_2 = R_3$ = impedances at ports 2 and 3. All external impedances, for the circuit as shown, must be pure real. The real terminating impedances on ports 2 and 3 must additionally be equal. Furthermore, the network of Figure 15.1 is only capable of splitting or combining two signals with equal amplitude and a phase difference of 180°. These are critical constraints that greatly limit widespread applicability of the circuit. However, the circuit topology of Figure 15.1 is actually capable of satisfying much more demanding design requirements than is evident from the limited balun example of Figure 15.1, with element values determined from (15.1). This becomes clear from a detailed circuit analysis of the topology.

To understand how the design flexibility of a three-port network, with the topology of Figure 15.1, may be expanded, it is instructive to first redraw the network of Figure 15.1 in the form shown in Figure 15.2(a). Note that the circuits in these two schematics are entirely equivalent. It is a little clearer now that the top arm has a series inductor followed by a shunt capacitor, with the elements interchanged in the lower arm. This still does not shed the light on the true basis for design and operation of the circuit. For that, we turn to Figure 15.2(b).

In Figure 15.2(b), additional shunt elements are shown, in a lighter tone, on both upper and lower arms. With these elements, the top arm comprises a lowpass π-network, while the lower arm comprises a highpass π-network. Both these

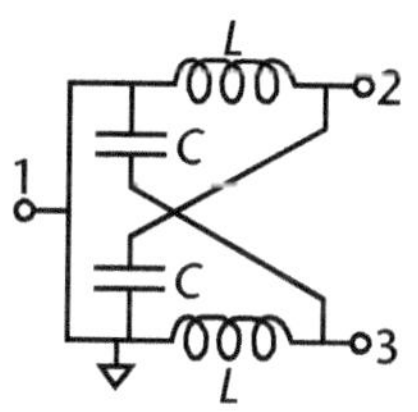

Figure 15.1 Lattice-balun splitter/combiner.

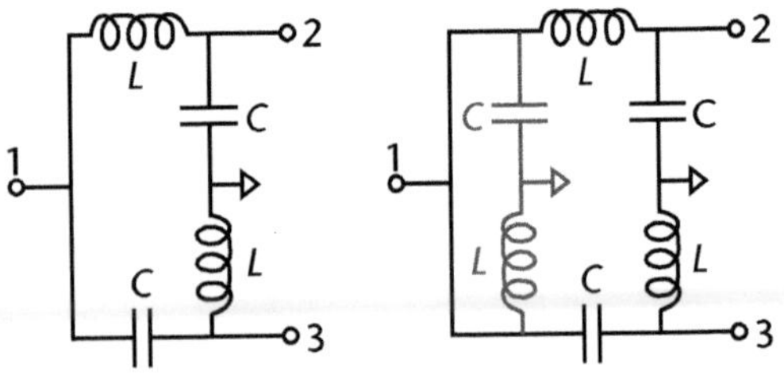

Figure 15.2 Lattice-balun splitter/combiner evolution: (a) lattice redrawn, and (b) lattice dual π-network equivalent.

networks can be designed to achieve a given insertion phase shift and match an arbitrary terminating complex impedance on one end to a second arbitrary complex impedance at the second end (see Chapter 11).

For the lattice balun of Figure 15.1, all external impedances are assumed to be real, and the impedances on ports 2 and 3 are assumed to be equal. Likewise, the voltages on ports 2 and 3 are taken to be equal but in anti-phase (i.e., differential). To function, the lowpass π-network must be designed to implement a phase lag of 90°, whereas the lowpass π-network must be designed to implement a phase advance of 90°. From these requirements, it follows, from the design equations of Chapter 11, that the elements in the top and bottom arms must be complex conjugates of each other. Because the first network elements, on the left, are connected to the same node 1 and are complex conjugates of each other, they self-cancel and thus can be excluded from the circuit. Hence, the circuit of Figure 15.2(b) simplifies to that of Figure 15.2(a).

The lattice-type balun can therefore be understood by viewing it as constituted from low and highpass π-networks in which the common node shunt elements of each are eliminated as a result of self-cancellation. In addition to minimizing the complexity of the network, eliminating the two elements from the circuit also has the additional benefit of improving performance for two reasons:

1. Resistive losses in the two elements are eliminated.
2. Reactive cancellation of the two elements is effectively achieved over infinite bandwidth, rather than just at the design frequency.

The lattice balun of Figure 15.1, and equivalently on Figure 15.2(a), with conjugate elements in the two arms, is limited to differential operation under the restrictive boundary conditions listed above. However, if the form is maintained, but more freedom is allowed in the design of the lattice elements, a much greater range of performance characteristics can be achieved.

Figure 15.3(a) shows a generalized form for a lattice coupler. The network series reactance elements jX_{n1}, and shunt susceptance elements jB_{n1}, are assumed independent of one another. Despite this, the design of the generalized form can still be approached as a combination of lowpass and highpass π-networks, following the design equations of Chapter 11. Complete synthesis and analysis equations for the coupler are given in Section 15.1.1.

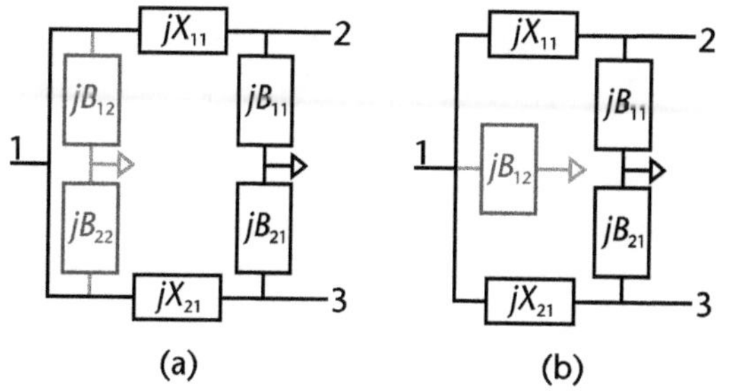

Figure 15.3 Generalized lattice coupler configuration: (a) dual π-network equivalent, and (b) resultant coupler elements.

As with the conventional lattice balun, regardless of the complex nature of the boundary conditions, complex impedances, and relative power levels, the generalized lattice coupler can continue to be designed such that the π-network shunt susceptances at the common node, jB_{12} and jB_{22}, are self-complex conjugates. In general, therefore, no resulting shunt susceptance, shown as jB_{12} in Figure 15.3(b), is required to implement the generalized form of the lattice coupler.

The generalized three-port coupler, with four independent variables, is capable of satisfying four independent boundary conditions or equations. As a result, the network can be designed to couple together three arbitrary complex external impedances, with an arbitrary power and phase relationship between the voltages on ports 2 and 3.

15.1 Generalized Lattice Splitter/Combiner Design Basics

The design of a generalized lattice three-port network, as in Figure 15.3(b), can simply be approached by analyzing the network requirements for the topology to operate as a power splitter. By reciprocity, the derived network can be expected to operate equally well in reverse mode as a power combiner. As described above, the analysis approach is based on viewing the three-port configuration as arising from the interconnection of lowpass and highpass π-networks, as in Figure 15.3(a).

For the conventional lattice-balun network at the beginning of the chapter, the phase shifts in the top and bottom arms of the network were $\pm 90°$, and the top and bottom elements were complex conjugates of each other. This resulted in very simple design criteria, at the price of very limited performance capabilities. If any of the simplified boundary conditions for the network are perturbed, in even a minor way, the design of the network becomes significantly more complicated and the top and bottom symmetry is broken.

To illustrate how the complexity arises, suppose that all the boundary conditions and performance requirements for the conventional lattice-balun of Figure 15.1 were maintained, save for a series capacitor added to the common port 1. This requires that at node 1, the "virtual" shunt elements of the two π-networks have a net susceptance equal to that of the now complex external admittance on port 1 to maintain an impedance match. To achieve this, the magnitudes of the phase shifts in the two arms of the network must now be unequal, although their phase difference must still be equal to 180°.

Likewise, if the terminating impedances on any of the three ports become complex, if an unequal power split between ports 2 and 3 is required, or a nondifferential phase is required, a difference in the magnitudes of the phase shifts in the two arms is required. In all cases, the required network will no longer have the top-to-bottom symmetry of Figure 15.1.

In the following section, design equations are presented that allow the design of a generalized lattice three-port network, capable of coupling arbitrary complex loads with arbitrary amplitude and phase splitting. Despite the greatly enhanced performance capabilities, the network still requires only four reactive elements.

15.1.1 Generalized Lattice Splitter/Combiner Design Equations

Figure 15.4(a) shows the general form of the three-port lattice splitter of Figure 15.3(b), with critical voltages and currents defined for operation as a splitter. Figure 15.4(b) shows an equivalent schematic for the coupler when used as a combiner.

In contrast to the assumptions made for the conventional lattice balun of Figure 15.1, the synthesis will be developed allowing for all external impedances Z_n to be complex. Likewise, no assumptions are made of the relative impedances and power levels at the two output splitter ports, or the two input combiner ports.

The three-port generalized lattice coupler can be used both for combining or splitting waveforms. However, it is sufficient to synthesize the coupler network elements for only one scenario. Reciprocity ensures that the derived network will also function equally well in the reverse mode. Here, for convenience, the synthesis equations will be presented for the generalized lattice coupler when used as a splitter.

For the lattice splitter synthesis, the following quantities are defined

$$Z_1 = R_1 + jX_1 = 1/(G_1 + jB_1) \tag{15.2}$$

$$Z_2 = R_2 + jX_2 = 1/(G_2 + jB_2) \tag{15.3}$$

$$Z_3 = R_3 + jX_3 = 1/(G_3 + jB_3) \tag{15.4}$$

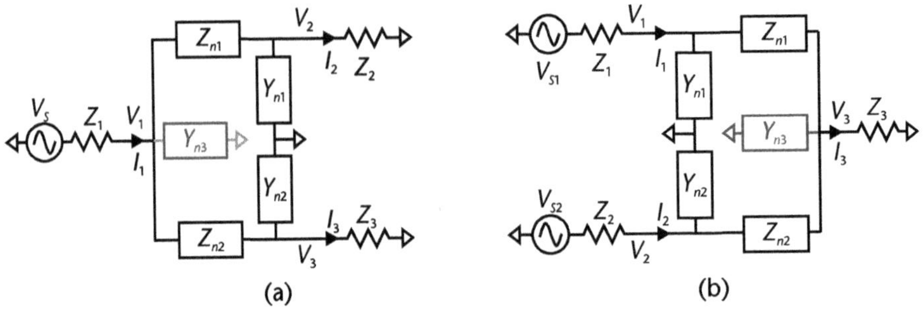

Figure 15.4 Generalized lattice splitter configurations: (a) splitter schematic, and (b) combiner schematic.

Let the phase shift required between the desired outputs on ports 2 and 3 be $\Delta\Phi$. In contrast to the conventional, but constrained, lattice balun, the magnitude of the phase shifts in each of the π-networks will not be assumed equal (i.e., $|\Delta\Phi|/2$). Instead, the phase shifts through the upper and lower π-networks will be taken as $\Delta\Phi_2$ and $\Delta\Phi_3$, respectively, which requires

$$|\Delta\phi_2 - \Delta\phi_3| = |\Delta\phi| \tag{15.5}$$

If the coupled power ratio between the outputs on ports 2 and 3 is

$$\text{dB\_power\_ratio} = \frac{P_2}{P_3} = \text{r\_dB} \tag{15.6}$$

consequently, the conductance looking into the upper lowpass π-network, G_{2S}, and the conductance looking into the lower highpass π-network, G_{3S}, must satisfy

$$G_{2S} = G_1/\left(1 + 10^{-\text{r\_dB}/10}\right) \tag{15.7}$$

and

$$G_{3S} = G_1/\left(1 + 10^{\text{r\_dB}/10}\right) \tag{15.8}$$

where G_1 is the source conductance on port 1 (Equation (15.2)).

An additional requirement, to ensure a match to the source on port 1, and simultaneously eliminate the need for either of the lattice shunt elements on port 1, is

$$B_{1\_\text{top}} + B_{1\_\text{bot}} = B_1 \tag{15.9}$$

where B_{1_top} and B_{1_bot} are the susceptances of the leftmost elements of the top and bottom π-networks, respectively, and B_1 is the susceptance of the source.

The key design variables for the network synthesis approach are the phase shifts in the two π-networks (i.e., $\Delta\Phi_2$ and $\Delta\Phi_3$), which are related to the desired phase split between the outputs on ports 2 and 3 by (15.5).

In addition to the phase shift through each of the π-networks, there is also a potential additional phase shift in each path due to the complex nature of the source and loads. To account for this, the modified net phase shifts required in the two paths are defined as $\Delta\Phi_2'$ and $\Delta\Phi_3'$, where

$$\Delta\phi_2' = -\left(\Delta\phi_n + \tan^{-1}(Z_2/Z_1)\right) \tag{15.10}$$

and

$$\Delta\phi_3' - \Delta\phi - \left(\Delta\phi_n + \tan^{-1}(Z_3/Z_1)\right) \tag{15.11}$$

Here, $\Delta\Phi_n$ is a phase-shift variable to be used in deriving the optimum element values for the coupler.

For network synthesis, the coupler elements will be assumed lossless, with

$$Z_{n\#} = jX_{n\#} \qquad \text{and} \qquad Y_{n\#} = jB_{n\#} \tag{15.12}$$

In terms of the phase shifts, $\Delta\Phi'_2$ and $\Delta\Phi'_3$, and design frequency ω_0, from analysis, the following relationships are derived

$$X_{n1} = \frac{R_2 \sin(\Delta\phi'_2) + X_2 \cos(\Delta\phi'_2)}{\sqrt{G_{2S}R_2}} \tag{15.13}$$

$$B_{n1} = \frac{G_2\left(R_2 - X_2 \tan(\Delta\phi'_2)\right) - \sqrt{G_{2S}R_2}\,\sec(\Delta\phi'_2)}{R_2 \tan(\Delta\phi'_2) + X_2} - B_2 \tag{15.14}$$

$$X_{n2} = -\frac{R_3 \sin(\Delta\phi'_3) + X_3 \cos(\Delta\phi'_3)}{\sqrt{G_{3S}R_3}} \tag{15.15}$$

$$B_{n2} = \frac{G_3\left(R_3 - X_3 \tan(\Delta\phi'_3)\right) - \sqrt{G_{3S}R_3}\,\sec(\Delta\phi'_3)}{R_3 \tan(\Delta\phi'_3) + X_3} - B_3 \tag{15.16}$$

Also, the net shunt susceptance on port 1 is given by

$$Y_{n3} = \frac{G_{2S}\left(R_2 - X_2 \tan(\Delta\phi'_2)\right) - \sqrt{G_{2S}R_2}\,\sec(\Delta\phi'_2)}{R_2 \tan(\Delta\phi'_2) + X_2} + \frac{G_{3S}\left(R_3 - X_3 \tan(\Delta\phi'_3)\right) - \sqrt{G_{3S}R_3}\,\sec(\Delta\phi'_3)}{R_3 \tan(\Delta\phi'_3) + X_3} - B_1 \tag{15.17}$$

In certain cases, it might be desirable to include a shunt reactive element, typically a capacitor, at the common node of the network. However, in most cases, in order to minimize the size of the module, a shunt element is not required, in which case, in (15.17), $Y_{n3} = 0$.

The above equations allow all the desired element values for the coupler to be determined in terms of the two phase shifts $\Delta\Phi'_2$ and $\Delta\Phi'_3$, which are, in turn, dependent on the single phase-shift variable $\Delta\Phi_n$ ((15.10) and (15.11)). The design challenge is thus reduced to determining the appropriate value for $\Delta\Phi_n$. A solution for the phase shift must be found that is consistent with satisfying $Y_{n3} = 0$ in (15.17). Unfortunately, in general, a closed-form expression for the appropriate value of $\Delta\Phi_n$, to set the value of (15.17) to zero, is not normally attainable. Therefore, a solution must typically be obtained from a numerical approach. Fortunately, this is typically not challenging.

Having determined the appropriate element values for the design of a generalized three-port lattice coupler, in order to evaluate its capabilities, expressions must be derived for the key electrical circuit parameters. Separate equations are required to characterize the network, when employed as a splitter or as a power combiner.

For the analysis, complex impedances for the network elements are defined as follows, to account for resistive dissipation in the elements

$$Z_{n\#} = R_{n\#} + jX_{n\#} \qquad \text{and} \qquad Y_{n\#} = G_{n\#} + jB_{n\#} \tag{15.18}$$

where $R_{n\#}$ denotes a resistive component and $G_{n\#}$ denotes a conductance component.

Equations to determine the key electrical circuit parameters for the lattice, in both the splitter and combiner modes, are given below.

Lattice splitter mode: key electrical parameters

$$\frac{V_2}{V_1} = \frac{1}{1 + Z_{n1}\left(Y_{n1} + 1/Z_2\right)} \tag{15.19}$$

$$\frac{V_3}{V_1} = \frac{1}{1 + Z_{n2}\left(Y_{n2} + 1/Z_3\right)} \tag{15.20}$$

$$I_1 = \left(1/Z_{nl} + 1/Z_{n2}\right)V_1 - V_2/Z_{n1} - V_3/Z_{n2} \tag{15.21}$$

Lattice combiner mode: key electrical parameters

$$V_3 = \frac{-\left(\dfrac{V_{S1}}{Z_1\left(1 + Z_{n1}\left(1/Z_1 + Y_{n1}\right)\right)} + \dfrac{V_{S2}}{Z_2\left(1 + Z_{n2}\left(1/Z_2 + Y_{n2}\right)\right)}\right)}{\left(\dfrac{1}{Z_{n1}\left(1 + Z_{n1}\left(1/Z_1 + Y_{n1}\right)\right)} + \dfrac{1}{Z_{n2}\left(1 + Z_{n2}\left(1/Z_2 + Y_{n2}\right)\right)} - \dfrac{1}{Z_{n1}} - \dfrac{1}{Z_{n2}} - \dfrac{1}{Z_3}\right)} \tag{15.22}$$

$$V_1 = \frac{V_3 + \left(Z_{n1}/Z_1\right)V_{S1}}{\left(1 + Z_{n1}\left(1/Z_1 + Y_{n1}\right)\right)} \tag{15.23}$$

$$V_2 = \frac{V_3 + \left(Z_{n2}/Z_2\right)V_{S2}}{\left(1 + Z_{n2}\left(1/Z_2 + Y_{n2}\right)\right)} \tag{15.24}$$

15.2 Generalized Lattice Design Examples

In this section, some design examples are presented to illustrate the flexibility and performance capabilities of the generalized three-port lattice in various scenarios. Its use in both splitter and combiner modes is examined.

15.2.1 Splitter Example: Conventional Lattice-Balun

First, for a baseline, the design of a conventional lattice-balun splitter, in the configuration of Figure 15.4(a), is evaluated, having performance specifications fully compatible with those for the conventional lattice-balun configuration of Figure 15.1. The required electrical specifications are given in Table 15.1.

Table 15.1 Conventional Lattice-Balun Splitter Specifications

Design Specifications	
f_{lo} 1,710 MHz	f_{hi} 2,025 MHz
Phase shift	$\Delta\Phi = 180°$
Impedance—port 1	50Ω
Impedance—port 2	8Ω
Impedance—port 3	8Ω
Inductors, $Q_L = 60$	Capacitors, $Q_C \approx 100$

To satisfy these specifications, applying the formulae of Section 15.1.1, the required schematic for the conventional lattice balun splitter, is as shown in Figure 15.5.

The required element values are determined as listed in Table 15.2.

Using these element values, the input reflection coefficient of the splitter on port 1 is as shown in Figure 15.6(a). A corresponding plot of the complex input impedance over frequency is shown in Figure 15.6(b).

In Figure 15.7(a), the net insertion gain for the splitter is shown, together with the associated power split to the two output ports 2 and 3 in Figure 15.7(b). The insertion loss is observed to be <~0.3 dB across the full design bandwidth. In addition, Figure 15.7(a) shows a maximum insertion loss delta of ~0.05 dB between the two insertion gain plots. This indicates that the insertion loss is primarily due to dissipation in the lattice elements, rather than to mismatch.

Figure 15.7(b) shows a maximum coupling imbalance between the two outputs across the band ~±1.5 dB. At the center band, the insertion gain to each of the outputs is ~ −3.25 dB, where the 0.25-dB loss is again attributable to resistive dissipation in the elements.

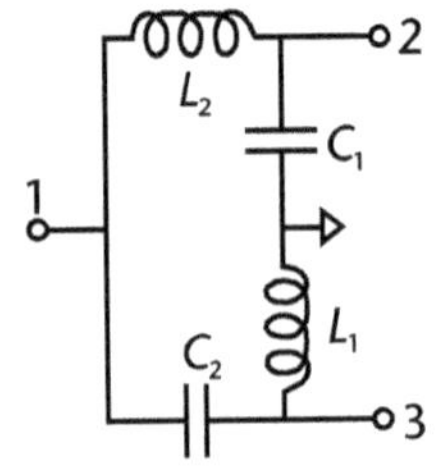

Figure 15.5 Conventional lattice-balun splitter schematic.

Table 15.2 Conventional Lattice-Balun Splitter Elements

Element Values	
$L_2 = 2.410$ nH	$C_1 = 3.013$ pF
$C_2 = 3.013$ pF	$L_1 = 2.410$ nH

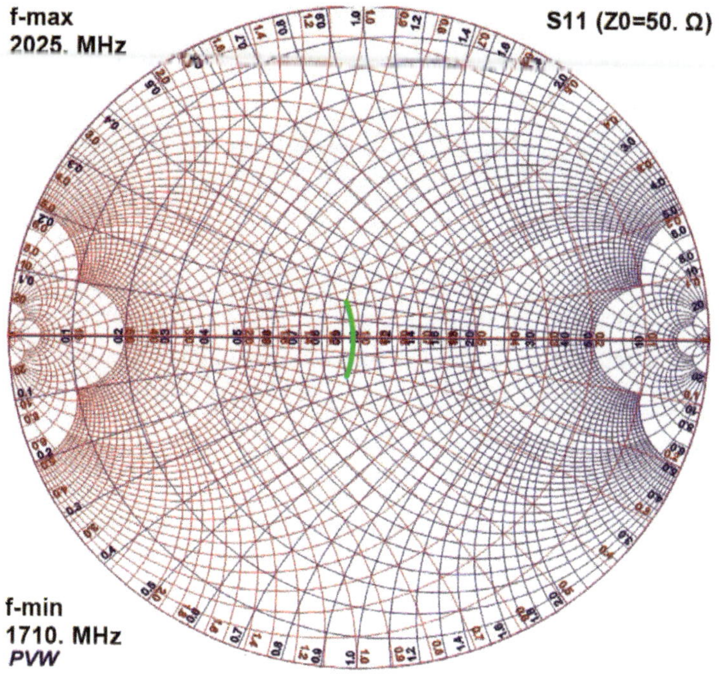

(a)

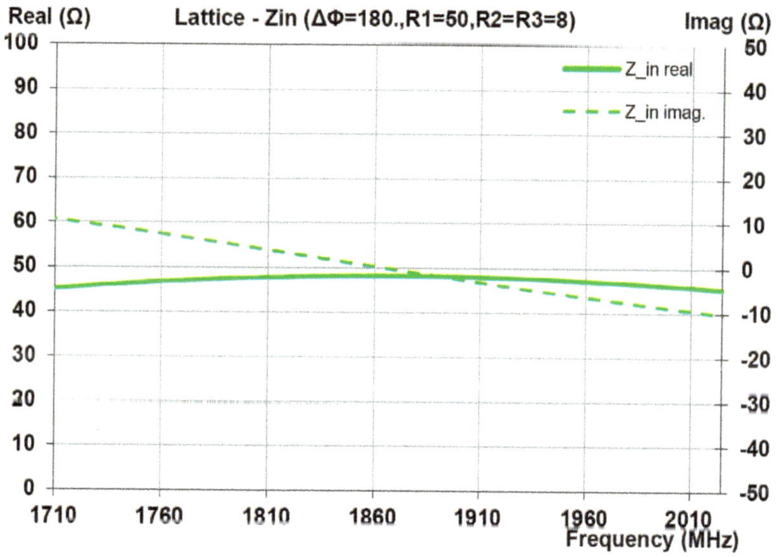

(b)

Figure 15.6 Conventional lattice-balun splitter impedance: (a) input reflection coefficient, and (b) input impedance.

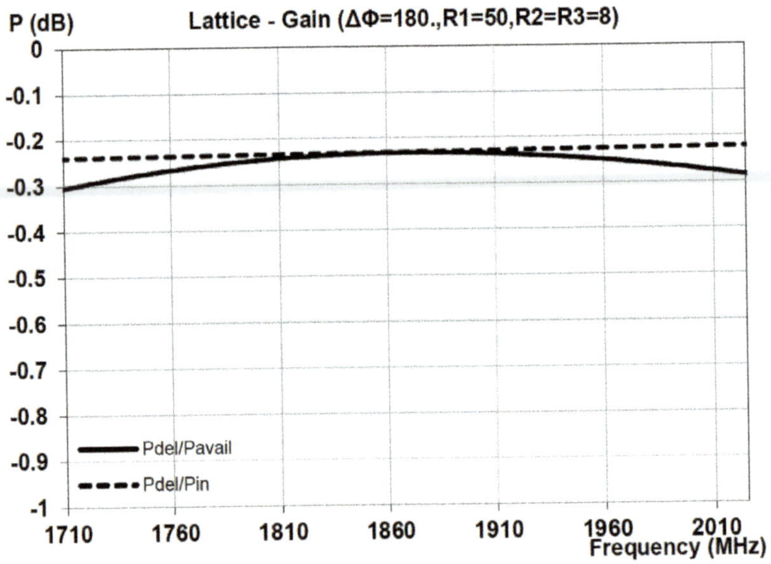

(a)

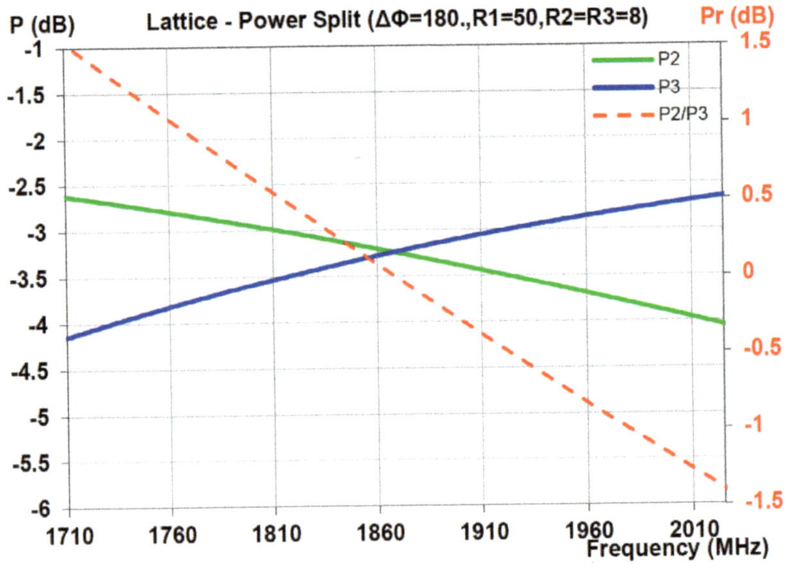

(b)

Figure 15.7 Conventional lattice-balun splitter gains: (a) net insertion gain, and (b) output power split.

Lastly, to complete the characterization of the key splitter performance characteristics, Figure 15.8 shows the insertion phases from port 1 to the two output ports 2 and 3.

The most important dependency is the relative phase shift $\Delta\Phi°$ between the two outputs, which is observed to be extremely flat across the full bandwidth, with a value of ~ −178.5°, very close to the differential target of 180°. Once more, the small phase deviation is attributable to resistive losses in the coupler elements.

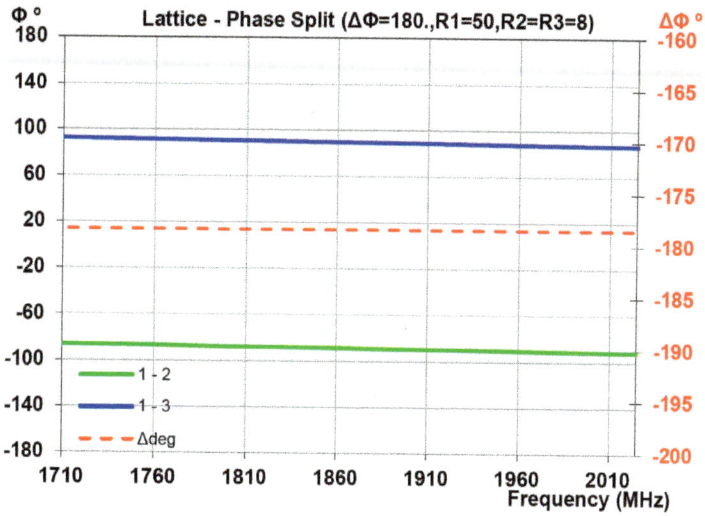

Figure 15.8 Conventional lattice-balun splitter insertion phase shifts.

15.2.2 Combiner Example: Conventional Lattice-Balun

In this section, again for a baseline, the design of the lattice-balun splitter of Section 15.2.1 is evaluated when used in the reverse mode as a power combiner. The equivalent combiner schematic is as shown in Figure 15.9.

Using the element values of Table 15.2, the input reflection coefficients of the combiner on the input ports 1 and 2, are shown in Figure 15.10(a). Corresponding plots of the complex input impedances over frequency are shown in Figure 15.10(b).

In Figure 15.11, the net insertion gain for the combiner is shown. The insertion loss is observed to be <~0.33 dB across the full design bandwidth. Figure 15.11 also shows a maximum insertion loss delta of ~0.1 dB between the two insertion gain plots. This indicates that the insertion loss is primarily due to dissipation in the lattice elements, rather than to mismatch.

Overall, the conventional lattice coupler/splitter has good wide-bandwidth characteristics for use as both a power splitter or combiner.

15.2.3 Splitter Example: Lattice-Balun with Complex Input Impedance

In practical circuits, a series capacitor is frequently required on the input port of a splitter, to function as a DC block. The presence of the capacitor results in the source impedance, as seen by the coupler on port 1, being complex.

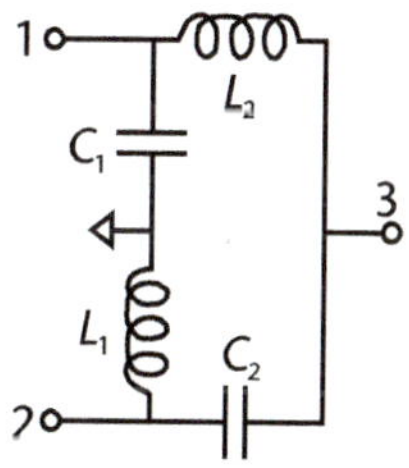

Figure 15.9 Conventional lattice balun combiner schematic.

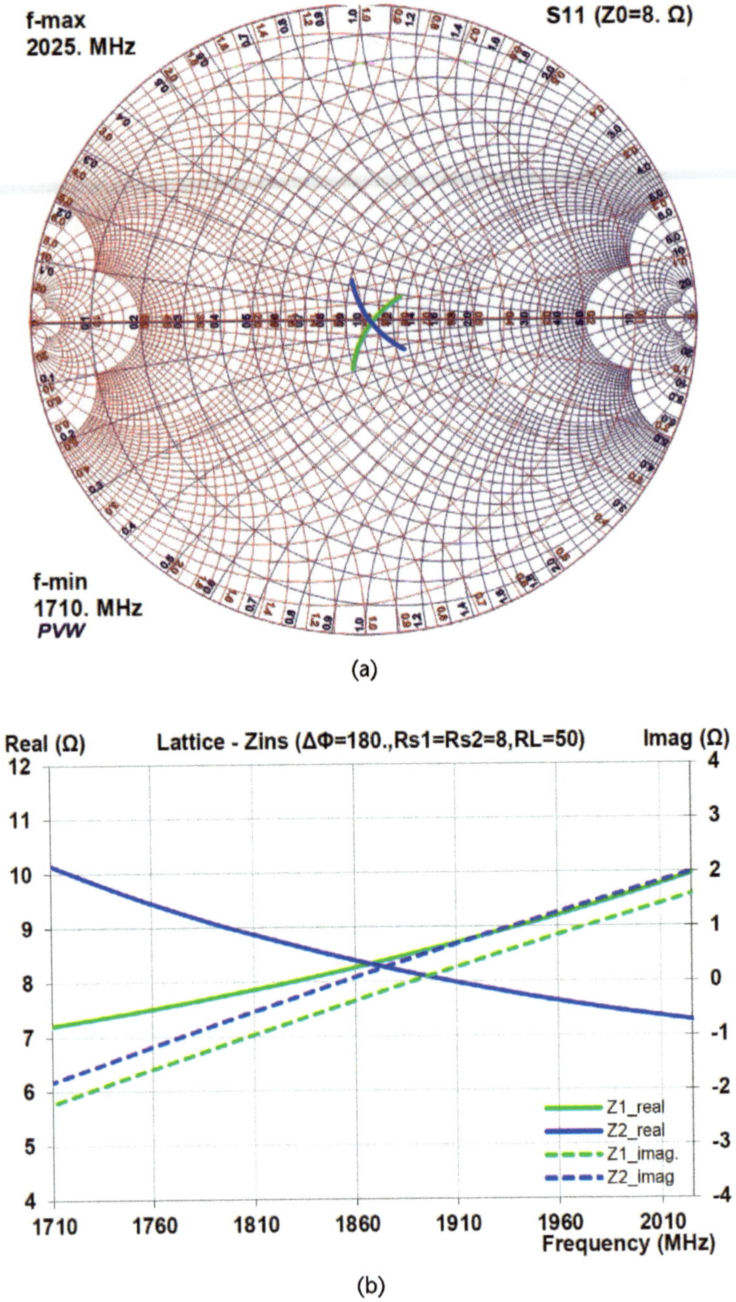

Figure 15.10 Conventional lattice-balun combiner impedances: (a) input reflection coefficients, and (b) input impedances.

In the conventional lattice-balun design, elements in the top and bottom arms are complex conjugates of each other. Therefore, with reference to the splitter of Figure 15.4(a), no residual shunt susceptance Y_3 is required at the input node. As such, the splitter presents a pure real input impedance at the port 1. If the source impedance on port 1 becomes complex, and the impedances on ports 2 and 3 remain equal and pure real, the network terminations still retain top-to-bottom symmetry. However,

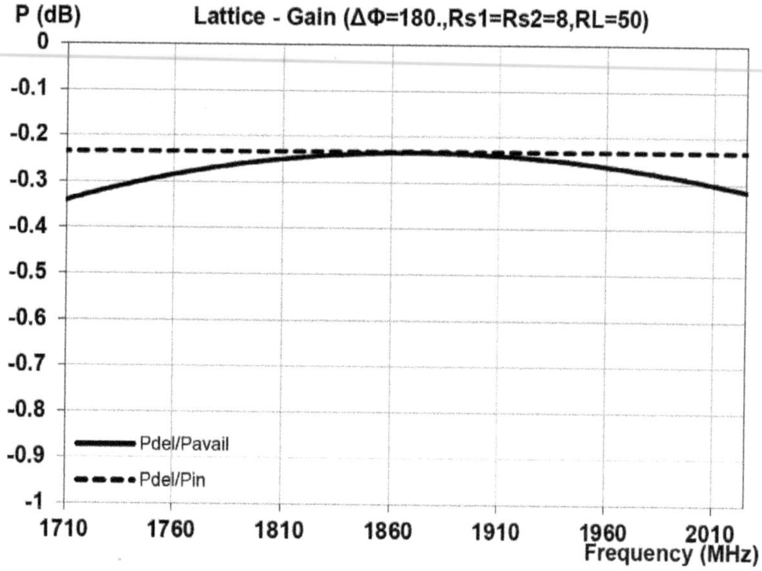

Figure 15.11 Conventional lattice-balun combiner gain.

in order to realize the required complex conjugate match on port 1, without the need for the shunt susceptance Y_3, in Figure 15.4(a), the complex conjugate duality of the two arms of the coupler can no longer be maintained.

The modified electrical specifications required for this example are given in Table 15.3. Note that the only difference between these specifications and those in Table 15.1 is the addition of the 22-pF series capacitor on port 1.

The splitter schematic to satisfy these specifications maintains the same form as in Figure 15.5, but with the modified element values listed in Table 15.4.

Note that these element values differ significantly from those for the conventional lattice-balun splitter in Table 15.2 without the series capacitor. The lattice with these element values is found to implement an insertion phase shift of −94.4° in the top path and 85.6° in the bottom path. This differs from the insertion phase shifts of ±90° in the conventional lattice-balun.

Using the element values of Table 15.4, the input reflection coefficient of the splitter on port 1, including the series capacitor, is shown in Figure 15.12(a). A

Table 15.3 Lattice-Balun Splitter Specifications with Complex Z_1

Design Specifications	
f_{lo} 1,710 MHz	f_{hi} 2,025 MHz
Phase shift	$\Delta\Phi = 180°$
Impedance—port 1	50Ω + series 22 pF
Impedance—port 2	8Ω
Impedance—port 3	8Ω
Inductors, $Q_L = 60$	Capacitors, $Q_C \approx 100$

Table 15.4 Lattice-Balun Splitter Elements with Complex Z_1

Element Values	
L_2 = 2.403 nH	C_1 = 3.841 pF
C_2 = 3.022 pF	L_1 = 3.297 nH

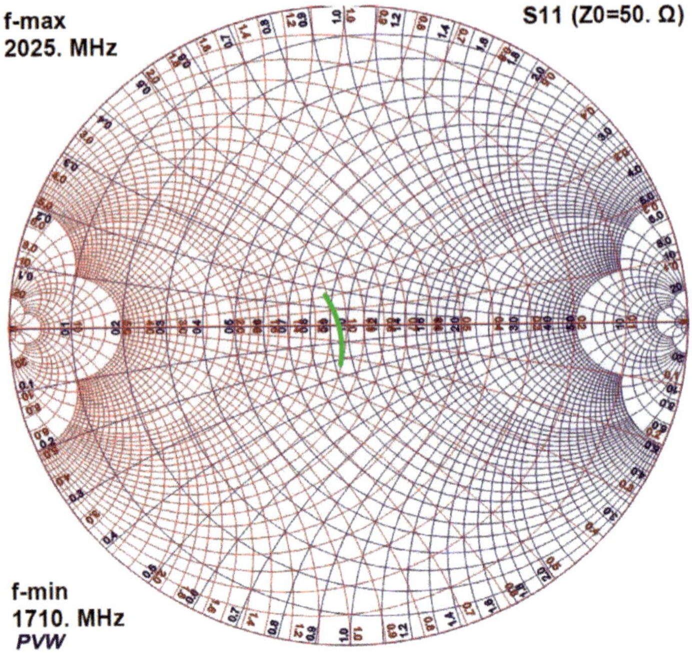

(a)

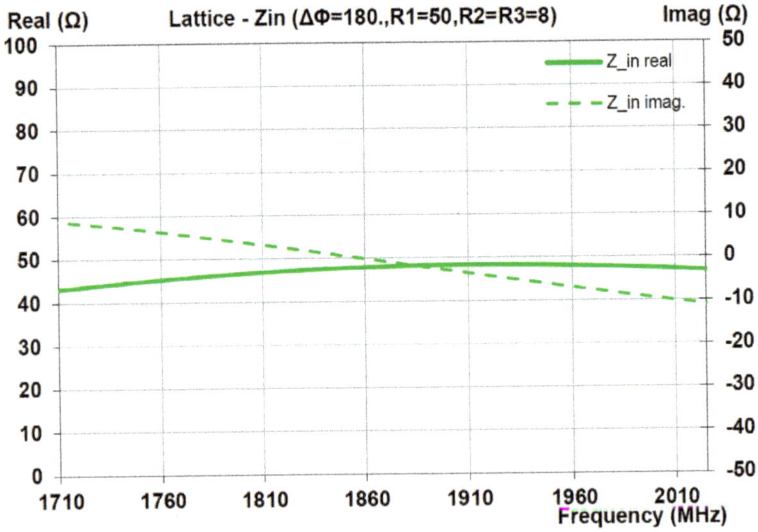

(b)

Figure 15.12 Lattice-balun splitter impedance with complex Z_1: (a) input reflection coefficient, and (b) input impedance.

corresponding plot of the complex input impedance over frequency is shown in Figure 15.12(b).

In Figure 15.13(a), the net insertion gain for the splitter is shown, together with the associated power split to the two output ports 2 and 3 in Figure 15.13(b). The insertion loss is observed to be <~0.3 dB across the full design bandwidth. In addition, Figure 15.13(a) shows a maximum insertion loss delta of ~0.06 dB between the two insertion gain plots.

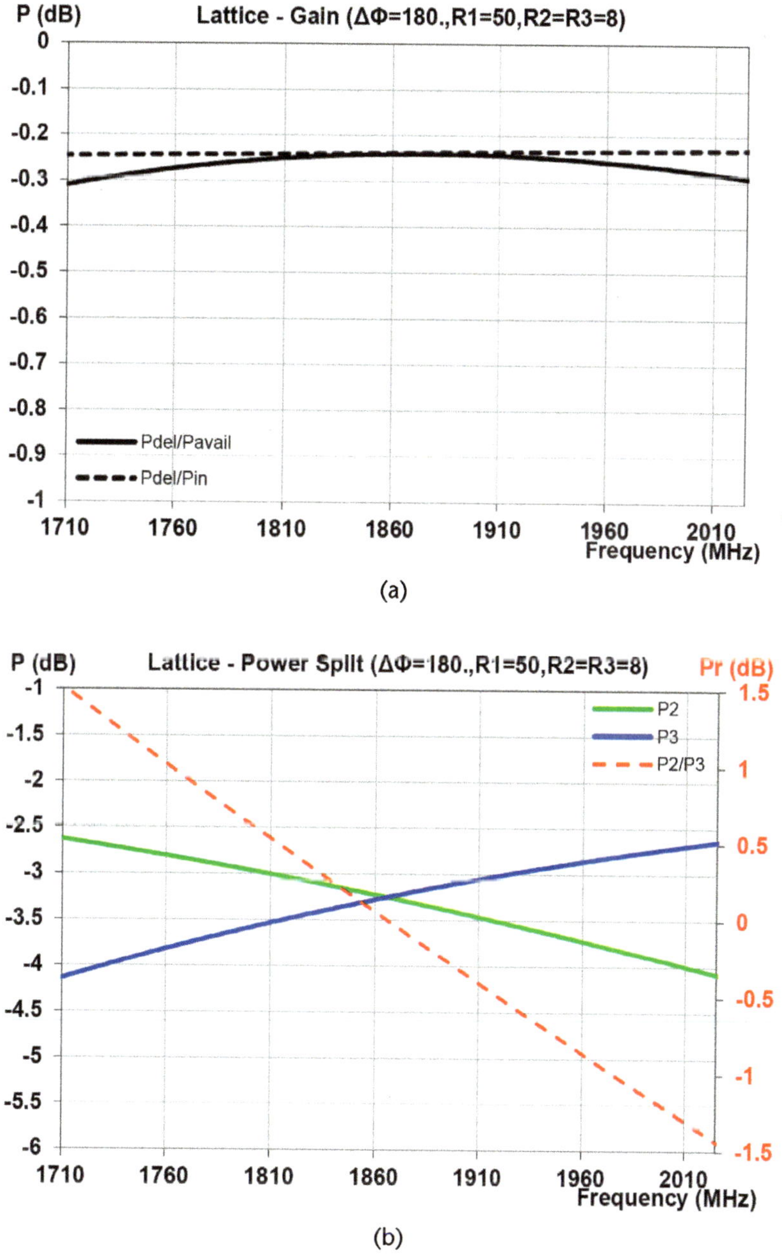

Figure 15.13 Lattice-balun splitter gains with complex Z_1: (a) net insertion gain, and (b) output power split.

Figure 15.13(b) shows a maximum coupling imbalance between the two outputs across the band ~ ±1.5 dB. At the center band, the insertion gains to each of the outputs is ~ −3.25 dB, where the 0.25-dB loss is again attributable to resistive dissipation in the elements.

Figure 15.14 shows the insertion phases from port 1 to the two output ports 2 and 3. The relative phase shift ΔΦ° between the two outputs is observed to be relatively flat, varying from ~ −178° to ~ −179° across the band. From these plots, it can be seen that the modified lattice splitter, designed in accordance with the equations of Section 15.1.1, correctly accommodates for the case of a complex input impedance. The splitter performance characteristics are little perturbed from those for the conventional lattice-balun with real impedances of Section 15.2.1.

15.2.4 Combiner Example: Lattice-Balun with a Complex Input Impedance

In this section, the design of the lattice-balun splitter of Section 15.2.3 is evaluated when used in the reverse mode as a power combiner. The coupler schematic remains that shown in Figure 15.9.

Using the element values of Table 15.4, the input reflection coefficients of the combiner on the input ports 1 and 2 are shown in Figure 15.15(a). Corresponding plots of the complex input impedances over frequency are shown in Figure 15.15(b).

In Figure 15.16, the net insertion gain for the combiner is shown. The insertion loss is observed to be <~0.34 dB across the full design bandwidth. In addition, Figure 15.16 also shows a maximum insertion loss delta of ~0.1 dB between the two insertion gain plots.

As with the splitter, the combiner performance characteristics are little perturbed from those for the conventional lattice-balun with real impedances of Section 15.2.2.

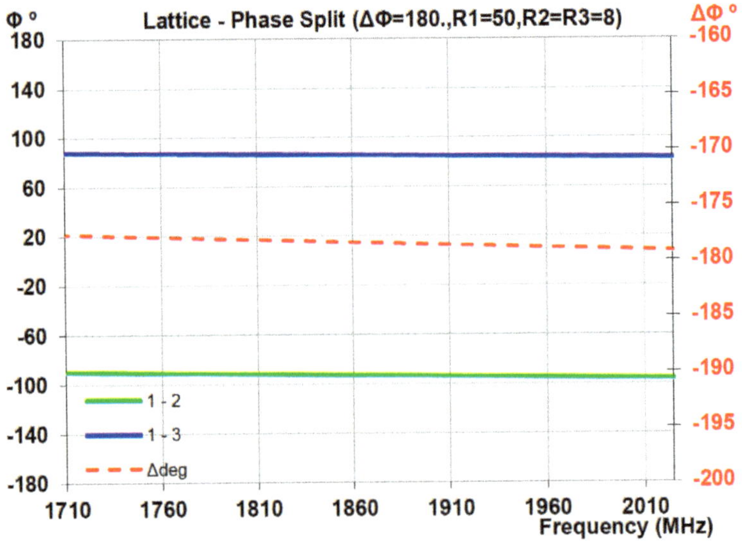

Figure 15.14 Lattice-balun splitter insertion phase shifts with complex Z_1.

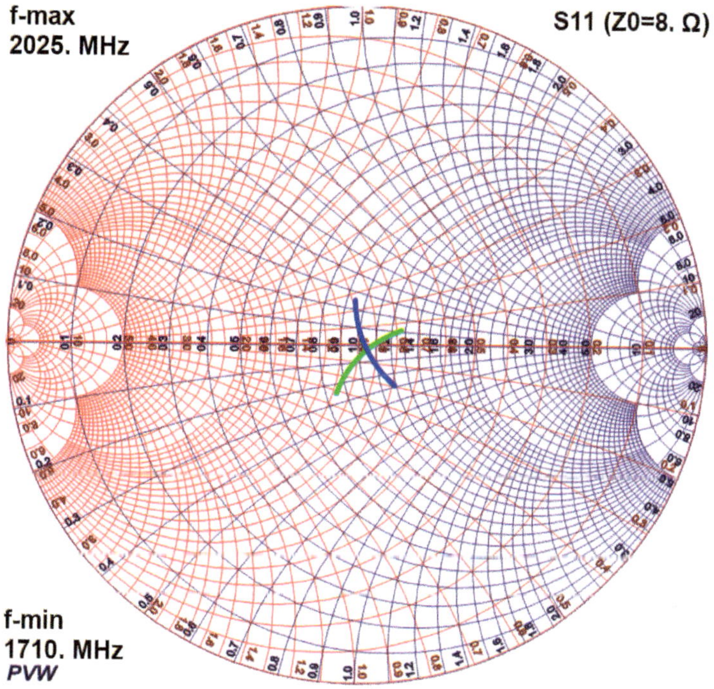

(a)

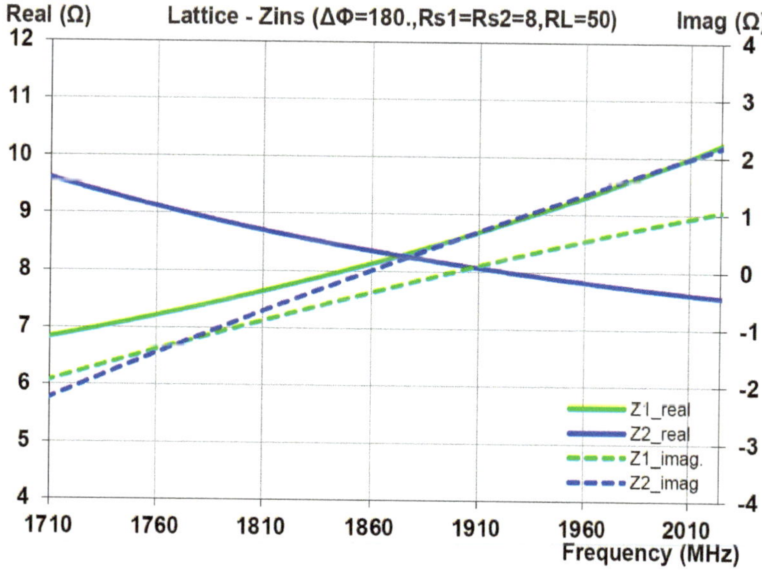

(b)

Figure 15.15 Lattice-balun combiner impedances with a complex load: (a) input reflection coefficients, and (b) input impedances.

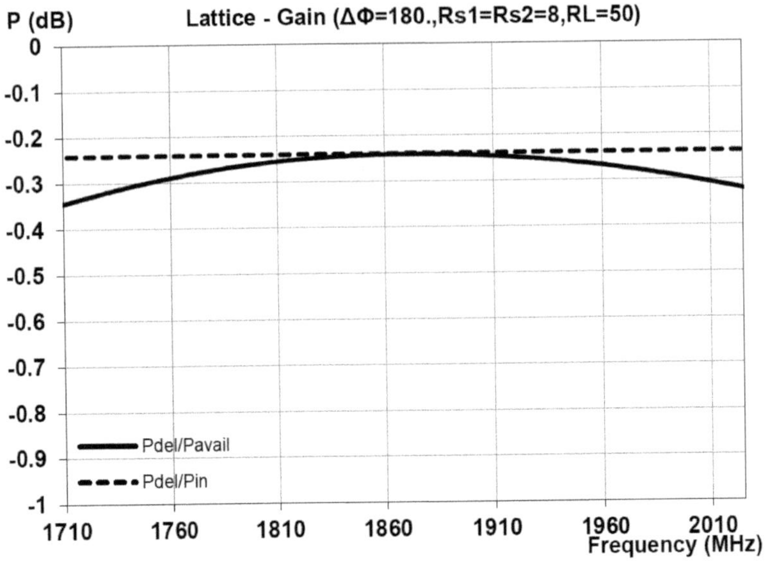

Figure 15.16 Lattice-balun combiner gain with a complex load.

15.2.5 Splitter Example: Lattice-Balun with a Complex Input Impedance and an Unequal Power Split

In many applications, a power splitter is required to implement an unequal power split of the output signals. The generalized lattice configuration is capable of achieving both arbitrary power and phase splitting. In this section, the design specifications of Section 15.2.3 are maintained, but with the additional requirement of a desired output power split ratio of 3 dB. With the series capacitor still included on port 1, a complex conjugate match is again required on port 1 of the lattice.

The full electrical specifications for this example are given in Table 15.3. Note that the specifications include both the 22-pF series capacitor on port 1 and an output power ratio of 3 dB.

In the conventional lattice-balun splitter schematic of Figure 15.5, and in the previous examples, the top-arm shunt element is a capacitor C_1, while the bottom-arm shunt element is an inductor L_1. However, in order to meet the required unbalanced

Table 15.5 Lattice-Balun Splitter Specifications with Complex Z_1 and 3-dB Power Split

Design Specifications	
f_{lo} 1,710 MHz	f_{hi} 2,025 MHz
Phase shift	$\Delta\Phi = 180°$
Output power ratio	3 dB
Impedance—port 1	50Ω + series 22 pF
Impedance—port 2	8Ω
Impedance—port 3	8Ω
Inductors, $Q_L = 60$	Capacitors, $Q_C \approx 100$

power split in Table 15.3, the lattice takes a different form, following with the design equations of Section 15.1.1. The modified configuration in this instance is shown in Figure 15.17, which shows also the input series capacitor C_S. For simplicity, the virtual elements are emitted in Figure 15.17.

The change from the previous configurations is that now the shunt elements in both arms are inductive. The shunt element C_1 in Figure 15.4 is replaced by the inductor L_3. The appropriate element values are listed in Table 15.6.

In this case, the lattice exhibits an insertion phase shift of −57.72° in the top arm and 122.28° in the bottom arm, far removed from the insertion phase shifts of ±90° in the conventional lattice-balun.

Using the element values of Table 15.6, the input reflection coefficient of the splitter on port 1, including the series capacitor, is shown in Figure 15.18(a). A corresponding plot of the complex input impedance over frequency is shown in Figure 15.18(b).

Figure 15.19(a) shows the net insertion gain for the splitter. The insertion loss is observed to be <~0.34 dB across the full design bandwidth. There is a maximum insertion loss delta of ~0.07 dB between the two insertion gain plots, again confirming the network incurs minimal mismatch loss across the full band.

These three plots are little changed from the corresponding ones in Section 15.2.3. This demonstrates that the modified lattice configuration is well suited to implementing wide power splits without significantly degrading the key performance characteristics of the splitter.

The associated power split between the two output ports 2 and 3 is shown in Figure 15.19(b). At the center of the band the network achieves almost exactly the 3-dB power split desired. Over the full bandwidth, the power ratio between the two outputs varies from ~1.5 dB to 4.8 dB.

Figure 15.20 shows the insertion phases from port 1 to the two output ports 2 and 3. The relative phase shift $\Delta\Phi°$ between the two outputs is observed to be slightly less flat than in the previous examples; however, it only varies from ~−176° to ~−180° across the band. This is likely to be more than adequate in most applications.

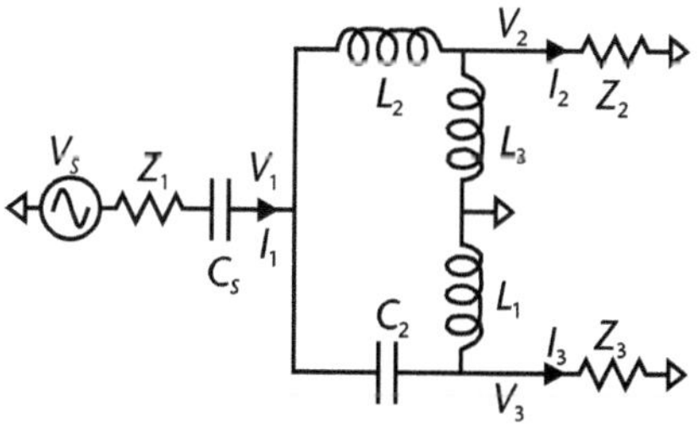

Figure 15.17 Lattice-balun splitter schematic with complex Z_1 and 3-dB power split.

Table 15.6 Lattice-Balun Splitter Elements with Complex Z_1 and 3-dB Power Split

Element Values	
L_2 = 1.766 nH	L_3 = 2.776 nH
C_2 = 2.912 pF	L_1 = 0.753 nH

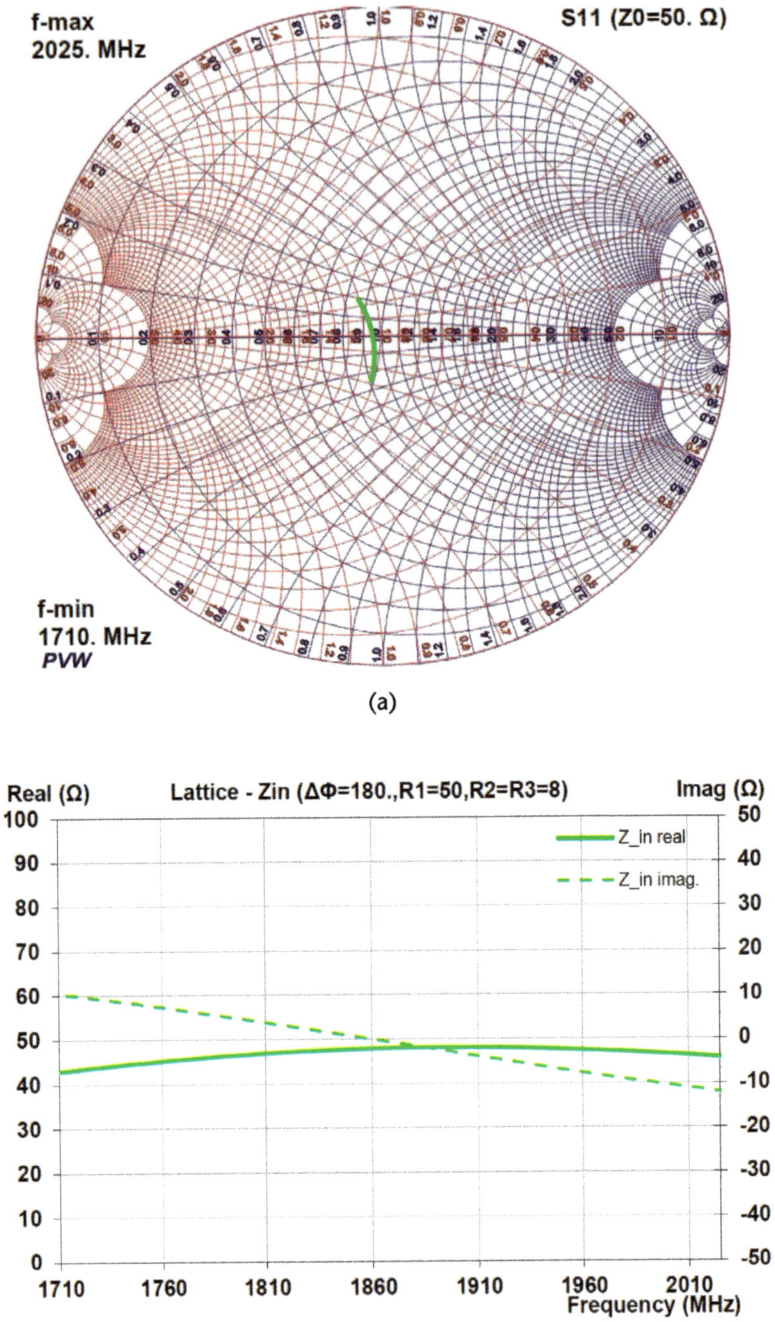

Figure 15.18 Lattice-balun splitter input impedance with complex Z_1 and 3-dB power split: (a) input reflection coefficient, and (b) input impedance.

This design example shows that the generalized modified lattice splitter architecture is very capable of accommodating a complex input impedance, simultaneously with a nonequal power split requirement. There is negligible degradation in the splitter performance characteristics from those of the conventional lattice-balun.

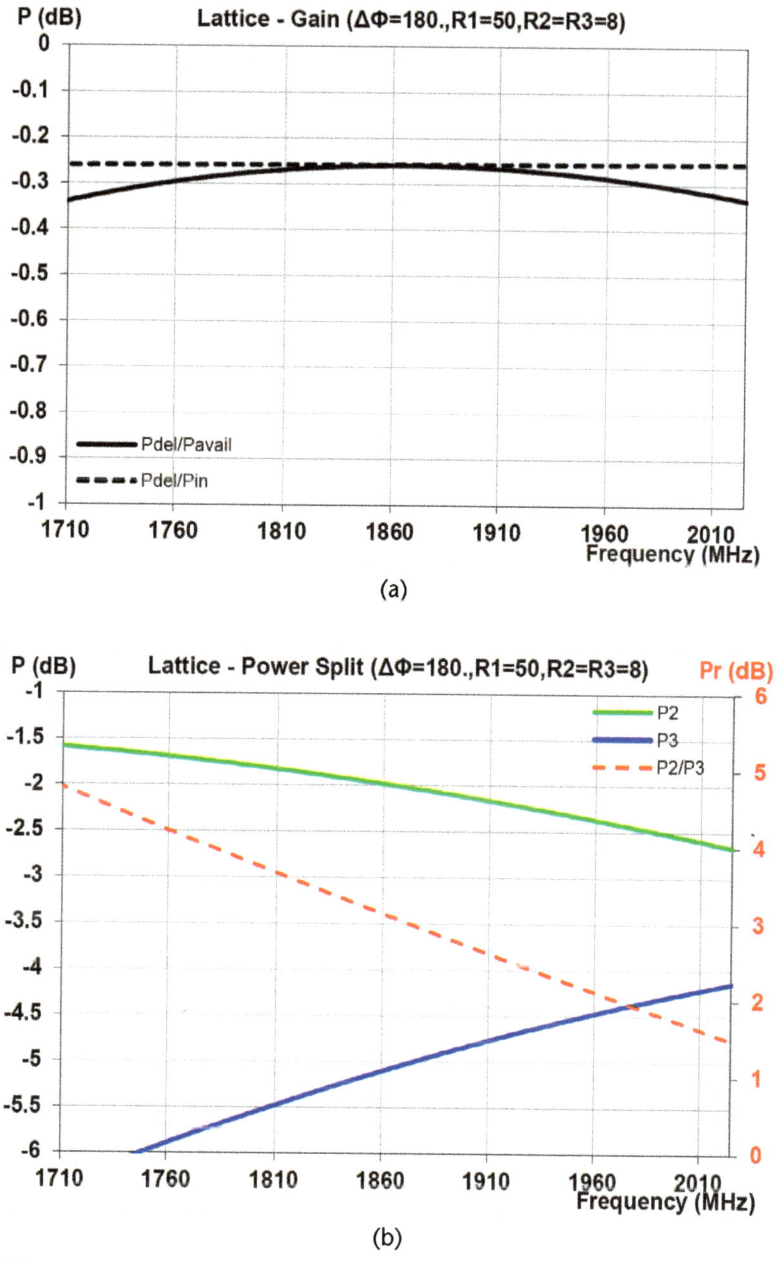

Figure 15.19 Lattice-balun splitter insertion gains with complex Z_1 and 3-dB power split: (a) net insertion gain, and (b) output power split.

15.2.6 Combiner Example: Lattice-Balun with a Complex Input Impedance and an Unequal Power Split

In this section, the design of the lattice-balun splitter of Section 15.2.5 is evaluated when used in reverse mode as a power combiner. Analogous with Section 15.2.5, the corresponding modified combiner schematic is shown in Figure 15.21.

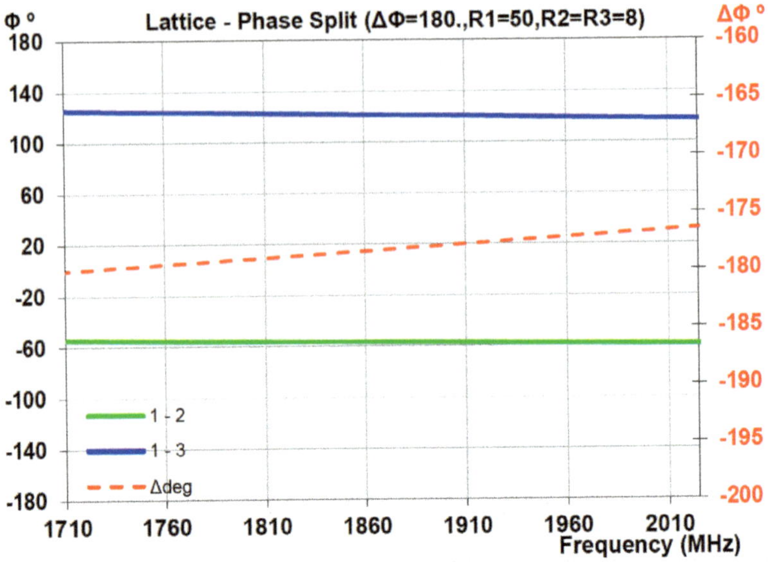

Figure 15.20 Lattice-balun splitter phase shifts with complex Z_1 and 3-dB power split.

Using the element values of Table 15.6, the input reflection coefficients of the combiner on the input ports 1 and 2 are shown in Figure 15.22(a). Corresponding plots of the complex input impedances over frequency are shown in Figure 15.22(b).

In this configuration, the real impedances on the two input ports track much more closely than do the imaginary components. This is a reversal of the impedance characteristics in Section 15.2.4 for the equal power combiner. Nonetheless, the magnitude of the reflection coefficients in both cases are very similar, resulting in no significant increase in insertion loss.

Figure 15.23 shows the net insertion gain for the combiner across the full band. The insertion loss is observed to be <~0.38 dB across the full design bandwidth. In addition, Figure 15.23 also shows a maximum insertion loss delta of ~0.1 dB between the two insertion gain plots. These characteristics are almost identical to those of the combiner in Section 15.2.4.

As with the splitter, the generalized lattice combiner architecture is very capable of accommodating a complex load impedance, together with unequal power inputs. There is negligible degradation in the combiner performance compared with the conventional lattice-balun combiner.

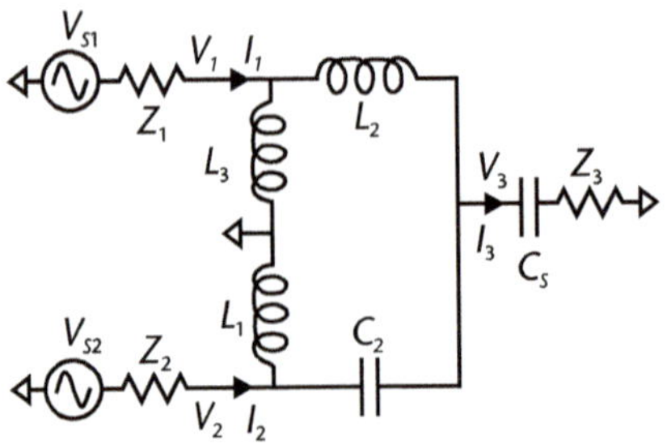

Figure 15.21 Lattice-balun combiner schematic with a 3-dB power split and a complex load.

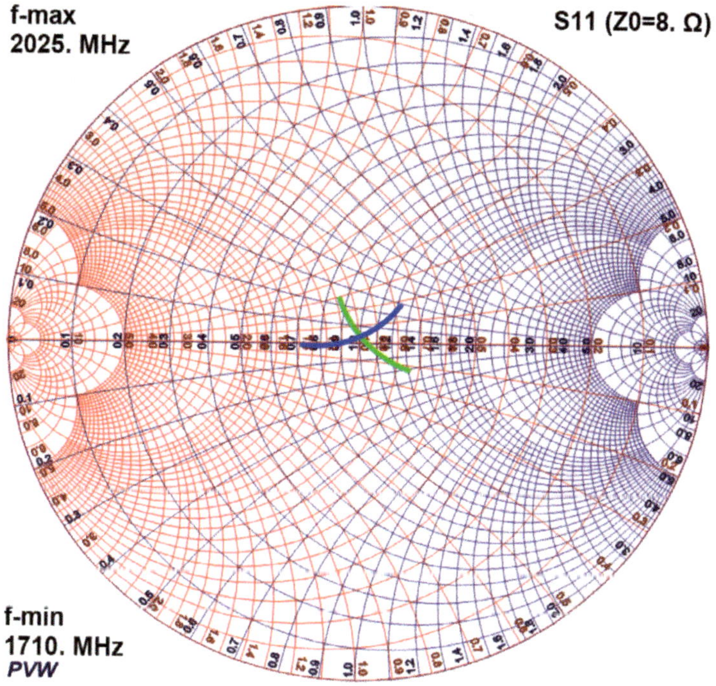

(a)

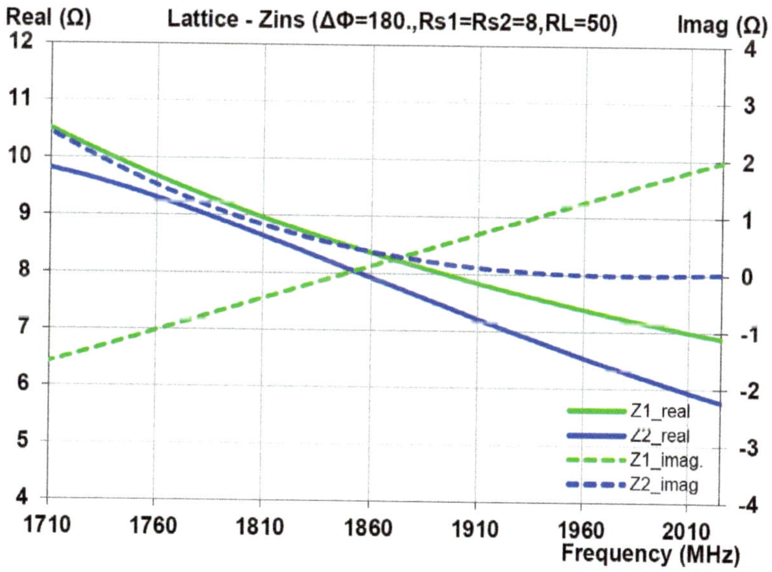

(b)

Figure 15.22 Lattice-balun combiner impedances with a 3-dB power split and a complex load: (a) input reflection coefficients, and (b) input impedances.

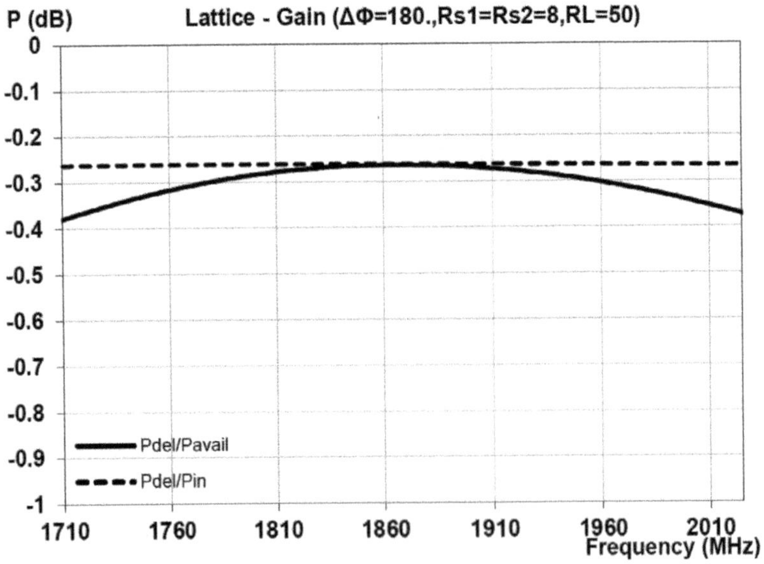

Figure 15.23 Lattice-balun combiner gain with a 3-dB power split and a complex load.

15.2.7 Splitter Example: Lattice with Multiple Mixed Specifications

This last splitter example is given to exemplify the broad capabilities of the lattice coupler. In the previous examples, the lattices were all designed to affect a differential signal splitting or combining. However, the generalized lattice architecture is capable of splitting or combining signals with an arbitrary phase relationship. To illustrate, in this example, a phase split $\Delta\Phi = 90°$ is specified. Thus, the lattice no longer performs as a balun.

For this example, a series capacitor C_S is again included on the input port, maintaining a complex impedance on port 1. As in the previous splitter example, an output power split of 3 dB is also specified. However, an additional complexity, in this example, is that the impedances on the two output ports are no longer identical. The detailed design specifications for the splitter are listed in Table 15.7.

The requirement for a quadrature phase split, in addition to the 3-dB power split, results in a further architectural change to the configuration of Figure 15.17.

Table 15.7 Lattice Splitter Multiple Mixed Specifications

Design Specifications	
f_{lo} 1,710 MHz	f_{hi} 2,025 MHz
Phase shift	$\Delta\Phi = 90\Omega$
Output power ratio	3 dB
Impedance—port 1	50Ω + series 22 pF
Impedance—port 2	8 + 2jΩ
Impedance—port 3	8°
Inductors, $Q_L = 60$	Capacitors, $Q_C \approx 100$

In order to satisfy the specifications in Table 15.7, the required lattice configuration takes the form shown in Figure 15.24. In this instance, both shunt element types are inverted relative to the conventional lattice-balun architecture of Figure 15.5.

The appropriate element values are listed in Table 15.8.

In this case, the lattice exhibits an insertion phase shift of −65.95° in the top arm and 38.08° in the bottom arm. Note that the insertion phase difference between the two arms ≠ 90° as there is an additional relative phase shift in the top arm due to the complex impedance on port 2.

Using the element values of Table 15.8, the input reflection coefficient of the splitter on port 1, including the series capacitor, is as shown in Figure 15.25(a). A corresponding plot of the complex input impedance over frequency is shown in Figure 15.25(b).

These dependencies are little changed from the previous lattice-balun splitter of Section 15.2.5, confirming that the lattice architecture is very capable of realizing the 90° phase shift with negligible impact on it input impedance.

Figure 15.26(a) shows the net insertion gain for the splitter. The insertion loss is observed to be <~0.27 dB across the full design bandwidth. The maximum insertion loss delta between the two insertion gain plots is again <0.07 dB, confirming that the network continues to exhibit only minimal mismatch loss across the full band. The insertion loss in this case is actually slightly less than that of the similar 180° splitter.

The associated power split between the two output ports 2 and 3 is shown in Figure 15.26(b). At the center of the band, the network achieves almost exactly the 3-dB power split desired. Over the full bandwidth, the power ratio between the two outputs varies from ~2.2 dB to 3.8 dB. This is also somewhat less than the coupling power ratio variation for the similar 180° splitter.

Figure 15.27 shows the insertion phases from port 1 to the two output ports 2 and 3.

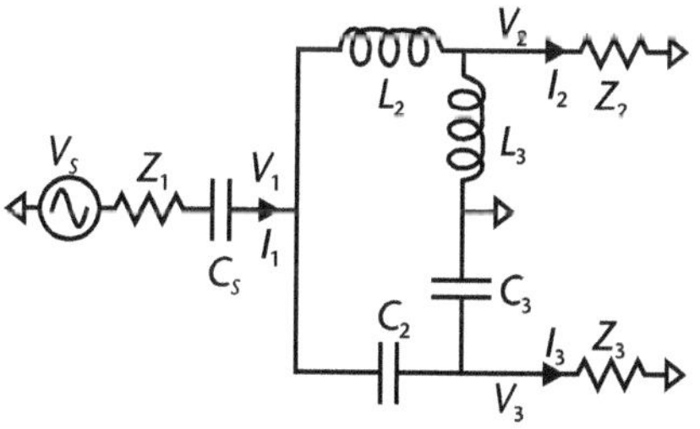

Figure 15.24 Lattice splitter schematic with multiple mixed specifications.

Table 15.8 Lattice Splitter Elements with Multiple Mixed Specifications

Element Values	
L_2 = 1.694 nH	L_3 = 6.830 nH
C_2 = 3.992 pF	C_3 = 9.603 nH

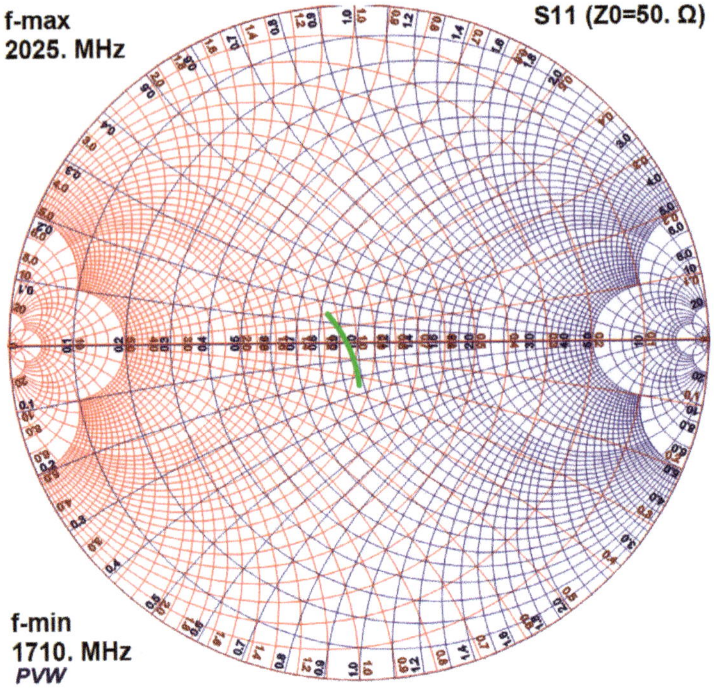

(a)

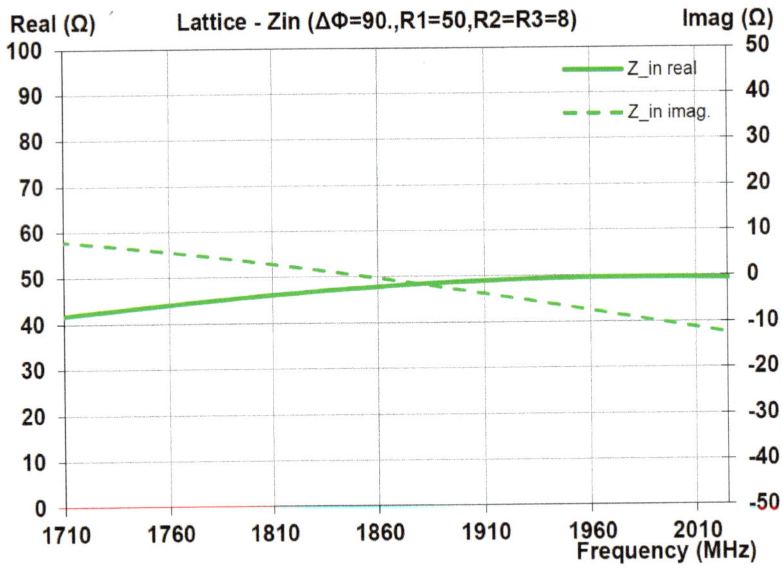

(b)

Figure 15.25 Lattice splitter impedance with multiple mixed specifications: (a) input reflection coefficient, and (b) input impedance.

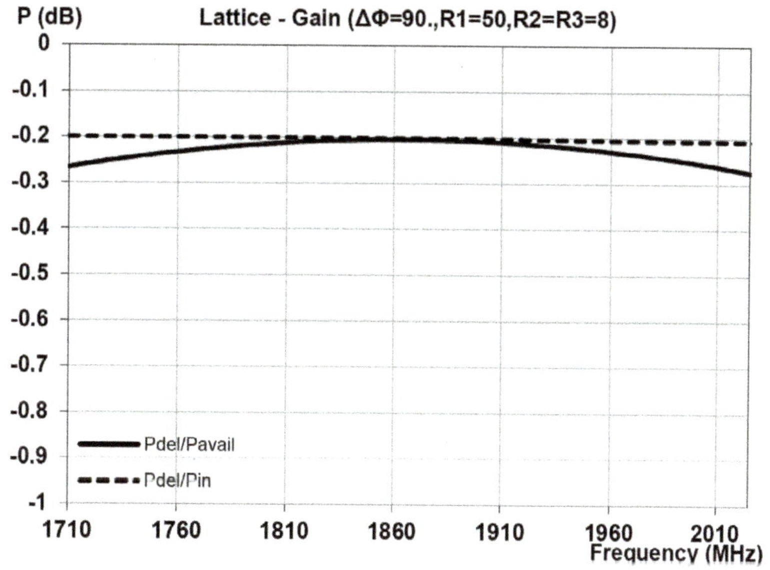

(a)

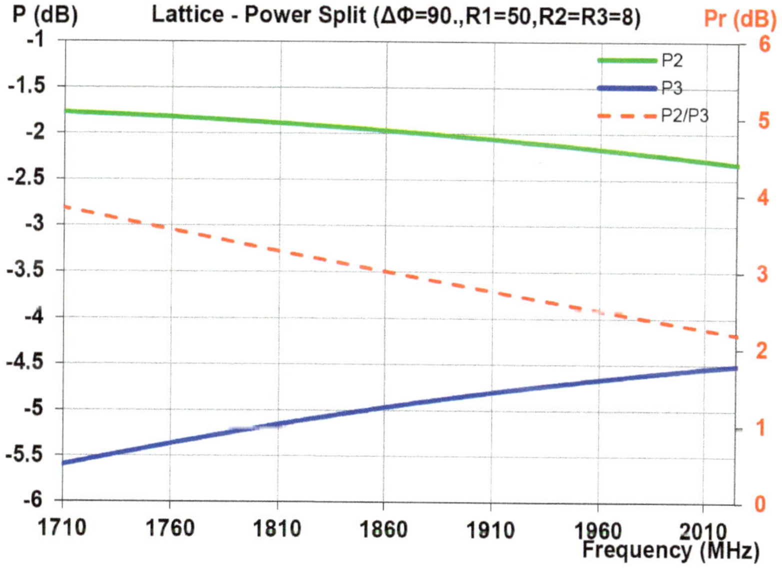

(b)

Figure 15.26 Lattice splitter insertion gains with multiple mixed specifications: (a) net insertion gain, and (b) output power split.

The relative phase shift $\Delta\Phi°$ between the two outputs is observed to be extremely flat and very close to the objective of 90°, varying from only ~ −89.5° to ~ −88.1° across the full band.

This design example demonstrates the wide capabilities of the generalized lattice splitter architecture. It offers excellent performance characteristics for power-splitting applications, regardless of source impedances, the power-splitting ratio, and the relative phase offset of the outputs. The splitter performs equally well under

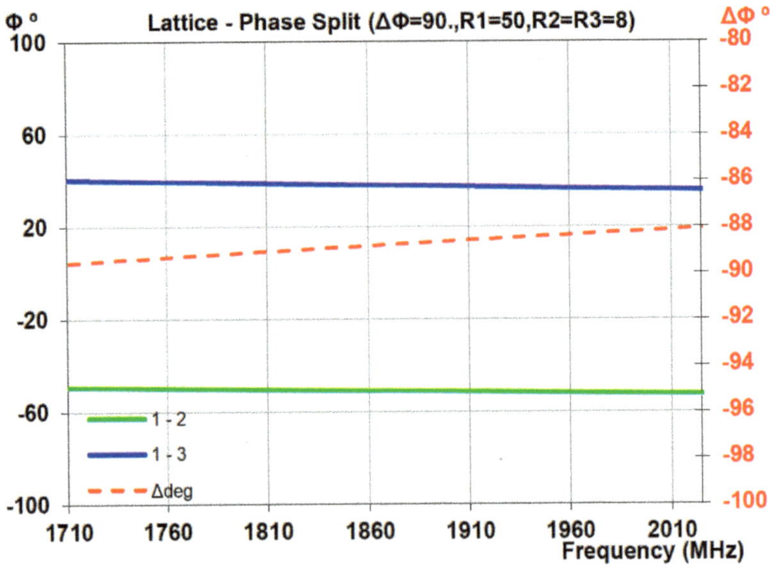

Figure 15.27 Lattice splitter insertion phase shifts with multiple mixed specifications.

all scenarios. In addition, the configuration is capable of meeting the varied specifications with only four components.

15.2.8 Combiner Example: Lattice with Multiple Mixed Specifications

In this section, the design of the mixed specifications lattice splitter of Section 15.2.7 is evaluated when used in reverse mode as a power combiner. Analogous with Section 15.2.7, the corresponding modified combiner schematic is shown in Figure 15.28.

Using the element values of Table 15.8, the input reflection coefficients of the combiner on the input ports 1 and 2 are shown in Figure 15.29(a). Corresponding plots of the complex input impedances over frequency are shown in Figure 15.29(b).

Note that the input impedance on port 1 is capacitive, consistent with a conjugate match to the inductive load. In this configuration, the real impedances on the two input ports vary across the band from ~6.5Ω to ~10Ω. The magnitude of the reflection coefficients on both ports are relatively low, resulting in no significant increase in insertion loss.

Figure 15.30 shows the net insertion gain for the combiner across the full band. The insertion loss is observed to be <~0.29 dB across the full design bandwidth.

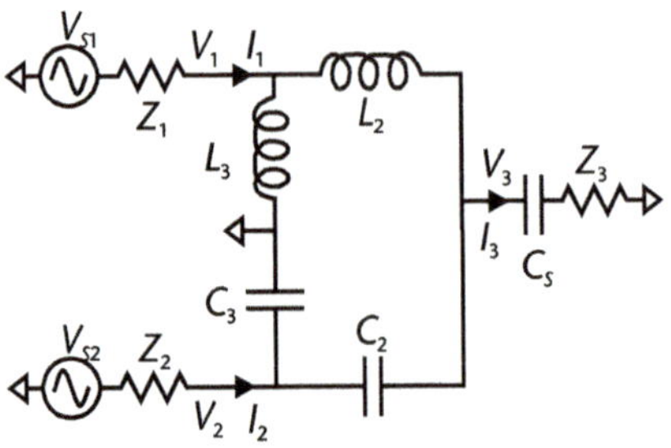

Figure 15.28 Lattice combiner schematic with multiple mixed specifications.

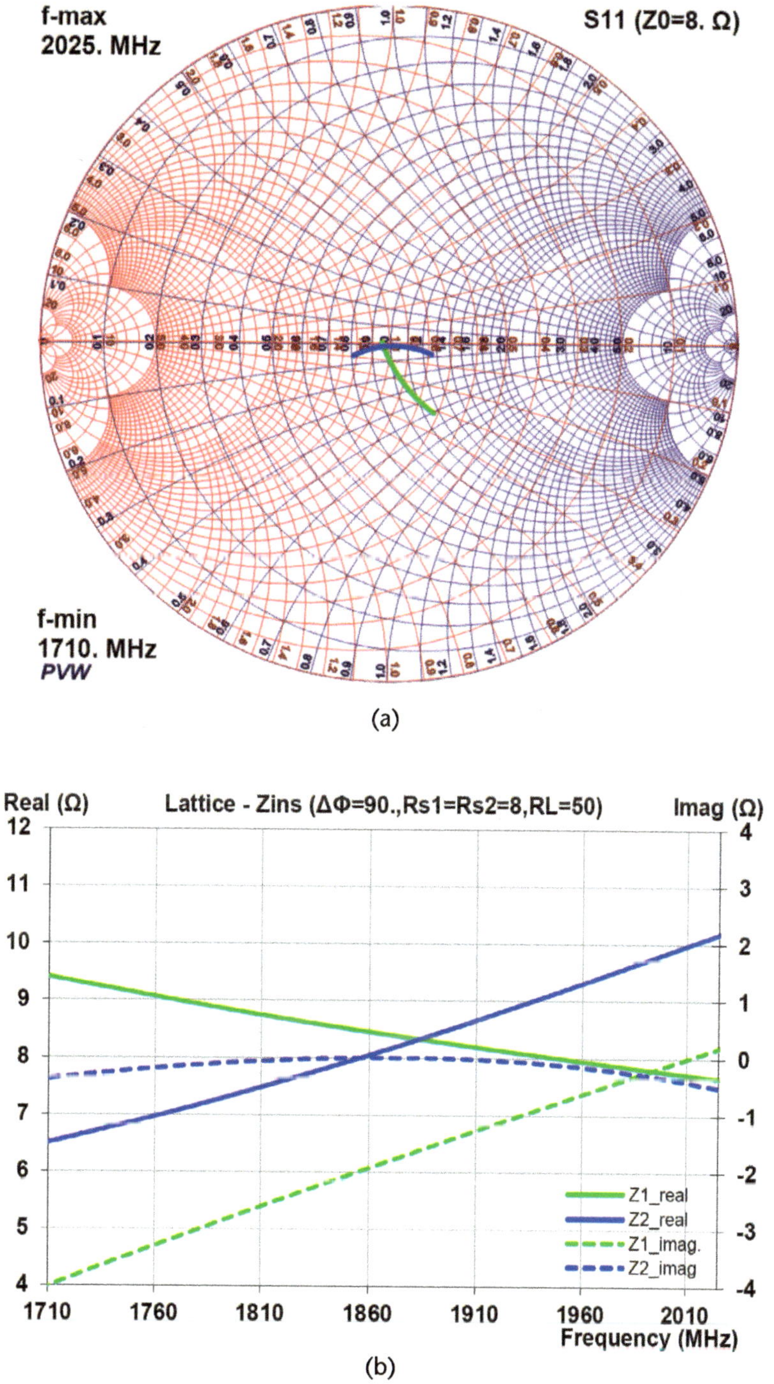

Figure 15.29 Lattice combiner impedances with multiple mixed specifications: (a) input reflection coefficients, and (b) input impedances.

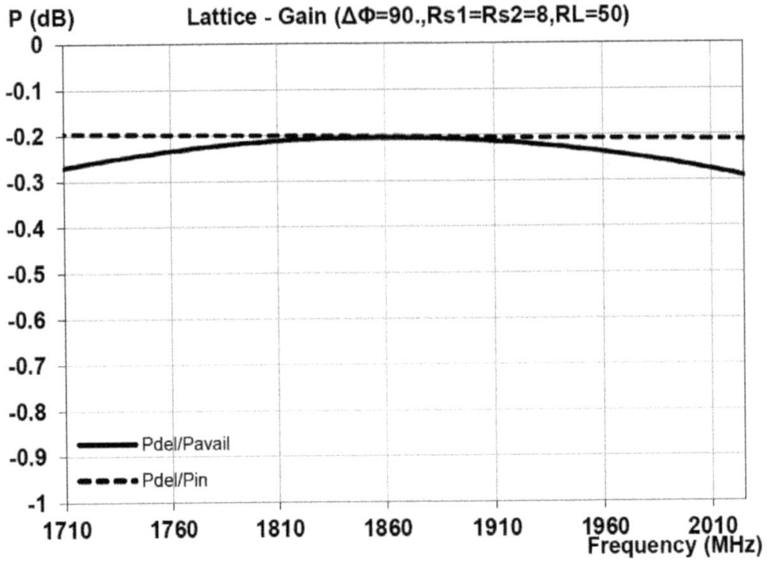

Figure 15.30 Lattice combiner gain with multiple mixed specifications.

In addition, Figure 15.30 also shows a maximum insertion loss delta of <0.07 dB between the two insertion gain plots. This is further confirmation that the combiner manifests only minimal insertion loss due to mismatch across the band.

As with the splitter, the generalized lattice combiner architecture maintains excellent performance characteristics with widely varying specifications. As with the splitter scenario, only four elements are required to meet the disparate design requirements.

15.2.9 Limited Case for Lattice with Additional Shunt Susceptance

In the previous sections, it was shown that the generalized lattice can be designed to meet multiple complex boundary conditions without the need for a shunt susceptance at the common node. The question arises as to whether improved performance for the lattice might be obtained if a finite value for the shunt element were allowed for in the design. The requirement for such an element can be generated by adjusting the relative phase shift between the two arms of the lattice according to (15.17). In all cases but one, the answer is that generally no significant performance improvement can be obtained by inclusion of such an element. To understand why, consider the insertion loss data shown in Figure 15.31.

Figure 15.31 is a composite of the lowpass and highpass π-network phase shifts in Figures 11.11(c) and 11.13(c); plotted as a function of $|\Delta\Phi|$, the net relative phase shift between the two arms. The assumption in the figure is that the magnitude of the phase shifts in each arm is equal. It can be seen that, across the range of typical values for $|\Delta\Phi|$, the insertion gain in each of the arms are close to the same. This implies that, if the phase shift in one arm was increased and the other necessarily decreased, no improvement in insertion gain could be expected. In consequence, the inclusion of a shunt element at the common node of the lattice generally offers no benefit.

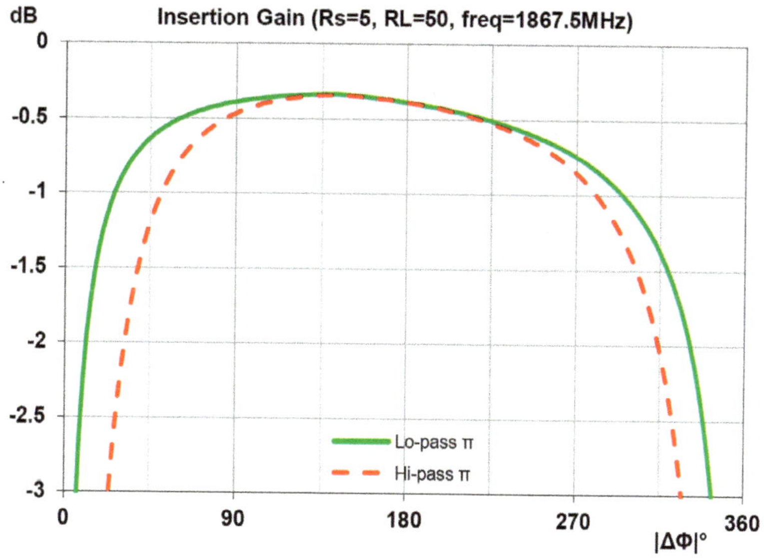

Figure 15.31 Lowpass and highpass π-network insertion gains versus phase shift.

There is one exceptional case in which the inclusion of an additional shunt element in the lattice network is required. That is when the coupler is required to implement specified insertion phases in each of the two arms. Adjusting the individual phase shifts, while retaining their desired offset, is the basis described in Section 15.1.1 for eliminating the undesired additional component. By specifying the individual phase shifts, this degree of freedom is removed and an additional shunt element may be required. In practice, for the composite lattice coupler of Figure 15.4, this is seldom a useful option as elucidated below.

For the conventional differential lattice-balun of Section 15.2.2, the phase shift in the upper arm is $\Delta\Phi_1 = -90°$, while that in the lower arm is $\Delta\Phi_2 = +90°$. Suppose that a constraint is added requiring $\Delta\Phi_1 = -120°$ and $\Delta\Phi_2 = +60°$. The reconfiguration of the coupler required to meet these constraints is shown in Figure 15.32.

The insertion gain for the conventional differential coupler design of Figure 15.9, previously shown in Figure 15.11, is duplicated in Figure 15.33(a) for comparison. Next to it, in Figure 15.33(b), the insertion loss is shown for the modified lattice coupler of Figure 15.32.

There is a noticeable increase in insertion loss for the coupler architecture required to meet the predefined phase shifts in the two arms. The reason for this can

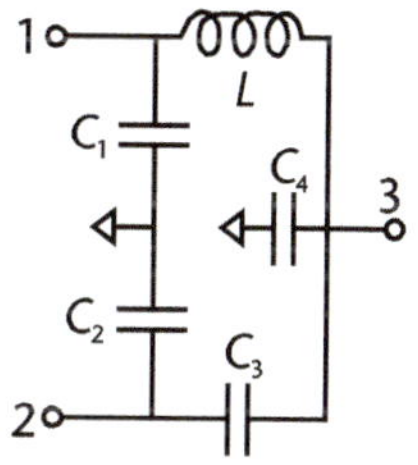

Figure 15.32 Lattice-balun combiner with defined $\Delta\Phi$'s schematic with the element values listed in Table 15.9.

Table 15.9 Lattice-Balun Combiner with Defined ΔΦ's Elements

Element Values	
$L = 2.088$ nH	$C_1 = 9.630$ pF
$C_2 = 2.671$ pF	$C_3 = 3.479$ pF
$C_4 = 0.984$ pF	

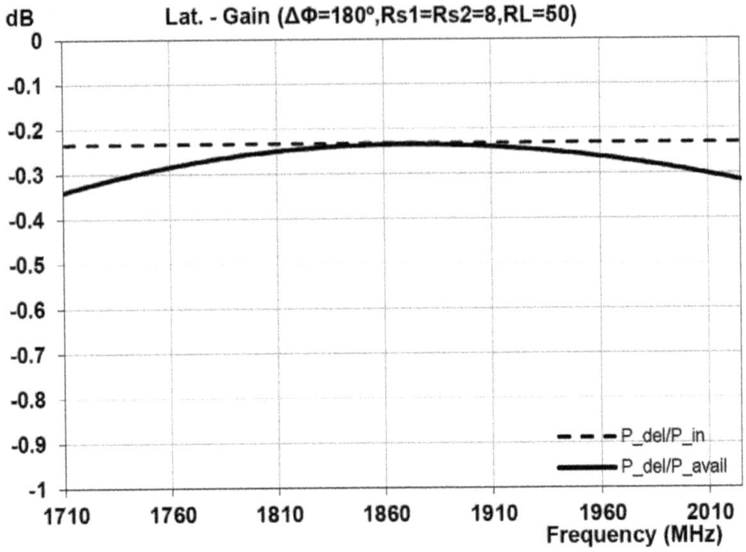

(a)

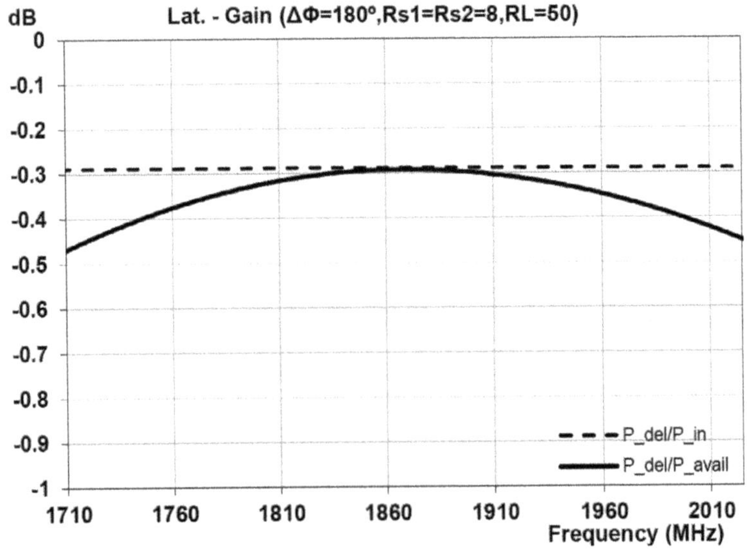

(b)

Figure 15.33 Comparison of conventional and modified lattice gains: (a) conventional lattice, and (b) lattice with defined ΔΦ.

be ascertained by reference to Figures 11.11(c) and 11.13(c). The former illustrates that increasing the insertion phase from −90° to the required −120° in the upper lowpass π-network is expected to result in a significant decrease in insertion gain for the section. On the contrary, Figure 11.13(c) shows that decreasing the insertion phase from +90° to the +60° in the lower highpass π-network is expected to result in higher insertion gain for that arm. However, this does not adequately compensate for the decreased insertion gain in the lowpass arm. In general, the greater the required difference in the magnitude of the phase offset in the two arms, the greater the insertion loss penalty.

If a significant difference in the magnitude of the phase offset between the two arms is required, the six-element lattice described in the following section is generally the preferred option.

15.3 Development of Six-Element Lattice Coupler

In the previous sections, it was described how two complementary π-networks could be combined to achieve a low-loss, three-port splitter/combiner architecture, typically referred to as a lattice. In combining the two networks, the two common-node elements can generally be eliminated. Thus, the overall configuration comprises only four elements, rather than six elements of the individual π-networks (Figure 15.2).

The question arises as to whether a lattice coupler with improved characteristics might be constructed from a combination of two T-networks, such as described in Section 11.6, rather than the two π-networks. An obvious disadvantage is that since a T-network comprises series elements at both ends, the opportunity for two of the elements to mutually cancel does not arise. Therefore, a lattice comprising two T-networks will necessarily require six elements. However, it is the focus of this section to demonstrate the many advantages that can accrue from the more complex dual T-based coupler network.

The four-element lattice coupler was constructed from a combination of two π-networks having conjugate shunt elements at the combining node. Schematics for the synthesis models used to design such a splitter and combiner are shown in Figures 15.2 and 15.3. As the elements at the combining node are self-conjugate, their physical representations are not required in the resulting coupler. Despite the absence of these virtual elements in the physical coupler configuration, their presence in the models used for synthesis significantly aids in the development of the design equations for the coupler.

Seeking to profit from this approach, a similar pair of virtual shunt elements is included in the synthesis models proposed for a dual T-based coupler. The synthesis models for the T-based couplers are shown in Figure 15.34. All components of the couplers are ideally pure reactive.

As with the π-based coupler designs, the shunt susceptive elements jB_V and $-jB_V$ will not be required in the physical implementation and are only present to aid in the synthesis. The splitter and combiner couplers can operate reciprocally, given the same boundary conditions; thus, without loss of generality, it is sufficient to develop the design equations for only one. We shall consider the design approach for the combiner coupler schematic shown in Figure 15.34(b).

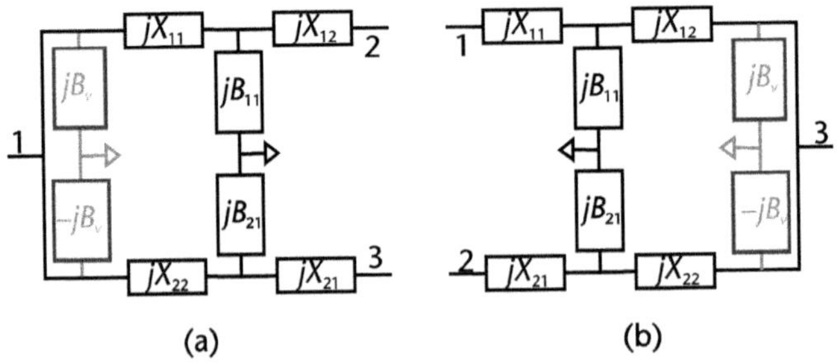

Figure 15.34 Six-element lattice configurations: (a) splitter schematic, and (b) combiner schematic.

To develop a generally applicable synthesis approach for the six-element coupler, as in Section 15.1.1, the power at the two paired input or output ports will be assumed unequal, having a power ratio defined by

$$dB_R = 10\log(P_1/P_2) \tag{15.25}$$

If the external load impedance at the combining node, port 3, is

$$Z_L = R_L + jX_L \tag{15.26}$$

then the corresponding external load admittance at port 3 is

$$Y_L = 1/Z_L = G_L + jB_L \tag{15.27}$$

where G_L is the load conductance and B_L is the load susceptance.

It follows that the effective loading conductance on the upper track 1, at the combining node, must be

$$G_{1L} = \frac{G_L}{1+10^{-dB_R/10}} \tag{15.28}$$

and the effective loading conductance on the lower track 2, at the combining node, must be

$$G_{2L} = \frac{G_L}{1+10^{dB_R/10}} \tag{15.29}$$

The most straightforward approach to synthesizing the appropriate values for the *T*-network elements would be to set the loading admittances on the two tracks at the combining node as

$$Y_{1L} = G_{1L} + jB_L/2 \tag{15.30}$$

and

$$Y_{2L} = G_{2L} + jB_L/2 \tag{15.31}$$

Referring to Figure 15.34(b), these equalities imply $B_V = 0$; thus, the virtual elements are excluded from the network. A valid configuration for the coupler would be obtained using these equalities, but, in general, the resulting network would be far from optimal in performance. To implement a coupler with more optimal performance characteristics, finite values for B_V must be evaluated.

In the case of the π-based lattice couplers, there was a unique value required for the virtual shunt susceptances in order to satisfy the boundary conditions. In such couplers, there were only three variables (i.e., elements) in each of the lattice arms of the conceptual models, as in Figure 15.34. For each arm, there are three boundary conditions to meet, namely, the real and imaginary parts of the impedance equation, plus the insertion phase through the network. Thus, only a unique solution is possible.

For the dual *T*-based couplers, each of the arms has the same three boundary conditions as with the π-based couplers. However, there are now four elements in each arm. Thus, the synthesis equations for the elements are underdetermined, implying an infinite set of solutions. These solutions can be obtained as a function of the virtual susceptance value B_V. By scanning through a range of values for B_V, a corresponding set of performance characteristics for the coupler may be obtained. The solution set that best optimizes one or more key performance parameters, such as insertion gain, input/output impedance, power balance, and maximum component value, can then be selected. In most cases, for synthesis, it is sufficient to assume that the magnitudes of the phase shifts through each arm are the same, that is,

$$|\Delta\phi_1| = |\Delta\phi_2| \tag{15.32}$$

although this is not required.

The *T*-based coupler synthesis equations are derived following the design equations for *T*-networks given in Section 11.5.

Equations are given in terms of the following variables

$$Z_n = R_n + jX_n \qquad Y_n = 1/Z_n = G_n + jB_n \tag{15.33}$$

$$Z_L = R_L + jX_L \qquad Y_L = 1/Z_L = G_L + jB_L \tag{15.34}$$

$$Y_{nL} = G_{nL} + jB_{nL} \qquad Z_{nL} = 1/Y_{nL} = R_{nL} + jX_{nL} \tag{15.35}$$

$$G_{1L} = \frac{G_L}{1 + 10^{-dB_R/10}} \qquad B_{1L} = B_V \tag{15.36}$$

$$G_{2L} = \frac{G_L}{1 + 10^{+dB_R/10}} \qquad B_{2L} = -B_V + B_L \tag{15.37}$$

Elemental parameters are determined by

$$B_{n1} = \frac{(B_n B_{nL} - G_n G_{nL})\sin\Delta\Phi_n - (B_n G_{nL} + G_n B_{nL})\cos\Delta\Phi_n}{\sqrt{G_n G_{nL}}} \tag{15.38}$$

$$X_{n1} = \frac{\left(1 - \sqrt{R_n / R_{nL} - (B_{n1} R_n)^2}\right)}{B_{n1} - X_n} \tag{15.39}$$

$$X_{n2} = \frac{\left(1 - \sqrt{R_{nL} / R_n - (B_{n1} R_{nL})^2}\right)}{B_{n1} - X_{nL}} \tag{15.40}$$

where $n = 1$ for the top arm and $n = 2$ for the bottom arm.

To compare the performance of the six-element lattice to the four-element lattice examined in detail earlier in the chapter, consider the target specifications for the four-element lattice coupler in Section 15.2.2, which are duplicated in Table 15.10.

The insertion loss and input impedances for the four-element combiner from Figure 15.11 and Figure 15.10(b) are repeated for comparison purposes in Figure 15.35.

The optimal configuration of the six-element lattice to meet the specifications of Table 15.12, in terms of maximum insertion gain, is shown in Figure 15.36.

The equivalent performance characteristics to those for the four-element lattice balun in Figure 15.35, are shown in Figure 15.37.

Both plots for the dual *T*-based combiner show noticeably improved characteristics over those for the four-element lattice combiner in Figure 15.35.

In the previous section, it was shown that imposing an offset in the magnitude of the insertion phases in the two arms of a four-element lattice results in an increased insertion loss for the coupler. This is shown in Figure 15.33. The increased loss was attributed to the steep drop-off in insertion gain of a π-network when the magnitude of the insertion phase moves outside the range

$$\sim 45^\circ \leq |\Delta\phi| \leq 90^\circ \tag{15.41}$$

T-networks have very similar insertion gain versus insertion phase characteristics to the π-networks, as can be seen from Figures 11.15(c) and 11.17(c). It might therefore be expected that the *T*-based lattice couplers would likewise suffer from increased insertion loss for insertion phases outside the range listed at the end of the previous paragraph. Fortuitously, this turns out not to be the case. The reason for this is that, rather than regarding each of the arms as comprising a single

Table 15.10 Target Six-Element Lattice Combiner Specifications

Design Specifications	
f_{lo} 1,710 MHz	f_{hi} 2,025 MHz
Phase shift	$\Delta\Phi = 180°$
Impedance—port 1	8Ω
Impedance—port 2	8Ω
Impedance—port 3	50Ω
Inductors, $Q_L = 60$	Capacitors, $Q_C \approx 100$

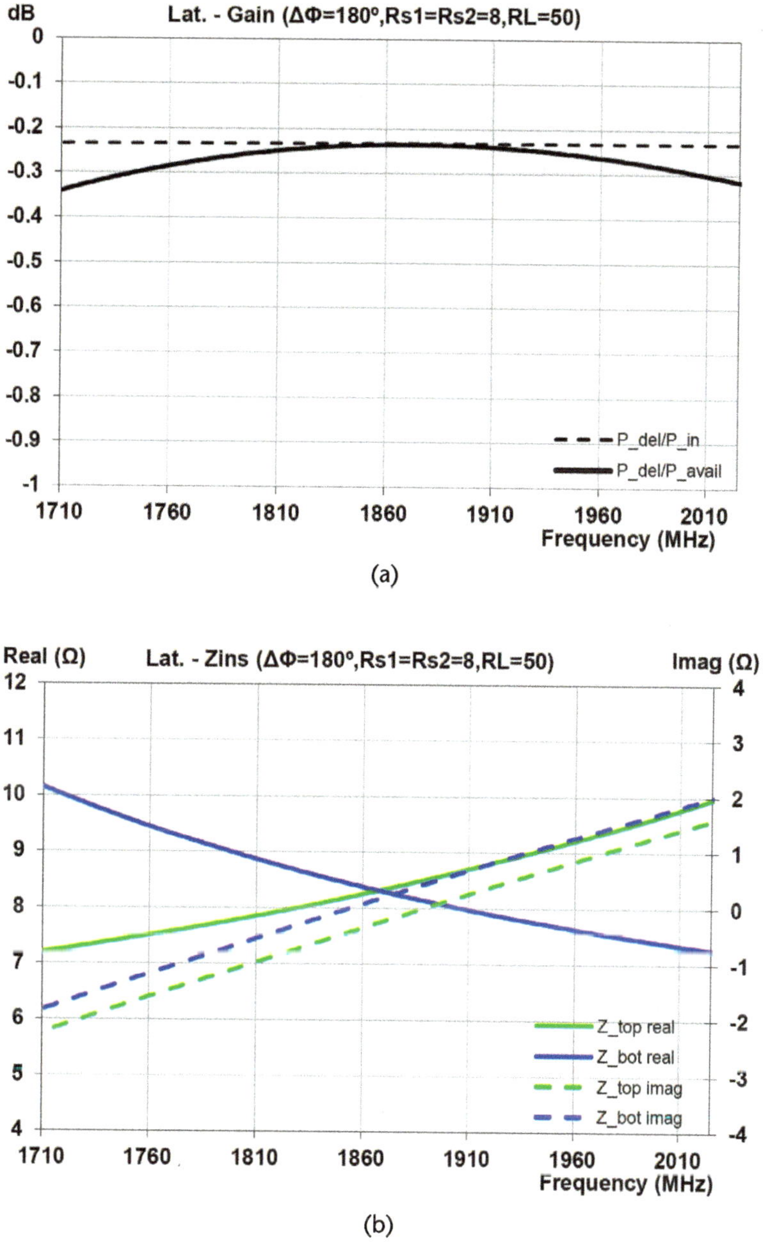

Figure 15.35 Four-element lattice-balun combiner characteristics: (a) insertion gain, and (b) input impedances.

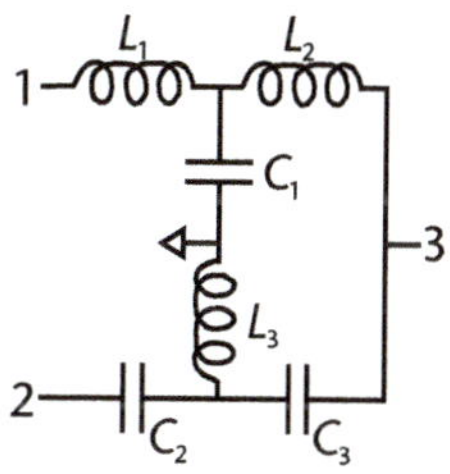

Figure 15.36 Six-element lattice-balun combiner configuration with the element values listed in Table 15.11.

Table 15.11 Six-Element Lattice Balun Elements

Element Values	
$L_1 = 0.793$ nH	$L_2 = 2.411$ nH
$L_3 = 2.411$ nH	$C_1 = 3.013$ pF
$C_2 = 9.157$ pF	$C_3 = 3.013$ pF

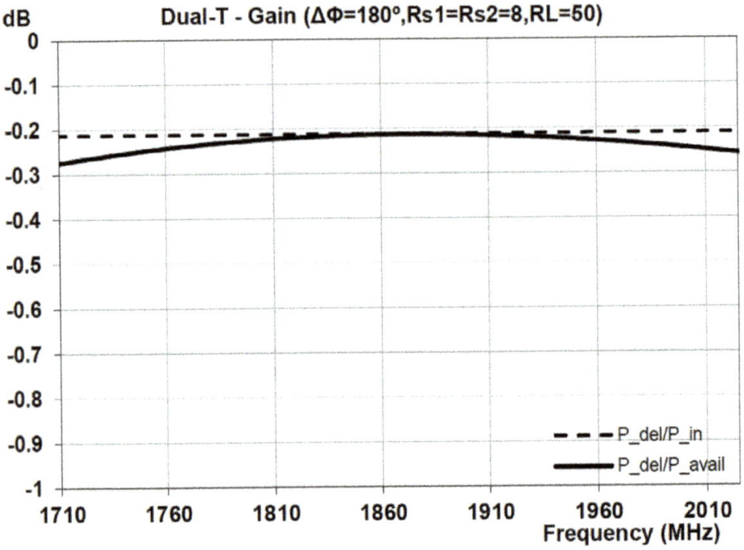

(a)

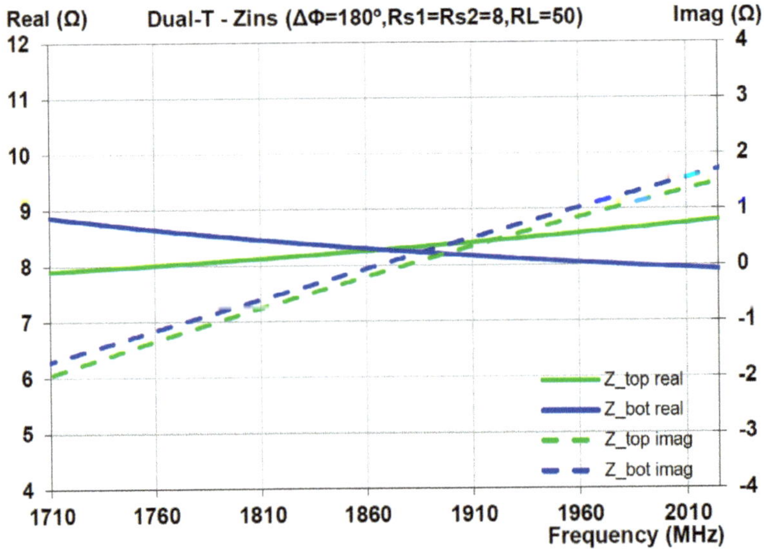

(b)

Figure 15.37 Six-element lattice-balun combiner characteristics: (a) insertion gain, and (b) input impedances.

T-network, they should instead be viewed as effectively comprising four-element networks, as shown in Figure 15.34. A four-element network does not have the same limitations as a three-element T-network. In fact, an impedance matching network with minimal phase shift and low loss can readily be implemented using only four elements. This is examined in detail in Section 5.1 of Volume 2 of this series. As a result, the six-element T-based coupler is particularly advantageous compared to the four-element configuration.

To compare the performance of the two networks, when used as coupler, if $\Delta\Phi$ denotes the difference in phase between the two input waveforms, that is,

$$\Delta\Phi = \phi_1 - \phi_2 \tag{15.42}$$

then the insertion phase shifts in the upper and lower arms of the coupler, $\Delta\phi_1$ and $\Delta\phi_2$, respectively, must satisfy the relationship

$$\Delta\Phi = \Delta\phi_2 - \Delta\phi_1 \tag{15.43}$$

All the couplers are designed to meet the following specifications.

The alternate coupler configurations are to be compared for two values of relative input phase offset $\Delta\Phi$, namely, 45° and 90°.

Figures 15.38 to 15.40 show alternative combiner architectures and element values for three network types designed to meet the specifications in Table 15.12. The first configurations, shown in Figure 15.38, are for four-element lattice designs. The second configurations, shown in Figure 15.39, are for six-element lattice designs optimized for insertion loss in the manner described earlier in this section. The third configurations, shown in Figure 15.40, are also for six-element lattice designs, but with the simplifying assumption that each of the arms is terminated in a pure real conductance at the common node. The purpose of these designs is to illustrate the enhanced performance characteristics of the six-element lattices, resulting from the optimization process.

In Figure 15.41, insertion gains for the three network types are shown for three relative input phase differences $\Delta\Phi$, as defined by (15.42). For all plots, the magnitudes of the phase shifts in the two arms of the combiner were assumed to be equal. For example, for $\Delta\Phi = 45°$, the assigned phase shift in the upper arm of the network $\Delta\phi_1 = -22.5°$, while that in the lower arm $\Delta\phi_2 = +22.5°$.

Table 15.12 Three-Port Combiner Networks Design Specifications

Design Specifications	
f_{lo} 1,710 MHz	f_{hi} 2,025 MHz
Impedance—port 1	8Ω
Impedance—port 2	8Ω
Impedance—port 3	50Ω
Inductors, $Q_L = 60$	Capacitors, $Q_C \approx 100$

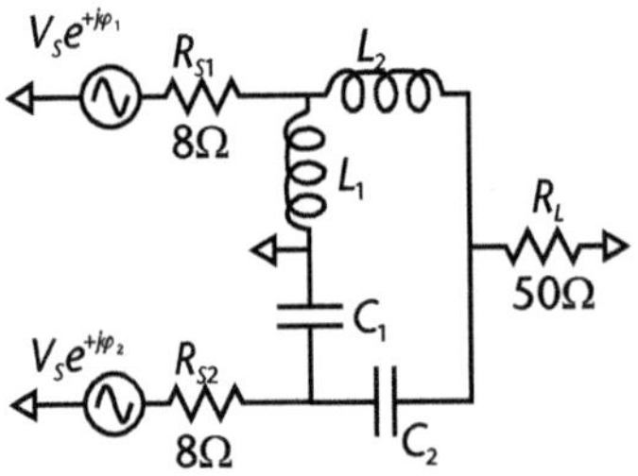

Design parameters	
$\Delta\Phi = 45°$	
$L_1 = 0.407$ nH	$L_2 = 0.923$ nH
$C_1 = 17.845$ pF	$C_2 = 7.874$ pF
$\Delta\Phi = 90°$	
$L_1 = 1.136$ nH	$L_2 = 1.705$ nH
$C_1 = 6.392$ pF	$C_2 = 4.261$ pF

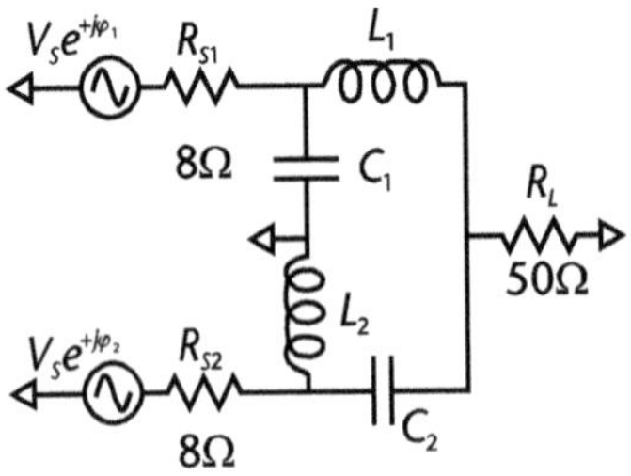

Design parameters	
$\Delta\Phi = 180°$	
$C_1 = 3.013$ pF	$L_1 = 2.411$ nH
$L_2 = 2.411$ nH	$C_2 = 3.013$ pF

Figure 15.38 Four-element lattice combiner architectures and element values.

Design parameters	
$\Delta\Phi = 45°$	
$C_1 = 10.496$ pF	$L_1 = 1.404$ nH
$L_2 = 3.183$ nH	$L_3 = 0.692$ nH
$C_2 = 5.172$ pF	$C_3 = 2.282$ pF
$\Delta\Phi = 90°$	
$C_1 = 19.144$ pF	$L_1 = 1.769$ nH
$L_2 = 2.653$ nH	$L_3 = 0.379$ nH
$C_2 = 4.107$ pF	$C_3 = 2.738$ pF

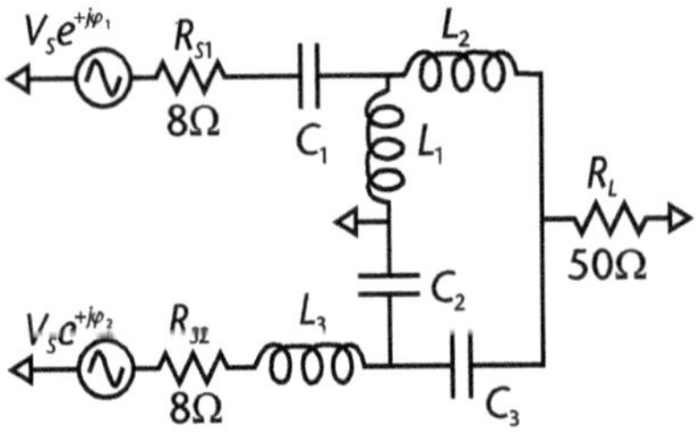

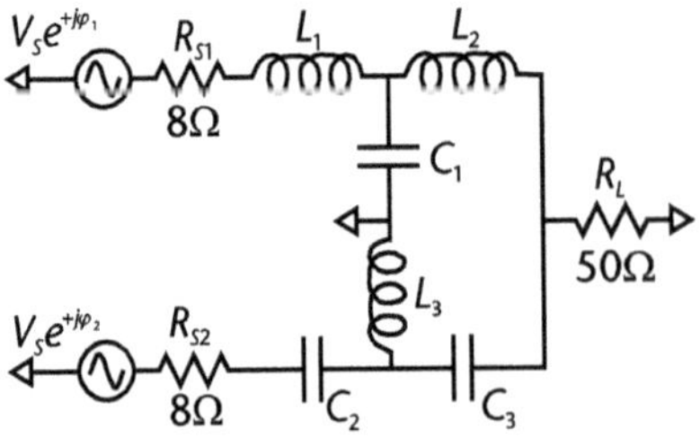

Design parameters	
$\Delta\Phi = 180°$	
$L_1 = 0.793$ nH	$C_1 = 3.013$ nH
$L_2 = 2.411$ nH	$C_2 = 9.157$ pF
$L_3 = 2.411$ nH	$C_3 = 3.013$ pF

Figure 15.39 Optimized *T*-network combiner architectures and element values.

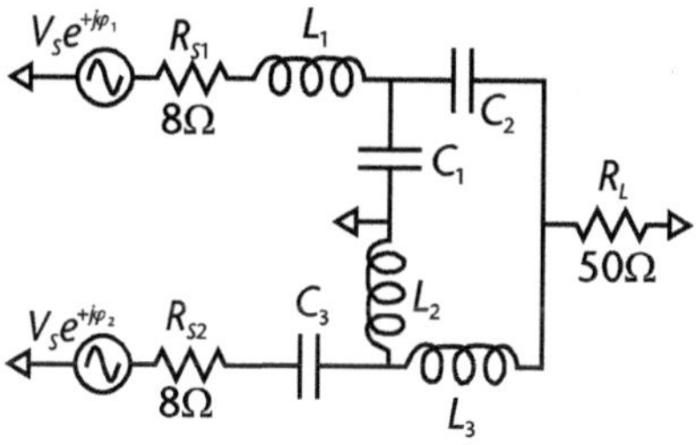

Design parameters	
$\Delta\Phi = 45°$	
$L_1 = 4.653$ nH	$C_1 = 1.153$ pF
$C_2 = 0.509$ pF	$C_3 = 1.561$ pF
$L_2 = 6.299$ nH	$L_3 = 14.276$ nH
$\Delta\Phi = 90°$	
$L_1 = 2.727$ nH	$C_1 = 2.131$ pF
$C_2 = 1.420$ pF	$C_3 = 2.663$ pF
$L_2 = 3.409$ nH	$L_3 = 5.113$ nH

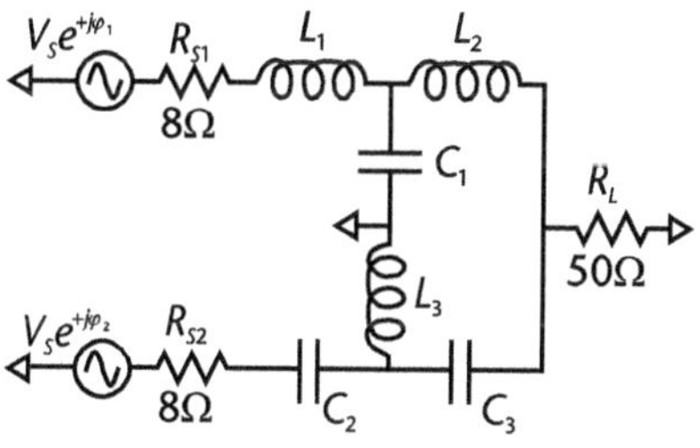

Design parameters	
$\Delta\Phi = 180°$	
$L_1 = 2.411$ nH	$C_1 = 3.013$ nH
$L_2 = 2.411$ nH	$C_2 = 3.013$ pF
$L_3 = 2.411$ nH	$C_3 = 3.013$ pF

Figure 15.40 *T*-networks combiner with real common node architectures and element values.

In Figure 15.41(a), for $\Delta\Phi = 45°$, the optimized six-element lattice combiner is seen to exhibit significantly higher gain across the passband than the alternative configurations. Its gain is much higher than the four element lattice, particularly at the band edges, due to the broader bandwidth characteristics of the virtual four-element networks in each arm. The insertion gain plot for the nonoptimized dual *T*-configuration (*T0 Elems*), designed assuming real terminating admittances on both arms at the common node, is significantly inferior and more dispersive than the other two.

In Figure 15.41(b), for $\Delta\Phi = 90°$, the insertion gain plots for the four-element lattice combiner, and the optimized six-element lattice combiner are near identical. The improved characteristics of the four-element lattice, compared to $\Delta\Phi = 45°$, are due to the magnitude of the insertion phases in each of the π-networks being within the range specified by (15.41). Hence, for a relative input phase offset of $|\Delta\Phi| = \sim 90°$, the optimized six-element lattice offers little advantage over the four-element configuration. Again, the nonoptimized dual *T*-configuration (*T0* Elems) response is significantly inferior.

In Figure 15.41(c), for $\Delta\Phi = 180°$, the optimized six-element lattice combiner has higher insertion gain than that of the four-element lattice combiner. However, the gain advantage is less than it was for $\Delta\Phi = 90°$. As in the previous plots, the nonoptimized dual *T*-configuration (*T0* Elems) has a greatly inferior gain response.

Given that the six-element lattice couplers are a composite of two four-element phase shift networks, they have considerably more design flexibility than do the conventional four-element lattice couplers. With the four-element designs, to accommodate a complex termination at the common node and avoid the need for an additional component, a trade-off must be made in the relative insertion phases through the two arms. With the six-element designs, this is unnecessary. The elements for the two arms are simply designed according to (15.38) to (15.40) for a given value

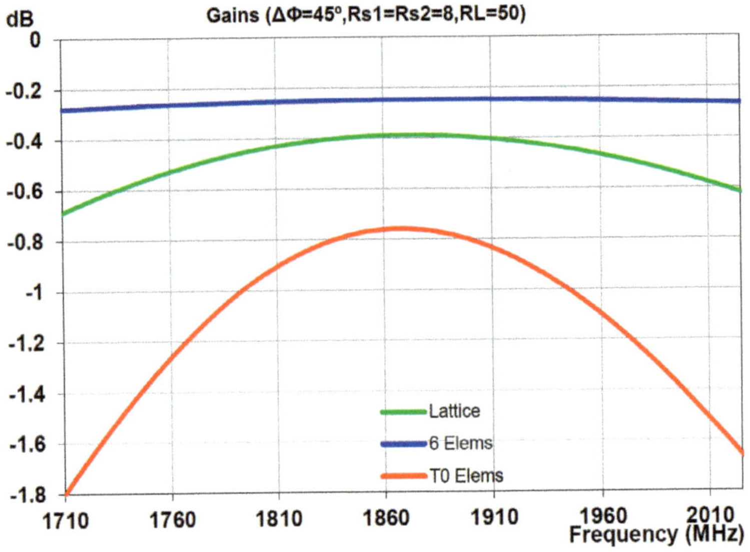

(a)

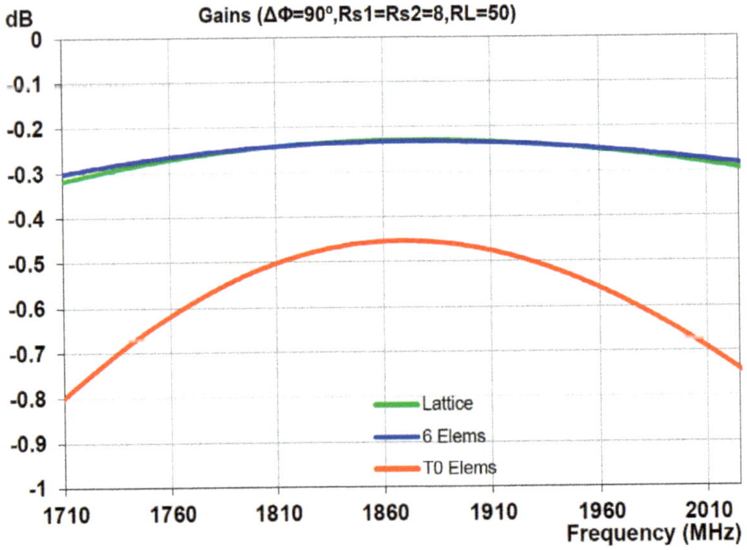

(b)

Figure 15.41 Combiner network gain variation and dependence on insertion phase delta: (a) $\Delta\Phi = 45°$, (b) $\Delta\Phi = 90°$, and (c) $\Delta\Phi = 180°$.

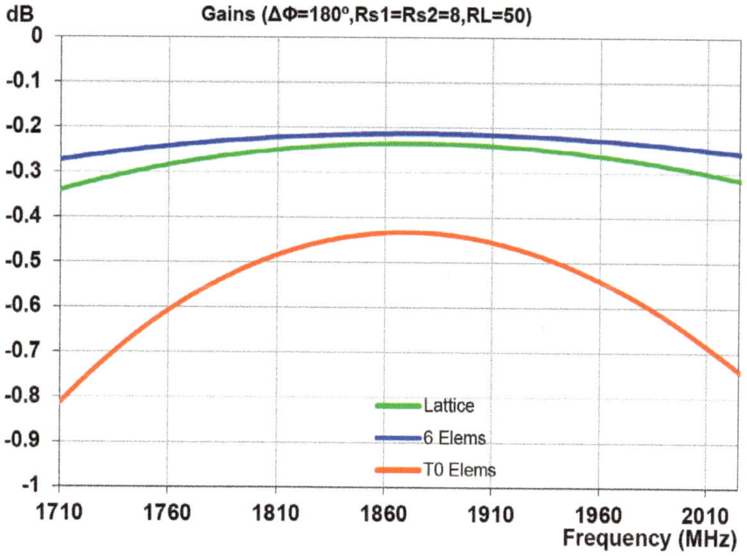

(c)

Figure 15.41 *(Continued)*

of the virtual susceptance B_V. The optimum values are determined that are associated with the best performance characteristics for the coupler.

In Section 15.2.9, the design of a four-element lattice coupler was considered, with the requirement for unequal magnitudes of insertion phase in the two arms. This required the addition of a fifth element to the network and, in addition, there was an increase in insertion loss. As explained earlier in this section, the further either of the insertion phases falls outside the range from (15.41), the higher the expected insertion loss. However, this is not an issue with the six-element lattice.

For the optimized six-element lattice combiner, the insertion phase shifts in the upper and lower arms of the coupler, $\Delta\phi_1$ and $\Delta\phi_2$, respectively, are solely constrained by (15.43). This allows for great flexibility in assigning an appropriate combination of the two phase shifts. In choosing the latter, it is important to determine how the overall coupler characteristics are affected by the choice.

Figure 15.42 shows the coupler gain dependencies versus upper-arm phase shift $\Delta\phi_1$ for three relative input phase differences $\Delta\Phi = 45°$, $90°$, and $180°$.

The fine structure in these plots is attributable to two factors: (1) the discrete nature of the optimization algorithm, and (2) discontinuities arising from circuit elements changing in type from inductive to capacitive, or vice versa.

For $\Delta\Phi = 45°$ and $90°$, it can be seen, from Figures 15.42(a, b), that the insertion gain plots are remarkedly unchanged for all values of $\Delta\Phi$. Essentially, the insertion gain value is near constant ~ -0.3 dB, irrespective of the chosen value $\Delta\phi_1$ for the insertion phase shift in the upper arm. For the case $\Delta\Phi = 180°$, the coupler gain is a maximum for $\Delta\phi_1 = -90°$ and $\Delta\phi_2 = +90°$. As the phase shifts move away from this point, the coupler gain decreases.

For the case in which the magnitudes of the phase shifts in the two arms are close to equal and $\Delta\Phi = 90°$, Figure 15.41(b) shows that there is negligible gain advantage to employing an optimized six-element coupler compared to using a conventional four-element design. Furthermore, Figure 15.42(b) shows the gain of the optimized six-element coupler phase shift is essentially independent of the value chosen for the upper arm insertion phase value $\Delta\phi_1$. Thus, an optimized six-element coupler design

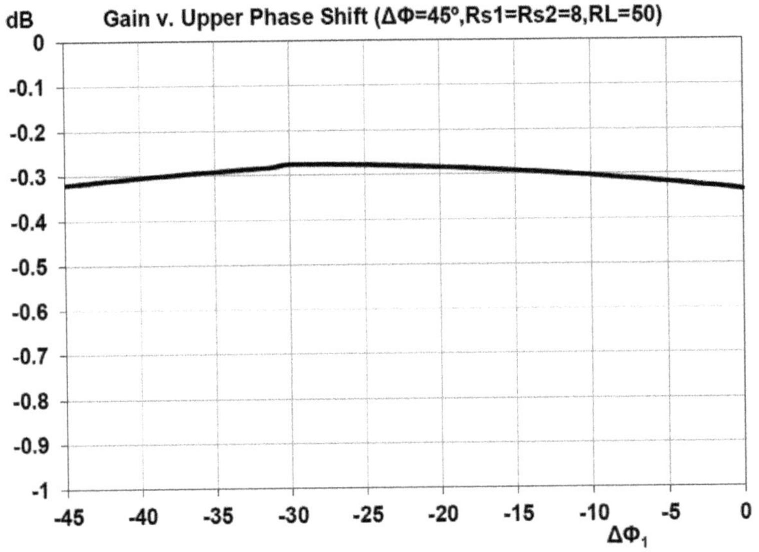

(a)

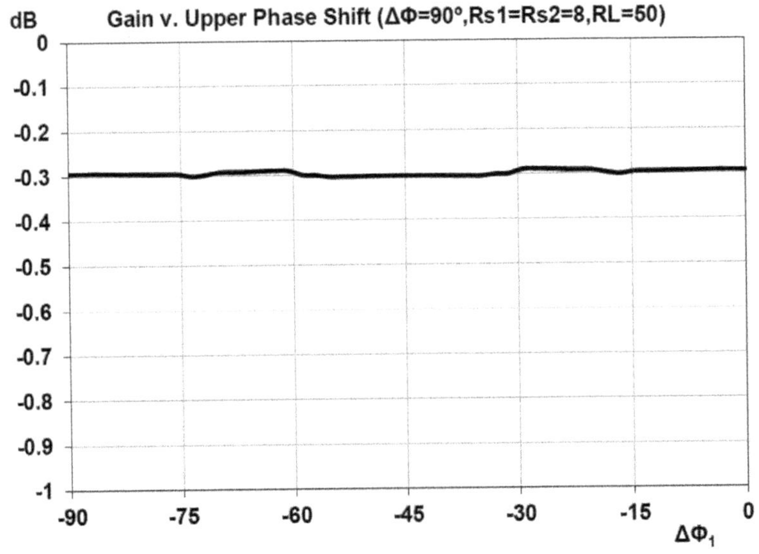

(b)

Figure 15.42 Six-element coupler gain dependencies versus upper-arm phase shift $\Delta\Phi_1$: (a) $\Delta\Phi = 45°$, (b) $\Delta\Phi = 90°$, and (c) $\Delta\Phi = 180°$.

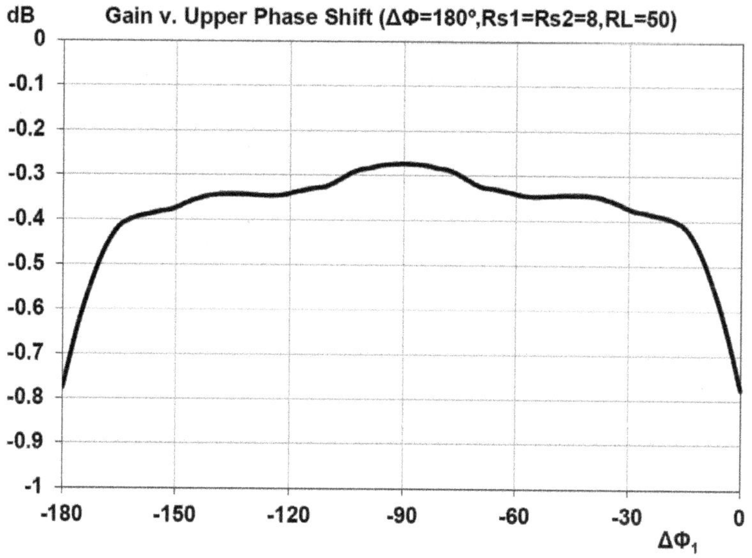

(c)

Figure 15.42 *(Continued)*

with $\Delta\phi_1 = -90°$ and $\Delta\phi_2 = 0°$, or $\Delta\phi_1 = 0°$ and $\Delta\phi_2 = +90°$, could be expected to have near the same gain characteristic as the four-element coupler. Given that there is no expected gain advantage (or disadvantage), there are other possible advantages to choosing these phase combinations. Three potential advantages are:

1. Real parts of input impedances are less dispersive across the passband.
2. The coupler can be used to implement an efficient switchable dual power state amplifier architecture (i.e., with high and low power output levels).
3. The coupler can be used to implement a Doherty-type amplifier.

The latter two advantages are examined fully in Chapter 4 of Volume 2 of this series. To illustrate the first point, Figure 15.43 shows the gain and input impedances for an optimized six-element coupler for $\Delta\Phi = 90°$ and $|\Delta\phi_1| = |\Delta\phi_2| = 45°$.

Figure 15.44 shows the gain and input impedances for an optimized six-element coupler for $\Delta\Phi = 90°$, $\Delta\phi_1 = -90°$, and $\Delta\phi_2 = 0°$.

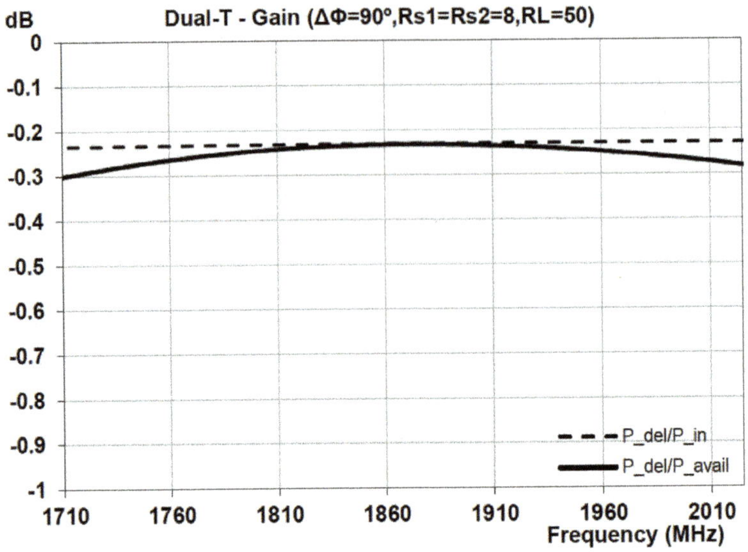

(a)

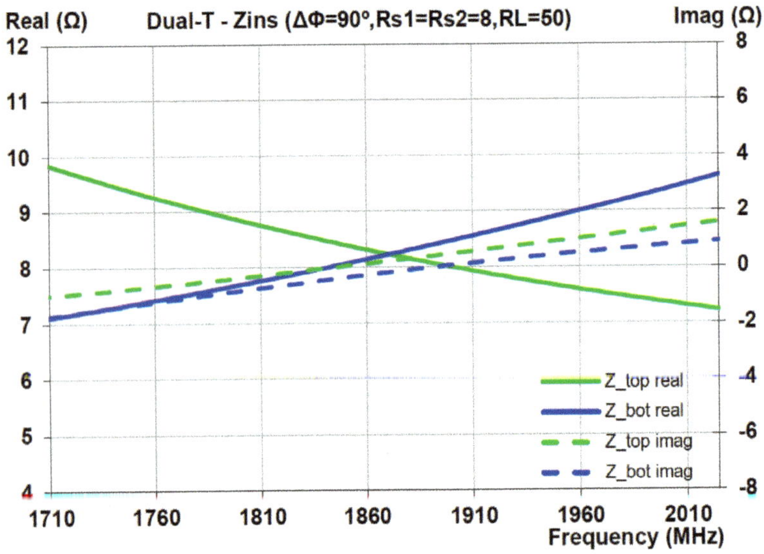

(b)

Figure 15.43 Six-element lattice configuration with $\Delta\Phi = 90°$, $|\Delta\Phi_1| = |\Delta\Phi_2| = 45°$; (a) gain, and (b) input impedances.

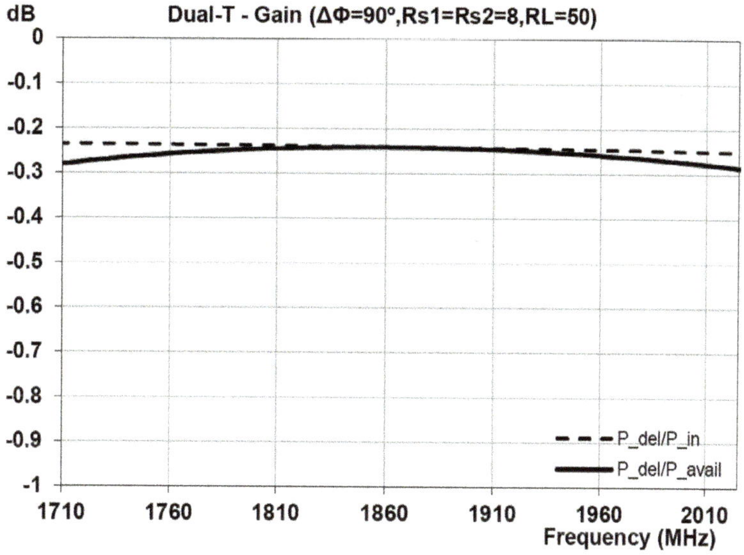

(a)

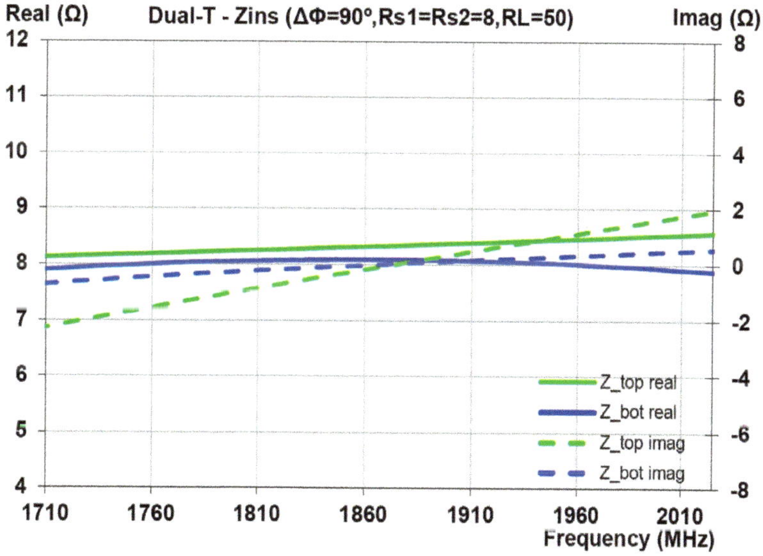

(b)

Figure 15.44 Six-element lattice configuration with $\Delta\Phi = 90°$, $\Delta\Phi_1 = -90°$, $\Delta\Phi_2 = 0°$: (a) gain, and (b) input impedances.

APPENDIX A

Miscellaneous Generic [M] = [ABCD] Network Parameters

In the expressions below, matrix coefficients have been expanded, where possible, to share common expressions in order to minimize computation.

$$[M]=\begin{bmatrix} 1+Y_2Z_1 & Z_1+Z_2(1+Y_2Z_1) \\ Y_2+Y_1(1+Y_2Z_1) & 1+Y_2Z_2+Y_1(Z_1+Z_2(1+Y_2Z_1)) \end{bmatrix} \tag{A.1}$$

$$[M]=\begin{bmatrix} 1+Y_2Z_2+Z_1(Y_1+Y_2(1+Y_1Z_2)) & Z_2+Z_1(1+Y_1Z_2) \\ Y_1+Y_2(1+Y_1Z_2) & 1+Y_1Z_2 \end{bmatrix} \tag{A.2}$$

$$[M]=\begin{bmatrix} 1+Y_2Z_1+Y_3(Z_1+Z_2(1+Y_2Z_1)) & Z_1+Z_2(1+Y_2Z_1) \\ Y_2+Y_1(1+Y_2Z_1)+Y_3(1+Y_2Z_2+Y_1(Z_1+Z_2(1+Y_2Z_1))) & 1+Y_2Z_2+Y_1(Z_1+Z_2(1+Y_2Z_1)) \end{bmatrix} \tag{A.3}$$

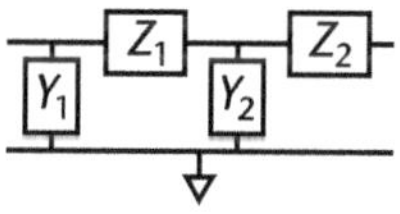

Figure A.1 Four-element topology.

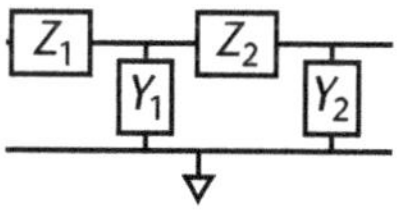

Figure A.2 Alternate four element topology.

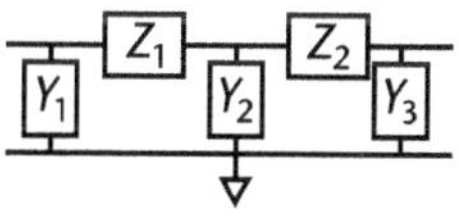

Figure A.3 Five-element topology.

$$[M] = \begin{bmatrix} 1 + Y_2Z_2 + Z_1(Y_2 + Y_1(1 + Y_2Z_2)) & Z_2 + Z_3(1 + Y_2Z_2) + Z_1(1 + Y_1Z_2 + Z_3(Y_2 + Y_1(1 + Y_2Z_2))) \\ Y_2 + Y_1(1 + Y_2Z_2) & 1 + Y_1Z_2 + Z_3(Y_2 + Y_1(1 + Y_2Z_2)) \end{bmatrix} \quad \text{(A.4)}$$

$$[M] = \begin{bmatrix} A & B \\ C & D \end{bmatrix}$$

with

$$\begin{aligned} A &= (1 + Y_2Z_2)(1 + Y_3Z_3) + Y_3Z_2 + CZ_1 \\ B &= Z_2 + Z_3(1 + Y_2Z_2) + DZ_1 \\ C &= Y_2 + Y_1(1 + Y_2Z_2)(1 + Y_3Z_3) + Y_3(1 + Y_1Z_2 + Y_2Z_3) \\ D &= Y_1Z_3 + (1 + Y_1Z_2)(1 + Y_2Z_3) \end{aligned} \quad \text{(A.5)}$$

$$[M] = \begin{bmatrix} A & B \\ C & D \end{bmatrix}$$

where

$$\begin{aligned} A &= 1 + Y_2Z_1 + Y_3(Z_1 + Z_2(1 + Y_2Z_1)) \\ B &= AZ_3 + Z_1 + Z_2(1 + Y_2Z_1) \\ C &= Y_2 + Y_1(1 + Y_2Z_1) + Y_3(1 + Y_2Z_2 + Y_1(Z_1 + Z_2(1 + Y_2Z_1))) \\ D &= CZ_3 + 1 + Y_2Z_2 + Y_1(Z_1 + Z_2(1 + Y_2Z_1)) \end{aligned} \quad \text{(A.6)}$$

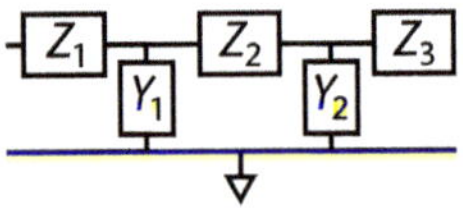

Figure A.4 Alternate five-element topology.

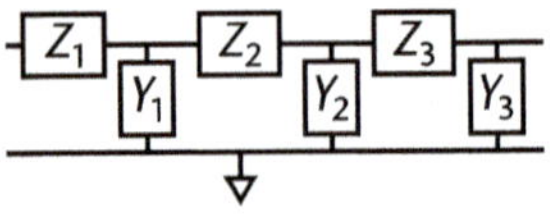

Figure A.5 Six-element topology.

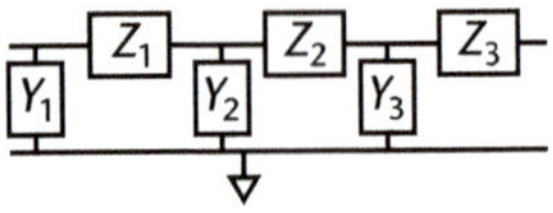

Figure A.6 Alternate six-element topology.

With $[ABCD]$ parameters from previous six element analysis

$$[M]=\begin{bmatrix} A+BY_4 & B \\ C+DY_4 & D \end{bmatrix} \tag{A.7}$$

$$[M]=\begin{bmatrix} A+CZ_2+Z_1\left(C+Y\left(A+CZ_2\right)\right) & B+DZ_2+Z_1\left(D+Y\left(B+DZ_2\right)\right) \\ C+Y\left(A+CZ_2\right) & D+Y\left(B+DZ_2\right) \end{bmatrix} \tag{A.8}$$

$$[M]=\begin{bmatrix} A\left(1+YZ_1\right)+BY & AZ_1+B+Z_2\left(A\left(1+YZ_1\right)+BY\right) \\ C\left(1+YZ_1\right)+DY & CZ_1+D+Z_2\left(C\left(1+YZ_1\right)+DY\right) \end{bmatrix} \tag{A.9}$$

$$[M]=\begin{bmatrix} A' & B' \\ C' & D' \end{bmatrix}$$

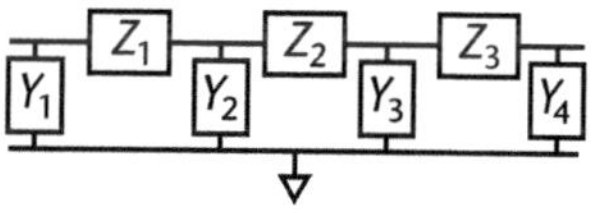

Figure A.7 Seven-element topology.

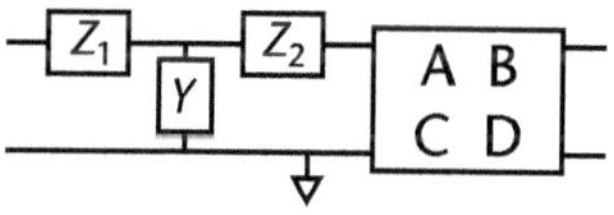

Figure A.8 Network with an input *T*-match.

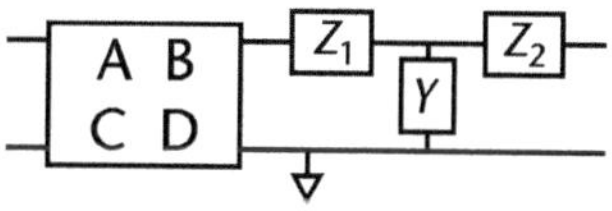

Figure A.9 Network with an output *T*-match.

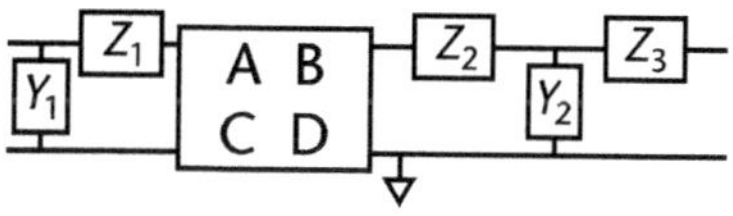

Figure A.10 Network with a two-element input match and output *T*-match.

where

$$\begin{aligned}
A' &= A(1+Y_2Z_2)+BY_2+Z_1(C(1+Y_2Z_2)+DY_2)\\
B' &= AZ_2+B+Z_1(CZ_2+D)\\
&\quad + Z_3(A(1+Y_2Z_2)+BY_2+Z_1(C(1+Y_2Z_2)+DY_2))\\
C' &= Y_1(A(1+Y_2Z_2)+BY_2)+(1+Y_1Z_1)(C(1+Y_2Z_2)+DY_2)\\
D' &= Y_1(AZ_2+B)+(1+Y_1Z_1)(CZ_2+D)\\
&\quad + Z_3(Y_1(A(1+Y_2Z_2)+BY_2)+(1+Y_1Z_1)(C(1+Y_2Z_2)+DY_2))
\end{aligned} \tag{A.10}$$

$$[M]=\begin{bmatrix} A' & B' \\ C' & D' \end{bmatrix}$$

where

$$\begin{aligned}
A' &= A+CZ_1+Y_2(AZ_2+B+Z_1(CZ_2+D))\\
B' &= AZ_2+B+Z_1(CZ_2+D)\\
C' &= AY_1+C(1+Y_1Z_1)+Y_2(CZ_2+D+Y_1(AZ_2+B+Z_1(CZ_2+D)))\\
D' &= CZ_2+D+Y_1(AZ_2+B+Z_1(CZ_2+D))
\end{aligned} \tag{A.11}$$

$$[M]=\begin{bmatrix} A+BY_1+Y_2(AZ+B(1+Y_1Z)) & AZ+B(1+Y_1Z) \\ C+DY_1+Y_2(CZ+D(1+Y_1Z)) & CZ+D(1+Y_1Z) \end{bmatrix} \tag{A.12}$$

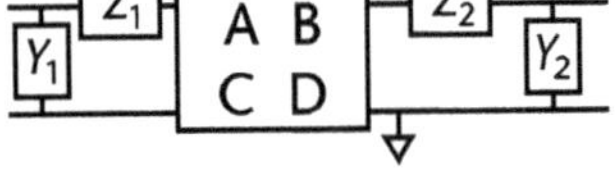

Figure A.11 Network with a two-element input and output matches.

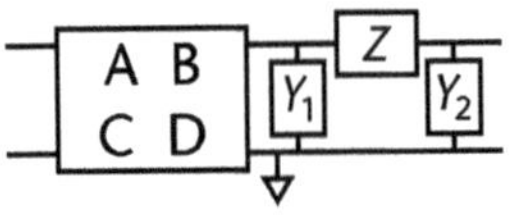

Figure A.12 Network with an output π-match.

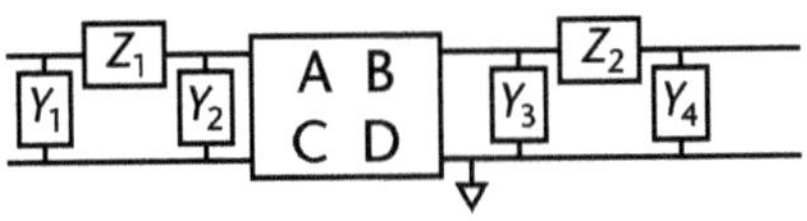

Figure A.13 Network with input and output π-matches.

$$[M] = \begin{bmatrix} 1+Y_2Z_1 & Z_1 \\ Y_1+Y_2(1+Y_1Z_1) & (1+Y_1Z_1) \end{bmatrix} \cdot \begin{bmatrix} A+BY_3+Y_4(AZ_2+B(1+Y_3Z_2)) & AZ_2+B(1+Y_3Z_2) \\ C+DY_3+Y_4(CZ_2+D(1+Y_3Z_2)) & CZ_2+D(1+Y_3Z_2) \end{bmatrix} \tag{A.13}$$

or

$$[M] = \begin{bmatrix} A(1+Y_2Z_1)+CZ_1 & B(1+Y_2Z_1)+DZ_1 \\ C+AY_2+Y_1(A(1+Y_2Z_1)+CZ_1) & D+BY_2+Y_1(B(1+Y_2Z_1)+DZ_1) \end{bmatrix} \cdot \begin{bmatrix} 1+Y_4Z_2 & Z_2 \\ Y_3+Y_4(1+Y_3Z_2) & (1+Y_3Z_2) \end{bmatrix} \tag{A.14}$$

$$[M] = \begin{bmatrix} A & B \\ C & D \end{bmatrix}$$

where

$$\begin{aligned} A &= A_2(A_1+B_1Y_2)+B_1C_2 \\ B &= B_2(A_1+B_1Y_2)+B_1D_2 \\ C &= A_2(C_1+D_1Y_2)+C_2D_1+Y_1(A_2(A_1+B_1Y_2)+B_1C_2) \\ D &= B_2(C_1+D_1Y_2)+D_1D_2+Y_1(B_2(A_1+B_1Y_2)+B_1D_2) \end{aligned} \tag{A.15}$$

$$[M] = \begin{bmatrix} A & B \\ C & D \end{bmatrix}$$

where

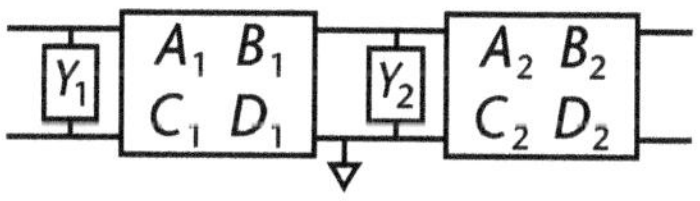

Figure A.14 Two networks with shunt elements.

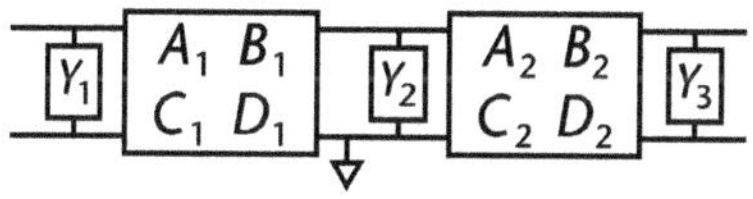

Figure A.15 Alternate two networks with shunt elements.

$$\begin{aligned}
A &= A_2(A_1 + B_1Y_2) + B_1C_2 + \big(B_2(A_1 + B_1Y_2) + B_1D_2\big)Y_3 \\
B &= B_2(A_1 + B_1Y_2) + B_1D_2 \\
C &= A_2(C_1 + D_1Y_2) + C_2D_1 + \big(A_2(A_1 + B_1Y_2) + B_1C_2\big)Y_1 \\
&\quad + \big(B_2(C_1 + D_1Y_2) + D_1D_2 + Y_1(B_2(A_1 + B_1Y_2) + B_1D_2)\big)Y_3 \\
D &= B_2(C_1 + D_1Y_2) + D_1D_2 + \big(B_2(A_1 + B_1Y_2) + B_1D_2\big)Y_1
\end{aligned} \tag{A.16}$$

APPENDIX B

Conversion Formulae Between Two-Port Network Parameters

Two-port network parametric, current, and voltage definitions are the following.

For S-parameters, the normalization impedances on ports 1 and 2 are $\underline{Z}_1$ and $\underline{Z}_2$, respectively.

From two-port $[ABCD]$ parameters:

$\Rightarrow S$

$$S_{11} = \frac{A + B/\underline{Z}_2 - C\underline{Z}_1 - D\underline{Z}_1/\underline{Z}_2}{A + B/\underline{Z}_2 + C\underline{Z}_1 + D\underline{Z}_1/\underline{Z}_2} \tag{B.1}$$

$$S_{12} = \frac{2(AD - BC)(\underline{Z}_1/\underline{Z}_2)}{A + B/\underline{Z}_2 + C\underline{Z}_1 + D\underline{Z}_1/\underline{Z}_2} \tag{B.2}$$

$$S_{21} = \frac{2}{A + B/\underline{Z}_2 + C\underline{Z}_1 + D\underline{Z}_1/\underline{Z}_2} \tag{B.3}$$

$$S_{22} = \frac{A + B/\underline{Z}_2 - C\underline{Z}_1 + D\underline{Z}_1/\underline{Z}_2}{A + B/\underline{Z}_2 + C\underline{Z}_1 + D\underline{Z}_1/\underline{Z}_2} \tag{B.4}$$

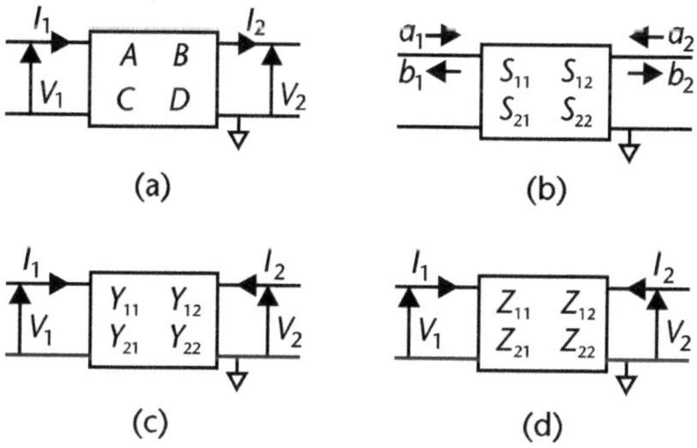

Figure B.1 Electrical parameters for network characterization: (a) matrix parameters, (b) scattering parameters, (c) admittance parameters, and (d) impedance parameters.

$\Rightarrow Y$

$$Y_{11} = D/B \tag{B.5}$$

$$Y_{12} = C - AD/B \tag{B.6}$$

$$Y_{21} = -1/B \tag{B.7}$$

$$Y_{22} = A/B \tag{B.8}$$

$\Rightarrow Z$

$$Z_{11} = A/C \tag{B.9}$$

$$Z_{12} = AD/C - B \tag{B.10}$$

$$Z_{21} = 1/C \tag{B.11}$$

$$Z_{22} = D/C \tag{B.12}$$

From two-port S-parameters:

$\Rightarrow [ABCD]$

$$A = \frac{(1+S_{11})(1+S_{22})+S_{12}S_{21}}{2S_{21}} \tag{B.13}$$

$$B = \frac{(1+S_{11})(1+S_{22})-S_{12}S_{21}}{2S_{21}} \underline{Z}_2 \tag{B.14}$$

$$C = \frac{(1-S_{11})(1-S_{22})-S_{12}S_{21}}{2S_{21}\underline{Z}_1} \tag{B.15}$$

$$D = \frac{(1-S_{11})(1+S_{22})+S_{12}S_{21}}{2S_{21}} \frac{\underline{Z}_2}{\underline{Z}_1} \tag{B.16}$$

$\Rightarrow Y$

$$Y_{11} = \frac{(1-S_{11})(1+S_{22})+S_{12}S_{21}}{((1+S_{11})(1+S_{22})-S_{12}S_{21})\underline{Z}_1} \tag{B.17}$$

$$Y_{12} = \frac{-2S_{12}}{((1+S_{11})(1+S_{22})-S_{12}S_{21})\underline{Z}_1} \tag{B.18}$$

$$Y_{21} = \frac{-2S_{21}}{((1+S_{11})(1+S_{22}) - S_{12}S_{21})\underline{Z}_2} \tag{B.19}$$

$$Y_{22} = \frac{(1+S_{11})(1-S_{22}) + S_{12}S_{21}}{((1+S_{11})(1+S_{22}) - S_{12}S_{21})\underline{Z}_2} \tag{B.20}$$

$\Rightarrow Z$

$$Z_{11} = \frac{(1+S_{11})(1-S_{22}) + S_{12}S_{21}}{(1-S_{11})(1-S_{22}) - S_{12}S_{21}}\underline{Z}_1 \tag{B.21}$$

$$Z_{12} = \frac{2S_{12}}{(1-S_{11})(1-S_{22}) - S_{12}S_{21}}\underline{Z}_2 \tag{B.22}$$

$$Z_{21} = \frac{2S_{21}}{(1-S_{11})(1-S_{22}) - S_{12}S_{21}}\underline{Z}_1 \tag{B.23}$$

$$Z_{22} = \frac{(1-S_{11})(1+S_{22}) + S_{12}S_{21}}{(1-S_{11})(1-S_{22}) - S_{12}S_{21}}\underline{Z}_2 \tag{B.24}$$

From two-port Y-parameters:

$\Rightarrow [ABCD]$

$$A = -Y_{22}/Y_{21} \tag{B.25}$$

$$B = -1/Y_{21} \tag{B.26}$$

$$C = Y_{12} - Y_{11}Y_{22}/Y_{21} \tag{B.27}$$

$$D = -Y_{11}/Y_{21} \tag{B.28}$$

$\Rightarrow S$

$$S_{11} = \frac{(1/\underline{Z}_1 - Y_{11})(1/\underline{Z}_2 - Y_{22}) + Y_{12}Y_{21}}{(1/\underline{Z}_1 + Y_{11})(1/\underline{Z}_2 - Y_{22}) - Y_{12}Y_{21}} \tag{B.29}$$

$$S_{12} = \frac{-(2/\underline{Z}_2)Y_{12}}{(1/\underline{Z}_1 + Y_{11})(Y_{22} + 1/\underline{Z}_2) - Y_{12}Y_{21}} \tag{B.30}$$

$$S_{21} = \frac{-(2/\underline{Z}_1)Y_{21}}{(1/\underline{Z}_1 + Y_{11})(1/\underline{Z}_2 + Y_{22}) - Y_{12}Y_{21}} \tag{B.31}$$

$$S_{22} = \frac{(1/\underline{Z}_1 + Y_{11})(1/\underline{Z}_2 - Y_{22}) + Y_{12}Y_{21}}{(1/\underline{Z}_1 + Y_{11})(Y_{22} + 1/\underline{Z}_2) - Y_{12}Y_{21}} \tag{B.32}$$

$\Rightarrow Z$

$$Z_{11} = \frac{Y_{22}}{Y_{11}Y_{22} - Y_{12}Y_{21}} \tag{B.33}$$

$$Z_{12} = \frac{Y_{12}}{Y_{11}Y_{22} - Y_{12}Y_{21}} \tag{B.34}$$

$$Z_{21} = \frac{Y_{21}}{Y_{11}Y_{22} - Y_{12}Y_{21}} \tag{B.35}$$

$$Z_{22} = \frac{Y_{11}}{Y_{11}Y_{22} - Y_{12}Y_{21}} \tag{B.36}$$

From two-port Z-parameters:

$\Rightarrow [ABCD]$

$$A = Z_{11}/Z_{21} \tag{B.37}$$

$$B = Z_{11}Z_{22}/Z_{21} - Z_{12} \tag{B.38}$$

$$C = 1/Z_{21} \tag{B.39}$$

$$D = -Z_{22}/Z_{21} \tag{B.40}$$

$\Rightarrow S$

$$S_{11} = \frac{(Z_{11} - \underline{Z}_1)(Z_{22} + \underline{Z}_2) - Z_{12}Z_{21}}{(Z_{11} + \underline{Z}_1)(Z_{22} + \underline{Z}_2) - Z_{12}Z_{21}} \tag{B.41}$$

$$S_{12} = \frac{2Z_{12}\underline{Z}_1}{(Z_{11} + \underline{Z}_1)(Z_{22} + \underline{Z}_2) - Z_{12}Z_{21}} \tag{B.42}$$

$$S_{21} = \frac{2Z_{21}\underline{Z}_2}{(Z_{11} + \underline{Z}_1)(Z_{22} + \underline{Z}_2) - Z_{12}Z_{21}} \tag{B.43}$$

$$S_{22} = \frac{(Z_{11} + \underline{Z}_1)(Z_{22} - \underline{Z}_2) - Z_{12}Z_{21}}{(Z_{11} + \underline{Z}_1)(Z_{22} + \underline{Z}_2) - Z_{12}Z_{21}} \tag{B.44}$$

$\Rightarrow Y$

$$Y_{11} = \frac{Z_{22}}{Z_{11}Z_{22} - Z_{12}Z_{21}} \tag{B.45}$$

$$Y_{12} = \frac{Z_{12}}{Z_{11}Z_{22} - Z_{12}Z_{21}} \tag{B.46}$$

$$Y_{21} = \frac{Z_{21}}{Z_{11}Z_{22} - Z_{12}Z_{21}} \tag{B.47}$$

$$Y_{22} = \frac{Z_{11}}{Z_{11}Z_{22} - Z_{12}Z_{21}} \tag{B.48}$$

APPENDIX C

Conversion of Four-Port S-Parameters to Y-Parameters

For S-parameters:

$$\begin{bmatrix} b_1 \\ b_2 \\ b_3 \\ b_4 \end{bmatrix} = \begin{bmatrix} S_{11} & S_{12} & S_{13} & S_{14} \\ S_{12} & S_{22} & S_{23} & S_{13} \\ S_{13} & S_{23} & S_{22} & S_{12} \\ S_{14} & S_{13} & S_{12} & S_{11} \end{bmatrix} \cdot \begin{bmatrix} a_1 \\ a_2 \\ a_3 \\ a_4 \end{bmatrix} \tag{C.1}$$

with Z_0 = characteristic impedance.

For Y-parameters:

$$\begin{bmatrix} I_1 \\ I_2 \\ I_3 \\ I_4 \end{bmatrix} = \begin{bmatrix} Y_{11} & Y_{12} & Y_{13} & Y_{14} \\ Y_{21} & Y_{22} & Y_{23} & Y_{24} \\ Y_{31} & Y_{32} & Y_{33} & Y_{34} \\ Y_{41} & Y_{42} & Y_{43} & Y_{44} \end{bmatrix} \cdot \begin{bmatrix} V_1 \\ V_2 \\ V_3 \\ V_4 \end{bmatrix} \tag{C.2}$$

in terms of:

$$E_1 = \frac{(S_{14}S_{23} - S_{13}S_{24})S_{32} + (1 + S_{22})(S_{13}S_{34} - (1 + S_{33})S_{14}) + S_{12}((1 + S_{33})S_{24} - S_{23}S_{34})}{(S_{13}S_{24} - S_{14}S_{23})S_{42} + (1 + S_{22})(S_{14}S_{43} - (1 + S_{44})S_{13}) + S_{12}((1 + S_{44})S_{23} - S_{24}S_{43})} \tag{C.3}$$

$$E_2 = \frac{(S_{13}S_{24} - S_{14}S_{23})S_{31} + (1 + S_{11})(S_{23}S_{34} - (1 + S_{33})S_{24}) + S_{21}((1 + S_{33})S_{14} - S_{13}S_{34})}{(S_{14}S_{23} - S_{13}S_{24})S_{41} + (1 + S_{11})(S_{24}S_{43} - (1 + S_{44})S_{23}) + S_{21}((1 + S_{44})S_{13} - S_{14}S_{43})} \tag{C.4}$$

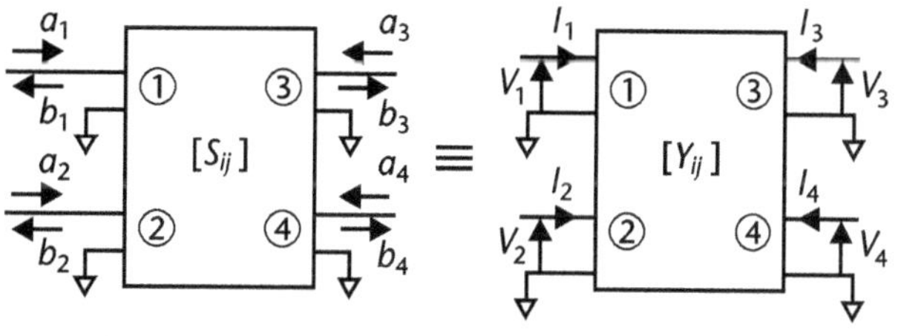

Figure C.1 Definition of S- and Y-matrix parameters.

$$E_3 = \frac{(S_{32}S_{41} - S_{31}S_{42})S_{14} + (1+S_{44})(S_{12}S_{31} - (1+S_{11})S_{32}) + S_{34}((1+S_{11})S_{42} - S_{12}S_{41})}{(S_{31}S_{42} - S_{32}S_{41})S_{24} + (1+S_{44})(S_{21}S_{32} - (1+S_{22})S_{31}) + S_{34}((1+S_{22})S_{41} - S_{21}S_{42})} \tag{C.5}$$

$$E_4 = \frac{(S_{31}S_{42} - S_{32}S_{41})S_{13} + (1+S_{33})(S_{12}S_{41} - (1+S_{11})S_{42}) + S_{43}((1+S_{11})S_{32} - S_{12}S_{31})}{(S_{32}S_{41} - S_{31}S_{42})S_{23} + (1+S_{33})(S_{21}S_{42} - (1+S_{22})S_{41}) + S_{43}((1+S_{22})S_{31} - S_{21}S_{32})} \tag{C.6}$$

we have:

$$Y_{11} = -\frac{1}{Z_0} + \frac{Y_{13}(S_{23}S_{34} - (1+S_{33})S_{24}) + Y_{14}((1+S_{44})S_{23} - S_{24}S_{43})}{S_{13}S_{24} - S_{14}S_{23}} \tag{C.7}$$

$$Y_{12} = \frac{Y_{13}((1+S_{33})S_{14} - S_{13}S_{34}) + Y_{14}(S_{14}S_{43} - (1+S_{44})S_{13})}{S_{13}S_{24} - S_{14}S_{23}} \tag{C.8}$$

$$Y_{13} = \frac{2(S_{13}S_{24} - S_{14}S_{23})/Z_0}{\begin{array}{l}(S_{13}S_{24} - S_{14}S_{23})S_{31} + (1+S_{11})(S_{23}S_{34} - (1+S_{33})S_{24}) + S_{21}((1+S_{33})S_{14} - S_{13}S_{34}) \\ +E_1((S_{13}S_{24} - S_{14}S_{23})S_{41} + (1+S_{11})((1+S_{44})S_{23} - S_{24}S_{43}) + S_{21}(S_{14}S_{43} - (1+S_{44})S_{13}))\end{array}} \tag{C.9}$$

$$Y_{14} = E_1 Y_{13}$$

$$Y_{21} = \frac{Y_{23}((1+S_{33})S_{24} - S_{23}S_{34}) + Y_{24}(S_{24}S_{43} - (1+S_{44})S_{23})}{S_{14}S_{23} - S_{13}S_{24}} \tag{C.10}$$

$$Y_{22} = -\frac{1}{Z_0} + \frac{Y_{23}(S_{13}S_{34} - (1+S_{33})S_{14}) + Y_{24}((1+S_{44})S_{13} - S_{14}S_{43})}{S_{14}S_{23} - S_{13}S_{24}} \tag{C.11}$$

$$Y_{23} = \frac{2(S_{14}S_{23} - S_{13}S_{24})/Z_0}{\begin{array}{l}(S_{14}S_{23} - S_{13}S_{24})S_{32} + (1+S_{22})(S_{13}S_{34} - (1+S_{33})S_{14}) + S_{12}((1+S_{33})S_{24} - S_{23}S_{34}) \\ +E_2((S_{14}S_{23} - S_{13}S_{24})S_{42} + (1+S_{22})((1+S_{44})S_{13} - S_{14}S_{43}) + S_{12}(S_{24}S_{43} - (1+S_{44})S_{23}))\end{array}} \tag{C.12}$$

$$Y_{24} = E_2 Y_{23}$$

$$Y_{31} = \frac{2(S_{31}S_{42} - S_{32}S_{41})/Z_0}{\begin{array}{l}(S_{31}S_{42} - S_{32}S_{41})S_{13} + (1+S_{33})(S_{12}S_{41} - (1+S_{11})S_{42}) + S_{43}((1+S_{11})S_{32} - S_{12}S_{31}) \\ +E_3((S_{31}S_{42} - S_{32}S_{41})S_{23} + (1+S_{33})((1+S_{22})S_{41} - S_{21}S_{42}) + S_{43}(S_{21}S_{32} - (1+S_{22})S_{31}))\end{array}} \tag{C.13}$$

$$Y_{32} = E_3 Y_{31}$$

$$Y_{33} = -\frac{1}{Z_0} + \frac{Y_{31}\left(S_{12}S_{41} - (1+S_{11})S_{42}\right) + Y_{32}\left((1+S_{22})S_{41} - S_{21}S_{42}\right)}{S_{31}S_{42} - S_{32}S_{41}} \quad \text{(C.14)}$$

$$Y_{34} = \frac{Y_{31}\left((1+S_{11})S_{32} - S_{12}S_{31}\right) + Y_{32}\left(S_{21}S_{32} - (1+S_{22})S_{31}\right)}{S_{31}S_{42} - S_{32}S_{41}} \quad \text{(C.15)}$$

$$Y_{41} = \frac{2(S_{32}S_{41} - S_{31}S_{42})/Z_0}{\begin{array}{l}(S_{32}S_{41} - S_{31}S_{42})S_{14} + (1+S_{44})\left(S_{12}S_{31} - (1+S_{11})S_{32}\right) + S_{34}\left((1+S_{11})S_{42} - S_{12}S_{41}\right) \\ +E_4\left((S_{32}S_{41} - S_{31}S_{42})S_{24} + (1+S_{44})\left((1+S_{22})S_{31} - S_{21}S_{32}\right) + S_{34}\left(S_{21}S_{42} - (1+S_{22})S_{41}\right)\right)\end{array}} \quad \text{(C.16)}$$

$$Y_{42} = E_4 Y_{41} \quad \text{(C.17)}$$

$$Y_{43} = \frac{Y_{41}\left((1+S_{11})S_{42} - S_{12}S_{41}\right) + Y_{42}\left(S_{21}S_{42} - (1+S_{22})S_{41}\right)}{S_{32}S_{41} - S_{31}S_{42}} \quad \text{(C.18)}$$

$$Y_{44} = -\frac{1}{Z_0} + \frac{Y_{41}\left(S_{12}S_{31} - (1+S_{11})S_{32}\right) + Y_{42}\left((1+S_{22})S_{31} - S_{21}S_{32}\right)}{S_{32}S_{41} - S_{31}S_{42}} \quad \text{(C.19)}$$

Four-Port Scattering Parameters for Basic Coupled-Inductors with Primary Shunt Capacitor

Four-port scattering matrix equations for the two coupled lines are:

$$\begin{bmatrix} b_1 \\ b_2 \\ b_3 \\ b_4 \end{bmatrix} = \begin{bmatrix} S_{11} & S_{12} & S_{13} & S_{14} \\ S_{21} & S_{22} & S_{23} & S_{24} \\ S_{31} & S_{32} & S_{33} & S_{34} \\ S_{41} & S_{42} & S_{43} & S_{44} \end{bmatrix} \cdot \begin{bmatrix} a_1 \\ a_2 \\ a_3 \\ a_4 \end{bmatrix} \tag{D.1}$$

If the normalization impedances on ports 1 to 4 are $\underline{Z}_1$ to $\underline{Z}_4$, respectively, and

k = mutual coupling coefficient
Y_C = capacitor admittance
Z_{L1} = impedance of primary inductor L_1
Z_{L2} = impedance of secondary inductor L_2

then S-parameters are:

$$S_{11} = \frac{(\omega k)^2 L_1 L_2 \left(1 - Y_C\left(\underline{Z}_1 - \underline{Z}_2\right)\right) + \left(Z_{L1} - \underline{Z}_1 + \underline{Z}_2 - Y_C Z_{L1}\left(\underline{Z}_1 - \underline{Z}_2\right)\right)\left(Z_{L2} + \underline{Z}_3 + \underline{Z}_4\right)}{(\omega k)^2 L_1 L_2 \left(1 + Y_C\left(\underline{Z}_1 + \underline{Z}_2\right)\right) + \left(Z_{L1} + \underline{Z}_1 + \underline{Z}_2 + Y_C Z_{L1}\left(\underline{Z}_1 + \underline{Z}_2\right)\right)\left(Z_{L2} + \underline{Z}_3 + \underline{Z}_4\right)} \tag{D.2}$$

$$S_{12} = \frac{2\left((\omega k)^2 L_1 L_2 Y_C + \left(1 + Y_C Z_{L1}\right)\left(Z_{L2} + \underline{Z}_3 + \underline{Z}_4\right)\right)\underline{Z}_1}{(\omega k)^2 L_1 L_2 \left(1 + Y_C\left(\underline{Z}_1 + \underline{Z}_2\right)\right) + \left(Z_{L1} + \underline{Z}_1 + \underline{Z}_2 + Y_C Z_{L1}\left(\underline{Z}_1 + \underline{Z}_2\right)\right)\left(Z_{L2} + \underline{Z}_3 + \underline{Z}_4\right)} \tag{D.3}$$

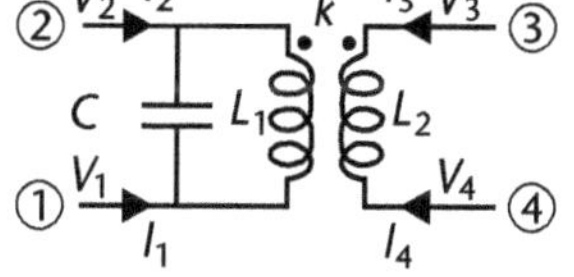

Figure D.1 Basic coupled inductors with primary shunt capacitor.

$$S_{13} = \frac{-2j\omega k\sqrt{L_1L_2}\underline{Z}_1}{(\omega k)^2 L_1L_2\left(1+Y_C\left(\underline{Z}_1+\underline{Z}_2\right)\right)+\left(Z_{L1}+\underline{Z}_1+\underline{Z}_2+Y_CZ_{L1}\left(\underline{Z}_1+\underline{Z}_2\right)\right)\left(Z_{L2}+\underline{Z}_3+\underline{Z}_4\right)} \tag{D.4}$$

$$S_{14} = \frac{2j\omega k\sqrt{L_1L_2}\underline{Z}_1}{(\omega k)^2 L_1L_2\left(1+Y_C\left(\underline{Z}_1+\underline{Z}_2\right)\right)+\left(Z_{L1}+\underline{Z}_1+\underline{Z}_2+Y_CZ_{L1}\left(\underline{Z}_1+\underline{Z}_2\right)\right)\left(Z_{L2}+\underline{Z}_3+\underline{Z}_4\right)} \tag{D.5}$$

$$S_{21} = \frac{2\left((\omega k)^2 L_1L_2Y_C+\left(1+Y_CZ_{L1}\right)\left(Z_{L2}+\underline{Z}_3+\underline{Z}_4\right)\right)\underline{Z}_2}{(\omega k)^2 L_1L_2\left(1+Y_C\left(\underline{Z}_1+\underline{Z}_2\right)\right)+\left(Z_{L1}+\underline{Z}_1+\underline{Z}_2+Y_CZ_{L1}\left(\underline{Z}_1+\underline{Z}_2\right)\right)\left(Z_{L2}+\underline{Z}_3+\underline{Z}_4\right)} \tag{D.6}$$

$$S_{22} = \frac{(\omega k)^2 L_1L_2\left(1+Y_C\left(\underline{Z}_1-\underline{Z}_2\right)\right)+\left(Z_{L1}+\underline{Z}_1-\underline{Z}_2+Y_CZ_{L1}\left(\underline{Z}_1-\underline{Z}_2\right)\right)\left(Z_{L2}+\underline{Z}_3+\underline{Z}_4\right)}{(\omega k)^2 L_1L_2\left(1+Y_C\left(\underline{Z}_1+\underline{Z}_2\right)\right)+\left(Z_{L1}+\underline{Z}_1+\underline{Z}_2+Y_CZ_{L1}\left(\underline{Z}_1+\underline{Z}_2\right)\right)\left(Z_{L2}+\underline{Z}_3+\underline{Z}_4\right)} \tag{D.7}$$

$$S_{23} = \frac{2j\omega k\sqrt{L_1L_2}\underline{Z}_2}{(\omega k)^2 L_1L_2\left(1+Y_C\left(\underline{Z}_1+\underline{Z}_2\right)\right)+\left(Z_{L1}+\underline{Z}_1+\underline{Z}_2+Y_CZ_{L1}\left(\underline{Z}_1+\underline{Z}_2\right)\right)\left(Z_{L2}+\underline{Z}_3+\underline{Z}_4\right)} \tag{D.8}$$

$$S_{24} = \frac{-2j\omega k\sqrt{L_1L_2}\underline{Z}_2}{(\omega k)^2 L_1L_2\left(1+Y_C\left(\underline{Z}_1+\underline{Z}_2\right)\right)+\left(Z_{L1}+\underline{Z}_1+\underline{Z}_2+Y_CZ_{L1}\left(\underline{Z}_1+\underline{Z}_2\right)\right)\left(Z_{L2}+\underline{Z}_3+\underline{Z}_4\right)} \tag{D.9}$$

$$S_{31} = \frac{-2j\omega k\sqrt{L_1L_2}\underline{Z}_3}{(\omega k)^2 L_1L_2\left(1+Y_C\left(\underline{Z}_1+\underline{Z}_2\right)\right)+\left(Z_{L1}+\underline{Z}_1+\underline{Z}_2+Y_CZ_{L1}\left(\underline{Z}_1+\underline{Z}_2\right)\right)\left(Z_{L2}+\underline{Z}_3+\underline{Z}_4\right)} \tag{D.10}$$

$$S_{32} = \frac{2j\omega k\sqrt{L_1L_2}\underline{Z}_3}{(\omega k)^2 L_1L_2\left(1+Y_C\left(\underline{Z}_1+\underline{Z}_2\right)\right)+\left(Z_{L1}+\underline{Z}_1+\underline{Z}_2+Y_CZ_{L1}\left(\underline{Z}_1+\underline{Z}_2\right)\right)\left(Z_{L2}+\underline{Z}_3+\underline{Z}_4\right)} \tag{D.11}$$

$$S_{33} = \frac{(\omega k)^2 L_1L_2\left(1+Y_C\left(\underline{Z}_1+\underline{Z}_2\right)\right)+\left(Z_{L1}+\underline{Z}_1+\underline{Z}_2+Y_CZ_{L1}\left(\underline{Z}_1+\underline{Z}_2\right)\right)\left(Z_{L2}-\underline{Z}_3+\underline{Z}_4\right)}{(\omega k)^2 L_1L_2\left(1+Y_C\left(\underline{Z}_1+\underline{Z}_2\right)\right)+\left(Z_{L1}+\underline{Z}_1+\underline{Z}_2+Y_CZ_{L1}\left(\underline{Z}_1+\underline{Z}_2\right)\right)\left(Z_{L2}+\underline{Z}_3+\underline{Z}_4\right)} \tag{D.12}$$

$$S_{34} = \frac{2\left(Z_{L1}+\underline{Z}_1+\underline{Z}_2+Y_CZ_{L1}\left(\underline{Z}_1+\underline{Z}_2\right)\right)\underline{Z}_3}{(\omega k)^2 L_1L_2\left(1+Y_C\left(\underline{Z}_1+\underline{Z}_2\right)\right)+\left(Z_{L1}+\underline{Z}_1+\underline{Z}_2+Y_CZ_{L1}\left(\underline{Z}_1+\underline{Z}_2\right)\right)\left(Z_{L2}+\underline{Z}_3+\underline{Z}_4\right)} \tag{D.13}$$

$$S_{41} = \frac{2j\omega k\sqrt{L_1L_2}\underline{Z}_4}{(\omega k)^2 L_1L_2\left(1+Y_C\left(\underline{Z}_1+\underline{Z}_2\right)\right)+\left(Z_{L1}+\underline{Z}_1+\underline{Z}_2+Y_CZ_{L1}\left(\underline{Z}_1+\underline{Z}_2\right)\right)\left(Z_{L2}+\underline{Z}_3+\underline{Z}_4\right)} \tag{D.14}$$

$$S_{42} = \frac{-2j\omega k\sqrt{L_1L_2}\underline{Z}_4}{(\omega k)^2 L_1L_2\left(1+Y_C\left(\underline{Z}_1+\underline{Z}_2\right)\right)+\left(Z_{L1}+\underline{Z}_1+\underline{Z}_2+Y_CZ_{L1}\left(\underline{Z}_1+\underline{Z}_2\right)\right)\left(Z_{L2}+\underline{Z}_3+\underline{Z}_4\right)} \tag{D.15}$$

$$S_{43} = \frac{2\left(Z_{L1}+\underline{Z}_1+\underline{Z}_2+Y_CZ_{L1}\left(\underline{Z}_1+\underline{Z}_2\right)\right)\underline{Z}_4}{(\omega k)^2 L_1L_2\left(1+Y_C\left(\underline{Z}_1+\underline{Z}_2\right)\right)+\left(Z_{L1}+\underline{Z}_1+\underline{Z}_2+Y_CZ_{L1}\left(\underline{Z}_1+\underline{Z}_2\right)\right)\left(Z_{L2}+\underline{Z}_3+\underline{Z}_4\right)} \tag{D.16}$$

$$S_{44} = \frac{(\omega k)^2 L_1 L_2 (1 + Y_C (\underline{Z}_1 + \underline{Z}_2)) + (Z_{L1} + \underline{Z}_1 + \underline{Z}_2 + Y_C Z_{L1} (\underline{Z}_1 + \underline{Z}_2))(Z_{L2} + \underline{Z}_3 - \underline{Z}_4)}{(\omega k)^2 L_1 L_2 (1 + Y_C (\underline{Z}_1 + \underline{Z}_2)) + (Z_{L1} + \underline{Z}_1 + \underline{Z}_2 + Y_C Z_{L1} (\underline{Z}_1 + \underline{Z}_2))(Z_{L2} + \underline{Z}_3 + \underline{Z}_4)} \quad \text{(D.17)}$$

Special Case: Common Normalization Impedance

$$Z_0 = \underline{Z}_1 = \underline{Z}_2 = \underline{Z}_3 = \underline{Z}_4 \quad \text{(D.18)}$$

$$S_{11} = S_{22} = \frac{(\omega k)^2 L_1 L_2 + (Z_{L2} + 2Z_0) Z_{L1}}{(\omega k)^2 L_1 L_2 (1 + 2Y_C Z_0) + (Z_{L1} + 2(1 + Y_C Z_{L1}) Z_0)(Z_{L2} + 2Z_0)} \quad \text{(D.19)}$$

$$S_{12} = S_{21} = \frac{2\left((\omega k)^2 L_1 L_2 Y_C + (1 + Y_C Z_{L1})(Z_{L2} + 2Z_0)\right) Z_0}{(\omega k)^2 L_1 L_2 (1 + 2Y_C Z_0) + (Z_{L1} + 2(1 + Y_C Z_{L1}) Z_0)(Z_{L2} + 2Z_0)} \quad \text{(D.20)}$$

$$\begin{aligned} -S_{13} &= -S_{31} = -S_{24} = -S_{42} = S_{14} = S_{41} = S_{23} = S_{32} \\ &= \frac{2j\omega k\sqrt{L_1 L_2} Z_0}{(\omega k)^2 L_1 L_2 (1 + 2Y_C Z_0) + (Z_{L1} + 2(1 + Y_C Z_{L1}) Z_0)(Z_{L2} + 2Z_0)} \end{aligned} \quad \text{(D.21)}$$

$$S_{33} = S_{44} = \frac{(\omega k)^2 L_1 L_2 (1 + 2Y_C Z_0) + (Z_{L1} + 2(1 + Y_C Z_{L1}) Z_0)(Z_{L2})}{(\omega k)^2 L_1 L_2 (1 + 2Y_C Z_0) + (Z_{L1} + 2(1 + Y_C Z_{L1}) Z_0)(Z_{L2} + 2Z_0)} \quad \text{(D.22)}$$

$$S_{34} = S_{43} = \frac{2(Z_{L1} + 2(1 + Y_C Z_{L1}) Z_0)(Z_0)}{(\omega k)^2 L_1 L_2 (1 + 2Y_C Z_0) + (Z_{L1} + 2(1 + Y_C Z_{L1}) Z_0)(Z_{L2} + 2Z_0)} \quad \text{(D.23)}$$

APPENDIX E

Quadratic Polynomial Fit Through Three Data Points

If dependencies at three data points are

at x_1 $$y_1 = a_0 + a_1x_1 + a_2x_1^2 \tag{E.1}$$

at x_2 $$y_2 = a_0 + a_1x_2 + a_2x_2^2 \tag{E.2}$$

at x_3 $$y_3 = a_0 + a_1x_3 + a_2x_3^2 \tag{E.3}$$

then coefficients are given by

$$a_0 = \frac{x_2x_3(x_3 - x_2)y_1 - x_1x_3(x_3 - x_1)y_2 + x_1x_2(x_2 - x_1)y_3}{(x_3 - x_2)x_1^2 - (x_3 - x_1)x_2^2 + (x_2 - x_1)x_3^2} \tag{E.4}$$

$$a_1 = \frac{(x_2^2 - x_3^2)y_1 + (x_3^2 - x_1^2)y_2 - (x_2^2 - x_1^2)y_3}{(x_3 - x_2)x_1^2 - (x_3 - x_1)x_2^2 + (x_2 - x_1)x_3^2} \tag{E.5}$$

$$a_2 = \frac{(x_3 - x_2)y_1 - (x_3 - x_1)y_2 + (x_2 - x_1)y_3}{(x_3 - x_2)x_1^2 - (x_3 - x_1)x_2^2 + (x_2 - x_1)x_3^2} \tag{E.6}$$

The location of the minimum/maximum value will occur at

$$x = -a_1/2a_2 \tag{E.7}$$

that is,

$$x = \frac{(x_2^2 - x_1^2)y_3 - (x_2^2 - x_3^2)y_1 - (x_3^2 - x_1^2)y_2}{2((x_3 - x_2)y_1 - (x_3 - x_1)y_2 + (x_2 - x_1)y_3)} \tag{E.8}$$

APPENDIX F

Analysis of Coupled Inductors Including Interwinding Capacitance

Coupled inductors are critical circuit elements widely employed in most compact RF modules in today's mobile communications devices. Typically, the inductors are implemented as metalized traces printed on one or more layers of a passive or active die. Magnetic coupling between the inductors is achieved principally through one of two ways. First, the inductors are wound tightly together in a coplanar topology; this is commonly referred to as proximity coupling. Second, the inductors are superposed on one another on different metallization layers; this is referred to as overlap coupling. In some structures, a combination of the two coupling methods may be employed. Whatever the case, the goal is generally to achieve as high a magnetic coupling coefficient (k) as possible.

Magnetic coupling between the two coils is maximized by minimizing the distance between the coils, either horizontally or vertically. Unfortunately, minimizing the distance between the coils also has the undesirable effect of maximizing the capacitance between the coils. This capacitance can significantly impact the performance characteristics of the coupled inductor pair. Thus, deriving a lumped-element model that accurately accounts for this capacitance is highly desirable. However, introducing the interwinding capacitance into a lumped-element model for a pair of coupled inductors presents significant complexity.

Figure F.1 shows a schematic representing a coupled-inductor pair having interwinding capacitance, and shunt reactive matching elements Y_S and Y_L. The driving sources on all four terminals are assumed unequal and arbitrary. In order to analyze this network, it is first necessary to derive the system equations for the coupled inductors in isolation. To accomplish this, an incremental model was developed, as detailed in Figure F.2.

The model assumes that the interwinding capacitance C_X is uniformly distributed along the length of the inductors. The primary inductor value is denoted by L_1, and that of the second by L_2. The magnetic coupling coefficient is denoted by k.

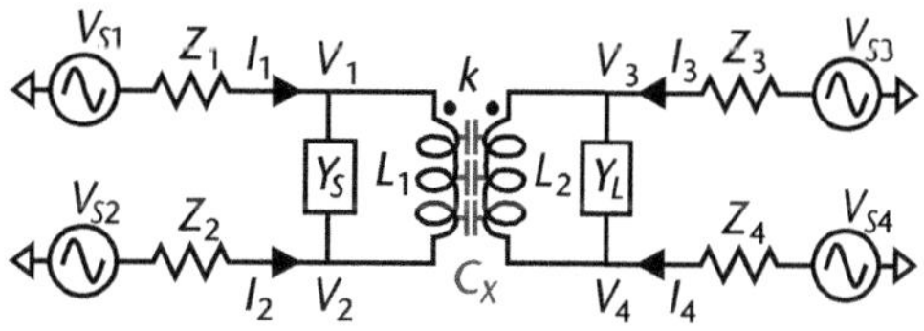

Figure F.1 Coupled inductors including interwinding capacitance and external shunt matching and sources.

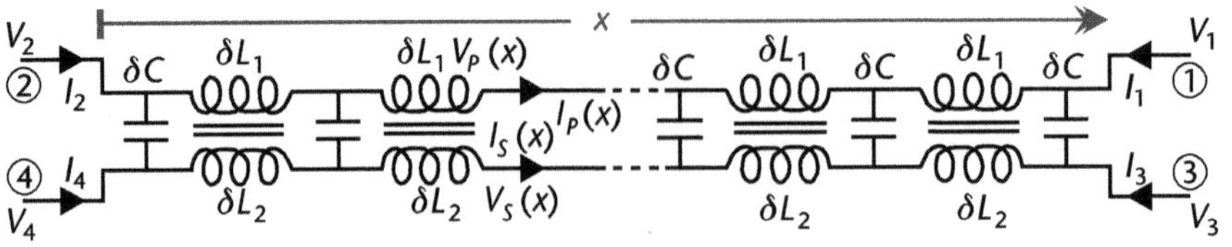

Figure F.2 Incremental model of coupled-inductor pair with interwinding capacitance.

Incremental circuit analysis yields

$$\frac{\delta I_P(x)}{\delta x} = -\left(V_P(x) - V_S(x)\right) Y_C / l \tag{F.1}$$

$$\frac{\delta I_S(x)}{\delta x} = \left(V_P(x) - V_S(x)\right) Y_C / l \tag{F.2}$$

$$\frac{\delta V_P(x)}{\delta x} = -I_P(x) Z_{L1}/l - j\omega k\left(\sqrt{L_1 L_2}/l\right) I_S(x) \tag{F.3}$$

$$\frac{\delta V_S(x)}{\delta x} = -I_S(x) Z_{L2}/l - j\omega k\left(\sqrt{L_1 L_2}/l\right) I_P(x) \tag{F.4}$$

where V_P and I_P represent the voltage and current along the primary inductor L_1, and V_S and I_S represent the voltage and current along the secondary inductor L_2. The equal lengths of the two inductors are denoted by l.

From these equations, we determine

$$\frac{\delta^2 V_P(x)}{\delta x^2} = \left(Z_{L1} - j\omega k\sqrt{L_1 L_2}\right)\left(\frac{Y_C}{l^2}\right)\left(V_P(x) - V_S(x)\right) \tag{F.5}$$

$$\frac{\delta^2 V_S(x)}{\delta x^2} = \left(Z_{L2} - j\omega k\sqrt{L_1 L_2}\right)\left(\frac{Y_C}{l^2}\right)\left(V_S(x) - V_P(x)\right) \tag{F.6}$$

Hence,

$$\frac{\delta^2 V_P(x)/\delta x^2}{\delta^2 V_S(x)/\delta x^2} = -\frac{Z_{L1} - j\omega k\left(\sqrt{L_1 L_2}\right)}{Z_{L2} - j\omega k\left(\sqrt{L_1 L_2}\right)} \tag{F.7}$$

Integration yields

$$\left(Z_{L2} - j\omega k\sqrt{L_1 L_2}\right) V_P(x) + \left(Z_{L1} - j\omega k\sqrt{L_1 L_2}\right) V_S(x) = Ax + B \tag{F.8}$$

where A and B are arbitrary constants. For simplification of the subsequent equations, the following substitutions are made

$$\eta_1 = Z_{L1} - j\omega k\sqrt{L_1 L_2} \qquad \text{and} \qquad \eta_2 = Z_{L2} - j\omega k\sqrt{L_1 L_2} \tag{F.9}$$

thus

$$\eta_2 V_P(x) + \eta_1 V_S(x) = A_X + B \tag{F.10}$$

substituting in (F.1)

$$\eta_1 \frac{\delta I_P(x)}{\delta x} = -(\eta_1 + \eta_2)(Y_C/l)V_P(x) + (Ax + B)(Y_C/l) \tag{F.11}$$

substituting in (F.2)

$$\eta_2 \frac{\delta I_S(x)}{\delta x} = -(\eta_1 + \eta_2)(Y_C/l)V_S(x) + (Ax + B)(Y_C/l) \tag{F.12}$$

substituting in (F.5)

$$\delta^2 V_P(x)/\delta x^2 = (\eta_1 + \eta_2)(Y_C/l^2)V_P(x) - (Ax + B)(Y_C/l^2) \tag{F.13}$$

substituting in (F.6)

$$\delta^2 V_S(x)/\delta x^2 = (\eta_1 + \eta_2)(Y_C/l^2)V_S(x) - (Ax + B)(Y_C/l^2) \tag{F.14}$$

Equations (F.13) and (F.14) are second-order nonhomogenous equations. To obtain a solution for $V_P(x)$ satisfying (F.13), first consider the homogeneous equation

$$\delta^2 V_P(x)/\delta x^2 - (\eta_1 + \eta_2)(Y_C/l^2)V_P(x) = 0 \tag{F.15}$$

with trial solution

$$V_P(x) = e^{rx}$$

substituting

$$r^2 e^{rx} - (\eta_1 + \eta_2)(Y_C/l^2)e^{rx} = 0 \tag{F.16}$$

thus

$$r = \pm\sqrt{(\eta_1 + \eta_2)Y_C}/l \tag{F.17}$$

leading to the complete homogeneous solution

$$V_P(x) = \xi_1 e^{-\sqrt{(\eta_1+\eta_2)Y_C}x/l} + \xi_2 e^{\sqrt{(\eta_1+\eta_2)Y_C}x/l} \tag{F.18}$$

where ξ_1 and ξ_2 are arbitrary constants.

For the particular solution to (F.13), assume

$$V_P(x) = ax^2 + bx + c \tag{F.19}$$

where a, b, and c are constants.

Substituting in (F.13),

$$2a - (\eta_1 + \eta_2)\left(\frac{Y_C}{l^2}\right)\left(ax^2 + bx + c\right) + (Ax + B)\left(\frac{Y_C}{l^2}\right) = 0 \tag{F.20}$$

$$\begin{aligned} &2a - a(\eta_1 + \eta_2)\left(\frac{Y_C}{l^2}\right)x^2 - ((\eta_1 + \eta_2)b - A)\left(\frac{Y_C}{l^2}\right)x \\ &+ (B - (\eta_1 + \eta_2)c)\left(\frac{Y_C}{l^2}\right) = 0 \end{aligned} \tag{F.21}$$

Hence,

$$a = 0 \qquad b = A/(\eta_1 + \eta_2) \tag{F.22}$$

$$c = B/(\eta_1 + \eta_2) \tag{F.23}$$

thus, the particular solution is

$$V_P(x) = (Ax + B)/(\eta_1 + \eta_2) \tag{F.24}$$

yielding a complete solution to (F.13) as

$$V_P(x) = \xi_1 e^{-\sqrt{(\eta_1+\eta_2)Y_C}x/l} + \xi_2 e^{\sqrt{(\eta_1+\eta_2)Y_C}x/l} + (Ax + B)/(\eta_1 + \eta_2) \tag{F.25}$$

Similarly, for a secondary inductor, the solution to (F.14) is

$$V_S(x) = \xi_3 e^{-\sqrt{(\eta_1+\eta_2)Y_C}x/l} + \xi_4 e^{\sqrt{(\eta_1+\eta_2)Y_C}x/l} + (Ax + B)/(\eta_1 + \eta_2) \tag{F.26}$$

Differentiating, we have

$$\delta^2 V_P(x)/\delta x^2 = (\eta_1 + \eta_2)\left(Y_C/l^2\right)\xi_1 e^{-\sqrt{(\eta_1+\eta_2)Y_C}x/l} + (\eta_1 + \eta_2)\left(Y_C/l^2\right)\xi_2 e^{\sqrt{(\eta_1+\eta_2)Y_C}x/l} \tag{F.27}$$

and

$$\delta^2 V_S(x)/\delta x^2 = (\eta_1 + \eta_2)\left(Y_C/l^2\right)\xi_3 e^{-\sqrt{(\eta_1+\eta_2)Y_C}x/l} + (\eta_1 + \eta_2)\left(Y_C/l^2\right)\xi_4 e^{\sqrt{(\eta_1+\eta_2)Y_C}x/l} \quad (F.28)$$

Substituting in (F.7),

$$\frac{\xi_1 e^{-\sqrt{(\eta_1+\eta_2)Y_C}x/l} + \xi_2 e^{\sqrt{(\eta_1+\eta_2)Y_C}x/l}}{\xi_3 e^{-\sqrt{(\eta_1+\eta_2)Y_C}x/l} + \xi_{41} e^{\sqrt{(\eta_1+\eta_2)Y_C}x/l}} = -\frac{\eta_1}{\eta_2}$$

from which it follows

$$\xi_3 = -\xi_1\eta_2/\eta_1 \quad (F.29)$$

and

$$\xi_4 = -\xi_2\eta_2/\eta_1 \quad (F.30)$$

For convenience, define new variables as

$$\xi_1 = \eta_1\psi_1 \quad (F.31)$$

$$\xi_2 = \eta_1\psi_2 \quad (F.32)$$

Equations (F.25) and (F.26) become

$$V_P(x) = \eta_1\psi_1 e^{-\sqrt{(\eta_1+\eta_2)Y_C}x/l} + \eta_1\psi_2 e^{\sqrt{(\eta_1+\eta_2)Y_C}x/l} + (Ax + B)/(\eta_1 + \eta_2) \quad (F.33)$$

and

$$V_S(x) = -\eta_2\psi_1 e^{-\sqrt{(\eta_1+\eta_2)Y_C}x/l} - \eta_2\psi_2 e^{\sqrt{(\eta_1+\eta_2)Y_C}x/l} + (Ax + B)/(\eta_1 + \eta_2) \quad (F.34)$$

Applying boundary conditions, $V_P(0) = V_2$, $V_S(0) = V_4$ gives

$$V_2 = \eta_1(\psi_1 + \psi_2) + B/(\eta_1 + \eta_2)$$

$$V_4 = -\eta_2(\psi_1 + \psi_2) + B/(\eta_1 + \eta_2)$$

Hence,

$$V_2 - V_4 = (\eta_1 + \eta_2)(\psi_1 + \psi_2)$$

and

$$B/(\eta_1 + \eta_2) = V_2 - \eta_1(\psi_1 + \psi_2) = V_4 + \eta_2(\psi_1 + \psi_2) \quad (F.35)$$

thus

$$\psi_1 + \psi_2 = V_2 - V_4/\eta_1 + \eta_2 \tag{F.36}$$

Hence,

$$B = \eta_2 V_2 + \eta_1 V_4 \tag{F.37}$$

Applying boundary conditions, $V_P(l) = V_1$, $V_S(l) = V_3$, we have

$$Al/(\eta_1 + \eta_2) = V_1 - V_2 + \eta_1\psi_1\left(1 - e^{-\sqrt{(\eta_1+\eta_2)Y_C}}\right) + \eta_1\psi_2\left(1 - e^{\sqrt{(\eta_1+\eta_2)Y_C}}\right) \tag{F.38}$$

$$Al/(\eta_1 + \eta_2) = V_3 - V_4 - \eta_2\psi_1\left(1 - e^{-\sqrt{(\eta_1+\eta_2)Y_C}}\right) - \eta_2\psi_2\left(1 - e^{\sqrt{(\eta_1+\eta_2)Y_C}}\right) \tag{F.39}$$

thus

$$\begin{aligned} &V_1 - V_2 - V_3 + V_4 + (\eta_1 + \eta_2)\psi_1\left(1 - e^{-\sqrt{(\eta_1+\eta_2)Y_C}}\right) \\ &+ (\eta_1 + \eta_2)\psi_2\left(1 - e^{\sqrt{(\eta_1+\eta_2)Y_C}}\right) = 0 \end{aligned} \tag{F.40}$$

from (F.36) and (F.40) determine

$$\begin{aligned} &V_1 - V_2 - V_3 + V_4 + (\eta_1 + \eta_2)\psi_1\left(1 - e^{-\sqrt{(\eta_1+\eta_2)Y_C}}\right) \\ &+ \left(V_2 - V_4 - (\eta_1 + \eta_2)\psi_1\right)\left(1 - e^{\sqrt{(\eta_1+\eta_2)Y_C}}\right) = 0 \end{aligned} \tag{F.41}$$

hence,

$$\psi_1 = \frac{V_1 - V_3 - (V_2 - V_4)e^{\sqrt{(\eta_1+\eta_2)Y_C}}}{(\eta_1 + \eta_2)\left(e^{-\sqrt{(\eta_1+\eta_2)Y_C}} - e^{\sqrt{(\eta_1+\eta_2)Y_C}}\right)} \tag{F.42}$$

also

$$\begin{aligned} &V_1 - V_2 - V_3 + V_4 + \left(V_2 - V_4 - (\eta_1 + \eta_2)\psi_2\right)\left(1 - e^{-\sqrt{(\eta_1+\eta_2)Y_C}}\right) \\ &+ (\eta_1 + \eta_2)\psi_2\left(1 - e^{\sqrt{(\eta_1+\eta_2)Y_C}}\right) = 0 \end{aligned} \tag{F.43}$$

hence,

$$\psi_2 = \frac{V_1 - V_3 - (V_2 - V_4)e^{-\sqrt{(\eta_1+\eta_2)Y_C}}}{(\eta_1 + \eta_2)\left(e^{-\sqrt{(\eta_1+\eta_2)Y_C}} - e^{\sqrt{(\eta_1+\eta_2)Y_C}}\right)} \tag{F.44}$$

from (F.38), (F.42), and (F.44),

$$Al = \eta_2(V_1 - V_2) + \eta_1(V_3 - V_4) \tag{F.45}$$

substituting in (F.33),

$$V_P(x) = \frac{1}{(\eta_1 + \eta_2)}\left(\begin{array}{l} \dfrac{\eta_1\left(\begin{array}{l}(V_1 - V_3)\left(e^{-\sqrt{(\eta_1+\eta_2)Y_C}x/l} - e^{\sqrt{(\eta_1+\eta_2)Y_C}x/l}\right) \\ +(V_2 - V_4)\left(e^{-(1-x/l)\sqrt{(\eta_1+\eta_2)Y_C}} - e^{(1-x/l)\sqrt{(\eta_1+\eta_2)Y_C}}\right)\end{array}\right)}{e^{-\sqrt{(\eta_1+\eta_2)Y_C}} - e^{\sqrt{(\eta_1+\eta_2)Y_C}}} \\ +(\eta_2 V_1 + \eta_1 V_3)(x/l) + (\eta_2 V_2 + \eta_1 V_4)(1 - x/l)\end{array}\right) \tag{F.46}$$

which, if $Y_C = 0$, reduces to

$$V_P(x) = V_2 + (V_1 - V_2)\left(\frac{x}{l}\right) \tag{F.47}$$

substituting in (F.34),

$$V_S(x) = -\frac{1}{(\eta_1 + \eta_2)}\left(\begin{array}{l} \dfrac{\eta_2\left(\begin{array}{l}(V_1 - V_3)\left(e^{-\sqrt{(\eta_1+\eta_2)Y_C}x/l} - e^{\sqrt{(\eta_1+\eta_2)Y_C}x/l}\right) \\ +(V_2 - V_4)\left(e^{-(1-x/l)\sqrt{(\eta_1+\eta_2)Y_C}} - e^{(1-x/l)\sqrt{(\eta_1+\eta_2)Y_C}}\right)\end{array}\right)}{e^{-\sqrt{(\eta_1+\eta_2)Y_C}} - e^{\sqrt{(\eta_1+\eta_2)Y_C}}} \\ -(\eta_2 V_1 + \eta_1 V_3)(x/l) + (\eta_2 V_2 + \eta_1 V_4)(1 - x/l)\end{array}\right) \tag{F.48}$$

from (F.11) and (F.33),

$$I_P(x) = \sqrt{(\eta_1 + \eta_2)Y_C}\left(\psi_1 e^{-\sqrt{(\eta_1+\eta_2)Y_C}x/l} - \psi_2 e^{\sqrt{(\eta_1+\eta_2)Y_C}x/l}\right) + \psi_3 \tag{F.49}$$

from (F.12) and (F.34),

$$I_S(x) = -\sqrt{(\eta_1 + \eta_2)Y_C}\left(\psi_1 e^{-\sqrt{(\eta_1+\eta_2)Y_C}x/l} - \psi_2 e^{\sqrt{(\eta_1+\eta_2)Y_C}x/l}\right) + \psi_4 \tag{F.50}$$

from (F.42), (F.44), and (F.49),

$$I_P(x) = \frac{\sqrt{(\eta_1 + \eta_2)Y_C}}{(\eta_1 + \eta_2)\left(e^{-\sqrt{(\eta_1+\eta_2)Y_C}} - e^{\sqrt{(\eta_1+\eta_2)Y_C}}\right)} \left(\begin{array}{l}(V_1 - V_3)\left(e^{-\sqrt{(\eta_1+\eta_2)Y_C}x/l} + e^{\sqrt{(\eta_1+\eta_2)Y_C}x/l}\right) \\ -(V_2 - V_4)\left(e^{-(1-x/l)\sqrt{(\eta_1+\eta_2)Y_C}} + e^{(1-x/l)\sqrt{(\eta_1+\eta_2)Y_C}}\right)\end{array}\right) + \psi_3 \tag{F.51}$$

from (F.42), (F.44), and (F.50),

$$I_S(x) = \frac{-\sqrt{(\eta_1+\eta_2)Y_C}}{(\eta_1+\eta_2)\left(e^{-\sqrt{(\eta_1+\eta_2)Y_C}} - e^{\sqrt{(\eta_1+\eta_2)Y_C}}\right)} \begin{pmatrix} (V_1 - V_3)\left(e^{-\sqrt{(\eta_1+\eta_2)Y_C}x/l} + e^{\sqrt{(\eta_1+\eta_2)Y_C}x/l}\right) \\ -(V_2 - V_4)\left(e^{-(1-x/l)\sqrt{(\eta_1+\eta_2)Y_C}} + e^{(1-x/l)\sqrt{(\eta_1+\eta_2)Y_C}}\right) \end{pmatrix} + \psi_4 \tag{F.52}$$

from (F.46),

$$\frac{\delta V_P(x)}{\delta x} = \frac{1}{(\eta_1+\eta_2)} \left(\frac{\eta_1\sqrt{(\eta_1+\eta_2)Y_C} \left(\begin{array}{l} -(V_1 - V_3)\left(e^{-\sqrt{(\eta_1+\eta_2)Y_C}x/l} + e^{\sqrt{(\eta_1+\eta_2)Y_C}x/l}\right) \\ +(V_2 - V_4)\left(e^{-(1-x/l)\sqrt{(\eta_1+\eta_2)Y_C}} - e^{(1-x/l)\sqrt{(\eta_1+\eta_2)Y_C}}\right) \end{array} \right)}{e^{-\sqrt{(\eta_1+\eta_2)Y_C}} - e^{\sqrt{(\eta_1+\eta_2)Y_C}}} + (\eta_2 V_1 + \eta_1 V_3 - \eta_2 V_2 + \eta_1 V_4) \right) / l \tag{F.53}$$

substituting in (F.3), with (F.51) and (F.52),

$$\begin{aligned} &(\eta_1+\eta_2)\psi_3 Z_{L1} + (\eta_1+\eta_2) j\omega k\sqrt{L_1L_2}\psi_4 \\ &+ \eta_2V_1 - \eta_2V_2 + \eta_1V_3 - \eta_1V_4 = 0 \end{aligned} \tag{F.54}$$

from (F.48),

$$\frac{\delta V_S(x)}{\delta x} = \frac{1}{\eta_1+\eta_2} \left(\frac{\eta_2\sqrt{(\eta_1+\eta_2)Y_C} \left(\begin{array}{l} -(V_1 - V_3)\left(e^{-\sqrt{(\eta_1+\eta_2)Y_C}x/l} + e^{\sqrt{(\eta_1+\eta_2)Y_C}x/l}\right) \\ +(V_2 - V_4)\left(e^{-(1-x/l)\sqrt{(\eta_1+\eta_2)Y_C}} + e^{(1-x/l)\sqrt{(\eta_1+\eta_2)Y_C}}\right) \end{array} \right)}{e^{-\sqrt{(\eta_1+\eta_2)Y_C}} - e^{\sqrt{(\eta_1+\eta_2)Y_C}}} - (\eta_2 V_1 + \eta_1 V_3 - \eta_2 V_2 + \eta_1 V_4) \right) / l \tag{F.55}$$

substituting in (F.4), with (F.51) and (F.52),

$$\begin{aligned} &(\eta_1+\eta_2) j\omega k\sqrt{L_1L_2}\psi_3 + (\eta_1+\eta_2)\psi_4 Z_{L2} \\ &+ \eta_2V_1 - \eta_2V_2 + \eta_1V_3 - \eta_1V_4 = 0 \end{aligned} \tag{F.56}$$

Hence from (F.54) and (F.56),

$$\psi_3 = \frac{\eta_2^2(V_2 - V_1) + \eta_1\eta_2(V_4 - V_3)}{(\eta_1+\eta_2)\left(Z_{L1}Z_{L2} + (\omega k)^2 L_1L_2\right)} \tag{F.57}$$

and

$$\psi_4 = \frac{\eta_1^2 (V_4 - V_3) + \eta_1\eta_2 (V_2 - V_1)}{(\eta_1 + \eta_2)\left(Z_{L1}Z_{L2} + (\omega k)^2 L_1 L_2\right)} \tag{F.58}$$

Substituting in (F.51) yields

$$\begin{aligned} I_P(x) = &\frac{\sqrt{(\eta_1 + \eta_2)Y_C}}{(\eta_1 + \eta_2)\left(e^{-\sqrt{(\eta_1+\eta_2)Y_C}} - e^{\sqrt{(\eta_1+\eta_2)Y_C}}\right)} \\ &\left(\begin{aligned} &(V_1 - V_3)\left(e^{-\sqrt{(\eta_1+\eta_2)Y_C}x/l} + e^{\sqrt{(\eta_1+\eta_2)Y_C}x/l}\right) \\ &- (V_2 - V_4)\left(e^{-(1-x/l)\sqrt{(\eta_1+\eta_2)Y_C}} + e^{(1-x/l)\sqrt{(\eta_1+\eta_2)Y_C}}\right) \end{aligned} \right) \\ &+ \frac{\eta_2^2 (V_2 - V_1) + \eta_1\eta_2 (V_4 - V_3)}{(\eta_1 + \eta_2)\left(Z_{L1}Z_{L2} + (\omega k)^2 L_1 L_2\right)} \end{aligned} \tag{F.59}$$

and in (F.52) yields

$$\begin{aligned} I_S(x) = &\frac{-\sqrt{(\eta_1 + \eta_2)Y_C}}{(\eta_1 + \eta_2)\left(e^{-\sqrt{(\eta_1+\eta_2)Y_C}} - e^{\sqrt{(\eta_1+\eta_2)Y_C}}\right)} \\ &\left(\begin{aligned} &(V_1 - V_3)\left(e^{-\sqrt{(\eta_1+\eta_2)Y_C}x/l} + e^{\sqrt{(\eta_1+\eta_2)Y_C}x/l}\right) \\ &- (V_2 - V_4)\left(e^{-(1-x/l)\sqrt{(\eta_1+\eta_2)Y_C}} + e^{(1-x/l)\sqrt{(\eta_1+\eta_2)Y_C}}\right) \end{aligned} \right) \\ &+ \frac{\eta_1^2 (V_4 - V_3) + \eta_1\eta_2 (V_2 - V_1)}{(\eta_1 + \eta_2)\left(Z_{L1}Z_{L2} + (\omega k)^2 L_1 L_2\right)} \end{aligned} \tag{F.60}$$

In summary, the spatially variant voltages and currents in a coupled inductor pair, with distributed capacitance, are governed by (F.46), (F.48), (F.59), and (F.60), with the variables η_1 and η_2 as defined in (F.9).

To apply the above equations to a circuit analysis of the coupled inductors, the spatially variant current equations must be related to the nodal currents at the four network terminals. If the currents flowing into the inductors are defined as I_{X1} to I_{X4}, as shown in Figure F.3, we have

$$I_{X1}(x) = -I_P(l) \tag{F.61}$$

$$I_{X2}(x) = I_P(0) \tag{F.62}$$

$$I_{X3}(x) = -I_S(l) \tag{F.63}$$

$$I_{X4}(x) = I_S(0) \tag{F.64}$$

Hence,

$$-I_{X1} = \frac{\sqrt{(\eta_1 + \eta_2)Y_C}}{(\eta_1 + \eta_2)\left(e^{-\sqrt{(\eta_1+\eta_2)Y_C}} - e^{\sqrt{(\eta_1+\eta_2)Y_C}}\right)}\begin{pmatrix}(V_1 - V_3)\left(e^{-\sqrt{(\eta_1+\eta_2)Y_C}} + e^{\sqrt{(\eta_1+\eta_2)Y_C}}\right) \\ -2(V_2 - V_4)\end{pmatrix} + \frac{\eta_2^2(V_2 - V_1) + \eta_1\eta_2(V_4 - V_3)}{(\eta_1 + \eta_2)\left(Z_{L1}Z_{L2} + (\omega k)^2 L_1L_2\right)} \tag{F.65}$$

$$I_{X2} = \frac{\sqrt{(\eta_1 + \eta_2)Y_C}}{(\eta_1 + \eta_2)\left(e^{-\sqrt{(\eta_1+\eta_2)Y_C}} - e^{\sqrt{(\eta_1+\eta_2)Y_C}}\right)}\begin{pmatrix}2(V_1 - V_3) - (V_2 - V_4) \\ \left(e^{-\sqrt{(\eta_1+\eta_2)Y_C}} + e^{\sqrt{(\eta_1+\eta_2)Y_C}}\right)\end{pmatrix} + \frac{\eta_2^2(V_2 - V_1) + \eta_1\eta_2(V_4 - V_3)}{(\eta_1 + \eta_2)\left(Z_{L1}Z_{L2} + (\omega k)^2 L_1L_2\right)} \tag{F.66}$$

$$-I_{X3} = \frac{-\sqrt{(\eta_1 + \eta_2)Y_C}}{(\eta_1 + \eta_2)\left(e^{-\sqrt{(\eta_1+\eta_2)Y_C}} - e^{\sqrt{(\eta_1+\eta_2)Y_C}}\right)}\begin{pmatrix}(V_1 - V_3)\left(e^{-\sqrt{(\eta_1+\eta_2)Y_C}} + e^{\sqrt{(\eta_1+\eta_2)Y_C}}\right) \\ -2(V_2 - V_4)\end{pmatrix} + \frac{\eta_1^2(V_4 - V_3) + \eta_1\eta_2(V_2 - V_1)}{(\eta_1 + \eta_2)\left(Z_{L1}Z_{L2} + (\omega k)^2 L_1L_2\right)} \tag{F.67}$$

$$I_{X4} = \frac{-\sqrt{(\eta_1 + \eta_2)Y_C}}{(\eta_1 + \eta_2)\left(e^{-\sqrt{(\eta_1+\eta_2)Y_C}} - e^{\sqrt{(\eta_1+\eta_2)Y_C}}\right)}\begin{pmatrix}2(V_1 - V_3) - (V_2 - V_4) \\ \left(e^{-\sqrt{(\eta_1+\eta_2)Y_C}} + e^{\sqrt{(\eta_1+\eta_2)Y_C}}\right)\end{pmatrix} + \frac{\eta_1^2(V_4 - V_3) + \eta_1\eta_2(V_2 - V_1)}{(\eta_1 + \eta_2)\left(Z_{L1}Z_{L2} + (\omega k)^2 L_1L_2\right)} \tag{F.68}$$

To facilitate subsequent analysis, the following variables are defined. Defining

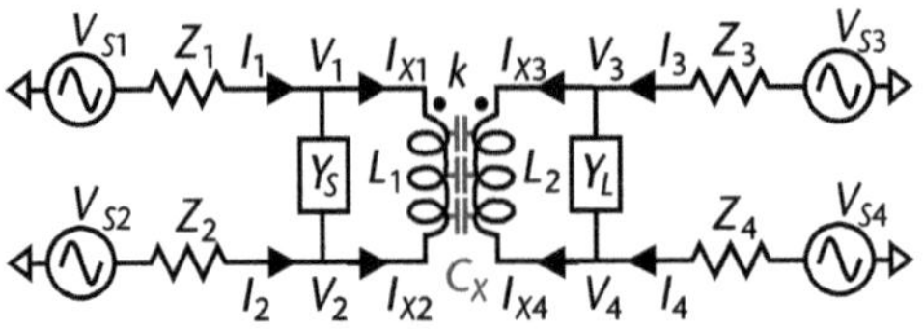

Figure F.3 Coupled-inductors including interwinding capacitance and nodal current relationships.

$$\eta_3 = \frac{\sqrt{(\eta_1 + \eta_2)Y_C}}{(\eta_1 + \eta_2)\left(e^{-\sqrt{(\eta_1+\eta_2)Y_C}} - e^{\sqrt{(\eta_1+\eta_2)Y_C}}\right)} \tag{F.69}$$

If $Y_C = 0$,

$$\eta_3 = \frac{-1}{2(\eta_1 + \eta_2)} \tag{F.70}$$

$$\eta_4 = \eta_3\left(e^{-\sqrt{(\eta_1+\eta_2)Y_C}} + e^{\sqrt{(\eta_1+\eta_2)Y_C}}\right) \tag{F.71}$$

$$\eta_5 = \frac{1}{(\eta_1 + \eta_2)\left(Z_{L1}Z_{L2} + (\omega k)^2 L_1L_2\right)} \tag{F.72}$$

In terms of these variables, the nodal currents are

$$I_{X1} = \left(\eta_2^2\eta_5 - \eta_4\right)V_1 + \left(2\eta_3 - \eta_2^2\eta_5\right)V_2 + (\eta_4 + \eta_1\eta_2\eta_5)V_3 - (2\eta_3 + \eta_1\eta_2\eta_5)V_4 \tag{F.73}$$

$$I_{X2} = \left(2\eta_3 - \eta_2^2\eta_5\right)V_1 + \left(\eta_2^2\eta_5 - \eta_4\right)V_2 - (2\eta_3 + \eta_1\eta_2\eta_5)V_3 + (\eta_4 + \eta_1\eta_2\eta_5)V_4 \tag{F.74}$$

$$I_{X3} = (\eta_4 + \eta_1\eta_2\eta_5)V_1 - (2\eta_3 + \eta_1\eta_2\eta_5)V_2 - \left(\eta_4 - \eta_1^2\eta_5\right)V_3 + \left(2\eta_3 - \eta_1^2\eta_5\right)V_4 \tag{F.75}$$

$$I_{X4} = -(2\eta_3 + \eta_1\eta_2\eta_5)V_1 + (\eta_4 + \eta_1\eta_2\eta_5)V_2 + \left(2\eta_3 - \eta_1^2\eta_5\right)V_3 - \left(\eta_4 - \eta_1^2\eta_5\right)V_4 \tag{F.76}$$

From the circuit analysis of Figure F.3, we have

Port 1 $$V_1 = V_{S1} - I_1Z_1 \tag{F.77}$$

Port 2 $$V_2 = V_{S2} - I_2Z_2 \tag{F.78}$$

Port 3 $$V_3 = V_{S3} - I_3Z_3 \tag{F.79}$$

Port 4 $$V_4 = V_{S4} - I_4Z_4 \tag{F.80}$$

Consequently,

$$I_{X1} = -Y_S(V_{S1} - V_{S2}) + (1 + Y_SZ_1)I_1 - Y_SZ_2I_2 \tag{F.81}$$

$$I_{X2} = -Y_S(V_{S2} - V_{S1}) - Y_SZ_1I_1 + (1 + Y_SZ_2)I_2 \tag{F.82}$$

$$I_{X3} = -Y_L(V_{S3} - V_{S4}) + (1 + Y_LZ_3)I_3 - Y_LZ_4I_4 \tag{F.83}$$

$$I_{X4} = -Y_L\left(V_{S4} - V_{S3}\right) - Y_L Z_3 I_3 + \left(1 + Y_L Z_4\right) I_4 \tag{F.84}$$

The above equations can be used to derive the terminal currents on all four ports of the network of Figure F.3 in terms of arbitrary external source voltages V_{S1} to V_{S4}, having associated source impedances Z_1 to Z_4.

Using these equations, closed-form expressions are developed below, for the slightly limited, although important, scenario in which the output inductor operates in single-ended mode. In this case, $V_{S4} = Z_4 = 0$. The input inductor may be driven either in single-ended or differential mode.

From (F.73) and (F.81)

$$\begin{aligned}&\left(1+\left(Y_S - \eta_4 + \eta_2^2\eta_5\right)Z_1\right)I_1 - \left(Y_S - 2\eta_3 + \eta_2^2\eta_5\right)Z_2 I_2 + \left(\eta_4 + \eta_1\eta_2\eta_5\right)Z_3 I_3\\ &= \left(Y_S - \eta_4 + \eta_2^2\eta_5\right)V_{S1} - \left(Y_S - 2\eta_3 + \eta_2^2\eta_5\right)V_{S2} + \left(\eta_4 + \eta_1\eta_2\eta_5\right)V_{S3}\end{aligned} \tag{F.85}$$

from (F.74) and (F.82),

$$\begin{aligned}&\left(Y_S - 2\eta_3 + \eta_2^2\eta_5\right)Z_1 I_1 - \left(1+\left(Y_S - \eta_4 + \eta_2^2\eta_5\right)Z_2\right)I_2 + \left(2\eta_3 + \eta_1\eta_2\eta_5\right)Z_3 I_3\\ &= \left(Y_S - 2\eta_3 + \eta_2^2\eta_5\right)V_{S1} - \left(Y_S - \eta_4 + \eta_2^2\eta_5\right)V_{S2} + \left(2\eta_3 + \eta_1\eta_2\eta_5\right)V_{S3}\end{aligned} \tag{F.86}$$

from (F.75) and (F.83),

$$\begin{aligned}&\left(\eta_4 + \eta_1\eta_2\eta_5\right)Z_1 I_1 - \left(2\eta_3 + \eta_1\eta_2\eta_5\right)Z_2 I_2 + \left(1+\left(Y_L - \eta_4 + \eta_1^2\eta_5\right)Z_3\right)I_3\\ &= \left(\eta_4 + \eta_1\eta_2\eta_5\right)V_{S1} - \left(2\eta_3 + \eta_1\eta_2\eta_5\right)V_{S2} + \left(Y_L - \eta_4 + \eta_1^2\eta_5\right)V_{S3}\end{aligned} \tag{F.87}$$

For simplification, define

$$\eta_6 = Y_S - \eta_4 + \eta_2^2\eta_5 \tag{F.88}$$

$$\eta_7 = -Y_S + 2\eta_3 - \eta_2^2\eta_5 \tag{F.89}$$

$$\eta_8 = \eta_4 + \eta_1\eta_2\eta_5 \tag{F.90}$$

$$\eta_9 = 2\eta_3 + \eta_1\eta_2\eta_5 \tag{F.91}$$

$$\eta_{10} = Y_L - \eta_4 + \eta_1^2\eta_5 \tag{F.92}$$

then equations are

$$\left(1 + \eta_6 Z_1\right)I_1 + \eta_7 Z_2 I_2 + \eta_8 Z_3 I_3 = \eta_6 V_{S1} + \eta_7 V_{S2} + \eta_8 V_{S3} \tag{F.93}$$

$$-\eta_7 Z_1 I_1 - (1+\eta_6 Z_2) I_2 + \eta_9 Z_3 I_3 = -\eta_7 V_{S1} - \eta_6 V_{S2} + \eta_9 V_{S3} \quad (F.94)$$

$$\eta_8 Z_1 I_1 - \eta_9 Z_2 I_2 + (1+\eta_{10} Z_3) I_3 = \eta_8 V_{S1} - \eta_9 V_{S2} + \eta_{10} V_{S3} \quad (F.95)$$

from (F.93),

$$\eta_8 Z_3 I_3 = \eta_6 V_{S1} + \eta_7 V_{S2} + \eta_8 V_{S3} - (1+\eta_6 Z_1) I_1 - \eta_7 Z_2 I_2 \quad (F.96)$$

substituting in (F.94),

$$\begin{aligned}&\left(\eta_7\eta_8 Z_1 + (1+\eta_6 Z_1)\eta_9\right) I_1 + \left(\eta_7\eta_9 Z_2 + (1+\eta_6 Z_2)\eta_8\right) I_2 \\ &= \left(\eta_6\eta_9 + \eta_7\eta_8\right) V_{S1} + \left(\eta_6\eta_8 + \eta_7\eta_9\right) V_{S2}\end{aligned} \quad (F.97)$$

substituting in (F.95),

$$\begin{aligned}&\left(\eta_8^2 Z_3 Z_1 - (1+\eta_6 Z_1)(1+\eta_{10} Z_3)\right) I_1 - \left(\eta_8\eta_9 Z_2 Z_3 + \eta_7 (1+\eta_{10} Z_3) Z_2\right) I_2 \\ &= \left(\eta_8^2 Z_3 - \eta_6 (1+\eta_{10} Z_3)\right) V_{S1} - \left(\eta_8\eta_9 Z_3 + \eta_7 (1+\eta_{10} Z_3)\right) V_{S2} - \eta_8 V_{S3}\end{aligned} \quad (F.98)$$

From which it is determined

$$\begin{aligned}&\begin{pmatrix}\eta_7^2 (1+\eta_{10} Z_3) Z_1 Z_2 + \eta_8^2 (1+\eta_6 Z_2) Z_1 Z_3 + \eta_9^2 (1+\eta_6 Z_1) Z_2 Z_3 \\ + 2\eta_7\eta_8\eta_9 Z_1 Z_2 Z_3 - (1+\eta_6 Z_1)(1+\eta_6 Z_2)(1+\eta_{10} Z_3)\end{pmatrix} I_1 \\ &= \begin{pmatrix}\eta_7^2 (1+\eta_{10} Z_3) Z_2 + \eta_8^2 (1+\eta_6 Z_2) Z_3 + \eta_9 (\eta_6\eta_9 + 2\eta_7\eta_8) Z_2 Z_3 \\ - \eta_6 (1+\eta_6 Z_2)(1+\eta_{10} Z_3)\end{pmatrix} V_{S1} \\ &+ \begin{pmatrix}\eta_6\eta_7 (1+\eta_{10} Z_3) Z_2 - \eta_8\eta_9 (1+\eta_6 Z_2) Z_3 + \eta_6\eta_8\eta_9 Z_2 Z_3 \\ - \eta_7 (1+\eta_6 Z_2)(1+\eta_{10} Z_3)\end{pmatrix} V_{S2} \\ &- \left(\eta_8 + (\eta_6\eta_8 + \eta_7\eta_9) Z_2\right) V_{S3}\end{aligned} \quad (F.99)$$

and

$$\begin{aligned}&\begin{pmatrix}\eta_7^2 (1+\eta_{10} Z_3) Z_1 Z_2 + \eta_8^2 (1+\eta_6 Z_2) Z_1 Z_3 + \eta_9^2 (1+\eta_6 Z_1) Z_2 Z_3 \\ + 2\eta_7\eta_8\eta_9 Z_1 Z_2 Z_3 - (1+\eta_6 Z_1)(1+\eta_6 Z_2)(1+\eta_{10} Z_3)\end{pmatrix} I_2 \\ &= \begin{pmatrix}\eta_6\eta_7 (1+\eta_{10} Z_3) Z_1 - \eta_8\eta_9 (1+\eta_6 Z_1) Z_3 + \eta_6\eta_8\eta_9 Z_1 Z_3 \\ - \eta_7 (1+\eta_6 Z_1)(1+\eta_{10} Z_3)\end{pmatrix} V_{S1} \\ &+ \begin{pmatrix}\eta_7^2 (1+\eta_{10} Z_3) Z_1 + \eta_9^2 (1+\eta_6 Z_1) Z_3 + \eta_8 (\eta_6\eta_8 + 2\eta_7\eta_9) Z_1 Z_3 \\ - \eta_6 (1+\eta_6 Z_1)(1+\eta_{10} Z_3)\end{pmatrix} V_{S2} \\ &+ \left(\eta_9 + (\eta_6\eta_9 + \eta_7\eta_8) Z_1\right) V_{S3}\end{aligned} \quad (F.100)$$

It should be noted that this pair of equations can also be applied in reverse to model the characteristics of a coupled inductor pair employed for differential splitting. In that case, the voltage source V_{S3} on inductor L_2 is treated as the input drive and voltages V_1 and V_2 represent the output amplitudes. Additionally, $V_{S1} = V_{S2} = 0$, and Z_{L1} and Z_{L2} are the output loads. The above equations then simplify to

$$\begin{pmatrix} \eta_7^2(1+\eta_{10}Z_3)Z_1Z_2 + \eta_8^2(1+\eta_6Z_2)Z_1Z_3 + \eta_9^2(1+\eta_6Z_1)Z_2Z_3 \\ +2\eta_7\eta_8\eta_9Z_1Z_2Z_3 - (1+\eta_6Z_1)(1+\eta_6Z_2)(1+\eta_{10}Z_3) \end{pmatrix} I_1 \\ = -\left(\eta_8 + (\eta_6\eta_8 + \eta_7\eta_9)Z_2\right)V_{S3} \tag{F.101}$$

and

$$\begin{pmatrix} \eta_7^2(1+\eta_{10}Z_3)Z_1Z_2 + \eta_8^2(1+\eta_6Z_2)Z_1Z_3 + \eta_9^2(1+\eta_6Z_1)Z_2Z_3 \\ +2\eta_7\eta_8\eta_9Z_1Z_2Z_3 - (1+\eta_6Z_1)(1+\eta_6Z_2)(1+\eta_{10}Z_3) \end{pmatrix} I_2 \\ = \left(\eta_9 + (\eta_6\eta_9 + \eta_7\eta_8)Z_1\right)V_{S3} \tag{F.102}$$

APPENDIX G

Analysis of Differential Coupled Inductors Including Interwinding Capacitance and Bias Network

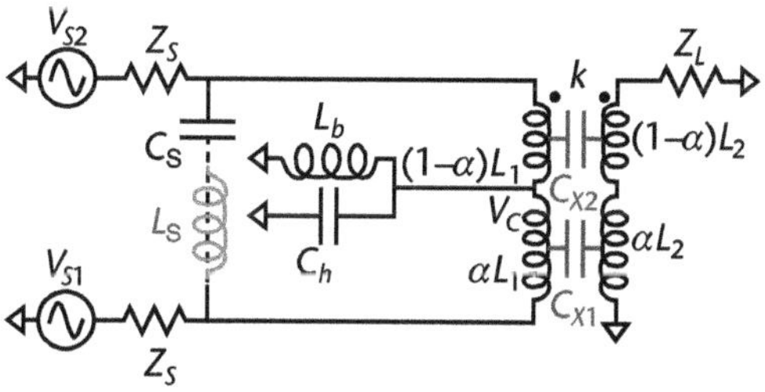

Figure G.1 Differential PA matching architecture with coupled inductors including interwinding capacitance and bias network.

Defining basic network parameters:

$$C_X = \text{total interwinding capacitance} \tag{G.1}$$

$$Z_S = R_S + jX_S \tag{G.2}$$

$$Y_S = 1/Z_S = G_S + jB_S \tag{G.3}$$

$$Z_L = R_L + jX_L \tag{G.4}$$

$$Z_{L1} = R_{L1} + j\omega L_1 \tag{G.5}$$

$$Z_{L2} = R_{L2} + j\omega L_2 \tag{G.6}$$

$$Z_{CS} = R_{CS} - j/\omega C_S \tag{G.7}$$

$$Z_{LS} = R_{LS} + j\omega L_S \tag{G.8}$$

$$Z_{Ct} = Z_{CS} + Z_{LS} \tag{G.9}$$

$$Z_{Lb} = R_{Lb} + j\omega L_b \tag{G.10}$$

$$Z_{Cb} = R_{Cb} - j/\omega C_b \tag{G.11}$$

$$Z_b = Z_{Lb}Z_{Cb}/(Z_{Lb} + Z_{Cb}) \tag{G.12}$$

and dependent variables:

$$\psi_{-L1} = Z_{L1} - j\omega k\sqrt{L_1L_2} \tag{G.13}$$

$$\psi_{+L1} = Z_{L1} + j\omega k\sqrt{L_1L_2} \tag{G.14}$$

$$\psi_{-L2} = Z_{L2} - j\omega k\sqrt{L_1L_2} \tag{G.15}$$

$$\psi_{+L2} = Z_{L2} + j\omega k\sqrt{L_1L_2} \tag{G.16}$$

$$\psi_1 = \frac{(1-\alpha)\psi_{-L1}}{2} \tag{G.17}$$

$$\psi_2 = \frac{1+\alpha(1-\alpha)\psi_{-L1}}{2Z_b} \tag{G.18}$$

$$\psi_5 = j\psi_2\left(-(1-\alpha)\omega k\sqrt{L_1L_2} + \alpha^2\psi_1\psi_{-L2}\omega C_X/2\right) \tag{G.19}$$

$$\begin{aligned}\psi_6 = {} & j\left(\psi_2\left(\alpha-(1-2\alpha)\psi_{-L1}/2Z_{Ct}\right)+\alpha^2(1-\alpha)\psi_{-L2}/2Z_b\right)\omega k\sqrt{L_1L_2} \\ & +\alpha^2(1+\psi_2Z_{L1}/Z_{Ct})\psi_{-L2}/2-(1-\alpha)(1+\psi_2Z_{L1}/Z_{Ct}+\psi_1/Z_b)\left(Z_L+(1-\alpha)\psi_{-L2}/2\right) \\ & -j\alpha\psi_1\left(\alpha\psi_2\psi_{-L1}Z_L/2Z_{Ct}+\left(Z_L+(1-\alpha)\psi_{-L2}/2\right)\left(1+\alpha(1-\alpha)(Z_{L1}+Z_{L2})/2Z_b\right)\right)\omega C_X\end{aligned} \tag{G.20}$$

$$\begin{aligned}\psi_7 = {} & j\left(\psi_2\left(1-\alpha+(1-2\alpha)\psi_{-L1}/2Z_{Ct}\right)+\alpha^3\psi_{-L2}/2Z_b\right)\omega k\sqrt{L_1L_2} \\ & -\alpha(1-\alpha)\left(Z_L+(1-\alpha)\psi_{-L2}/2\right)\psi_{-L2}/2Z_b \\ & +\left(1+\alpha\psi Z_{+L1}/2Z_b+\psi_2Z_1/Z_{Ct}\right)\left((1-\alpha)Z_L+(1-2\alpha)\psi_{-L2}/2\right) \\ & +j\alpha^2\psi_1\left(\psi_2\left(\psi_{-L1}Z_L/Z_{Ct}-\psi_{-L2}\right)-\alpha\left(Z_L+(1-\alpha)\psi_{-L2}/2\right)\psi_{-L2}/Z_b\right)\omega C_X/2\end{aligned} \tag{G.21}$$

$$\begin{aligned}\psi_8 = {} & \psi_1\left(Z_L+(1-\alpha)\psi_{-L2}/2\right)\left(\psi_2+\alpha(1+\psi_1/Z_b)\right)-\alpha^3\psi_{-L1}\psi_{-L1}/4 \\ & +j\alpha^2\left(\psi_2(\psi_{-L1}+\psi_{-L2})+\alpha\psi_1\psi_{-L2}/Z_b\right)\left(\omega k\sqrt{L_1L_2}-\psi_1\left(Z_L+(1-\alpha)\psi_{-L2}/2\right)\omega C_X\right)/2\end{aligned} \tag{G.22}$$

$$\begin{aligned}\psi_9 = &-j\left(\psi_1\psi_2(1-\alpha)+\alpha^2\psi_{-L2}(\psi_2+\alpha\psi_1/Z_b)/2\right)\omega k\sqrt{L_1L_2}\\&+\left(Z_L+(1-\alpha)\psi_{-L2}/2\right)\left(\psi_1\left(\psi_2+\alpha(1+\psi_1/Z_b)\right)-\psi_2(1-\alpha)Z_{L1}\right)\\&+\alpha^2\left(\psi_2 Z_{L1}-\alpha\psi_{-L1}/2\right)\psi_{-L2}/2\\&+j\alpha^2\psi_1\psi_{-L2}\left(\psi_1\psi_2+\left(Z_L+(1-\alpha)\psi_{-L2}/2\right)(\psi_2+\alpha\psi_1/Z_b)\right)\omega C_X/2\end{aligned} \tag{G.23}$$

$$\psi_{10} = j(1-\alpha)\psi_{-L2}\omega k\sqrt{L_1L_2}/2+\psi_1\left(Z_{L2}-(1-\alpha)\psi_{-L2}/2\right) \tag{G.24}$$

$$\begin{aligned}\psi_{11} = &\,\psi_1\psi_2\left(j\psi_1(Z_L+Z_{L2})-\left(Z_L+(1-\alpha)\psi_{-L2}\right)\omega k\sqrt{L_1L_2}\right)/Z_{Ct}\\&+j(1-\alpha)\left(Z_L+(1-\alpha)\psi_{-L2}/2\right)\left((1+\psi_2 Z_{L1}/Z_{Ct})\psi_{-L2}-\alpha\psi_1\psi_{+L2}/Z_b\right)/2\end{aligned} \tag{G.25}$$

$$\begin{aligned}\psi_{12} = &\,\psi_2\left(\psi_1\left((1-\alpha)\psi_{-L2}+Z_L\right)/Z_{Ct}+(1-\alpha)\psi_{-L2}/2\right)\omega k\sqrt{L_1L_2}\\&-j\left(Z_L+(1-\alpha)\psi_{-L2}/2\right)\left(\psi_2\psi_{-L2}\left(1+(1-\alpha)Z_{L1}/Z_{Ct}\right)-\alpha\left(\psi_{-L2}-\alpha\psi_1\psi_{+L2}/Z_b\right)\right)/2\\&-j\psi_1\psi_2(Z_{L2}+Z_L)(1+\psi_1/Z_{Ct})\end{aligned} \tag{G.26}$$

$$\psi_{13} = j\alpha(1-\alpha)\psi_1\left(Z_L+(1-\alpha)\psi_{-L2}/2\right)\left(\psi_{-L1}\psi_{+L2}+\psi_{+L1}\psi_{-L2}\right)/2Z_b \tag{G.27}$$

$$\begin{aligned}\psi_{14} = &\,\psi_1\psi_2\left(j\psi_1(Z_L+Z_{L2})-\left((1-\alpha)(4-\alpha)\psi_{-L2}/2+(2-\alpha)Z_L\right)\omega k\sqrt{L_1L_2}/2\right)\\&+j\left(Z_L+(1-\alpha)\psi_{-L2}/2\right)\left((1-\alpha)\psi_2\psi_{-L2}Z_{L1}+\alpha\psi_1\left(\psi_2 Z_{L2}-\psi_{-L2}+\alpha\psi_1\psi_{+L2}/Z_b\right)\right)/2\end{aligned} \tag{G.28}$$

$$\psi_{15} = \frac{\psi_5\omega C_X}{\left(\psi_2+\alpha(Z_S+\psi_1)/Z_b\right)\omega k\sqrt{L_1L_2}+(\psi_7 Z_S-\psi_9)\omega C_X} \tag{G.29}$$

$$\psi_{16} = \frac{\left(\psi_2+\left((1-\alpha)Z_S+\alpha\psi_1\right)/Z_b\right)\omega k\sqrt{L_1L_2}+(\psi_6 Z_S+\psi_8)\omega C_X}{\left(\psi_2+\alpha(Z_S+\psi_1)/Z_b\right)\omega k\sqrt{L_1L_2}+(\psi_7 Z_S-\psi_9)\omega C_X} \tag{G.30}$$

$$\psi_{17} = -\begin{pmatrix}\psi_2 Z_S(\omega k)^2 L_1L_2/Z_{Ct}+j\alpha\left(\psi_2\psi_{-L2}+\psi_{+L2}\left((1-\alpha)Z_S+\alpha\psi_1\right)/Z_b\right)\omega k\sqrt{L_1L_2}/2\\+(Z_{L2}+Z_L)\left(\alpha\psi_{-L1}/2+Z_S(1+\psi_2 Z_{L1}/Z_{Ct})\right)+(\psi_{11}Z_S+\alpha\psi_{13}/2)\omega C_X\end{pmatrix} \tag{G.31}$$

$$\begin{aligned}\psi_{18} = &\,\psi_2(1-\alpha/2+Z_S/Z_{Ct})(\omega k)^2 L_1L_2\\&-j\alpha\left(\alpha\psi_{+L2}(Z_S+\psi_1)/Z_b+\psi_2 Z_{L2}\right)\omega k\sqrt{L_1L_2}/2\\&+(Z_{L2}+Z_L)\left(Z_S(1+\alpha\psi_{+L1}/2Z_b+\psi_2 Z_{L1}/Z_{Ct})+\psi_2 Z_{L1}-\alpha\psi_{-L1}/2\right)\\&-(\psi_{12}Z_S+\psi_{14})\omega C_X\end{aligned} \tag{G.32}$$

$$\xi_1 = \begin{pmatrix} \alpha(1-\alpha)\left((Z_{L2}+Z_L)Z_{L1}+(\omega k)^2 L_1L_2\right)(\psi_{+L1}+2\psi_2 Z_{L1})/Z_{Ct} \\ +\psi_2\begin{pmatrix} (Z_{L2}+Z_L)(1+Z_S/Z_{Ct}-2\alpha)Z_{L1}-j\alpha Z_{L2}(1-2\alpha)\omega k\sqrt{L_1L_2}/2 \\ +\left(1+Z_S/Z_{Ct}+\alpha^2-5\alpha/2\right)(\omega k^2)L_1L_2 \end{pmatrix} \\ +j\alpha(1-2\alpha)\psi_{+L2}\left(1+\alpha((1-\alpha)Z_{L1}-\psi_1)/Z_b\right)\omega k\sqrt{L_1L_2}/2 \\ +(Z_{L2}+Z_{L1})\begin{pmatrix} Z_S+\alpha(\psi_{+L1}-(1-2\alpha)\psi_{-L1})/2 \\ +\alpha(1-\alpha)(2Z_{L1}+\psi_{+L1})(Z_S+\alpha Z_{L1})/2Z_b \end{pmatrix} \end{pmatrix}\omega k\sqrt{L_1L_2}/Z_b \tag{G.33}$$

$$\xi_2 = \begin{pmatrix} \begin{pmatrix} (\psi_{11}-\psi_{12})(1+\alpha(1-\alpha)Z_{L1}/Z_b)+(\alpha\psi_{11}-(1-\alpha)\psi_{12})Z_S/Z_b \\ +(1-2\alpha)\psi_{14}/Z_b \end{pmatrix}\omega k\sqrt{L_1L_2} \\ +\psi_6\begin{pmatrix} (Z_{L2}+Z_L)(Z_S+\psi_2(1+Z_S/Z_{Ct})Z_{L1}+\alpha(\psi_{+L1}Z_S/Z_b-\psi_{-L1})/2) \\ -(\psi_{12}Z_S-\psi_{14})\omega C_X+\psi_2(Z_S/Z_{Ct}+1-\alpha/2)(\omega k)^2 L_1L_2 \\ -j\alpha(\psi_2 Z_{L2}+\alpha\psi_{+L2}(\psi_1+Z_S)/Z_b)\omega k\sqrt{L_1L_2}/2 \end{pmatrix} \\ +\psi_7\begin{pmatrix} (Z_{L2}+Z_L)(Z_S-\psi_2(1-Z_S/Z_{Ct})Z_{L1}+\alpha\psi_{-L1}/2)+(\psi_{11}Z_S-\psi_{14})\omega C_X \\ +\psi_2(Z_S/Z_{Ct}-1+\alpha/2)(\omega k)^2 L_1L_2 \\ +j\alpha(\psi_2 Z_{L2}+\psi_{+L2}(\alpha\psi_1+(1-\alpha)Z_S)/Z_b)\omega k\sqrt{L_1L_2}/2 \end{pmatrix} \\ -\psi_9\begin{pmatrix} (Z_{L2}+Z_L)(2+2\psi_2 Z_{L1}/Z_{Ct}+\alpha\psi_{+L1}/2Z_b)-(\psi_{12}-\psi_{11})\omega C_X \\ +2\psi_2(\omega k)^2 L_1L_2/Z_{Ct}+j\alpha\psi_{+L2}(1-2\alpha)\omega k\sqrt{L_1L_2}/2Z_b \end{pmatrix} \end{pmatrix}\omega C_X \tag{G.34}$$

$$\begin{aligned} \xi_3 = {} & \alpha\left((\omega k)^2 L_1L_2+(Z_{L2}+Z_L)Z_{L1}\right)\left((1-\alpha)\psi_{+L1}/2+\psi_2(Z_S+(1-\alpha)Z_{L1})\right)\omega k\sqrt{L_1L_2}/Z_b \\ & + \begin{pmatrix} (1+\alpha(Z_S+(1-\alpha)Z_{L1})/Z_b)(\alpha\psi_{13}/2+\psi_{14})\omega k\sqrt{L_1L_2} \\ +(\psi_7 Z_S-\psi_9)\left(\psi_2(\omega k)^2 L_1L_2+(Z_{L2}+Z_L)\psi_2 Z_{L1}+(\alpha\psi_{13}/2+\psi_{14})\omega C_X\right) \\ +(\psi_9-\psi_8)\begin{pmatrix} (Z_{L2}+Z_L)(Z_S+\psi_2(1+Z_S/Z_{Ct})Z_{L1}+\alpha(\psi_{+L1}Z_S/Z_b-\psi_{-L1})/2) \\ -(\psi_{12}Z_S-\psi_{14})\omega C_X+\psi_2(Z_S/Z_{Ct}+1-\alpha/2)(\omega k)^2 L_1L_2 \\ -j\alpha(\psi_2 Z_{L2}+\alpha\psi_{+L2}(\psi_1+Z_S)/Z_b)\omega k\sqrt{L_1L_2}/2 \end{pmatrix} \end{pmatrix}\omega C_X \end{aligned} \tag{G.35}$$

$$\xi_4 = \left(\begin{array}{l} \left(\alpha(1-\alpha)\psi_{+L1} + \left(Z_S + 2\alpha(1-\alpha)Z_{L1}\right)\psi_2\right)\left(\omega k^2\right)L_1L_2/Z_{Ct} \\ + \left(Z_{L2} + Z_L\right)\left(\begin{array}{l} \alpha\left(\psi_{+L1} - (1-2\alpha)\psi_{-L1}\right)/2 \\ + \left(1 + \psi_2 Z_{L1}/Z_{Ct}\right)\left(Z_S + 2\alpha(1-\alpha)Z_{L1}\right) \\ - \left(Z_S + \alpha Z_{L1}\right)\left(1 - \psi_2 + \alpha(1-\alpha)\psi_{+L1}Z_{L1}/Z_{Ct}\right) \end{array} \right) \\ - j\alpha(1-2\alpha)\left(\begin{array}{l} \psi_2\psi_{-L2} - \psi_{+L2} \\ + \alpha\psi_{+L2}\left(\psi_1 - (1-\alpha)Z_{L1}\right)/Z_b \end{array} \right)\omega k\sqrt{L_1L_2}/2 \end{array} \right)\omega k\sqrt{L_1L_2}/Z_b \qquad (G.36)$$

$$\xi_5 = \left(\begin{array}{l} \left(\psi_6 + \psi_7\right)\left(\psi_2(\omega k)^2 L_1L_2Z_S/Z_{Ct} + \left(Z_{L2} + Z_L\right)Z_S\left(1 + \psi_2 Z_{L1}/Z_{Ct}\right)\right) \\ - \left(\psi_6 - \psi_7\right)\left(\begin{array}{l} \left(Z_{L2} + Z_L\right)\alpha\psi_{-L1}/2 \\ + j\alpha\left(\psi_2\psi_{-L2} + \alpha\psi_{+L2}\left(\psi_1 + Z_S\right)/Z_b\right)\omega k\sqrt{L_1L_2}/2 \end{array} \right) \\ - \psi_8\left(2\psi_2(\omega k)^2 L_1L_2/Z_{Ct} + \left(Z_{L2} + Z_L\right)\left(2\left(1 + \psi_2 Z_{L1}/Z_{Ct}\right) + \alpha\psi_{+L1}/2Z_b\right)\right) \\ + \left(\begin{array}{l} \left(1 + \alpha\left(Z_S + (1-\alpha)Z_{L1}\right)/Z_b\right)\left(\psi_{11} - \psi_{12}\right) \\ + (1-2\alpha)\left(j\alpha\psi_{+L2}\left(\psi_7 Z_S - \psi_8\right)/2 - \psi_{12}Z_S - \alpha\psi_{13}/2\right)/Z_b \end{array} \right)\omega k\sqrt{L_1L_2} \\ + \psi_6\left(Z_{L2} + Z_L\right)Z_S\alpha\psi_{+L1}/2Z_b \\ - \left(\psi_6\left(\psi_{12}Z_S + \alpha\psi_{13}/2\right) - \psi_7\left(\psi_{11}Z_S + \alpha\psi_{13}/2\right) - \psi_8\left(\psi_{12} - \psi_{11}\right)\right)\omega C_X \end{array} \right)\omega C_X \quad (G.37)$$

$$\begin{array}{l} \xi_6 = (1-\alpha)\left((\omega k)^2 L_1L_2 + \left(Z_L + Z_{L2}\right)Z_{L1}\right)\left(\alpha\psi_{+L1}/2 + \psi_2\left(Z_S + \alpha Z_{L1}\right)\right)\omega k\sqrt{L_1L_2}/Z_b \\ + \left(\begin{array}{l} +\left(\psi_8 - \psi_9\right)\left(\begin{array}{l} \left(Z_{L2} + Z_L\right)\left(Z_S\left(1 + \psi_2 Z_{L1}/Z_{Ct}\right) + \alpha\psi_{-L1}/2\right) + \left(\psi_{11}Z_S + \alpha\psi_{13}/2\right)\omega C_X \\ + j\alpha\left(\psi_2\psi_{-L2} + \psi_{+L2}\left(\alpha\psi_1 + (1-\alpha)Z_S\right)/Z_b\right)\omega k\sqrt{L_1L_2}/2 \\ + \psi_2(\omega k)^2 L_1L_2Z_S/Z_{Ct} \end{array} \right) \\ + \left(1 + (1-\alpha)\left(Z_S + \alpha Z_{L1}\right)/Z_b\right)\left(\alpha\psi_{13}/2 + \psi_{14}\right)\omega k\sqrt{L_1L_2} \\ + \left(\psi_6 Z_S - \psi_8\right)\left(\psi_2(\omega k)^2 L_1L_2 + \left(Z_L + Z_{L2}\right)\psi_2 Z_{L1} + \left(\alpha\psi_{13}/2 + \psi_{14}\right)\omega C_X\right) \end{array} \right)\omega C_X \end{array}$$

(G.38)

G.1 Odd-Mode Network Solutions

For these modes,

$$V_{S1} = V_S = -V_{S2} \tag{G.39}$$

In terms of

$$\text{denom} = \left((\omega k)^2 L_1L_2(1+2Z_S/Z_{Ct}) + (Z_{L2}+Z_L)\left(Z_{L1}+2Z_S(1+Z_{L1}/Z_{Ct})\right)\right)\omega k\sqrt{L_1L_2} \tag{G.40}$$

Network solutions are

$$\frac{I_1}{V_S} = \frac{2\left((\omega k)^2 L_1L_2/Z_{Ct} + (Z_{L2}+Z_L)(1+Z_{L1}/Z_{Ct})\right)\omega k\sqrt{L_1L_2} + \xi_1 + \xi_2}{\text{denom} + (\xi_1+\xi_2)Z_S + \xi_3} \tag{G.41}$$

$$\frac{V_1}{V_S} = \frac{\left((\omega k)^2 L_1L_2 + (Z_{L2}+Z_L)Z_{L1}\right)\omega k\sqrt{L_1L_2} + \xi_3}{\text{denom} + (\xi_1+\xi_2)Z_S + \xi_3} \tag{G.42}$$

$$Z_1 = \frac{(V_1/V_S)}{(I_1/V_S)} \tag{G.43}$$

$$\frac{I_2}{V_S} = \frac{2\left((\omega k)^2 L_1L_2/Z_{Ct} + (Z_{L2}+Z_L)(1+Z_{L1}/Z_{Ct})\right)\omega k\sqrt{L_1L_2} + \xi_4 + \xi_5}{\text{denom} + (\xi_4+\xi_5)Z_S + \xi_6} \tag{G.44}$$

$$\frac{V_2}{V_S} = \frac{\left((\omega k)^2 L_1L_2 + (Z_{L2}+Z_L)Z_{L1}\right)\omega k\sqrt{L_1L_2} + \xi_6}{\text{denom} + (\xi_4+\xi_5)Z_S + \xi_6} \tag{G.45}$$

$$Z_2 = \frac{(V_2/V_S)}{(I_2/V_S)} \tag{G.46}$$

$$I_L = \frac{\begin{pmatrix} \left(1+\psi_2\left(Z_{L1}+j\psi_1^2\omega C_X\right)/Z_{Ct}\right)V_1 \\ -\left(1+\alpha\psi_{+L1}/2Z_b + \psi_2 Z_{L1}/Z_{Ct} + j(1+\psi_1/Z_{Ct})\psi_1\psi_2\omega C_X\right)V_2 \\ -\alpha\psi_{-L1}I_1/2 + \left(\psi_2 Z_{L1} - \alpha\psi_{-L1}/2 + j\psi_1^2\psi_2\omega C_X\right)I_2 \end{pmatrix}}{j\psi_2\left(\omega k\sqrt{L_1L_2} - (Z_L + (1-\alpha)\psi_{-L2}/2)\psi_1\omega C_X\right)} \tag{G.47}$$

$$V_L = I_L Z_L \tag{G.48}$$

$$V_C = V_2 + (1-\alpha)\left(j\omega k\sqrt{L_1L_2}I_L + (V_2 - V_1)Z_{L1}/Z_{Ct} - Z_{L1}I_2\right) \\ + j(1-\alpha)^2\psi_{-L1}\left(V_2 - V_L - (1-\alpha)\left(\psi_{-L1}(V_1 - V_2)/Z_{Ct} + \psi_{-L1}I_2 + \psi_{-L1}I_L\right)/2\right)\omega C_X/2 \tag{G.49}$$

G.2 Output Impedance on Port 3

Driving port 3 with voltage V_{S3}, we have

$$\frac{I_1}{V_{S3}} = -\frac{\psi_{15}\psi_{18} + j\psi_2\left(\omega k\sqrt{L_1L_2} + \psi_{10}\omega C_X\right)}{\psi_{17} + \psi_{16}\psi_{18}} \tag{G.50}$$

$$\frac{I_2}{V_{S3}} = \psi_{15} + \psi_{16}I_1/V_{S3} \tag{G.51}$$

$$\frac{I_L}{V_{S3}} = \frac{\left(\begin{array}{l} j\psi_1\psi_2\omega C_X - \left(\alpha\psi_{-L1}/2 + \left(1 + \psi_2\left(Z_{L1} + j\psi_1^2\omega C_X\right)/Z_{Ct}\right)Z_S\right)I_1/V_{S3} \\ + \left(\begin{array}{l} -\alpha\psi_{-L1}/2 + \psi_2(1 + Z_S/Z_{Ct})\left(Z_{L1} + j\psi_1^2\omega C_X\right) \\ + (1 + \alpha\psi_{+L1}/2Z_b + j\psi_1\psi_2\omega C_X)Z_S \end{array}\right)I_2/V_{S3} \end{array}\right)}{j\psi_2\left(\omega k\sqrt{L_1L_2} - \left(Z_L + (1-\alpha)\psi_{-L2}/2\right)\psi_1\omega C_X\right)} \tag{G.52}$$

$$V_L/V_{S3} = 1 + (I_L/V_{S3})Z_L \tag{G.53}$$

$$Z_{out} = -V_L/I_L = (V_{S3}/I_L + Z_L) \tag{G.54}$$

G.3 Design Synthesis

For lossless case

$$Z_{L1} = j\omega L_1 \qquad \& \qquad Z_{L2} = j\omega L_2 \tag{G.55}$$

and for $L_s = 0$

$$1/Z_{Ct} = j\omega C_t \tag{G.56}$$

G.3.1 For Input Match on Ports 1 and 2

For input match, we require

$$Y_{in} = G_S - jB_S$$

Defining

$$a = \left(1 - k^2\right)^2 \left(\omega L_1 G_S / R_L\right) \tag{G.57}$$

$$b = 2\left(\left(1 - k^2\right)^2 \omega L_1 G_S X_L / R_L - k^2\right) \tag{G.58}$$

and for port 1

$$c = \omega L_1 G_S \left(R_L + X_L^2 / R_L\right) + \begin{pmatrix} G_S \Im(\xi_3) - B_S \Re(\xi_3) - \Im(\xi_1 + \xi_2) \\ -\left(G_S \Re(\xi_3) + B_S \Im(\xi_3) - \Re(\xi_1 + \xi_2)\right)\left(X_L + \left(1 - k^2\right)\omega L_2\right)/R_L \end{pmatrix} / \omega k \sqrt{L_1 L_2} \tag{G.59}$$

and for port 2

$$c = \omega L_1 G_S \left(R_L + X_L^2 / R_L\right) + \begin{pmatrix} G_S \Im(\xi_6) - B_S \Re(\xi_6) - \Im(\xi_4 + \xi_5) \\ -\left(G_S \Re(\xi_6) + B_S \Im(\xi_6) - \Re(\xi_4 + \xi_5)\right)\left(X_L + \left(1 - k^2\right)\omega L_2\right)/R_L \end{pmatrix} / \omega k \sqrt{L_1 L_2} \tag{G.60}$$

where $\Re(\,)$ represents the real part of the argument, and $\Im(\,)$ represents the imaginary part. Solutions for the secondary inductance L_2 can be expressed in the standard quadratic form

$$L_2 = \frac{-b - \sqrt{b^2 - 4ac}}{2a} \tag{G.61}$$

Also,

$$\omega C_S = \frac{1}{\omega L_1} + \begin{pmatrix} G_S\left(\left(1 - k^2\right)\omega L_2 + X_L\right) - B_S R_L \\ -\left(G_S \Re(\xi_3) + B_S \Im(\xi_3) - \Re(\xi_1 + \xi_2)\right)/\omega^2 L_1 k \sqrt{L_1 L_2} \end{pmatrix} / 2R_L \tag{G.62}$$

Solution of (G.61) and (G.62) requires a recursive approach, as with the single-ended coupled-resonators synthesis with interwinding capacitance of Section 12.3. A recursive solution is required since the ξ-terms are themselves $f(L_2)$. Since the ξ-terms all have first-order dependencies on C_X and $1/Z_b$, and the latter are assumed to be small, and the initial seed solution for L_2 is generated with $\xi_n = 0$, for all n. This is used to obtain a first solution for C_S, which, in turn, generates first solutions for all ξ_n. An improved estimate for L_2 is then generated and the procedure is repeated recursively. Typically, ~20 recursions are required to arrive at relatively precise values for L_2 and C_S.

In lieu of a single shunt capacitor C_S, if a low-input impedance is required at the third harmonic, as in Section 14.1, denoting the frequency at the center of the passband by ω_0 and that at the center of the third-harmonic band by ω_3, we must replace it by a series trap with

$$C_S' = \left(1 - (\omega_0/\omega_3)^2\right)C_S \tag{G.63}$$

$$L_S = \frac{1}{\left(\omega_3^2 + \omega_0^2\right)C_S} \tag{G.64}$$

Appendix H

Some Useful Materials Data

Materials	*Symbol*	*Elec. Rest.* ρ $(\Omega \times 10^{-8} m)$ $(\mu\Omega\ cm)$	*CTE* $(\times 10^{-6}$ *m/mK)*	*Density* ρ $(kg \times 10^3/m^3)$	*Elec. Conduct.* σ $(S \times 10^7/m)$	*Relative Permit.* ε_r	*Relative Perm.* μ_r	*Thermal conduct.* σ *(W/mK)*	*Melting Point (°C)*	*Process difficulty*	*Cost ($/oz)*
Air				0.00		1	1				
Aluminium	Al	2.65	23.1	2.70	3.77		1		660.3	Low melting pt.	
Aluminum nitride	AlN		–25	3.4055		9.66	1				
Alumina	Al_2O_3			3.97							
Barium titanate	$BaTiO_3$		6	6.02		1200					
Cobalt	Co	6.34	13	8.71	1.58		600		1,495		0.9
Copper	Cu	1.69	16.6	8.79	5.92		1		1,085	Marginal melting pt.	
GaAs	GaAs		7								
Gold	Au	2.35	14.2	19.29	4.26		1		1,064	Marginal melting pt.	
Indium phosphide	InP		4.6								
Iridium	Ir	5.3	6.4	22.50	1.89		1		2,466	Oxide	1,085.0

Materials	*Symbol*	*Elec. Rest. ρ (Ω × 10^{-8}m) (μΩ cm)*	*CTE (×10^{-6} m/mK)*	*Density ρ (kg × 10^3/m^3)*	*Elec. Conduct. σ (S × 10^7/m)*	*Relative Permit. ε_r*	*Relative Perm. μ_r*	*Thermal conduct. σ (W/mK)*	*Melting Point (°C)*	*Process difficulty*	*Cost ($/oz)*
Molybdenum	Mo	5.7	4.8	10.30	1.75		1	139	2,623	Oxide	0.9
Nickel	Ni	6.84	13.4	8.90	1.46		200		1,455		0.5
Palladium	Pd	10.8	11.8	12.00	0.93		1	72	1,555		610.0
Platinum	Pt	10.58	8.8	21.10	0.95		1	73	1,768		1,477.0
Rhodium	Rh	4.7	8.2	12.30	2.13		1		1,963		1,450.0
Ruthenium	Ru	7.1	6.4	12.36	1.41		1	120	2,334	Oxide	120.0
Sapphire	Al_2O_3		6	3.97		9.6	1	148			
Silicon	Si		2.618	2.33		12	1	1.4			
Silica	SiO_2		0.55	2.15		3.9	1	15			
Silicon nitride	Si_3N_4			3.247							
Silver	Ag	1.59	18.9	10.50	6.29		1		962	Low melting pt.	
Strontium titanate	$SrTiO_3$		9.4	4.81		310					
Tantalum	Ta			16.60							
Titanium	Ti	54	8.6	4.50	0.19				1,668		
Titanium dioxide	TiO_2		9	4.23		100					
Tungsten	W	5.5	4.5	19.20	1.82		1	197	3,422	Oxide	1.5

Selected Bibliography

Abrie, P. L. D., *The Design of Impedance-Matching Networks for Radio-Frequency and Microwave Amplifiers*, Dedham, MA: Artech House, 1985.

Bachi, J., "Design and Implementation of High Efficiency Power Amplifiers for 5G Applications," Thèse de Doctorat, Institut Polytechnique de Paris, Palaiseau, November 28, 2022.

Bakalski, W., W. Simburger, and H. Knapp, "Lumped and Distributed Lattice-Type LC-Baluns," *IEEE MTT-S International Microwave Symposium Digest*, 2002, pp. 209–212.

Bi, J., "Chireix's/LINC Power Amplifier for Base Station Applications Using GaN Devices with Load Compensation," Master of Science Thesis, Delft University, September 2008.

Chen, W., et al., "Radio Frequency Power Amplifier for Wireless Communication," *Microwave Wireless Communications*, December 2016, pp. 261–300.

Chireix, H., "High Power Outphasing Modulation," *Proc. IRE*, Vol. 23, No. 11, November 1935, pp. 1370–1392.

Chireix, H., "Means for Radio Communication," US Patent 1,882,119, October 11, 1932.

Cripps, S. C., *Advanced Techniques in RF Power Amplifier Design*, Norwood, MA: Artech House, 2002, pp. 58–72.

Cripps, S. C., "Pushing and Pulling," *IEEE Microwave Magazine*, October 2020.

Cripps, S. C., *RF Power Amplifiers for Wireless Communications*, Norwood, MA: Artech House, 2006.

Doherty, W. H., "Amplifier," US Patent 2,210,028, August 6, 1940.

Fagot, J., "Method and Means for the Linear Transmission or Amplification of Amplitude-Modulated Carrier Waves," US Patent 2,282,714, May 12, 1938.

Ghannouchi, F. M., and M. S. Hashmi, *Load-Pull Techniques with Applications to Power Amplifier Design*, New York: Springer, 2013.

Gonzalez, G., *Microwave Transistor Amplifiers: Analysis and Design*, Boston, MA: Pearson, 1996.

Grebennikov, A., *RF and Microwave Power Amplifier Design*, 2nd ed., New York: McGraw-Hill Education, 2016.

Grebennikov, A., et al., "High-Efficiency Doherty Power Amplifiers: Historical Aspect and Modern Trends," *IEEE Proc.*, Vol. 100, No. 12, September 13, 2012, pp. 3190–3219.

Grebennikov, A., N. Kumar, and B. S. Yarman, *Broadband RF and Microwave Amplifiers*, Boca Raton, FL: CRC Press, 2017.

Grebennikov, A., N. Sokal, and M. Franco, "High-Efficiency Doherty Power Amplifiers," Chapter 10 in *Switchmode RF and Microwave Power Amplifiers*, 2nd ed., 2012, pp. 529–573.

Gustrau, F., *RF and Microwave Engineering: Fundamentals of Wireless Communications*, New York: Wiley, 2012.

Hakala, I., L. Gharavi, and R. Kaunisto, "Chireix Power Combining with Saturated Class-B Power Amplifiers," *12th GaAs Symposium*, Amsterdam, 2004.

Hou, J. A., and Y. H. Wang, "A Compact Quadrature Hybrid Based on High-Pass and Low-Pass Lumped Elements," *IEEE Microw. Wireless Compon. Lett.*, Vol. 17, No. 8, August 2007, pp. 595–597.

Hou, J. A., and Y. H. Wang, "Design and Analysis of Novel Quadrature Hybrids with Compact Lumped Elements," *Int. Symp. on Microw. and Optical Tech.*, December 2009, pp. 536–539.

Hoverstein, J., "Efficient and Linear Microwave Transmitters for High Peak-to-Average Ratio Signals," PhD submission, University of Colorado, Dept. Elec. and Computer Eng., April 12, 2010.

Imran, T., T. Zahid, and W. Ahmad, *X-Band Low Noise Amplifier & Ultra Broad Band Wilkinson Power Divider*, London, UK: Lambert Academic Publishing, 2011.

Jarry, P., and J. Beneat, "Scattering Parameters and ABCD Matrices" in *Advanced Design Techniques and Realizations of Microwave and RF Filters*, New York: Wiley-IEEE Press, 2008.

Kim, B., "Introduction to Doherty Power Amplifier," Chapter 1 in *Doherty Power Amplifiers*, 2018, pp. 1–30.

Makimoto, M., and S. Yamashita, *Microwave Resonators and Filters for Wireless Communication: Theory, Design and Application*, New York: Springer, 2001.

Mavaddat, R., *Network Scattering Parameters*, New York: World Scientific Publishing Company, 1996.

Pozar, D., *Microwave Engineering*, New York: Wiley, 2011.

Qureshi, J. H., et al., "A 90-W Peak Power GaN Outphasing Amplifier with Optimum Input Signal Conditioning," *IEEE Transactions on Microwave Theory and Techniques*, Vol. 57, No. 8, August 8, 2009, pp. 1925–1935.

Raab, F., et al., "Power Amplifiers and Transmitters for RF and Microwave," *IEEE Transactions on Microwave Theory and Techniques*, Vol. 50, No. 3, March 2002.

Raab, F., et al., "RF and Microwave Power Amplifier and Transmitter Technologies—Part 1," *High Frequency Electronics*, May 2003.

Raab, F., et al., "RF and Microwave Power Amplifier and Transmitter Technologies—Part 3," *High Frequency Electronics*, September 2003.

Ramella, C., et al., "High Efficiency Power Amplifiers for Modern Mobile Communications: The Load-Modulation Approach," *Electronics Review*, MDPI, Basel, Switzerland, 2017, pp. 1–29.

Rogers, J., and C. Plett, *Radio Frequency Integrated Circuit Design*, Norwood, MA: Artech House, 2003.

Sorrentino, R., G. Bianchi, and K. Chang, *Microwave and RF Engineering*, New York: Wiley, 2010.

Wang, H., P. Asbeck, and C. Fager, "Millimeter-Wave Power Amplifier Integrated Circuits for High Dynamic Range Signals," *IEEE J. of Microwaves, MTT Symp.*, January 2021, pp. 299–316.

White, J. F., *High Frequency Techniques: An Introduction to RF and Microwave Design and Computer Simulation*, New York: IEEE Press, 2004.

Wong, J., A. Grebennikov, and N. Watanabe, "Doherty Amplifier Combines High Power and Efficiency," *Microwaves and RF*, October 7, 2017.

Wright, P., J. Zhao, and L. Fei, "Innovative Architectures for Advanced Handset Power Amplifier Performance," *2012 IEEE Bipolar/BiCMOS Circuits and Technology Meeting (BCTM)*, 2012.

Wright, P., "Three-Port Shunted-Inductor Lattice Coupler," US Patent 11,742,819, August 2023.

Wright, P. V., "Quadrature Lattice Matching Network," US Patent 8,811,531, August 19, 2014.

About the Author

Peter V. Wright graduated with an engineering degree from Cambridge University, United Kingdom, and began working as a microwave engineer for Marconi Communications, Chelmsford, United Kingdom. Following that, he moved to the United States to assume an engineering position with Microwave Associates, Burlington, Massachusetts. After a few years, anxious to advance his theoretical understanding of RF component design and theory, he left to enter the PhD program in the Electrical Sciences Department at the Massachusetts Institute of Technology (MIT). His thesis work was in the areas of microwave, acoustics, and optical component design and modeling. Working under Professor Haus, he pioneered the application of the coupling-of-modes (COM) theory to the design of surface acoustic wave (SAW) devices.

Upon graduating from MIT, he became a staff member at Lincoln Laboratories, Bedford, Massachusetts. There he worked on superconducting signal processing circuits, and acousto-optic spectrum analyzers. Seeing an opportunity to apply his COM theory to the design of practical SAW devices, he left to join RF Monolithics in Dallas, Texas. During his time at RF Monolithics, Dr. Wright wrote most of the software for designing the company's wide range of SAW resonators and filters. As an outcome of this work, he received multiple patents for numerous innovative architectures.

Dr. Wright also worked in the acoustic field for Schlumberger in Clamart, France. There, he created an innovative acoustic signal processing algorithm to rapidly process logs for evaluating image cement integrity on the outside of a borehole casing. This algorithm was widely employed by the company in its tools used in the oil industry.

Dr. Wright also worked for Thomson Microsonics in Sophia-Antipolis, France. There, he returned to his microwave focus to work on RF cellphone modules. In turn, this led him to join TriQuint Semiconductor in Hillsboro, Oregon, where he worked on the design of RF modules for mobile communications devices. He was with the company (later Qorvo) for 18 years. During that time, he expanded on the device modeling capabilities that he had previously developed and applied them predominantly to cellphone power amplifier RF modules. These techniques pointed the way to many innovative and advantageous device architectures.

Dr. Wright was awarded around 50 issued patents and is currently retired and living in Cascais, Portugal. There he enjoys writing historical and science fiction novels. He is also an avid gardener.

Index

E

F

H

I

M

N

O

P

Q

R

S

T

V

Y

Z

Artech House Microwave Library

Behavioral Modeling and Linearization of RF Power Amplifiers, John Wood

Chipless RFID Reader Architecture, Nemai Chandra Karmakar, Prasanna Kalansuriya, Randika Koswatta, and Rubayet E-Azim

Chipless RFID Systems Using Advanced Artificial Intelligence, Larry M. Arjomandi and Nemai Chandra Karmakar

Control Components Using Si, GaAs, and GaN Technologies, Inder J. Bahl

Design of Linear RF Outphasing Power Amplifiers, Xuejun Zhang, Lawrence E. Larson, and Peter M. Asbeck

Design Methodology for RF CMOS Phase Locked Loops, Carlos Quemada, Guillermo Bistué, and Iñigo Adin

Design of CMOS Operational Amplifiers, Rasoul Dehghani

Design of RF and Microwave Amplifiers and Oscillators, Second Edition, Pieter L. D. Abrie

Digital Filter Design Solutions, Jolyon M. De Freitas

Discrete Oscillator Design Linear, Nonlinear, Transient, and Noise Domains, Randall W. Rhea

Distortion in RF Power Amplifiers, Joel Vuolevi and Timo Rahkonen

Distributed Power Amplifiers for RF and Microwave Communications, Narendra Kumar and Andrei Grebennikov

Electric Circuits: A Primer, J. C. Olivier

Electronics for Microwave Backhaul, Vittorio Camarchia, Roberto Quaglia, and Marco Pirola, editors

An Engineer's Guide to Automated Testing of High-Speed Interfaces, Second Edition, José Moreira and Hubert Werkmann

Envelope Tracking Power Amplifiers for Wireless Communications, Zhancang Wang

Essentials of RF and Microwave Grounding, Eric Holzman

Frequency Measurement Technology, Ignacio Llamas-Garro, Marcos Tavares de Melo, and Jung-Mu Kim

FAST: Fast Amplifier Synthesis Tool—Software and User's Guide, Dale D. Henkes

Feedforward Linear Power Amplifiers, Nick Pothecary

Filter Synthesis Using Genesys S/Filter, Randall W. Rhea

Foundations of Oscillator Circuit Design, Guillermo Gonzalez

Frequency Synthesizers: Concept to Product, Alexander Chenakin

Fundamentals of Nonlinear Behavioral Modeling for RF and Microwave Design, John Wood and David E. Root, editors

Generalized Filter Design by Computer Optimization, Djuradj Budimir

Handbook of Dielectric and Thermal Properties of Materials at Microwave Frequencies, Vyacheslav V. Komarov

Handbook of RF, Microwave, and Millimeter-Wave Components, Leonid A. Belov, Sergey M. Smolskiy, and Victor N. Kochemasov

High-Efficiency Load Modulation Power Amplifiers for Wireless Communications, Zhancang Wang

High-Linearity RF Amplifier Design, Peter B. Kenington

High-Speed Circuit Board Signal Integrity, Second Edition, Stephen C. Thierauf

Integrated Microwave Front-Ends with Avionics Applications, Leo G. Maloratsky

Intermodulation Distortion in Microwave and Wireless Circuits, José Carlos Pedro and Nuno Borges Carvalho

Introduction to Modeling HBTs, Matthias Rudolph

An Introduction to Packet Microwave Systems and Technologies, Paolo Volpato

Introduction to RF Design Using EM Simulators, Hiroaki Kogure, Yoshie Kogure, and James C. Rautio

Introduction to RF and Microwave Passive Components, Richard Wallace and Krister Andreasson

Klystrons, Traveling Wave Tubes, Magnetrons, Crossed-Field Amplifiers, and Gyrotrons, A. S. Gilmour, Jr.

Lumped Element Quadrature Hybrids, David Andrews

Lumped Elements for RF and Microwave Circuits, Second Edition, Inder J. Bahl

Microstrip Lines and Slotlines, Third Edition, Ramesh Garg, Inder Bahl, and Maurizio Bozzi

Microwave Component Mechanics, Harri Eskelinen and Pekka Eskelinen

Microwave Differential Circuit Design Using Mixed-Mode S-Parameters, William R. Eisenstadt, Robert Stengel, and Bruce M. Thompson

Microwave Engineers' Handbook, Two Volumes, Theodore Saad, editor

Microwave Filters, Impedance-Matching Networks, and Coupling Structures, George L. Matthaei, Leo Young, and E. M. T. Jones

Microwave Imaging Methods and Applications, Matteo Pastorino and Andrea Randazzo

Microwave Material Applications: Device Miniaturization and Integration, David B. Cruickshank

Microwave Materials and Fabrication Techniques, Second Edition, Thomas S. Laverghetta

Microwave Materials for Wireless Applications, David B. Cruickshank

Microwave Mixer Technology and Applications, Bert Henderson and Edmar Camargo

Microwave Mixers, Second Edition, Stephen A. Maas

Microwave Network Design Using the Scattering Matrix, Janusz A. Dobrowolski

Microwave Power Amplifier Design with MMIC Modules, Howard Hausman

Microwave Radio Transmission Design Guide, Second Edition, Trevor Manning

Microwave and RF Semiconductor Control Device Modeling, Robert H. Caverly

Microwave Transmission Line Circuits, William T. Joines, W. Devereux Palmer, and Jennifer T. Bernhard

Microwave Techniques in Superconducting Quantum Computers, Alan Salari

Microwaves and Wireless Simplified, Third Edition, Thomas S. Laverghetta

Millimeter-Wave GaN Power Amplifier Design, Edmar Camargo

Modern Microwave Circuits, Noyan Kinayman and M. I. Aksun

Modern Microwave Measurements and Techniques, Second Edition, Thomas S. Laverghetta

Modern RF and Microwave Filter Design, Protap Pramanick and Prakash Bhartia

Neural Networks for RF and Microwave Design, Q. J. Zhang and K. C. Gupta

Noise in Linear and Nonlinear Circuits, Stephen A. Maas

Nonlinear Design: FETs and HEMTs, Peter H. Ladbrooke

Nonlinear Microwave and RF Circuits, Second Edition, Stephen A. Maas

On-Wafer Microwave Measurements and De-Embedding, Errikos Lourandakis

Parameter Extraction and Complex Nonlinear Transistor Models, Günter Kompa

Passive RF Component Technology: Materials, Techniques, and Applications, Guoan Wang and Bo Pan, editors

PCB Design Guide to Via and Trace Currents and Temperatures, Douglas Brooks with Johannes Adam

Practical Analog and Digital Filter Design, Les Thede

Practical Microstrip Design and Applications, Günter Kompa

Practical Microwave Circuits, Stephen Maas

Practical RF Circuit Design for Modern Wireless Systems, Volume I: Passive Circuits and Systems, Les Besser and Rowan Gilmore

Practical RF Circuit Design for Modern Wireless Systems, Volume II: Active Circuits and Systems, Rowan Gilmore and Les Besser

Principles of RF and Microwave Design, Matthew A. Morgan

Production Testing of RF and System-on-a-Chip Devices for Wireless Communications, Keith B. Schaub and Joe Kelly

Q Factor Measurements Using MATLAB, Darko Kajfez

Radio Frequency Integrated Circuit Design, Second Edition, John W. M. Rogers and Calvin Plett

Radio Frequency Machine Learning: A Practical Deep Learning Perspective, Scott Kuzdeba

Relativistic Field Theory for Microwave Engineers, Matthew A. Morgan

Reflectionless Filters, Matthew A. Morgan

RF Bulk Acoustic Wave Filters for Communications, Ken-ya Hashimoto

RF Circuits and Applications for Practicing Engineers, Mouqun Dong

RF Design Guide: Systems, Circuits, and Equations, Peter Vizmuller

RF Linear Accelerators for Medical and Industrial Applications, Samy Hanna

RF Measurements of Die and Packages, Scott A. Wartenberg

RF and Microwave Power Amplifiers, Frederick H. "Fritz" Raab

The RF and Microwave Circuit Design Handbook, Stephen A. Maas

RF and Microwave Coupled-Line Circuits, Rajesh Mongia, Inder Bahl, and Prakash Bhartia

RF and Microwave Oscillator Design, Michal Odyniec, editor

RF Passive Network Design and Synthesis for Mobile Communications, Volume 1, Peter V. Wright

RF Power Amplifiers for Wireless Communications, Second Edition, Steve C. Cripps

RF Systems, Components, and Circuits Handbook, Ferril A. Losee

Scattering Parameters in RF and Microwave Circuit Analysis and Design, Janusz A. Dobrowolski

The Six-Port Technique with Microwave and Wireless Applications, Fadhel M. Ghannouchi and Abbas Mohammadi

Solid-State Microwave High-Power Amplifiers, Franco Sechi and Marina Bujatti

Stability Analysis of Nonlinear Microwave Circuits, Almudena Suárez and Raymond Quéré

Substrate Integrated Suspended Line Circuits and Systems, Kaixue Ma and Yongqiang Wang

Substrate Noise Coupling in Analog/RF Circuits, Stephane Bronckers, Geert Van der Plas, Gerd Vandersteen, and Yves Rolain

System-in-Package RF Design and Applications, Michael P. Gaynor

Technologies for RF Systems, Terry Edwards

Terahertz Metrology, Mira Naftaly, editor

Understanding Quartz Crystals and Oscillators, Ramón M. Cerda

Vertical GaN and SiC Power Devices, Kazuhiro Mochizuki

The VNA Applications Handbook, Gregory Bonaguide and Neil Jarvis

Wideband Microwave Materials Characterization, John W. Schultz

*Wired and Wireless Seamless Access Systems for Public Infrastructur*e, Tetsuya Kawanishi

For further information on these and other Artech House titles, including previously considered out-of-print books now available through our In-Print-Forever® (IPF®) program, contact:

Artech House
685 Canton Street
Norwood, MA 02062
Phone: 781-769-9750
Fax: 781-769-6334
e-mail: artech@artechhouse.com

Artech House
16 Sussex Street
London SW1V 4RW UK
Phone: +44 (0)20 7596 8750
Fax: +44 (0)20 7630 0166
e-mail: artech-uk@artechhouse.com

Find us on the World Wide Web at: www.artechhouse.com